SCOTT W. STARRATT

I0820435

Interpreting Pre-Quaternary Climate from the Geologic Record

Perspectives in Paleobiology and Earth History

The Perspectives in Paleobiology and Earth History Series
David J. Bottjer and Richard K. Bambach, Editors

Anthony Hallam, *Phanerozoic Sea-Level Changes*
Ronald E. Martin, *One Long Experiment: Scale and Process in Earth History*

Interpreting Pre-Quaternary Climate from the Geologic Record

Judith Totman Parrish

Columbia University Press □ *New York*

Columbia University Press
Publishers Since 1893
New York Chichester, West Sussex
Copyright © 1998 Judith Totman Parrish-Jones
All rights reserved

Library of Congress Cataloging-in-Publication Data

Parrish, Judith Totman.
Interpreting pre-Quaternary climate from the geologic record /
Judith Totman Parrish.
p.   cm. — (Perspectives in paleobiology and earth history
series)
Includes bibliographical references and index.
ISBN 0–231–10206–2
1. Paleoclimatology.   2. Geology, Stratigraphic.   3. Sedimentology.
4. Paleonotology.   I. Title.   II. Series.
QC884.P37   1998
551.69'09'01—DC21   98–6659

∞

Casebound editions of Columbia University Press books
are printed on permanent and durable acid-free paper.
Printed in the United States of America

10 9 8 7 6 5 4 3 2 1

This book is dedicated to sedimentary geologists and paleontologists everywhere. Your disciplines are now too often viewed as old-fashioned, static, and dispensable when curricula and organizations are downsized. I hope this book shows that nothing could be further from the truth.

Contents

Acknowledgments

I don't pretend to be an expert on all or even most of the subjects treated in this book. However, I have had the good fortune to work with many wonderful scientists who *are* experts and who have been kind enough, first, to attribute to me some ability to interpret climate in the geologic record in ways they felt they could not, and, second, to invite me to work with them. This book quite literally could not have been written without them, for many of them directly inspired me to write a book that could be used by any expert in one type of rock or fossil as a reference for paleoclimatic interpretation of other types of rocks and fossils. Any ability on my part to address the various types of paleoclimatic indicators comes directly from those experiences. These generous collaborators include David Bottjer, Ian Daniel, Mary Droser, Russell Dubiel, Donald Gautier, Steven Good, Eric Gyllenhaal, James Hein, Howard Hutchison, Peter McCabe, Michael Parrish, Fred Peterson, Thomas Rich, Robert Spicer, Christine Turner, Patricia Vickers-Rich, Jack Wolfe, and, especially, A. M. Ziegler. Many, many others have, in one way or another, contributed greatly to my education about various paleoclimate indicators. I owe my paleoclimate-modeling colleagues a debt for the time they have spent helping me understand various types of numerical climate models—Eric Barron, Mark Chandler, Thomas Crowley, Brian Farrell, John Kutzbach, Bette Otto-Bliesner, and Paul Valdes.

I have also been inspired by my students, who are experts in their own right. I have learned a great deal from the truly scholarly work in some term papers written for my classes in paleoclimatology, specifically those of Michael Cassiliano, Lois Roe, Dena Smith, Steve Van der Hoven, and Mou Yun, whose term papers were masterful. I thank my graduate students, Gavin Lawson, Timothy Demko, Carlos González-León, Thomas Moore, Danielle Montague-Judd, and Glen Tanck, and some of the other University of Arizona graduate students with whom I have been fortunate to be closely associated, namely Katherine Gregory, Michał Kowalewski, Mark Rigali, Gerilyn Soreghan, and John Yarnold, all of whom contributed in one way or another to my understanding of paleoclimate. Stanford University graduate students Tom Hickson and Lisa Lamb were wonderful companions and sounding boards during my sabbatical there.

I am especially grateful to the many friends and colleagues who contributed specifically to this book by reading various sections and making innumerable helpful suggestions for its improvement: Warren Allmon (Paleontological Research Institute), György Bárdossy (Hungarian Academy of Sciences), John Barron (USGS), Timothy Bralower (University of North Carolina), Elizabeth Brouwers (USGS), Robyn Burnham (University of Michigan), Steven Calvert (University of British Columbia), Blaine Cecil (USGS), Ronald Charpentier (USGS), Emily CoBabe (University of Massachusetts), Andrew Cohen (University of Arizona), Peter DeCelles (University of Arizona), David Dettman (University of Arizona), Russell Dubiel (USGS), Nicholas Eyles

(University of Toronto), Karl Flessa (University of Arizona), Karl Föllmi (University of Neuchatel), Eric Force (USGS), Ethan Grossman (Texas A&M University), Pamela Hallock-Muller (University of South Florida), Steven Hasiotis (University of Colorado), James Kennett (University of California, Santa Barbara), Gary Kocurek (University of Texas at Austin), Greg Mack (New Mexico State University), Bette Otto-Bliesner (National Center for Atmospheric Research), David Rea (University of Michigan), G. I. Smith (USGS), Gary Upchurch (SW Texas State University), and Jack Wolfe (University of Arizona). In addition, Warren Allmon, Thomas Cronin (USGS), and David Bottjer put a tremendous amount of work into reviewing the entire manuscript. As always, of course, errors are my own and not theirs.

Special thanks to Professor Stephan Graham and the sedimentary geology graduate students of the Department of Geological and Environmental Sciences, Stanford University, as well as the staffs of Branner Library and the GES office, for their hospitality and encouragement during my sabbatical there; to Ed Lugenbeel of Columbia University Press for his persistence; to my parents, Bob and Dru Totman, for their unwavering and generous support, both emotional and material; and to my husband, Will Jones, for his love and patience.

Finally, I would like to express particular gratitude to A. M. (Fred) Ziegler, University of Chicago, with whom I spent four years as a research associate and under whose tutelage I began my career in paleoclimatology. In my years at Chicago, I was constantly in awe of Fred's ability to foster creativity in others. I never quite figured out how he did it, other than in the obvious way—by example. Serious science was never more fun than in Fred's lab.

This work was made possible in part by NSF Award EAR 9023558.

Judith Totman Parrish
Dept. of Geosciences
University of Arizona
Tucson, Arizona 85721
parrish@geo.arizona.edu

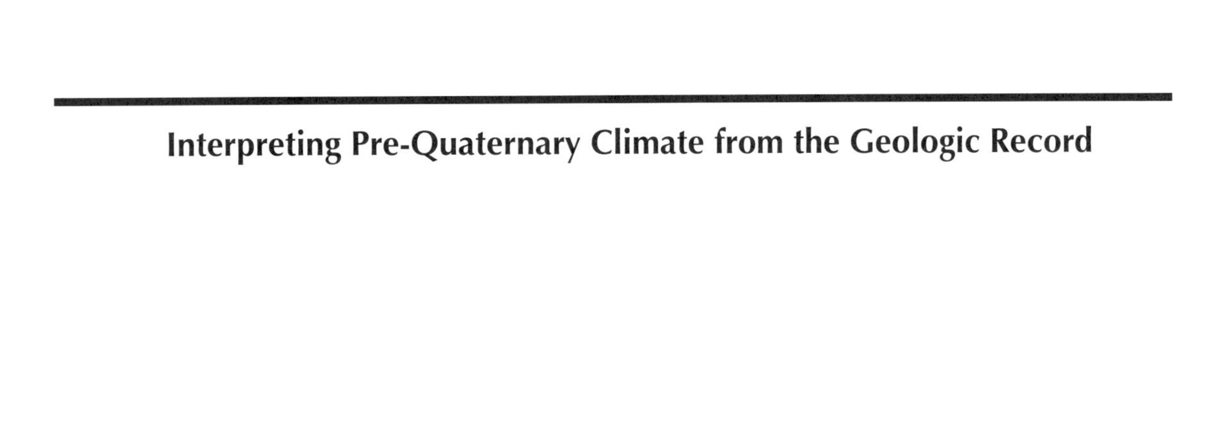

Interpreting Pre-Quaternary Climate from the Geologic Record

Perspectives in Paleobiology and Earth History

1

Introduction

Paleoclimatology—the study of ancient climates—has been practiced at least since the time of Agassiz and Darwin, both of whom were particularly interested in evidence for glaciers where none exist today (Agassiz 1828; Darwin 1842), and in what the presence of those glaciers might have implied about climatic change. The glaciation they postulated was quite recent in geologic history, dating from the Quaternary Period, and with the exception of a few notable treatments (Köppen and Wegener 1924; Brooks 1949), "paleoclimatology" was nearly synonymous with "Quaternary climatology" until the 1960s. At that time, geologists began to take seriously the notion that the continents have changed position on the surface of the Earth and, with that breakthrough, pre-Quaternary climatology came into its own. Since then, and particularly since about 1980, when continental reconstructions by various research groups began to closely resemble each other even for the distant past, the field of pre-Quaternary climatology has grown quickly.

Growth of the field has been on two fronts—models and data. The systematic and widespread use of climate models to understand the distribution of paleoclimatic indicators started with the numerical models for the Cretaceous by Barron and Washington (1982a,b), and with the conceptual models for the entire Phanerozoic by Parrish (Parrish 1982; Parrish and Curtis 1982), which augmented earlier efforts by Ross (1975), Nairn and Smithwick (1976), and Ziegler et al. (1977). The use of numerical models started to accelerate in the late 1980s, when groups in addition to Barron's began publishing models of pre-Quaternary climates (Crowley, Hyde, and Short 1989; Kutzbach and Gallimore 1989; Chandler, Rind, and Ruedy 1992; Valdes and Sellwood 1992). Equalling the growth in modeling efforts have been attempts to understand the climatic significance of rocks and fossils—the paleoclimatic indicators. For example, starting with the work of Retallack (1981, 1983), understanding of the paleoclimatic significance of various pre-Quaternary paleosols was broadened from earlier studies, which had concentrated almost solely on bauxite, laterite, and coal. All in all, the field—and its literature—has grown very fast and has become increasingly specialized. This book is aimed at students and geologists who are interested in learning more about how paleoclimates are interpreted from the geologic record and who want to find someplace to start in the vast literature on the subject.

The theme of this book is *Paleoclimate interpretations for every time and place are strengthened by the use of as many different indicators as possible and by consideration of those indicators in their regional and global contexts.*

This book provides a summary of methods in pre-Quaternary climatology—that is, it is a sourcebook that provides an entry into the literature. How various indicators are used to interpret paleoclimate is presented, not analytical details. In general, I tried to avoid grand summaries, as they are usually so generalized as to be rather useless. I make extensive use of examples, and it is not uncommon for the examples presented to be the only ones I could find in the literature at the time of writing. Many of the methods and indicators described herein are still controversial, underexploited, and/or poorly constrained. I hope this book will inspire colleagues and students to go forth and rectify this situation. Tremendous potential exists for cross-fertilization among the various subjects with regard to approaches, methods of interpretation, understanding of strengths and limitations, and even graphical presentation of information. For this reason, some examples were chosen not so much because they had spectacular conclusions or illustrated the method better than others but because they had points of view or methods of presentation that might inspire geologists and paleontologists working on other topics. A few of the paleoclimatic indicators discussed in this book have now been discounted. They are included because attempts to use them are widespread in the literature, but I caution readers to be aware that their inclusion is for the purpose of publicizing their demise.

I discuss only in passing the current grand themes of paleoclimatologic research—mechanisms of changing CO_2 levels, causes of glacial episodes, orogenic effects on climate, and so on. Rather, the book is focused on the data used to interpret these processes. Climate models are treated very lightly. Information about climate models has been synthesized in numerous volumes, for example, Hecht (1985), Crowley and North (1991), and Barron and Moore (1994), and here I concentrate on the benefits and some of the problems of model-data comparisons. I have intentionally ignored methods that are useful only in Quaternary geology, and length limitations forced me to concentrate on pre-Quaternary examples. Although this perpetuates the divide between Quaternary and pre-Quaternary paleoclimatology, I do not feel that divide will ever disappear, although it may become fuzzier. Quaternary climatology has been treated splendidly in the book *Quaternary Paleoclimatology* by Bradley (1985), which in some ways served as a model for this book. A book that provides a good link between modern and ancient climate studies is *Climate, Earth Processes and Earth History* (Huggett 1991).

The usefulness of stable isotopes of carbon and oxygen has been dealt with extensively in the literature (e.g., Arthur et al. 1983; Faure 1986). In this book, the use of stable isotopes for paleoclimate interpretation is integrated with the treatment of the rocks and fossils from which isotope data are taken; only a brief summary of the general principles is presented in this chapter. Similarly, issues concerning levels and changes in CO_2 through Earth history are addressed in the context of the paleoclimatic indicators from which the relevant data are derived.

A word is in order at this point regarding the use of the term "paleoclimatic indicator"[1]. Strictly speaking, a paleoclimatic indicator should be something that indicates a particular climate. A good example is evaporites, which are very clearly indicative of dry climates, however strictly one might want to define "dry." In the literature, however, the discussion of paleoclimatic indicators is commonly focused on the consideration of whether something is controlled by a particular climate. An observation might be explained by climate, but the connection might not be well-enough established for similar observations to be taken as indicative of similar climates. This is particularly true of events in evolution and extinction. For example, Tucker and Benton (1982) attributed the widespread extinctions of terrestrial vertebrates in the Late Triassic to climatic change. The extinction was possibly a *consequence* of climatic change, but it is not necessarily *indicative* of climatic change because extinctions of terrestrial vertebrates at other times have been caused by other types of processes. In this book, most of the paleoclimatic indicators discussed do indicate a particular climate.

GUIDE TO THIS BOOK

Most of this book (chapters 2 to 5) is devoted to the description and analysis of paleoclimatic indicators, divided into biotic and lithologic indicators and further divided into marine and terrestrial indicators. Three indicators merit special mention with regard to the organization of the book. Evaporites, even when precipitated from marine waters, are indicative of climate on the continents and thus are grouped with the other terrestrial climatic indicators. Indicators of gla-

1. I avoid the term "proxy" primarily because it is less descriptive. Furthermore, the dictionary definition of "proxy" is, in a word, "substitute," and the way the word has been used—for example, "oxygen isotopes as a proxy for temperature"—is not really correct; strictly, oxygen isotopes would be a proxy for a thermometer. "Proxy" should never be used as an adjective (e.g., a "proxy indicator").

ciation are likewise grouped with terrestrial climatic indicators, even though many, perhaps most, indicators of glacial conditions are found in the marine record. In contrast, clay minerals in the deep sea, which reflect climate on land, are included in the chapter on marine lithologic indicators because, paradoxically, the relationship between climate and their distribution in the deep sea is more straightforward than the relationship between climate and their distribution on land.

Chapter 6 deals with different kinds of climate models and how they are applied in paleoclimatology. It also contains a discussion of data-model comparisons. In the final chapter, I discuss integrative studies of pre-Quaternary climates, using some case histories to illustrate my points. Paleoclimatology is a highly interdisciplinary field and, as stated in the theme, interpretation of paleoclimate from the geologic record benefits from the use of all available information.

GEOGRAPHY OF ATMOSPHERIC CIRCULATION

An emphasis in this book is on the *geography of climate,* that is, patterns of climate on the surface of the Earth and how they change through time. Studies of paleoclimate approach the problem in two ways. One is to map paleoclimatic patterns, which is a natural way for geologists to think about climate. The other is to document climatic changes through time. Such changes are usually studied in a limited number of localities, and implicit in many such studies is the notion that changes must be at least regional if not global. However, in many cases, the temporal changes in climate probably are not global, but local, and thus the emphasis here on the geography of climate is intended to enforce thinking about the heterogeneity of climate in space.

Zonal Circulation

The Earth is a heat engine, and circulation of the atmosphere is controlled primarily by the uneven distribution of energy on the surface. More solar energy per unit area arrives at the Earth's surface at lower latitudes than at higher latitudes, creating a thermal gradient and the tendency for air to rise at the equator, flow poleward, sink at the poles, and return equatorward. The other primary control on atmospheric circulation is the rotation of the Earth. Circulation on an Earth with a homogeneous surface that is rotating at the same velocity that it has today is illustrated in figure 1.1. The circulation is zonal, that is, parallel to latitude. The warmest air, which rises at the equator and flows poleward, is deflected from a straight path by the rotation of the Earth, and, likewise, the coldest air, sinking at the poles and flowing equatorward, is deflected from a straight path to the equator. The heat exchange between equator and poles is effected by three circulation cells in each hemisphere, and the dynamics of the system force air to rise at about 60° north and south and to sink at about 30° north and south. The corresponding distribution of atmospheric pressure at the surface is high pressure where the air is sinking and low pressure where it is rising, regardless of the cause of the vertical circulation. The vertical circulation in low latitudes is the Hadley circulation, a term familiar to many (see figure 1.1). If the surface of the Earth were all ocean, the belts of low pressure at the equator and at high mid-latitudes would be rainy belts, and the belt of high pressure at low mid-latitudes and the polar high pressure cells would be dry (see figure 1.1).

The speed of the rotation of the Earth determines the number of zones. Because the moon is receding from the Earth, the energy of the Earth-Moon system is decreasing, and one effect is a slowing of the rotation of the Earth. Modeling of the Earth-Moon system has led to the conclusion

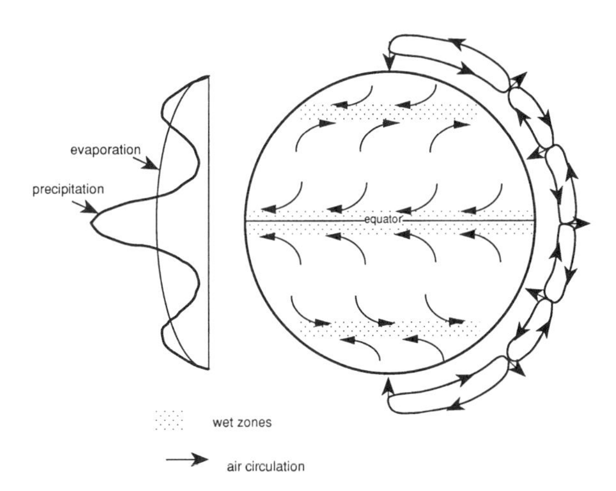

Figure 1.1. Idealized zonal circulation, with upper atmosphere and surface winds, zonal evaporation/precipitation, and wet zones.

that the rotation rate of the Earth has been faster in the past, and this prediction has been supported by studies of growth rings in fossils, which give estimates of about 400 days per year in the Ordovician-Silurian and 440 days per year in the late Precambrian (Scrutton 1978). Higher rotation rates for the Earth imply a different organization of the zonal circulation from that seen today, with more zones. The effect of higher rotation rates has been modeled by Hunt (1979), who found that the number of zones shown in figure 1.1 would be maintained to rotational rates of about 700 days per year, above which the circulation would reorganize to a greater number of zones.

Meridional Circulation

The idealized zonal circulation in figure 1.1 is for an Earth with a thermally homogeneous surface, but this circulation is an important component of atmospheric circulation today and is expressed well over the oceans, particularly over the Pacific Ocean. However, the Earth's surface is not thermally homogeneous, and anything that affects the exchange of heat will modify the zonal circulation. The factors that modify the zonal circulation are distributions of land, sea, and mountains. Atmospheric chemistry partially controls the overall thermal budget of the Earth but probably does not change the spatial circulation pattern. Likewise, the latitudinal temperature gradient, which is controlled by the dynamics of heat exchange in ways that are not fully understood, does affect the surficial expression of climate but has not been shown to affect the pattern of circulation. Two factors strongly affect the heat budget, even on a local scale. These are albedo, that is, the reflectivity of the Earth's surface, and the vertical circulation in a temperature-stratified ocean, which affects the horizontal temperature field at the surface. However, despite their strong effects on the global heat budget, neither has been shown to affect atmospheric circulation patterns except locally or over time scales that are unlikely to be detectable in the pre-Quaternary record.

The principal modifying feature to the zonal circulation is the distribution of land and sea, which largely controls the distribution of the large-scale circulation cells, such as the subtropical high-pressure cells over the oceans (figure 1.2). The modification of circulation comes about because of the thermal contrast between land, which has a relatively low heat capacity, and sea, which has a relatively high heat capacity. The effect on circulation will be strongest where the thermal contrast between land and sea is strongest, which means that the strongest cells tend to be in mid-latitudes. In summer (figure 1.2, bottom), the subtropical high-pressure cells are strongest because the thermal contrast between the relatively cool oceans and the warm adjacent continents is strongest. In winter (figure 1.2, top), the high mid-latitude low-pressure cells are strongest because the continents are cold and the adjacent oceans relatively much warmer. The vertical circulation that originates from the dynamics of the zonal circulation is enhanced by thermal processes. Thermal contrast at very high and very low latitudes is less than at mid-latitudes but can nevertheless be important. For example, the surface pressure over Antarctica is always relatively high even in the winter when the thermal contrast with the surrounding oceans is the lowest in the annual cycle. In contrast, the surface pressure over the North Pole is high relative to the surrounding continents only in the summer, when the continents are warm. Despite the insulating layer of ice, the Arctic Ocean is warmer than the surrounding continents in the winter, and the relative surface pressure is lower over the Arctic than over the adjacent continents.

The effect of mountains, particularly high plateaus, on circulation patterns is both thermal and mechanical. Standing waves in the upper atmospheric circulation in the Northern Hemisphere occur downstream (east) of the mid-latitude Colorado and Tibetan Plateaus (e.g., Ruddiman and Raymo 1988). The creation of the standing waves and consequent reorganization of the circulation thought to have been caused by the uplift of these regions has been cited as responsible for climatic changes during the Cenozoic Era (e.g., Ruddiman and Kutzbach 1991). The thermal effects of the plateaus also strongly affect circulation by enhancing monsoonal circulation.

Monsoonal circulation is defined by climatologists as any climate that has a seasonal alternation between high and low pressure such that most of the rain falls during the summer months (e.g., Hayden 1988). However, some such climates are extreme in that total annual rainfall is high and the seasonality of rainfall is very strong. The strongest monsoon is the Asian monsoon, and Dubiel et al. (1991) argued that only the strongest monsoonal climates are likely to leave a

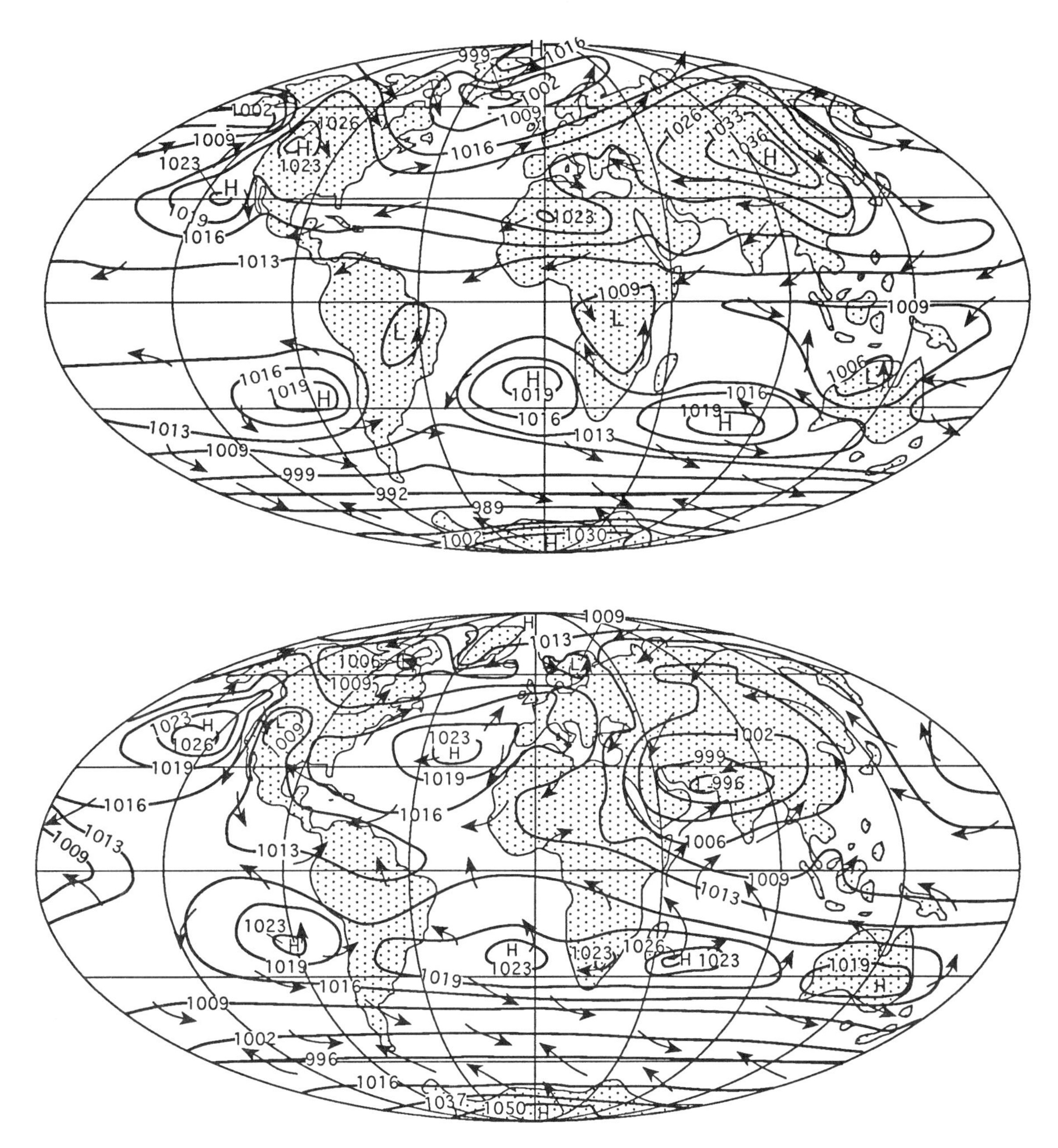

Figure 1.2. General circulation of the atmosphere in the northern winter *(top)* and northern summer *(bottom)*. *Contours* are sea-level pressure in millibars. *Arrows* are surface winds. Modified from Trewartha, G.T., Robinson, A.H., and Hammond, E.H. (1967) *Elements of Geography,* 5th edition, McGraw-Hill Book Company. By permission of The McGraw-Hill Companies.

recognizable signature in the geologic record. Because strong monsoonal circulation might have been important in the past, particularly on the supercontinent Pangea, the characteristics of the Asian monsoon will be further described here.

Monsoonal Circulation

The characteristics of the Asian monsoon are alternating very low pressure in the summer, averaging 999 millibars (mb), and very high pressure, averaging 1035 mb, in the winter. The annual variation in rainfall can be extreme, with the average monthly rainfall ranging from 0 to 2 cm mean monthly precipitation in the driest months to > 100 cm in the wettest months (Rumney 1968).

Four factors contribute to the strength of the Asian monsoon. These are the size of Asia, the Tibetan Plateau, the cross-equatorial pressure contrast with the southern Indian Ocean circulation, and the supply of warm, moist air from the Indian Ocean. The larger a continent is, the more isolated is its interior from the ameliorating effects of the surrounding oceans, and interior Asia has a very large temperature contrast between summer and winter mean temperatures, as much as 44°C in southeastern Russia. The Tibetan Plateau contributes to the thermal contrast between summer and winter by isolating the interior of the continent from the influence of the warm Indian Ocean in the winter and by blocking the influx of cold air from the north in the summer. The summer monsoon is further strengthened by the Tibetan Plateau because the feature acts as a high-altitude heat source. The low pressure that would form over the continent in the summer as it heats up is further enhanced because the warm ground is high in the atmosphere on the Tibetan Plateau. The effect is illustrated in figure 1.3.

The summer monsoon is also strengthened by the effects of latent heat[2] release from the warm, moist air that originates over the Indian Ocean. As the air is drawn into the low pressure over northern India, it begins to rise and release its moisture. The release of latent heat contributes to the energy of the system. Finally, the cross-equatorial thermal and pressure contrast with the southern Indian Ocean augments, and indeed may be the primary driving force behind, the summer monsoon (Webster 1987; Young 1987). The contrast increases the intensity of the summer monsoon low pressure and also affects the southern Indian Ocean subtropical high-pressure cell, which is strongest in the southern winter (figure 1.2); other subtropical high pressure cells are strongest in the summer, when the thermal contrast between ocean and adjacent continents is greatest. The large, warm Asian continent in the Northern Hemisphere affects the southern Indian Ocean subtropical high pressure cell more than do southern Africa or Australia. A consequence of the thermal contrast is that atmospheric and oceanic flow, particularly in the summer, cross the equator.

Precipitation and Evaporation

Perhaps the most common practice in paleoclimatology is to use the zonal component of rainfall (figure 1.1) as a predictor for the distribution of paleoclimatic indicators on the continents. The practice persists despite the obvious fact that rainfall is not zonal in distribution, even in the present world where, with the exception of Asia, the continents are small and modify the zonal circulation only minimally. A map of the distribution of rainfall today emphasizes this fact (figure 1.4). The asymmetry of rainfall across continents is especially striking in South America, where the zonal component of rainfall is disrupted by mountains. The zonal component of rainfall is even more strongly modified by mountains than the general circulation, because rain falls mostly from clouds in the lower atmosphere. Mountains can even cause rain to fall in regions that might otherwise be dry, because the lifting action of mountains on air moving across the surface can cause the air to reach saturation. The effects of mountains can be summarized as follows:

1. the higher the starting humidity and, to a lesser extent, the higher the starting temperature of the air mass, the greater will be the influence of mountains as the air mass flows over them, and
2. the higher the starting temperature and humidity, the higher the temperature will be in the lee of mountains.

In paleoclimatology, the general practice has been to concentrate on precipitation instead of evaporation, but precipitation is only half the story if climates are to be characterized in any biologically or sedimentologically meaningful way. Evaporation is equally important. Evaporation is controlled by temperature and relative humidity. Because temperature is strongly dependent on latitude, evaporation also is dependent on latitude. However, it is the balance between evaporation and precipitation that controls climate. Referring again to an idealized zonal climate (see figure 1.1), evaporation is highest in low latitudes, where temperatures are the warmest. Evaporation decreases away from the equator

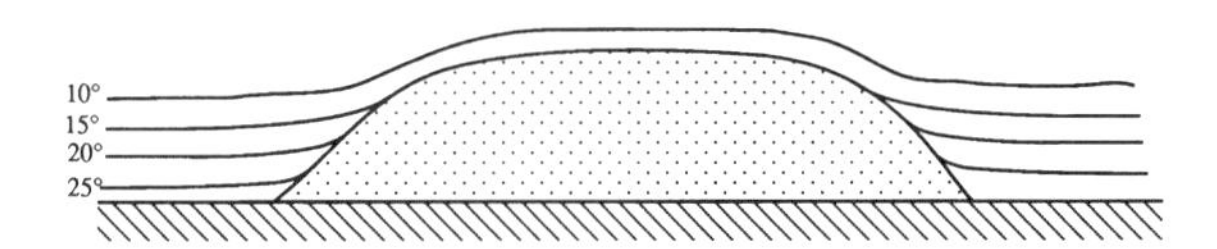

Figure 1.3. Schematic diagram of the high-altitude heat effect of the Tibetan Plateau. Note the elevation of isotherms over the plateau relative to the surrounding lowlands.

2. Latent heat is the heat released during condensation. Sensible heat is that which is measured with a thermometer.

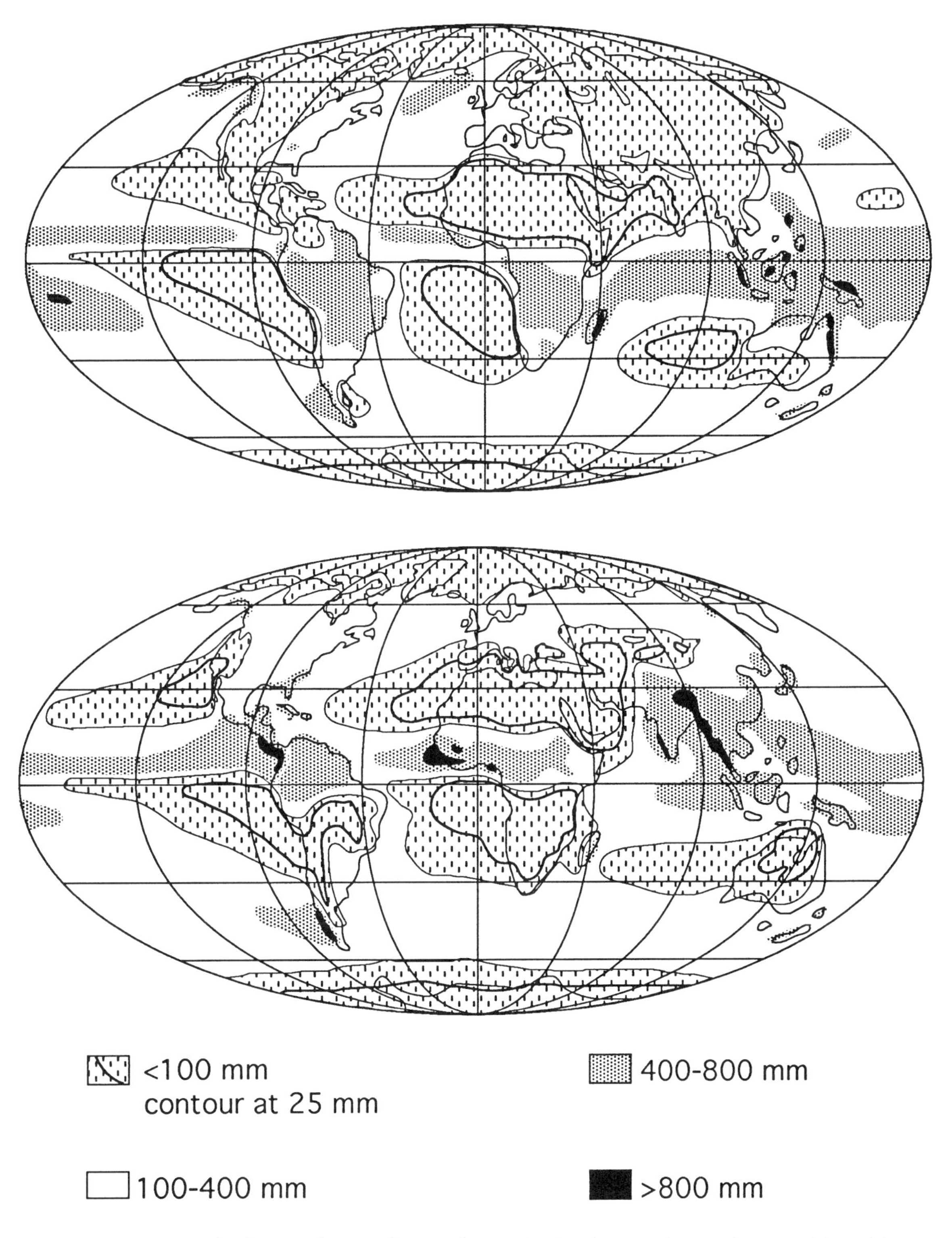

Figure 1.4. Precipitation for the periods December to February *(top)* and June to August *(bottom)*. Adapted from Barry and Chorley (1987) *Atmosphere, Weather, and Climate*. Methuen & Co.

and the balance is determined by the zonal rainfall. An interesting feature of these curves is the balance at high latitudes. The polar regions are among the driest, but they are also the coldest. Thus even though precipitation is low, evaporation is so low that the overall effect is that climate may be regarded to be relatively wet.

A useful parameter for understanding the balance between evaporation and precipitation is the Thornthwaite index, which is an expression of potential evapotranspiration, that is, the amount of water that would be evaporated from the soil surface and transpired by plants if the water supply to surface and plants were continuous. It is calculated from monthly temperatures and includes corrections for latitude and for the length of each month. The index is widely available in weather station data (Gyllenhaal 1991), and combined data from Asia, Africa, and Central and South America describe a relatively narrow envelope by latitude in the index (figure 1.5a), despite the wide variations at the same sites in both evaporation and mean annual precipitation (figure 1.5b,c). Gyllenhaal (1991) used this index to good effect in interpreting the climatic significance of certain terrestrial indicators.

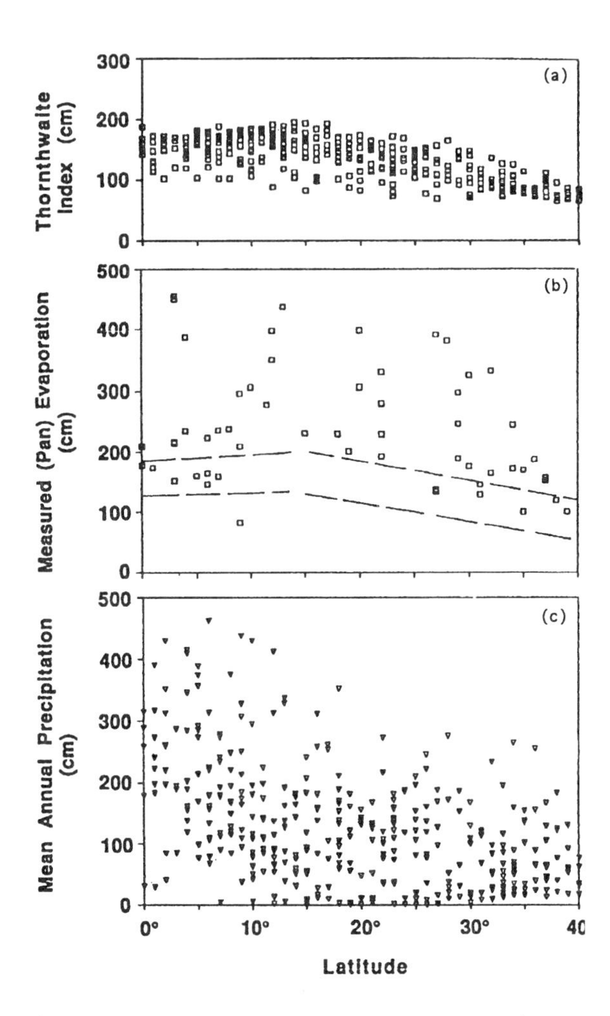

Figure 1.5. Evapotranspiration, evaporation, and mean annual precipitation by latitude for sites in Asia, Africa, and Central and South America. *(a)* Thornthwaite index of evapotranspiration; *(b)* measured evaporation; *(c)* mean annual precipitation. See text for further explanation. From Gyllenhaal (1991).

PALEOGEOGRAPHY

An understanding of pre-Quaternary paleoclimates requires an understanding of paleogeography. The most important elements of paleogeography for paleoclimate modeling are the positions of the continents and shorelines and the positions, altitudes, and orientations of mountain ranges. For interpretation of paleoclimate from the geologic record, a knowledge of the geographic setting of the locality studied—coastal plain, intermontane basin, and so on—is also important.

Continental positions and paleotopography are determined using the types of information listed in table 1.1. The use of paleomagnetic and tectonic information is the most desirable from the point of view of a paleoclimatologist because these data are independent of climate. In contrast, the use of biogeographic similarity and the distribution of paleoclimatic indicators to position continents in global reconstructions is undesirable, for two reasons. First, using those types of data to position the continents and then using reconstructions of the continental positions to model or study global climate introduces circularity into the process (Rosen 1992). Second, use of the distribution of paleoclimatic indicators (including biogeographic patterns) assumes a certain relationship between climate and those indicators and a particular type of climate, usually zonal (e.g., Scotese and Barrett 1990; Witzke 1990). If the actual climatic patterns departed from the zonal model (or whatever model is assumed), studying that departure and the reasons for it is rendered impossible. This is the principal reason that studies of Paleozoic, particularly pre-Carboniferous, climates have lagged behind studies of younger climates. The other information used to reconstruct continental positions for those times is commonly unreliable, so much so that paleogeographers are forced to use biogeographic and paleoclimate information in their reconstructions.

Table 1.1
Types of information used to reconstruct paleogeography

Continental positions
- Paleomagnetic data
  - Sea-floor spreading (Jurassic to Recent only)
  - Polar-wander paths
- Timing and geometry of rifting and collisional events
- Biogeographic similarity/dissimilarity between continental plates
- Distribution of paleoclimatic indicators

Paleotopography—altitudes of mountains in different tectonic settings, used to estimate minimum and maximum paleoaltitudes (Ziegler et al. 1985); locally, recent examples may be higher or lower than indicated
- Continent-continent collisions: >4000–10000 m
- Ocean-continent collisions: 2000–4000 m
- Island arcs, rift shoulders: 1000–2000 m
- Some forearc ridges: 200–1000 m

The spatial and temporal scales of paleogeographic reconstructions must be matched to the scale of climatic processes to be examined, but it may not be possible to study climate rigorously at all spatial and temporal scales. The larger the spatial scale, the more difficult it is to resolve short events in time. On the other hand, the longer the time scale studied, the more difficult it is to resolve small-scale variations in space. Global reconstructions are commonly made for each geologic age, but they are generalized even at that scale of resolution (e.g., Ziegler et al. 1985), because, for example, large changes in the positions of shorelines can take place within that length of time.

For local paleogeographic reconstructions, on a scale of 10^1 to 10^2 kilometers, lithostratigraphic correlation may be adequate. For larger-scale paleogeographic reconstructions, other types of data are used to correlate. Biostratigraphic time scales are generally consistent on each continental plate. Correlation from plate to plate is more difficult but is helped by the practice of referring local biostratigraphic scales to global standards, such as those established by Harland et al. (1982), through reference to radiometric dates, where possible.

The importance of mountains to climate has not escaped the attention of climate modelers, who have repeatedly attempted to assess the effects of mountains (Barron and Washington 1985; Ruddiman and Raymo 1988; Ruddiman and Kutzbach 1991; Otto-Bliesner 1993). Although some studies have suggested that topography has a minimal effect on global climate (Barron and Washington 1985), others have shown that topography has a profound effect, at least over large regions if not globally.

Reconstructing paleotopography, especially paleoaltitude, is a challenging task for the paleogeographer because little direct evidence may be preserved in the geologic record (Chase et al., in press). Predominately erosional settings, that is, "highlands" (Ziegler et al. 1985; Scotese and Golonka 1992), are reconstructed by uplift ages, ages of metamorphism, and sedimentary gradient analysis from surrounding areas of deposition. The plate tectonic setting of the mountain range can be a general guide to paleoaltitude (see table 1.1). Paleoaltitude has also been estimated from plant fossils (e.g., Gregory 1994 and references therein) and estimates of the oxygen isotopic composition of runoff (e.g., Norris et al. 1996). An important point to bear in mind is that "uplift" means different things in different contexts, and that uplift in the geological sense of the word, that is, raising of rocks at depth to the surface, is not necessarily accompanied by the generation of relief (Molnar and England 1990). Processes that can be studied to determine changes in uplift of the surface, such as incision rates, changes in sedimentation rate and style, and changes in flora, are also affected by climatic changes. For example, higher sedimentation rates could be the result of increased uplift rates, increased rainfall rates (Bull 1991), or changes in resistance of the rock being weathered as the weathered terrane is unroofed and different geology exposed. In addition, changes in climate and uplift are not necessarily independent (e.g., Ruddiman and Raymo 1988).

TIME AND CORRELATION

Paleogeography provides the spatial context for paleoclimate studies. Of equal importance is the temporal context. Paleoclimate records must be dated in order to relate them to each other and to establish the sequence and timing of climatic change. Resolution in the pre-Late Cretaceous record is commonly relatively poor—rarely a few tens of thousands, to as poor as a few millions of years in the early Paleozoic—especially globally. From about the Late Cretaceous onward, temporal resolution is better. Efforts to construct a framework for correlation are sometimes referred to as construction of an age model, to emphasize the level of inference that goes into them. Age models are based on biostratigraphy, radiometric age dates, magnetostratigraphy, and event stratigraphy. The latter includes the use of data such as variations in the stable isotope record, the record of volcanic ash beds (which are commonly also dated radiometrically), and inferences about the influence of the Earth's orbital (Milankovitch) cycles on the stratigraphic record. On relatively small spatial scales, lithostratigraphic methods can be used. Excellent discussions of age-model construction in the service of paleoclimatology can be found in Sancetta, Heuser, and Hall (1992) and Berger et al. (1993). An example of the effect of different age models, in this case on calculations of carbonate sedimentation rates through the Tertiary, is illustrated in figure 1.6.

One of the challenges of interpreting paleoclimate from the pre-Quaternary record is understanding how the climate signal has been preserved. Short-term changes may be difficult to resolve not only because the age model lacks resolution but also because the signal might be time-averaged, for example, by bioturbation. Alternatively, short-term changes may be observable but difficult to put into context. For example, many cyclic sequences in the older geologic record appear to reflect Milankovitch-scale cyclicity, especially the 400 Ka eccentricity cycle. But whether the cyclicity is actually related to Milankovitch cycles or simply happens to fit more or less into a 400 Ka framework because of lack of resolution in dating can be difficult to prove. Even in the younger geologic record, different dating methods can give discrepant results (Sancetta, Heusser, and Hall 1992). In short, interpretation of the climate signal, as recorded in the geologic record, is subject to the same sources of error as other kinds of interpretation of the stratigraphic record.

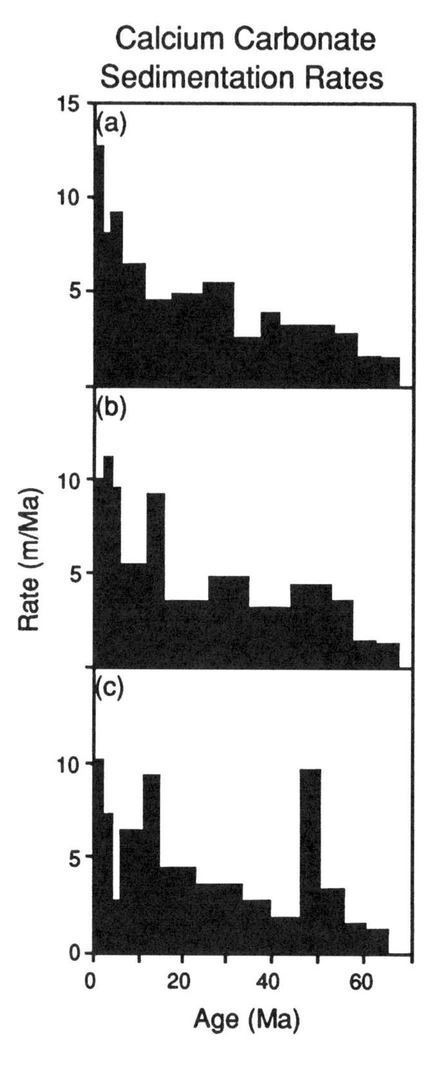

Figure 1.6. Effect of different age models on calculations of calcium carbonate sedimentation rates through the Tertiary. *(a)* Berggren, Kent, and Flynn (1985) and Berggren, Kent, and Van Couvering (1985); *(b)* Harland et al. (1982); *(c)* Davies et al. (1977). From Delaney and Boyle (1988) *Paleoceanography* 3:137–56, copyright by the American Geophysical Union.

INTRODUCTION TO PALEOCLIMATIC INDICATORS

Paleoclimatic indicators can be divided into three groups, among which some overlap exists. These are paleontological, lithological, and geochemical indicators. Paleontological indicators, discussed in chapters 2 and 4, can be divided loosely into four types. These are direct indicators, nearest-living-relative (NLR) indicators, empirical-morphological indicators, and biogeography. Direct indicators can be viewed as paleoclimate index taxa, that is, taxa that had very specific climatic tolerances. NLR indicators are those whose value as paleoclimatic indicators comes from their evolutionary relationship to modern taxa that have specific climatic tolerances. Empirical-morphological indicators are those whose morphology has been determined empirically to be correlated with climate. Finally, the distributions of organisms are commonly controlled at least partly by climate, and thus biogeography of extinct organisms is commonly used to interpret broad paleoclimatic patterns.

Lithologic indicators, discussed in chapters 3 and 5, come from the marine and terrestrial realms and have different meanings for climate depending on the realm. Marine lithologic indicators record ocean temperatures, variations in biologic productivity, and water chemistry, particularly oxygen content. All of these are influenced to some extent by atmospheric climate. On land, lithologic indicators contain more direct information about atmospheric climate because the land surface is in direct contact with the atmosphere. Terrestrial paleoclimatic indicators contain information principally about temperature and moisture, but some also have information about wind direction and strength.

Geochemical indicators are principally the stable isotopes of oxygen and carbon, although isotopes of strontium and other elements also have been exploited for climate information. This geochemical information is collected in conjunction with lithologic and paleontologic information.

The paleoclimatic indicators discussed in this book are listed in table 1.2, arranged by the paleoclimatic parameters they indicate. Many indicators discussed here, such as changes in depth zonation of brachiopods as indicative of sea level, are just examples; changes in depth zonation of a number of benthic invertebrates would probably be equally useful to that of brachiopods. It is apparent from the table that temperature is the paleoclimatic parameter that either has received the most attention or is the most amenable to interpretation from the geologic record, probably both.

General Principles of Biotic Indicators of Paleoclimate

Plants and animals can be exquisitely sensitive to climate, and, through what Dodd and Stanton (1981) called taxonomic uniformitarianism, fossils can be excellent paleoclimatic indicators. Organisms respond to climate in four ways that can be applied to the geologic record of climate. First, the distribution of organisms on a regional or global basis is commonly controlled at least partly by climate. *Paleobiogeography* has thus been an important method of paleoclimate interpretation, and compilations such as *Atlas of Palaeobiogeography* (Hallam 1973) and *Palaeozoic Palaeogeography and Biogeography* (McKerrow and Scotese 1990) are good sources of information on global paleoclimatic patterns. Second, the relationship of an organism to its physical and biological environment can be controlled principally by climate, and thus the study of *paleoecology* has proved useful. Third, adaptations of organisms for certain climates may be preserved in the fossil record, and thus the *morphology* of a fossil can hold important clues to paleoclimate. Finally, although an organism's *physiology*, that is, the internal chemical environment, is not preserved, it can sometimes be inferred, and arguments about the physiology of certain organisms have played an important role in the interpretation of those organisms as paleoclimatic indicators. Most discussions of the relationship between fossils and paleoclimate begin with paleobiogeography, and for many organisms, their entire usefulness as paleoclimatic indicators comes from biogeographic studies. Paleoecological, morphological, and physiological arguments are restricted to certain taxa. The principles of paleobiogeography will be discussed at length here because they apply to all taxa. Only brief discussion of the other methods of using fossils as paleoclimatic indicators is included here, with the specifics deferred to the sections on individual taxa.

Paleobiogeography

Biogeography is based on three facts about the distribution of organisms:

1. Not all taxa live everywhere.
2. Taxa are not randomly distributed.
3. Some taxa tend to occur together, whereas others tend not to occur together.

The distribution of organisms is controlled by their physical and biotic environments. A general principle of biogeography is that the more severe the physical environment, the more it exerts control over the distribution of organisms. In contrast, the distribution of organisms in less severe environments may be more dependent on biotic, rather than physical, environments. In addition, the relationship between biotic and physical control is at least partly dependent on geographic scale. Biotic interactions may be more important

Table 1.2
Paleoclimatic indicators discussed in this book and the climate parameters indicated by each

Paleoclimatic parameter	Paleoclimatic indicator(s)
Temperature	General: biomarkers, photic-zone reefs, carbonate platforms, nontropical limestones, kaolinite, vermiculite, sepiolite, palygorskite, attapulgite,[a] ikaite, taphonomy of freshwater fishes, Oxisols, sand wedges
	Morphology: foraminifera, silicoflagellates, terrestrial vertebrates
	"Index" fossils[b]: foraminifera, nannofossils, marine ostracodes, *Bolboforma,* marine mollusks
	"Faunal indices"[b]: foraminifera, diatoms, radiolarians, dinoflagellates, silicoflagellates, nannofossils, terrestrial vertebrates (all quantitatively defined except nannofossils)
	"Traditional" biogeography: conodonts, brachiopods, marine mollusks, graptolites, trilobites, terrestrial plant megafossils, terrestrial vertebrates? (see text)
	Latitudinal migration of assemblages[c]: planktonic foraminifera, brachiopods, marine mollusks, diatoms, multiple taxa integrated
	Diversity gradients: benthic foraminifera, larger foraminifera, dinoflagellates, multiple taxa integrated
	Transfer functions: marine ostracodes, dinoflagellates
	Nearest-living-relative method: marine ostracodes, pollen, terrestrial plant megafossils, crocodilians
	$\delta^{18}O$ in foraminifera, marine diatoms, planktonic foraminifera + diatoms, conodonts, brachiopods, marine mollusks, corals, marine and terrestrial vertebrates, chalk, phosphorites, Al- and Fe-bearing paleosol minerals; $\delta^{18}O$ gradients in foraminifera, nannofossils, marine mollusks
	Li/Cd in brachiopods
Seasonality of temperature	Radiolarians, wood, leaf taphonomy
Vertical temperature structure of oceans	$\delta^{13}C$ in biomarkers; $\delta^{18}O$ in foraminifera, diatoms, brachiopods, marine mollusks, marine vertebrates
Glaciers, ice, ice volume	General: loess (see text), tillite, glaciomarine deposits, abraded bedrock, ice-modified clasts, dropstones (see text), ice-rafted debris, sea-ice scour marks, glacial depositional systems
	Sea-level changes (physical effects): brachiopod depth zonation, cyclothems
	$\delta^{18}O$ in foraminifera, marine vertebrates, chalk, brachiopods
Productivity, changes in nutrient status of oceans	General: foraminifera[d], nannofossils, radiolarians, tasmanitids, graptolites, marine mollusks, marine reptiles and whales, chalk, phosphorite, bedded biogenic chert, disseminated phosphate, organic-rich rocks, drowned reefs and carbonate platforms, barite, magnetic susceptibility
	"Index" fossils[b]: radiolaria
	$\delta^{13}C$ in foraminifera, mollusks
	$\delta^{15}N$ in marine diatoms (Quaternary only), bulk sediment
	Cd/Ca and Ba in planktonic foraminifera
O_2 content of marine waters	Organic-rich rocks[e], marine trace fossils, laminated rocks, iron-bearing minerals, manganese, other trace elements? (see text)
	Morphology: marine ostracodes, marine bivalves
	Biogeography: benthic foraminifera
Global carbon budget, CO_2	General: leaf cuticle
	$\delta^{13}C$ in foraminifera, chalk, biomarkers, terrestrial vertebrates, paleosol carbonates, goethite

Table 1.2 *(continued)*

Paleoclimatic parameter	Paleoclimatic indicator(s)
Continental weathering	General: clay minerals in ocean and epeiric sea sediments, oolitic iron minerals, fluvial sandstone composition Sr isotopes in foraminifera, conodonts, marine mollusks, disseminated phosphate Ge in marine diatoms
Storms	Flat-pebble conglomerates, hummocky cross-stratification, storm-graded beds
Wind direction and strength	Patch reefs, eolian sandstones, volcanic ash deposits, eolian dust in the deep sea and in lake sediments, clay pellet dunes
Rainfall, paleohydrology, evapotranspiration	General: siderite? (see text), charophytes, terrestrial trace fossils, loessite, clay pellet dunes, evaporites, terrestrial zeolite assemblages, paleosols, karst, coal beds, lake levels, lake deposits, fluvial/alluvial deposits, large-scale erosion surfaces Biogeography: pollen, terrestrial plant megafossils, terrestrial and freshwater mollusks "Index fossils"[b]: *Classopollis,* conchostracans Morphology: terrestrial plant megafossils, leaf cuticle $\delta^{18}O$, $\delta^{13}C$ in platform carbonates, terrestrial mollusks, paleosol carbonates, terrestrial vertebrates, lacustrine carbonates Paleoecology: terrestrial plants, terrestrial vertebrates
Seasonality of moisture	Wood, lungfish, Aridisols, clay pellet dunes, Oxisols, Vertisols, coal beds, ephemeral streams
Meteoric and freshwater chemistry	Nearest-living-relative method: freshwater ostracodes, freshwater diatoms Morphology: freshwater ostracodes $\delta^{18}O$ in freshwater mollusks, freshwater vertebrates, paleosol carbonates, lacustrine carbonates
Snow cover	Niveo-eolian deposits
Climate cycles	Chalk (both lithologic and isotopic), clastic sedimentary rocks, eolian depositional systems, evaporites, coal beds? (see text), lake deposits

This table should not be taken to mean that these are the only useful indicators for each climate parameter, but they are representative. Other indicators are found in the literature but were not discussed here because of length considerations. The table does not include modeling studies. In addition, as stated in the theme of this book, the strongest interpretations of paleoclimate are those based on more than one type of information.

[a]Sepiolite, palygorskite, and attapulgite are indicative of aridity as well as warmth. Other clays in the deep sea can provide some clues to temperature but are more indicative of general weathering conditions on the continents.

[b]"Index fossils" are individual taxa. "Faunal indices" are statistical assemblages of taxa that are indicative of a certain climate.

[c]Assemblages defined by ranges of taxa or factor analysis (planktonic foraminifera), peak abundances of taxa (planktonic foraminifera), peak diversities (planktonic foraminifera), or as temperature indices (diatoms).

[d]Productivity is determined from foraminifera using their biogeography, dominance/diversity in planktonic foraminifera (also used in marine mollusks), diversity coupled with factor analysis of assemblages (also used in nannofossils and radiolarians), and planktonic/benthic ratios.

[e]See discussion in text.

in determining the distribution of taxa when studied at small scales than at large scales. In contrast, the physical environment, for example, geography or climate, may be more important at large scales.

Biogeographic units. Biogeographers and paleobiogeographers attempt to divide biotas into biogeographic units of various scales. No fixed hierarchy of biogeographic units exists. The smallest units are generally called communities or assemblages. The term "community" has ecological implications that some workers wish to avoid, and these workers tend to use "assemblages." "Faunule" and "florule" also are used to indicate small biogeographic units, and a number of other terms exist. "Assemblage," "faunule," and "florule," however, also may be used to designate the organisms from a single locality, which has no particular biogeographic meaning, whereas "community" implies a group of organisms of a certain taxonomic composition that is found in more than one locality. "Endemic center" was used by Kauffman (1973) in nearly the same sense that other paleobiogeographers have used "community." The next higher biogeographic unit is the "formation," a term that is uncommon in the paleobiogeographic literature. "Region" is a biogeographic unit that is sometimes used in a sense that implies that it is larger than a "formation" but smaller than a province, but it has also been used to describe the next level larger than province. This term also is not common, and the term "subprovince" tends to be used for the next level smaller than province. The most common terms are those applied to the next two largest biogeographic units: "province" and "realm." Provinces are generally of a size such that the globe will have between four and twenty provinces, depending on the taxon studied, the time studied, and whether the unit is defined on the basis of one or more higher taxa (that is, just bivalve mollusks or all mollusks, for example). "Realm" is usually reserved for the largest biogeographic units, and this usage is different from the use of the term to distinguish marine from terrestrial environments. Realms generally contain from zero to several provinces. In the former case, global biogeographic differentiation is interpreted as low, and thus the global biota can be divided into only two or three biogeographic units. Provinces and subprovinces are the units most likely to be related to either climate or continental positions.

Although some attempts have been made to set standard criteria for defining biogeographic units, in practice provinces are defined by geographic breaks in diversity or by obvious changes in biotic composition, where they are defined at all. This qualitative method of defining biogeographic units is what I occasionally refer to as "traditional paleobiogeography." This approach has been criticized (Rosen 1984, 1992), mostly because criteria for establishing provinces are not stated or because arbitrary adjustments are made even to quantitatively defined provinces.

The most detailed global paleobiogeographic study was of Cretaceous bivalves, undertaken by Kauffman (1973) partly to establish a uniform method of defining biogeographic units, in order to bring about continuity in biogeographic analysis through time and to facilitate the study of the evolution of biogeographic patterns. A notable feature of Kauffman's biogeography is that he strictly followed set limits in percent endemics at each level (table 1.3), and this resulted in a number of biogeographic units that were displaced in the hierarchy. For example, the North American Province divided in the Cenomanian into the Gulf-Atlantic coast subprovince and the Western Interior endemic center. Standard practice would have the province divided into two subprovinces, regardless of the levels of endemism, but by adhering to strict quantitative limits, Kauffman preserved potentially valuable information. Indeed, he simply formalized observations made by other paleobiogeographers, that the global paleobiota cannot always be neatly divided into equivalent biogeographic units (e.g., Charpentier 1984).

Unfortunately, few, if any, paleobiogeographers followed Kauffman's (1973) lead in these approaches, so his study failed to standardize paleobiogeographic analysis, and it remains the least comparable to others. One can speculate on the reasons why his approach was not embraced. One reason is likely the preference for a strictly hierarchical system, despite the evident problems. Another is the fact that strict numerical criteria cannot be used to distinguish even modern biogeographic units (e.g., Campbell and Valentine 1977), and many paleobiogeographers do not attempt to work at the species level on a global scale. Finally, Kauffman assumed that modern biogeography is a model for the past,

Table 1.3
Limits of endemism for the hierarchy of biogeographic units established by Kauffman (1973)

Biogeographic unit	Percent endemic species[a]
Realms	>75%
Regions	50–75%
Provinces	25-50%
Subprovinces	10-25%
Endemic centers	5-10%

[a]Percentages are calculated excluding cosmopolitan taxa (that is, taxa that occur in every biogeographic unit)

which did not allow for the possibility that global faunas might at times be so depauperate or the taxa so widespread that biogeographic differentiation was lower than at present.

Numerous sources of error can affect paleobiogeographic interpretations (good discussions in Charpentier 1984; Rosen 1992). These are misdating of samples, taxonomic misidentification, paleogeographic mislocation of small continental fragments, preservational bias, sampling bias toward North America and Europe, and publishing biases of various sorts, such as lack of availability of industry publications. Additional problems arise from the use of form taxa, such as leaves, or multi-element organisms, such as conodonts and plants, in which different body parts may have different taxonomic names. Provinciality is defined on the basis of biologic taxa, so the use of form taxa could introduce error, especially in estimates of diversity.

Diversity gradients. In many groups of organisms, diversity follows a latitudinal gradient, independent of other aspects of their biogeography, such that diversity is highest in low latitudes. Diversity, also called species richness by paleobiogeographers (Hallam 1994), is the number of different taxa in a specified geographic area. In the oceans, diversity gradients are almost entirely controlled by temperature; salinity gradients that are strong enough to affect biogeography are very local. Figure 1.7 illustrates diversity gradients for modern bivalve mollusks in the Pacific Ocean. Note that the gradient is different on either side of the Pacific. Sharp breaks in the gradient can indicate strong temperature gradients between water masses, but diversity in the marine realm, particularly among microorganisms, is complicated by variations in productivity (e.g., Ottens and Nederbragt 1992). Diversity gradients also occur in terrestrial organisms but are complicated by the fact that global land masses are not as contiguous as the global ocean and by the factors in addition to temperature that affect terrestrial diversity, most notably rainfall. Diversity gradients are widely used in paleoclimatology.

As global diversity changes, chances are that the interpreted paleobiogeography, including diversity gradients, will also change. Ecological and physical space are also thought to control diversity, although the relationship is by no means straightforward (Valentine and Moores 1970; Flessa 1975; Flessa and Sepkoski 1978; Schopf 1979; see Haydon, Radtkey, and Pianka 1993, and Schluter and Ricklefs 1993 for more recent discussions). Diversity is also affected by different philosophies in establishing the taxonomy of organisms (e.g., see Leckie 1989), by erection of parallel taxonomies by workers in different countries (Berggren 1977), and by sampling (Hanski, Kouki, and Halkka 1993). Rarely are these problems addressed in studies that use diversity gradients or raw diversity data in paleoclimate interpretation.

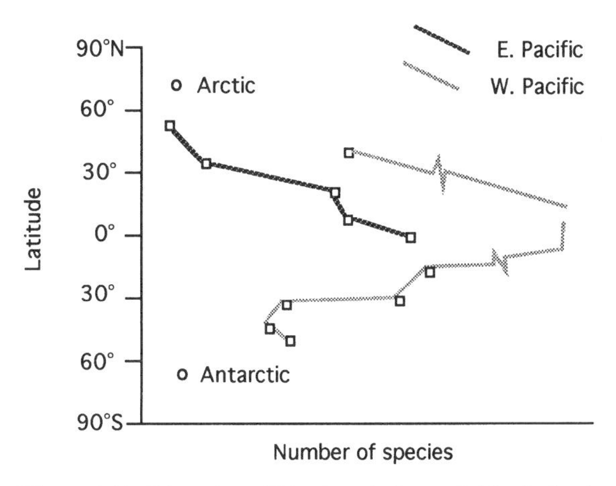

Figure 1.7. Diversity of bivalves (schematic) by latitude in the eastern and western Pacific Ocean. From data in Schopf (1970).

Numerical methods in paleobiogeography. To use biogeographic units in paleoclimate studies, we must be able to define them. The most objective way to define paleobiogeographic units is statistically, using similarity coefficients. Similarity coefficients are a class of statistics that measure the similarity of the composition of two samples. Thirty-nine similarity coefficients have been used in biogeography (Shi 1993a), and different coefficients treat the data quite differently. In practice, similarity coefficients are not used as extensively as one might expect, at least not for paleoclimate interpretation, although they are commonly used to determine biogeographic affinity among different continental plates (e.g., Flessa et al. 1979). Part of the reason is illustrated by an example from modern biogeography, in which the similarities of the faunas in modern provinces were compared. Campbell and Valentine (1977) used both Simpson's and Jaccard's coefficients ($a/(a + b)$ and $a/(a + b + c)$, respectively, where a is the number of taxa in both faunas and b and c the number of taxa in each fauna). At the generic level, modern provinces have similarities as low as 0.23 (Simpson's) and

0.06 (Jaccard's) for those that are distant and as high as 0.91 (Simpson's) and 0.7 (Jaccard's) for adjacent provinces. At the family level, the similarities are 0.73 to 0.97 (Simpson's) and 0.47 to 0.85 (Jaccard's). Obviously, establishing a single similarity threshold below which two faunas might be regarded to be in different provinces is not possible, although such attempts have been made. A graphical method using similarity coefficients is called a similarity net.

Multivariate statistical methods are commonly used in analysis of paleobiogeography and paleoclimate. These can be divided into transfer functions, which identify assemblages and link them to physical parameters, and various forms of factor analysis, which use similarity matrices of samples or taxa. Transfer functions statistically define the composition of assemblages from modern biotas in certain environmental conditions; those assemblages are then used to identify similar environmental conditions in the past (Imbrie and Kipp 1971; Sachs, Webb, and Clark 1977; Cronin and Dowsett 1990; Pisias, Roelofs, and Weber 1997). The method is based on two assumptions, that the biology of the organisms is systematically and linearly linked to physical parameters and that the environmental requirements of the extant forms is identical to the extinct forms in all ways. Indirect ecological linkages to environment are problematic. Examples of circumstances that would affect the analyses are changes in the ecological structure of the biota across several groups, for example, during changes in total diversity; nonequilibrium between the present-day climate and biota; ranges of past climatic conditions that are not equivalent to present-day ranges or conditions with no modern analogs; or variations in selective preservation further back in time (Kipp 1976; Ruddiman 1977; Sachs, Webb, and Clark 1977; Bartlein and Whitlock 1993). An example of a no-analog condition would be the proposed low-salinity layer at the ocean surface above 40° latitude in the North Atlantic at the last glacial maximum. An example of changes in ecological structure of the biota would be the blooms of the coccolithophorid *Braarudosphaera* and the radiolarian *Cycladophora davisiana* at times during the Tertiary. Further complicating matters is a poor understanding of the tolerances of modern forms. For microfossils, this is commonly determined from core-top samples related to present-day climate parameters, but the core-top samples probably include some fossils as old as 5000 Ka (Ruddiman 1977).

Several statistical methods are used to construct transfer functions, including regression analysis, principal components analysis, temperature indices with regression, response-surface analysis (which correlates a single taxon with several climatic variables), and others. All these methods give approximately the same results (Mix 1989; Bartlein and Whitlock 1993). Semiquantitative, graphical transfer functions have also been used (Hazel 1988).

The principal limitation to transfer functions from the point of view of this book is that they are applicable to only very young biotas, Miocene or Pliocene at the oldest (e.g., Cronin and Dowsett 1990) and perhaps even younger for some groups, although transfer function-like statistics have been developed for older fossils (e.g., Kovach and Spicer 1996). The reason is that the method is developed from modern species and their climate parameters and the further back in time, the smaller is the component of modern species in the biota. Although the method could conceivably be extended back in time indefinitely by reference to successive biotic associations, in practice, the assumptions that are inherent in the method become more and more problematic.

Factor analysis is performed in two modes, called R mode and Q mode. In R-mode analysis, taxa are grouped by co-occurrence, whereas in Q-mode analysis, samples are grouped by their similarity in taxonomic composition. Two types of factor analysis are common in paleobiogeographic studies, cluster analysis and ordination. Cluster analysis shows links among taxa or samples, the strengths of which are indicated by the statistical rank of the links. The principal advantages of cluster analysis are that the graphical display of relationships is easily interpreted and that the method does not require assumptions about the data, such as linear or normal distribution, that are common to ordination techniques. The disadvantages are that a dendrogram is two-dimensional and unordered and therefore does not reflect environmental gradients, and that distortions can be introduced because the data are forced into discrete groups where none might exist.

Ordination techniques produce groupings of taxa or samples according to their similarities to other groups such that the first axes explain most of the variability in the data set. For example, a Q-mode analysis of shallow-water marine localities might show groupings along three axes, of which the first would likely be temperature, the second salinity, and the third substrate characteristics. Along the first axis, localities occurring in warm water would cluster at one end, those in cold water at the other end, and those from intermediate temperatures in the middle, and a similar pattern might be seen along the second and third axes for low to high salinity and soft to hard substrates, respectively. The analyses produce correlation matrices that, in Q mode, can be mapped geographically. Climate parameters are usually not part of the analysis, so interpretation of the axes of variability in factor analysis is subjective. Some techniques, such as canonical correspondence analysis, do include climate parameters as part of the ordination.

Common ordination techniques are polar ordination, various types of correspondence analysis, and principal components analysis; principal coordinate analysis and nonmetric multidimensional scaling are used less commonly (Shi 1993b). Raw data are either presence/absence or abundance data (Shi 1993b), although a few workers have used trinary, ranked abundances, that is, 0 (absent), 1 (present), 2 (common) (Corrège 1993). Because abundance data may be biased by taphonomic effects and not reflect true original abundance, Shi (1993b) recommended limiting analysis to binary data sets, especially for global studies. In practice, this is usually the only way such studies can be approached because many published data sets include only presence/absence anyway.

All ordination methods are conceptually similar except polar ordination, the characteristics of which are summarized in table 1.4, and were comprehensively reviewed by Shi (1993b). In polar ordination (PO), two points from the data set represent poles and the others are arrayed along the ordination axis between them. Principal components analysis (PCA) has the advantage of performing dual ordination, both R- and Q-mode, so that the results can be illustrated on a single plot, one mode as vectors and one as points, displaying the relationship between samples and variables (taxa). Principal coordinate analysis (PCO) is similar to PCA but less restrictive for binary data. Unlike PCA and PCO, which give equal weight to all data, correspondence analysis (CA) weights the data to maximize the correlation between taxon and sample scores. CA works well if the data are dominated by one or two strong gradients, but otherwise it does poorly. Detrended correspondence analysis (DCA) was designed to overcome artifacts of the other methods. Most comparison studies have found DCA to be superior to CA, but it can mask curvilinear relationships that are real (Shi 1993b). Canonical correspondence analysis (e.g., Kovach and Spicer 1996, in a paleoclimate application) allows direct incorporation of climate parameters.

As with all statistics, numerical paleobiogeographical techniques are useful for discerning patterns in the data. However, in no case can the statistical result be taken without consideration of the actual physical properties and limitations of the data set itself and the context of its construction. In cases throughout this book in which one of these techniques is used, the importance of examining the context of the data will be abundantly evident.

Limitations of paleobiogeographic methods. A number of problems exist in interpreting paleobiogeography (Rosen 1992). First, paleobiogeographers implicitly use modern biogeographic patterns, which are based on species, as reference for how paleobiogeographic patterns "should" look (e.g., Kauffman 1973). However, paleontologists commonly cannot work at the species level; hard parts may be diagnostic only at higher levels. The higher the taxonomic level, the larger will be the smallest definable biogeographic unit. Second, biogeographers studying modern faunas have the advantage of being able to study the distribution of organisms in relation to known barriers to distribution, and boundaries between biogeographic units may be defined partly on physical barriers that may or may not represent discontinuities in biotic composition. Paleobiogeographers, on the other hand, may not know what the physical barriers were and therefore must rely more heavily on clues from biogeographic differentiation. In addition, continents have split and sutured and slivers of continental crust have moved long distances compared with the major plates. This means that the paleobiogeographer must first identify paleogeographic boundaries, in order to avoid lumping faunas that were not near each other or splitting those that were.

Third, sampling of a fossil biota is almost certainly incomplete because certain organisms are not preserved easily or at all. Selective and differ-

Table 1.4
Properties and mutual comparisons of multivarate statistical methods used in biogeography

	Technique						
Properties	Cluster analysis	Polar ordination (PO)	Principal component analysis (PCA)	Principal coordinate analysis (PCO)	Correspondence analysis (CA)	Detrended correspondence analysis (DCA)	Nonmetric multidimensional scaling (NMDS)
Purposes	To forcibly divide objects into groups	To summarize large data sets by reducing multidimensionality into a lower dimension, to display interrelationships of data points (samples or taxa in the lower dimension, and to extract major directions of data variation					
Assumptions	Data expected to be heterogenous	Data sets are expected to be continuous and linear				No assumptions made about data normality or linearity	
Raw data	Binary or quantitative	Binary or quantitative	Mainly quantitative	Binary or quantitative	Binary or quantitative	Binary or quantitative	
Input format	Similarity or distance matrix	Similarity or distance matrix	Correlation or covariance matrix (Euclidean distance)	Similarity or distance matrix	Distance matrix (Chi-square distance)	Distance matrix (Chi-square distance)	Rank-order dissimilarity matrix
Output fomrat	Two-dimensional hierarchical plot	Usually two- or three-dimensional scatter plot (ordination space)					
Characteristics	Both Q-mode and R-mode analysis	Dual ordination	Q-mode only	Samples and taxa dually ordinated	Samples and taxa dually ordinated	Samples and taxa dually ordinated	Q- and R-mode axes chosen in advance
Distortion	Between-group relationships distorted, but within-group distortion minimized	Within-group relationships distored, but distortion of between-groups relationships minimized Moderately tolerant to horseshoe effect	Most sensitive to horseshoe effect	More tolerant to horseshoe effect than PCA	Less affected by horseshoe effect (compared with PCA, PO, PCO)	Both methods are resistant to the arch effect and effects of non-linearity and heterogeneity of data	
Testing means	1. Cophenitic correlation coefficient; 2. Stopping-rule methods; 3. Trellis diagram	1. Applying ordination methods to either simulated data or real data with known results and comparing the arrangement of the data points in the ordination space with known relationships of the data points; 2. Calculating the Spearman correlation coefficient between ordination positions of data points and their original positions in the raw data; 3. Using a statistic measure; 4. Applying Procrustes analysis; 5. Using percentage of eigen values of stress values, but these values must be used with caution					

Reprinted from *Palaeogeography, Palaeoclimatology, Palaeoecology* 105, Shi, G.R., "Multivariate data analysis in palaeoecology and palaeobiogeography—a review," pp. 199–234 (1993) with kind permission of Elsevier Science-NL, Sara Burgerhartstraat 25, 1055 KV Amsterdam, The Netherlands. Shi also provided numerous references to applications of the various methods.

ential preservation are common, and climate itself can control preservation.

Fourth, how taxa are defined varies from group to group and among different workers; this is the monographic effect. For example, dinosaur paleobiogeography is well known, within the limitations of the available study sites, because dinosaurs have been popular animals for study, whereas smaller terrestrial vertebrates are relatively poorly known. Ironically, among terrestrial animals, the largest ones also were the most mobile and, perhaps, the least affected by (and least indicative of) climate. Marine taxa, which are well studied, are not studied with equal vigor. A scan of the paleontological literature shows, for example, that brachiopods have been much more popular subjects than crinoids. Because no set procedure exists for defining biogeographic units, paleobiogeographers for the various groups of organisms have differed in their interpretations of biogeography, which, in turn, means that paleoclimate can be interpreted very differently for the same time from biogeographic patterns of different groups. Monographic effect includes bias in preferential sampling on certain continents (Charpentier 1984).

Rosen (1992) pointed out that there are six quite different approaches that are commonly used in paleobiogeographic studies, of which two are particularly relevant here. Both start with collection of raw paleobiogeographic data and determination of the paleobiogeographic patterns. The first approach combines the interpreted paleobiogeographic patterns with assumptions about paleobiogeographic processes to examine geologic processes. The second approach combines the interpreted paleobiogeographic patterns with assumptions about geologic processes to examine paleobiogeographic processes. Both approaches are used in paleoclimate studies, particularly on a global scale. The first approach treats paleoclimate as if it were a paleobiogeographic process, usually assumes a zonal pattern for paleoclimate and therefore for large-scale biogeographic patterns, and goes on to make arguments about the distribution of continents. The second approach assumes that the continental reconstructions are accurate and interprets paleoclimate from the resulting paleobiogeographic patterns. As noted by Rosen (1992), these approaches are commonly used in the same study with no acknowledgment of resulting methodological confusion. Perhaps not in a single paper, but certainly over a series of papers, this lack of rigor can lead to circular results (Rosen 1992). Rarely, if ever, do paleobiogeographers question the assumption that climate influenced paleobiogeographic patterns (Rosen 1992).

The heart of this discussion is that how paleobiogeographic patterns are defined will influence how paleoclimate is interpreted from them. Nowhere has this been more evident than in the study of Cretaceous bivalves by Kauffman (1973). He identified 17 provinces and subprovinces. His evaluation of the biogeographic patterns led him to conclude that climate zonation was strong during the Cretaceous; by comparison with paleobiogeographic studies from earlier in the Mesozoic, temperature would have had to drop precipitously. In fact, the Cretaceous was one of the warmest intervals in Earth history (Fischer and Arthur 1977).

Paleoecology

Individual taxa may be direct indicators of paleoclimate or may be indicators of paleoclimate through their relationship to extant taxa. *Direct* indicators of paleoclimate are those taxa whose climatic tolerances are well known because they survive today, and whose presence in the fossil record is thus indicative of specific climatic conditions. Few direct indicators of paleoclimate exist in the pre-Quaternary fossil record because most species that ever lived are extinct (Raup 1986:47). Of the small proportion of fossil species that are still extant, few have climatic tolerances that are narrow enough to be useful in paleoclimatology. Individual species within the same genus can have widely differing climatic tolerances, so the use of direct indicators of climate is applicable only at the species level and is dependent on the paleontologist's confidence in identifying fossil organisms to that level. Transfer functions, in which groups of taxa are linked to climate and then the groups are used in paleoclimate interpretation, have rarely been extended further back in time than the Pliocene.

The use of *nearest living relatives* as paleoclimatic indicators is based on the assumption that a fossil taxon has the same climatic tolerances as its closest extant relative (see discussion in Wolfe 1993). A familiar example of this method is the use of dinosaurs as indicators of warm climates. Before paleontologists concluded that dinosaurs were probably not cold-blooded, it was thought that dinosaurs, like many extant reptiles, must have been dependent on the environment for warmth. The use of the NLR method has ranged from within-genus to within-class comparisons, and even within-genus comparisons can be problematic. For example, the plant genera *Metasequoia* and *Glyptostrobus* are now mutually exclusive in their biogeographic ranges, but in the Tertiary they commonly occurred together (Wolfe 1979, 1994a). An additional problem with the NLR method is that fossil taxa are sometimes not identified correctly (e.g., Wolfe 1971, 1993). The NLR method, if used judiciously, can, however, be helpful in paleoclimate interpretation.

Physiology and morphology

Some fossils might be termed *empirical-morphological* indicators of climate. This method is based on the observation that organisms respond to climatic variation in a limited number of ways and that, within broad groups of organisms, those that live in similar climates tend to have similar features. The features can sometimes be empirically and quantitatively linked to climate. The assumption is that the morphological and/or physiological responses to climate in these groups of organisms have not changed through time. The power of this method in some groups is that the morphological and/or physiological response is independent of taxon.

General Principles of Lithologic Indicators of Paleoclimate

Almost all lithologic indicators of paleoclimate are sedimentary rocks; the only exceptions are paleosols that formed on solid rock rather than on sediments, and these are relatively uncommon in the geologic record. Information about paleoclimate from rocks comes from their specific environments of formation and from their distribution in space and time on both global and local scales. The environments in which the indicators form determine their paleoclimatic significance. Their distribution determines how they are used to interpret spatial and temporal patterns of paleoclimate.

Lithologic indicators of paleoclimate are subject to most of the same limitations as biotic indicators, including sampling, dating, and the vagaries of the geologic record. A problem with the interpretation of lithologic indicators, particularly at a global scale, is absence of data in the sense that absence of data on lithologic indicators has, on occasion, been taken to indicate that the conditions for their formation did not exist. For example, workers might conclude that dry or wet zones on the continents have expanded or contracted, when all the data really show is a change in the distribution of relevant data. Because so many factors other than climate can determine whether a particular indicator formed and/or was preserved, using absence of data to interpret paleoclimate is not recommended.

Raw data for the use of rocks as indicators of paleoclimate is almost always binary, that is, presence/absence. This is because thicknesses and abundances are commonly controlled by factors unrelated to climate, such as basin subsidence rates. Only in the context of basin analysis, where all aspects of basin history are documented, can abundance and thickness data begin to be used to interpret changes in climate. However, such studies are unusual and, with respect to paleoclimate interpretation, full of caveats.

Methods of quantifying deposition, which is particularly common in studies of marine lithologic indicators of climate, are sedimentation rate (SR), mass accumulation rate (MAR), and component mass accumulation rate (CMAR, usually abbreviated according to the component measured, such as organic carbon, i.e., OCAR). Sedimentation rate is simply the thickness of rock divided by the time represented by that thickness. Sedimentation rate is a poor indicator of sedimentological processes except over the coarsest of scales, for two reasons. First, the finest time resolution possible in any given stratigraphic sequence is usually much longer than the time scale of the depositional processes by which the sequence formed. Thus many variations in sedimentation and even in environment can have occurred within the smallest time interval discernible by most dating techniques. Second, sedimentation in most if not all environments is discontinuous on all time scales. Although the amount of "missing" *time* can be estimated (e.g., Sadler and Strauss 1990), "missing" *events* cannot be divined. This means that climatic change, for example, might appear to be less or, more commonly, more abrupt than it really was.

Multivariate analysis is less commonly used in the study of lithologic indicators than in the study of biotic indicators of paleoclimate. The reason for this is that the data are not as easily quantified. Fossils occur as assemblages with characteristics that can be tabulated for each sample. No comparable "assemblage" of lithologic indicators exists, so statistical treatments are limited. In general, trends in space and time are documented in terms of changes in the distribution of indicators (e.g., Hallam 1984a; Witzke 1990) and compared informally and qualitatively with results of paleoclimate models (e.g., Parrish, Ziegler, and Scotese 1982; Otto-Bliesner 1993).

Two examples of statistical studies that have been done are the following. Parrish (1982; Parrish and Curtis 1982) used a simple chi-square test for a comparison of the distribution of upwelling indicators and the distribution of upwelling zones predicted from climate models. In this test, the measured area of continental shelf covered by predicted upwelling zones for each time period was taken as the probability that an upwelling indicator would correspond to a predicted upwelling zone simply by chance. The chi-square test revealed if more indicators corresponded with predicted upwelling than would be expected. Sageman and Hollander (in press) recommended use of moving-window cross-correlation, a method that had been used for dating but also can be applied to interpretation of patterns in time-series data. By using different window sizes, that is, fragments of the time series consisting of three to twenty data points, patterns and strengths of positive and negative correlation between paleoclimatic indicators through time can be documented. The significance of this method is that it can provide objective evidence of complex processes wherein, for example, two indicators might be correlated under certain conditions but not under others.

Stable Isotopes and Paleoclimate

The stable isotopes of oxygen, carbon, strontium, hydrogen, and nitrogen all have been utilized in paleoclimate studies or in paleoecological studies that have paleoclimatic implications. Of these, the stable isotopes of oxygen and carbon have been by far the most important and, in many types of studies, the analysis and interpretations of these isotopes has become almost routine. Stable isotope data provide quantitative information about paleotemperature; paleoprecipitation and paleoevaporation; glacial ice volume; production, burial, and weathering of organic carbon; and other processes that are related to paleoclimatic change. With the exception of strontium, all the isotopes directly record paleoclimatic or paleoecologic processes. Strontium, on the other hand, reflects the composition of rocks at the Earth's surface. It is related to climate indirectly in that the ratios of strontium isotopes in water are affected by weathering.

Stable isotopes are reported as ratios of heavy to light isotopes, and with the exception of strontium, the notation is usually in the form $\delta^{a}B$, where a is the atomic weight of the heavier isotope, B is the element, and δ is the change in the ratio of the heavy isotope to the light isotope relative to some standard. Changes in the ratio are brought about by fractionation of the isotopes during biologic and physical processes. Strontium isotope values are presented directly as a measured ratio, although some workers now use $\Delta^{87}Sr$ (e.g., Ingram and DePaolo 1993). One basic assumption is common to all studies using stable isotopes. That assumption is that the isotopic composition of the rock or fossil was controlled by the isotopic composition of the fluid in which the mineral (including organic matter) or shell was formed and by the mineral phase (Faure 1986). The relationships among the isotopic composition of the fluid, the mineral phase, and the isotopic composition of rock, organic matter, and shell are demonstrated by empirical studies and are the subject of continual refinement. In general, the key to paleoclimate interpretation using isotopes is determination of the controls on the isotopic composition of the fluid. In the following sections, the general principles of analysis and application of isotopes of each element are discussed briefly. Specific information on the application of stable isotopes is integrated into the sections on the rocks and fossils that are used for isotope analysis. For example, oxygen isotopes have been studied extensively in foraminifera. Because interpretation of the information on isotopes depends on interpretations of the ecology and biogeography of foraminifera, studies of the oxygen isotope information derived from these organisms is integrated with the other information derived from them.

Oxygen

Paleotemperature determinations rely on the observation that as temperature rises, the activity of ^{16}O goes up relative to ^{18}O, that is, if carbonate is formed at 20°C, it will contain a higher proportion of ^{16}O than if it formed at 10°C. Proportions of the isotopes are expressed as per mil (‰) $\delta^{18}O$, calculated as follows:

$$\delta^{18}O = \left[\frac{(^{18}O/^{16}O)_{sample} - (^{18}O/^{16}O)_{std}}{(^{18}O/^{16}O)_{std}}\right] \times 10^3 \quad (1.1)$$

where *std* refers to an analytical standard. The standards used are PDB (from the Pee Dee Belemnite) for carbonates, and SMOW (standard mean ocean water) for water and for minerals other than carbonates (e.g., Hudson and Anderson 1989). PDB is +30.86‰ on the SMOW scale (Arthur et al. 1983).

The numbers resulting from equation 1.1 can be either negative or positive. Negative values are referred to as light because they indicate a higher proportion of the light isotope ^{16}O, whereas positive values are referred to as heavy. If data shift toward more positive values, even if they are still negative, the data are described as becoming heavier or higher. If data shift toward more negative values, even if they are still positive, the data are described as becoming lighter or lower.

For low-magnesium calcite, generally regarded as the ideal source for oxygen isotope data because of its stability and abundance, paleotemperature is calculated from the isotopic composition using the following equation (Arthur et al. 1983):

$$t(°C) = 16.0 - 4.14(\delta_{carbonate} - \delta_{water}) + 0.13(\delta_c - \delta_w)^2. \quad (1.2)$$

Paleotemperature determinations can also be made from aragonite, apatite, and silica. The precipitation kinetics of these minerals are different, requiring different equations for the calculation of paleotemperature. For aragonite, the equation is

$$t(°C) = 20.6 - 4.34(\delta_a - \delta_w) \quad (1.3)$$

(see also Faure 1986; Grossman and Ku 1986). The general formula for apatite is $Ca_5(PO_4)_3(OH, F, Cl)$, although Cl is rare in biogenic apatite. Isotope determinations are made from the oxygen in the PO_4, which is diagenetically stable. For apatite, the paleotemperature equation is (Kolodny, Luz, and Navon 1983)

$$t(°C) = 111.4 - 4.3(\delta_p - \delta_w). \quad (1.4)$$

Paleotemperature determinations on biogenic silica are made with the following equation (Juillet-Leclerc and Labeyrie 1987):

$$t(°C) = 17.2 - 2.4(\delta^{18}O_{Si} - \delta^{18}O_w - 40) - 0.2(\delta^{18}O_{Si} - \delta^{18}O_w - 40)^2. \quad (1.5)$$

For all equations, δ_c and δ_a are relative to PDB; δ_p, δ_{Si}, and δ_w are relative to SMOW; and the more negative the sample, that is, the less ^{18}O it has, the higher the temperature.

Some variation in these equations appears in the literature as a result of refinements of the fractionation determinations (e.g., Erez and Luz 1983). However, the differences among the equations over the temperature range of interest to paleoclimatologists are insignificant (Hudson and Anderson 1989).

A number of processes must be considered before interpreting variations in oxygen isotopes as indicative of temperature changes (Arthur et al. 1983). These are

1. fractionation within the atmosphere and during crystallization,
2. changes in the original isotopic composition of the water,
3. biological fractionation, and
4. diagenesis, including differential dissolution (Wu, Herguera, and Berger 1990).

Fractionation of oxygen isotopes occurs within the atmosphere as water is evaporated, transported, and precipitated by atmospheric processes. This fractionation occurs on both global and local scales (Yurtesever 1975). Rainout, that is, the increasing fractionation of atmospheric moisture as an air parcel moves over land, condenses, and loses water in precipitation, is a problem mostly for interpreting terrestrial isotopic signatures; fractionation on a larger scale potentially affects the interpretation of marine isotopic signatures as well.

Because light water ($H_2{}^{16}O$) preferentially evaporates, saline water tends to have higher $\delta^{18}O$. Elevated salinity has the same effect on isotopic signatures as do cold temperatures. Because saline water is sometimes also warm, produced by evaporation at high temperatures, the effects of temperature and salinity on isotopic composition work against each other. Determining the relative effects of salinity and temperature can be difficult (Tourtelot and Rye 1969). However, many workers ignore the possible effects of salinity because they are relatively small.

The so-called vital effect means that the hard parts of some organisms are not useful for isotope studies, and the effect can be responsible for subtle variations even in organisms that are commonly used. Echinoderms (Weber and Raup 1966) and corals (e.g., McConnaughey 1989) deposit their tests out of isotopic equilibrium with seawater. They also form skeletons of high-Mg calcite, which, in addition to being diagenetically unstable (e.g., Popp, Anderson, and Sandberg 1986a), concentrates ^{18}O relative to low-Mg calcite. In groups that are commonly used for isotope analysis, for example, foraminifera, distinguishing vital from other effects can be difficult. Zachos et al. (1992) illustrated one way to do this (figure 1.8). The oxygen isotope values for ostracodes and the benthic foraminifer *Gyroidina* were plotted against those for the benthic foraminifer *Cibicidoides* from the same samples. The straight-line correlations indicate that the organisms all precipitate carbonate in isotopic equilibrium with seawater and the offsets are indicative of the relative vital effects.

The isotopic signature of biogenic or nonbiogenic minerals can be affected by diagenesis, which includes infiltration of the microstructure by diagenetic minerals, recrystallization, and dissolution. In shells, the criteria for the recognition of a lack of diagenetic effects are

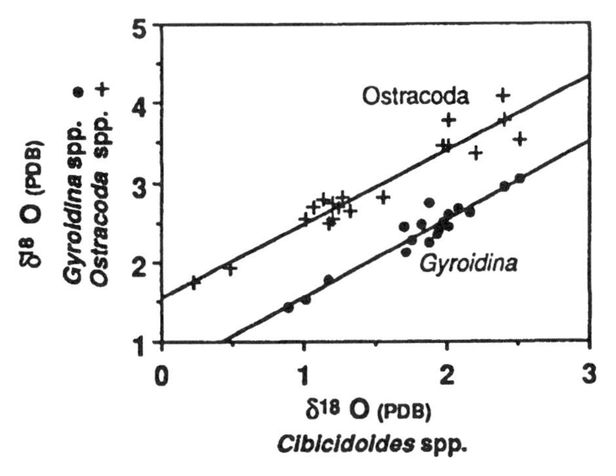

Figure 1.8. Comparison of the $\delta^{18}O$ composition of the benthic foraminiferan *Gyroidina* and benthic ostracodes, plotted against *Cibicidoides*. From Zachos et al. (1992).

1. 100% aragonitic composition (this will be true only in unaltered shells originally composed of aragonite),
2. preservation of the details of the microscopic shell structure,
3. absence of iron- and/or manganese-rich calcite,
4. lack of cathodoluminescence, and
5. preservation of original trace element ratios, such as Mg, Na, S, and Sr

(Popp, Anderson, and Sandberg 1986b; Veizer, Fritz, and Jones 1986; Grossman 1994; Mii and Grossman 1994; but see Land 1995). In addition, $\delta^{18}O$ and $\delta^{13}C$ are lower in diagenetically altered shells, which can be useful for comparing shells from the same taxon and within a single geologic unit. Discerning diagenesis in phosphate is more difficult, and, early on, phosphates were regarded as immune to diagenesis with respect to their $\delta^{18}O$ signatures (Kolodny, Luz, and Navon 1983).

The greatest challenge in studies using oxygen isotopes to estimate temperature is knowing the original isotopic composition of the water. The original isotopic composition of the source water for oxygen-bearing minerals has varied substantially through time, and this can affect paleotemperature determinations. The principal source of variation, especially in the oceans, appears to have been the volume of continental ice. Ocean water during the Ice Ages was heavier (enriched in ^{18}O) than is modern ocean water (Shackleton 1967). The ice-volume effect was variable through time as the ice caps waxed and waned, but it was around 1.2‰. A comparable variation caused by temperature alone would require a temperature change of about 7°C. On the continents, rainout and latitudinal effects may have been as or more important than the ice-volume effect in determining the original isotopic composition of the water.

Determining the original isotopic composition is particularly difficult for continental waters in temperate latitudes, where $\delta^{18}O$ is strongly affected by both temperature and precipitation changes (Lawrence and White 1991). Dettman and Lohmann (1993) showed how $\delta^{18}O$ in freshwater bivalve shells ($\delta^{18}O_s$) might vary in constant-$\delta^{18}O$ water ($\delta^{18}O_w$), where temperature is the dominant control on shell $\delta^{18}O$. In the example in figure 1.9, $\delta^{18}O_s$ variability is a direct measure of the *minimum* temperature variability (figure 1.9a–c). Variation of $\delta^{18}O_s$ in waters that vary in both temperature and $\delta^{18}O$ is shown in figure 1.9d,e. Variation in $\delta^{18}O_w$ results in greater or lesser dampening of the temperature signal because warm temperatures cause both rainfall and surface waters to be enriched in ^{18}O but shell material to be depleted in ^{18}O. Environments dominated by seasonally variable water can be the most difficult to interpret because the temperature effects can cancel out the $\delta^{18}O$ variations of the water. This would be obvious in the fossil record only where the $\delta^{18}O$ variations exceed the possible temperature effect. Another complication is identifying the growth season for the organism studied. Dettman and Lohmann (1993) used dark bands to identify the winter intervals of mollusk shell growth ("biological shutdown" on figure 1.9); these had the heaviest isotopic values. However, dark bands are not consistently indicative of such conditions (Jones and Quitmyer 1996).

A feature of the global $\delta^{18}O$ record that has puzzled geologists at least since the 1960s is the very light values obtained from Paleozoic rocks and fossils (Knauth and Epstein 1976; Karhu and Epstein 1986; Popp, Anderson, and Sandberg 1986b; Veizer, Fritz, and Jones 1986; Hudson and Anderson 1989; Railsback 1990). These values are consistent with temperatures as high as 30°C or more. If an ice-volume effect is included, to adjust for the known Carboniferous glaciation and the purported Devonian glaciation, the estimated temperatures would be even higher. Four explanations have been offered for these observations: (1) changing isotopic composition of seawater, (2) higher temperatures, (3) sequestering of warm, saline, $\delta^{18}O$-enriched waters in the deep ocean, and (4) increased diagenetic alteration of older samples. It is regarded as unlikely that the isotopic composition of seawater has changed much through time (Veizer, Fritz, and Jones 1986). None of the mechanisms for such a change could act on the short time scales indicated by some of the fluctuations, and evidence from meteoric cements (Hays and Grossman 1991) and ophiolites indicate that $\delta^{18}O$ has remained the same (Muehlenbachs and Clayton 1976; Gregory 1991). High temperatures seem inconsistent with current knowledge of the temperature tolerances of shallow-water marine organisms; such high temperatures for early parts of the Paleozoic would require parallel evolution at a high taxonomic level among all groups (Veizer, Fritz, and Jones 1986). The same problem applies to evolutionary changes in biological fractionation of isotopes (Veizer, Fritz, and Jones 1986). Railsback and colleagues (1990; Railsback et al. 1990) noted that the $\delta^{18}O$ data for the Paleozoic are all from shallow-water carbonates and fossils. If bottom-water generation for the early Paleozoic was in low latitudes and the water warm and saline, as seems likely given the geography, shallow waters could have been significantly depleted in ^{18}O (see also Luz, Kolodny, and Kovach 1984). None of these explanations has been embraced by all workers, and some argue that Paleozoic, especially early Paleozoic, isotope data must be used with extreme caution, regardless of evidence for lack of diagenesis or the organism from which the data come. Grossman's (1992) recommendation was "know your samples," which can apply to isotope studies of any organism. In referring to isotope analysis of brachiopods, Grossman (1992) specifically mentioned the need to understand sedimentological, local paleogeographic, and tectonic histories of the samples.

Carbon

Carbon tracks changes in the carbon cycle; changes caused by temperature are only 0.035‰ per °C. The carbon cycle is complex and fractionation strong among the various reservoirs except ocean water and carbonate (figure 1.10). Atmospheric and dissolved oceanic CO_2 is taken up by plants, with consequent isotopic fractionation, which is different for different groups of plants. Fractionation is slight between the ocean carbon reservoir and carbonates, so that deposition and dissolution of large amounts of carbonate have little effect on the carbon isotopic composition of the oceans, despite the great disparity in carbon reservoir size. In contrast, fractionation is great between the atmosphere, ocean, and plants. Thus deposition and weathering of organic carbon, particularly in marine sediments, has a strong effect on the oceanic and atmospheric carbon isotopic composition. Carbon isotope studies have been applied in various ways to the study of variations in CO_2 in the atmosphere (Hayes 1993).

Strontium

Strontium isotopes are not directly indicative of climate. Rather, strontium isotopes reveal changes in processes that can either cause or be

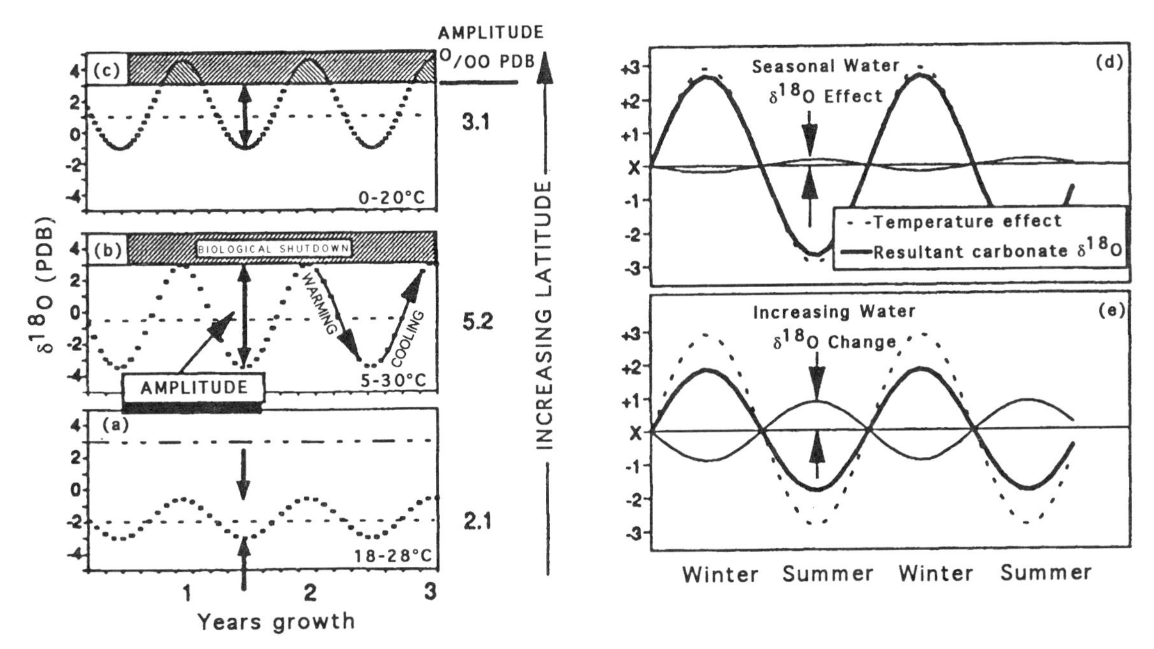

Figure 1.9. *(a–c)* Schematic patterns of the variation in $\delta^{18}O$ in the growth of bivalves in different temperature regimes. In all cases, the $\delta^{18}O$ of the water is constant. *Dots* are $\delta^{18}O$ of the bivalve shells, which varies with the annual cycle, and the *cross-hatch area* is the period during which growth ceases, so that in (c), the upper parts of the $\delta^{18}O$ curves are not represented in the bivalve shell $\delta^{18}O$. *(d,e)* Models of seasonal change in shell $\delta^{18}O$ as a deviation from a mean value (X). The effects of temperature and varying water $\delta^{18}O$ are combined. The temperature effects are the same in both examples, but the change in water $\delta^{18}O$ during the year is greater in (e) than in (d). From Dettman and Lohman (1993) in *Climate Change in Continental Isotopic Records*, pp. 153–63. Geophysical Monograph 78, copyright by the American Geophysical Union.

the result of climatic change. The strontium isotopic composition of seawater, which is recorded without fractionation in marine organisms, is controlled by riverine input of chemical weathering products from the continents, exchange with the upper mantle/crust at submarine hydrothermal vents, and diagenesis and dissolution of deep-sea carbonates:

$$N\, dR_{SW}/dt = J_r(R_r - R_{SW}) + J_h(R_h - R_{SW}) + J_c(R_c - R_{SW}) \qquad (1.6)$$

where N is the total number of moles of Sr in the ocean; J_r, J_h, J_c are fluxes of riverine, hydrothermal, and carbonate (diagenesis or dissolution) Sr, respectively; and R_r, R_h, R_c, R_{sw} are the $^{87}Sr/^{86}Sr$ ratios of river water, hydrothermal water, carbonate, and seawater, respectively. Carbonates have a small effect because the fluxes are small and the isotopic ratio is not very different from seawater, so they are commonly ignored.

The use of strontium isotopes as climatic indicators results from changes in $^{87}Sr/^{86}Sr$ in marine organisms that can be traced as changes in riverine influx and the extent to which the changes in riverine influx can be attributed to climate. Distinguishing climate effects from other effects is currently a subject of intense research (Berner and Rye 1992; Clemens, Farrell, and Gromet 1993).

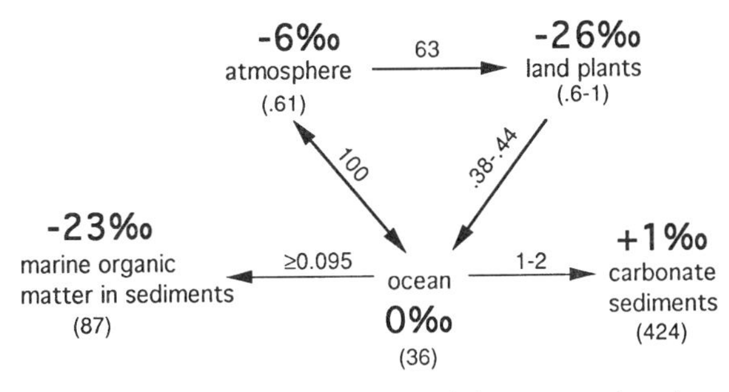

Figure 1.10. Simplified diagram of the principal carbon reservoirs and fluxes among them and the mass of carbon in each reservoir. The *large numbers* indicate the isotopic composition of carbon in each reservoir. The numbers paralleling the *arrows* are average values for the fluxes, in 10^{15} gC/yr; recycling between sediments and ocean is excluded. Numbers in parentheses are estimated masses in 10^{18} gC. From information in Arthur et al. (1983).

Nitrogen

Organisms take up light nitrogen preferentially. Therefore where productivity is high, $\delta^{15}N$ in sedimentary organic matter remains low, but under low productivity conditions, $\delta^{15}N$ increases. The equation for determining $\delta^{15}N$ is

$$\delta^{15}N = \left[\frac{(^{15}N/^{14}N)_{sample} - (^{15}N/^{14}N)_{atm.\ N2}}{(^{15}N/^{14}N)_{atm.\ N2}} - 1\right] \times 1000 \qquad (1.7)$$

The standard is the isotopic composition of atmospheric nitrogen.

2

Marine Biotic Indicators of Paleoclimate

Marine biotic indicators of paleoclimate include microfossils and macrofossils, and paleoclimate information comes from the biogeography of the organisms and their chemistry. Microfossils have provided more information by far than macrofossils about paleoclimatic patterns and change for the Cretaceous and Tertiary Periods, and the chemistry of calcareous microfossils, in particular, has been extremely important. For Earth history before the Cretaceous, most information about paleoclimate in the marine realm has come from the biogeography of macrofossils, although microfossils and fossil chemistry also have been important in interpretations of climate for some intervals.

Most microorganisms lived in the water column and their remains were deposited across the ocean basins, so their fossils are indicators of open-ocean conditions. Near the continents, the record of microfossils tends to be diluted by clastic sediments. In contrast, most of the macroorganisms lived on the continental margins and in epicontinental seaways, so their fossils are better indicators of paleoclimate on the continental shelves. Contributing to the differential utility of micro- and macrofossils is the fact that preserved ocean-basin sediments Cretaceous and older are sparse.

MARINE MICROFOSSILS

Microfossils are the hard elements of microscopic, usually single-celled, organisms, or microscopic hard elements of larger, soft-bodied, multicellular organisms. Microfossils represent three types of habits—benthic, planktonic, and nektonic[1]. Benthic organisms live on the sea floor or in the sediments. Planktonic organisms live in the water column and usually have limited or no ability to move through the water except by passive means (for example, by changing their specific gravity). Nektonic organisms are capable of active motion through the water column. Most paleoclimate information comes from planktonic microfossils, but benthic groups have provided important clues to climate and the evolution of the oceans (Kennett 1982).

The major groups of marine microfossils that have been used in paleoclimate interpretation are listed in table 2.1, and their stratigraphic ranges are in figure 2.1. The composition of these microfossils is diverse. The most important group, the foraminifera, is calcareous; these organisms are both planktonic and benthic. Other calcareous groups are nannofossils (predominately the remains of calcareous algae) and ostracodes (bivalved crustaceans). Groups that have siliceous tests are diatoms, radiolarians, and silicoflagellates. Dinoflagellates and tasmanitids

1. A note about terminology: These terms also have the form planktic, benthic, and nektic. For reasons that are not obvious, it has become customary to use both "planktic" and "planktonic," but with a distinct preference for the latter. "Benthic" is used far more commonly in the literature than "benthonic," and "nektic" is used only rarely. Therefore in this book, the terms "planktonic," "benthic," and "nektonic" are used, even though the form is inconsistent.

Table 2.1.
Marine microfossils used in pre-Quaternary climate studies

Benthic	
	Foraminifera[ab]
	Ostracodes[ab]
Nektonic and/or Benthic	
	Conodonts[ab]
Planktonic	
	Foraminifera[ab]
	Diatoms[ac]
	Coccoliths[ac]
	Radiolarians[ab]
	Dinoflagellates[d]
	Silicoflagellates[d]
	Tasmanitids[c]
	Bolboforma[e]

[a]Most important.
[b]Heterotrophs.
[c]Autotrophs.
[d]Flagellates.
[e]Possibly heterotroph.

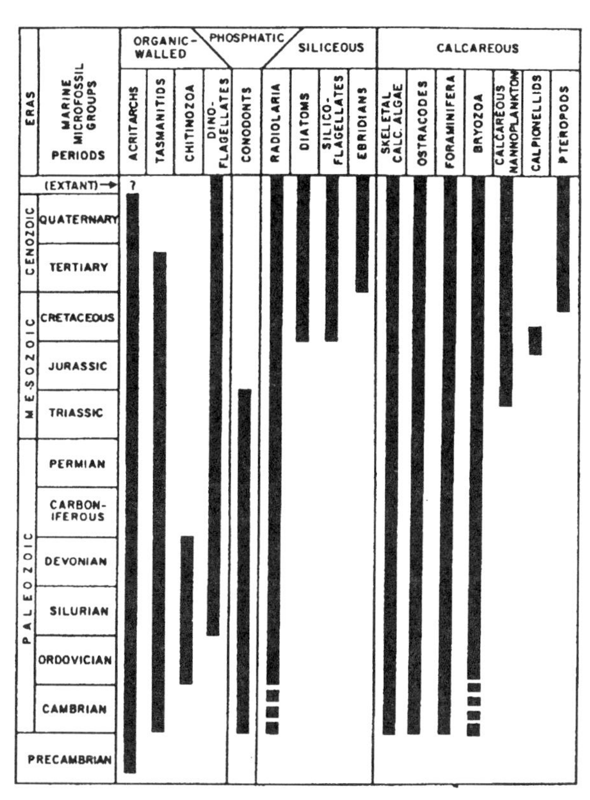

Figure 2.1. Stratigraphic ranges of microfossil groups. Modified from Berggren (1978).

have resistant, organic-walled tests. Conodonts are the phosphatic elements of a larger, soft-bodied organism.

Some microfossil groups, such as pteropods (planktonic snails whose remains are reduced, conical shells) and acritarchs, have, for a number of reasons, so far proved to be of limited use in pre-Quaternary climatology. Some are abundant enough for paleoclimate studies only in the Quaternary, others have been used mostly in biostratigraphy or in biogeographic studies principally concerned with paleogeographic reconstruction (e.g., Colbath 1990). These taxa are not treated here.

In paleobiologic studies, no formal depth divisions exist for planktonic organisms other than the distinction between photic and aphotic zones. The photic zone is the upper portion of the water column, which is illuminated by sunlight; it can vary in depth depending on the latitude and turbidity of the water. Paleodepth ranges for some groups of planktonic organisms are referred to as shallow, intermediate, and deep. The depths implied by these terms vary from taxon to taxon and through time, but they encompass at most the upper 200 m of water. By far the highest concentrations of most organisms are within the photic zone.

Foraminifera

Based on the sheer volume of literature, planktonic foraminifera have been the most important fossils in paleoclimate studies, because biogeographic and geochemical techniques are combined effectively in the study of these organisms. Foraminifera are benthic or planktonic heterotrophs (Culver 1993). Benthic foraminifera evolved first, in the Cambrian, and the earliest forms were agglutinated or pseudochitinous (Culver 1993); most other benthic foraminifera construct their tests from calcite. Benthic foraminifera occur in brackish to normal-salinity waters at all depths and latitudes (Kennett 1982). The highest diversity of benthics is in the equatorial regions, but the latitudinal diversity gradient can be complicated by local conditions, especially high productivity. The so-called larger foraminifera, which are 300 μm to 16 mm long and which were particularly diverse in the mid-Paleozoic, occur only in shallow tropical waters (Kennett 1982).

Planktonic foraminifera evolved in the Middle Jurassic and began to diversify in the late Early Cretaceous (Leckie 1989). They are stenohaline and shallow dwelling, found within the upper 200 m of the water column. Three depth-related assemblages have been identified in modern faunas (Bé 1977): shallow dwellers, at <50 m; intermediate dwellers, at 50 to 100 m; and deep dwellers, at >100 m. However, foraminifera do not occupy rigid depth zones and many species occupy relatively broad depth ranges (Bé 1977; Fairbanks et al. 1982). In addition, the preferred depths can be related to the thermocline (that is, above, within, or below it), and if the position of the thermocline relative to the surface changes, so might the depths at which the foraminifera live (Kennett, Keller, and Srinivasan 1985). Planktonic foraminifera also show a general diversity gradient with latitude, but also with complications (Ottens and Nederbragt 1992). Modern planktonic foraminifera are distributed in latitudinal zones closely related to temperature, and these species are bipolar. This facilitates the study of global climates because the southern representatives of a species have temperature requirements similar to those of the northern representatives, and such a relationship is assumed for past assemblages as well.

Foraminiferal morphology and paleoclimate

Some species of planktonic foraminifera exhibit a morphological response to temperature. For example, *Neogloboquadrina pachyderma* [which also appears in the literature as *Globigerina pachyderma*, e.g., Ruddiman (1977), or *Globoquadrina pachyderma*, e.g., Bé (1977)] is a coiled species that changes coiling direction with temperature. Most specimens coil to the right, but in waters colder than about 8°C, >90% of specimens in a sample will be left-coiling (Kellogg 1976; Bé 1977; Kennett and Venz 1995). *N. pachyderma* did not evolve until the Quaternary, and coiling directions have not been suggested as paleotemperature indicators in other species, but Cretaceous planktonic foraminifera include forms that are highly ornamented, with ribs, nodular growths, or double keels. These were tropical, open-ocean forms (Eicher and Diner 1989; Leckie et al. 1991; Huber 1992), and the ornamentation aided buoyancy in the warm, less viscous waters (Leckie 1989). Latitudinal shifts in the distribution of Cretaceous keeled foraminifera have been used to trace movement of the polar front in much the same way as coiling direction in *N. pachyderma*.

Benthic foraminiferal morphology, particularly size and test-wall thickness, has also been interpreted as indicative of variations in oxygenation levels in the oceans (Kaiho 1994). However, the method, based on modern taxa, is at least partly taxon based, so its applicability to the older record might be limited. In addition, the reliability of some of the taxa as oxygen-level indicators has been questioned (e.g., Sen Gupta and Machain-Castillo 1993).

Paleobiogeography of foraminifera

The global biogeographic distribution of foraminiferal taxa has not been used very much in paleoclimate studies because the distribution of calcareous sediments in the ocean basins is now and has in the past been quite patchy (e.g., Ruddiman 1977). The major exception to this general observation is the biogeography of large, benthic foraminifera, which has been exploited in climate studies of the late Paleozoic, in particular. Most biogeographic studies of foraminifera have concentrated on parts of oceans.

Planktonic foraminifera. Because foraminifera live throughout the oceans, studies of the biogeography of foraminifera and changes caused by paleoclimatic or other processes should, in principle, be a simple matter of generating detailed maps of the distribution of taxa through time in ocean cores. However, foraminifera, like all oceanic microfossils, are subject to dissolution as their tests sink through the water column. The depth at which a significant proportion of carbonate dissolves, resulting in a high rate of change in the species assemblage as the less robust tests dissolve, is called the lysocline (Berger 1968); the depth at which virtually all carbonate is gone is called the carbonate compensation depth (CCD). Sediments deposited below the lysocline will have fossil foraminiferal faunas that are quite unlike the fauna living at the surface (figure 2.2). Moreover, the depths of the lysocline and CCD vary in space and through time. Another complication is that warmer-water forms are larger, which contributes to greater accumulation of foraminiferal carbonate above the CCD in the tropics (Bé 1977).

Application of foraminiferal biogeography to the Tertiary used latitudinal migration of assemblages to define paleoclimatic change. Excursions in the latitudinal positions of warm-, cool-, and cold-water groups, defined using factor analysis, were called acme events (Haq and Lohmann 1976; Haq 1982). A summary of acme events in the North Atlantic in the early Tertiary is shown

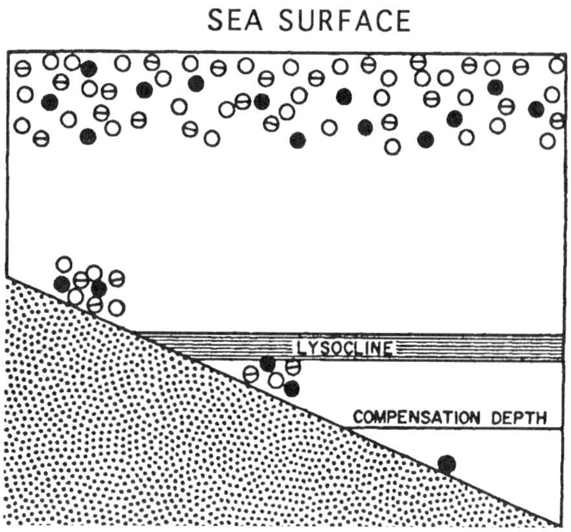

Figure 2.2. Schematic illustration of the changes in preserved foraminiferal faunas with depth and relationship to the lysocline and CCD. *Filled, hatched,* and *open circles* represent different species of foraminifera in the living fauna (at top) and on sediments (at bottom). Adapted from Bé (1977, figure 12).

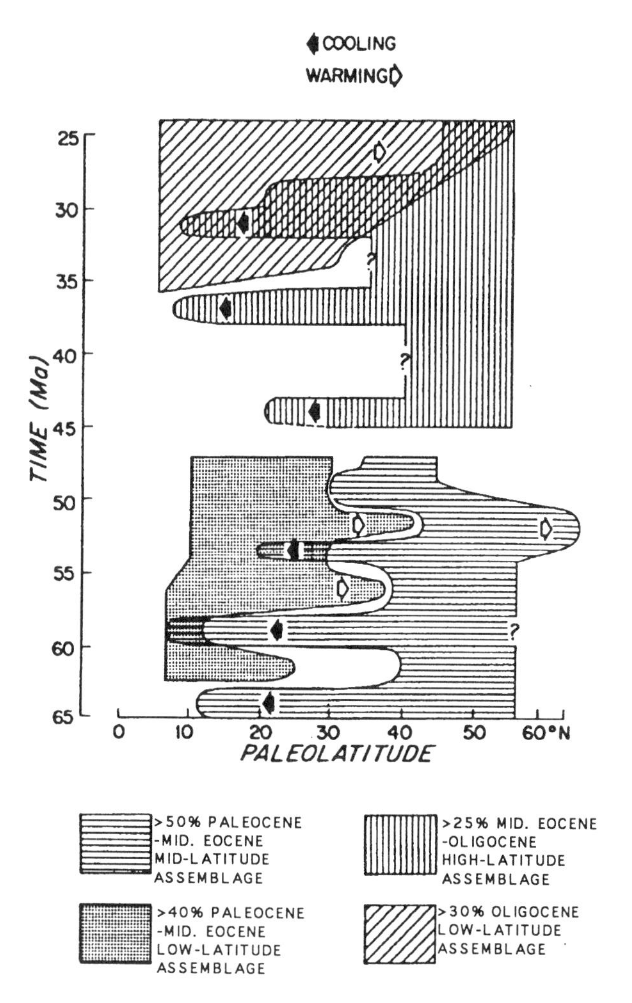

Figure 2.3. Summary of acme events in North Atlantic, early Tertiary foraminifera (Haq 1982). *Arrows* indicate direction of latitudinal migration of foraminiferal assemblages. From Haq, B.U., et al. (1977) *Journal of Geophysical Research* 82:3861–76, copyright by the American Geophysical Union.

in figure 2.3. Note that the assemblages changed not only with latitude but also with time. The most useful parts of this data set are where migration in one group was accompanied by a complementary migration in the other. This is particularly well expressed between 65 and 47 Ma. Overlap between two assemblages (for example, at about 30 Ma) is likely to reflect the nature of the latitudinal temperature gradient. A high degree of overlap would indicate an even gradient, whereas a low degree of overlap would indicate a sharp break in the gradient.

Relatively few foraminiferal taxa are endemic, so identifying paleobiogeographic units on the basis of degrees of endemism is not possible. Haq and Lohmann (1982) and Haq (1982) used factor analysis to identify assemblages. An alternative method is to demarcate paleobiogeographic units on the basis of peak abundances of taxa (Ruddiman and McIntyre 1976). Boersma and Premoli-Silva (1991) used this approach for the Paleogene of the Atlantic Ocean, and Spezzaferri (1995) used it for the Oligocene and Miocene of the Atlantic, Indian, and southwestern Pacific Oceans. They termed the paleoclimatically diagnostic faunal units "indices." Variations in the percent of each index in time and space provided detailed information about climatic change.

Similar information was used for interpreting Cretaceous climatic change by Huber (1992), who also took particular note of the morphology of the taxa. He plotted assemblage data in two ways, as latitudinal diversity plots (figure 2.4) and as biogeographic maps (figure 2.5). The realms defined by Huber (1992) were based on latitudinal changes in diversity of both total and keeled taxa; presence of endemic, high-latitude taxa; and absence of tropical taxa in high latitudes. Huber (1992) hypothesized that the decreasing diversity with latitude was a response not only to lower temperatures but also to a decrease in habitat from shallowing of the photic zone (from the lower incidence angle of sunlight) and greater seasonality.

Huber (1992) went on to reconstruct surface circulation for the Late Cretaceous Southern Ocean based on the biogeographic boundaries

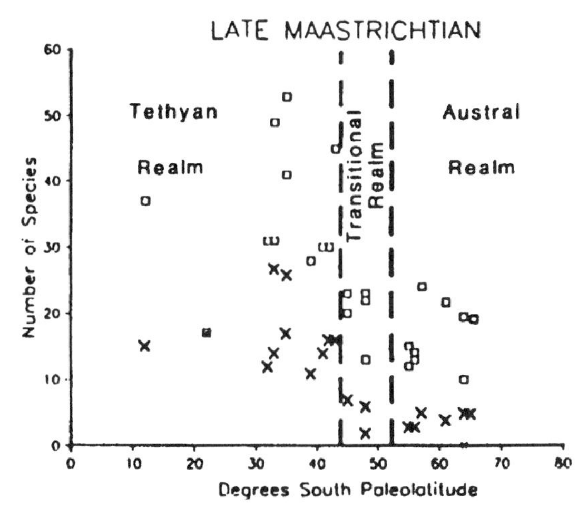

Figure 2.4. Example of a diversity plot of planktonic foraminifera in the Southern Ocean in the late Maastrichtian (latest Cretaceous). *Squares,* total diversity; *crosses,* diversity of keeled forms. Reprinted from *Palaeogeography, Palaeoclimatology, Palaeoecology* 92, Huber, B.T., "Paleobiogeography of Campanian-Maastrichtian foraminifera in the southern high latitudes," pp. 325–60 (1992) with kind permission of Elsevier Science-NL, Sara Burgerhartstraat 25, 1055 KV Amsterdam, The Netherlands.

(figure 2.5) and on combined data for the entire Campanian-Maastrichtian interval. The reconstructed currents reflect the relatively higher-diversity faunas on the western side of what is now Australia versus the eastern side in the late Maastrichtian. Huber's (1992) reconstruction implies influx of warm water from the low-latitude Indian Ocean to the western side and cooling of this water in higher latitudes as it circumvented Australia and flowed towards the South Pole.

Huber (1992) cautioned that a number of potential problems might exist with the data and therefore these interpretations. First, he pointed out that local paleoceanographic and preservational conditions can skew the data because the sample sites are so scattered. Reconstructions, even on the scale seen in figure 2.5, can be wrong if a preserved fauna lived in local conditions not representative of the regional conditions. Second, differential preservation of taxa, which in this case would most likely be preferential loss of smaller, thinner-walled forms, is likely a problem. Another common problem in studies of foraminifera is that faunal lists are commonly limited to the larger planktonic foraminifera, >150 μm; it is known for benthic foraminifera that smaller and larger taxa may respond differently to changing environmental conditions (e.g., Kennett and Stott 1991). Fourth, there is a lack of uniformity in taxonomy. For example, late Maastrichtian diversity is twice as high as early Campanian diversity as a result of the greater number of studies on Maastrichtian faunas and differing taxonomic philosophies, not as a result of climate. Finally, Huber (1992) noted inconsistencies in the definition of stages and substages of the Cretaceous. These types of concerns apply generally to biogeographic studies.

Some foraminifera, e.g., species of the family Guembelitrinidae (Kroon and Nederbragt 1990), may be indicative of variable conditions, including upwelling. Relationships of fossil planktonic foraminiferal taxa to productivity can be established by paleobiogeographical and/or isotope analysis. An ideal place to perform such an analysis, with the hope of establishing a standard that can be applied elsewhere, is someplace where other evidence, for example, phosphate or biogenic chert, can be used to estimate variations in productivity. Almogi-Labin, Bein, and Sass (1993) performed just such an analysis, studying the foraminiferal faunas in a well-studied, Late Cretaceous upwelling zone in Israel. They identified four planktonic assemblages based on observations about abundance, diversity, and dominance among genera (table 2.2). One of the assemblages was a typical, low-latitude Tethyan oceanic assemblage, whereas the other three were increasingly different from the typical Tethyan fauna. The authors argued that differences among the assemblages were related to variations in productivity and thus could be used to estimate productivity.

Almogi-Labin, Bein, and Sass (1993) studied two sections, Zin and Shefela, that were thought to be near the core of the upwelling zone and on the margins of the upwelling zone, respectively. The example presented here is from Zin. Keeled globotruncanids, which are thought to have been deeper dwellers, were gradually eliminated (figure 2.6f). Almogi-Labin, Bein, and Sass (1993) interpreted this as an indication of intensification of the oxygen minimum zone, which would tend to eliminate the deeper dwellers first (see Caron and Homewood 1983; but also van Eijden 1995). Changes in dominance between heterohelicids and globigerinelloids (figure 2.6c,d) were interpreted to indicate changes in productivity. The heterohelicids tended to occur where planktonic foraminifera were relatively abundant and the assemblage highly diverse, including keeled forms. In contrast, globigerinelloids showed more of a

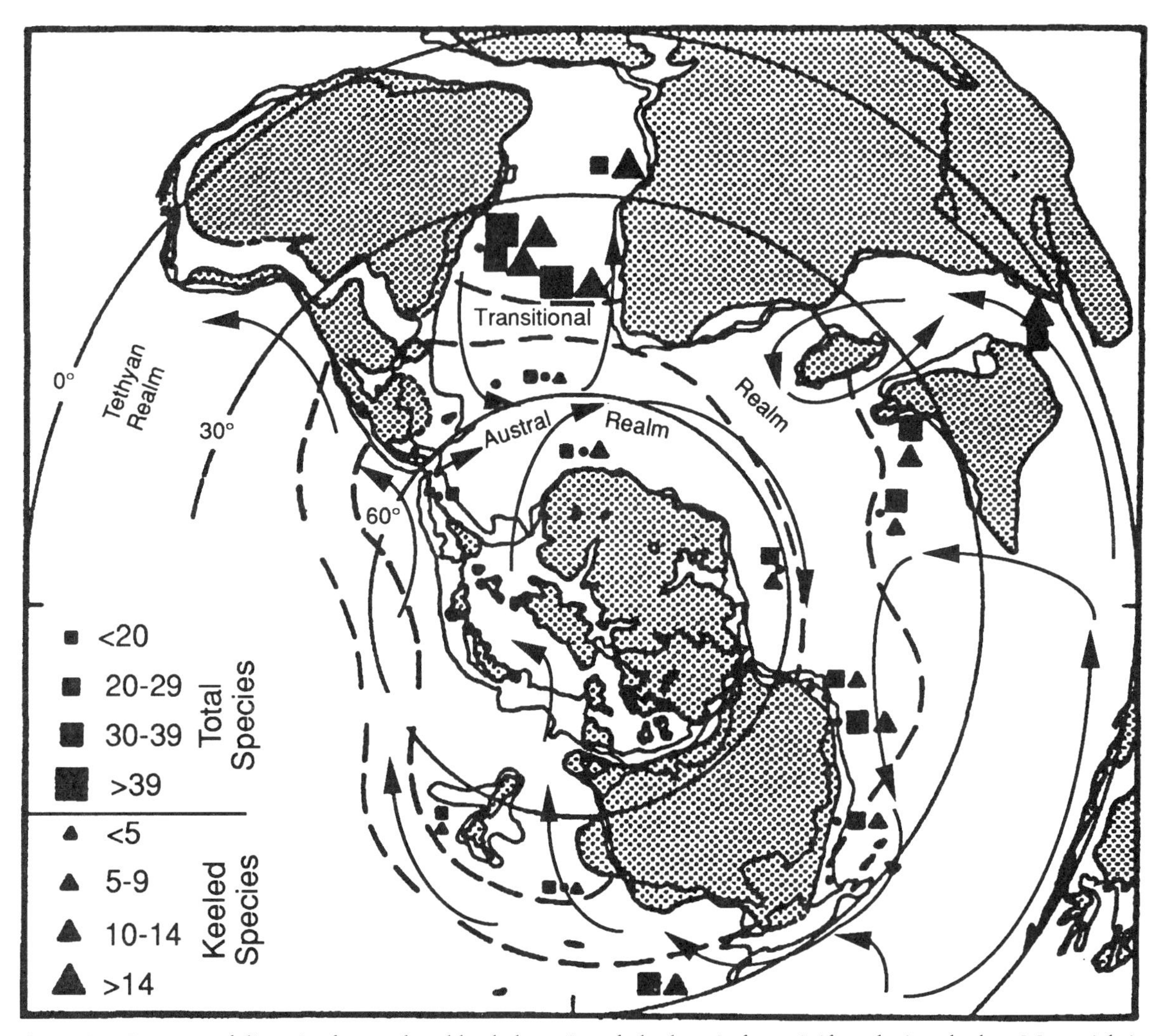

Figure 2.5. Patterns of diversity for total and keeled species of planktonic foraminifera during the late Maastrichtian (latest Cretaceous) and reconstruction of surface currents in the Campanian-Maastrichtian (Late Cretaceous) Southern Ocean (*arrows*). *Shaded areas* are exposed land. Reprinted from *Palaeogeography, Palaeoclimatology, Palaeoecology* 92, Huber, B.T., "Paleobiogeography of Campanian-Maastrichtian foraminifera in the southern high latitudes," pp. 325–60 (1992) with kind permission of Elsevier Science-NL, Sara Burgerhartstraat 25, 1055 KV Amsterdam, The Netherlands.

tendency to occur with low abundances of planktonic foraminifera and low-diversity assemblages containing few globotruncanids (see table 2.2). The heterohelicid assemblage, which is the more typical Tethyan assemblage, was interpreted to indicate lower-productivity waters, whereas the globigerinelloid assemblage was interpreted to indicate higher-productivity waters. Almogi-Labin, Bein, and Sass (1993) interpreted the changes in the foraminiferal faunas as a strengthening followed by a weakening of the upwelling system. Note the relationship of the productivity estimates (figure 2.7a), based on the assemblages of planktonic foraminifera, to the bottom-water oxygen level (figure 2.7b), based on the assemblages of benthic foraminifera, discussed in the next section.

Almogi-Labin, Bein, and Sass's (1993) method relied on peak diversities and thus is somewhat different from the methods applied to Tertiary deep-sea cores, which use peak abundances. Diversity here is simple diversity, that is, the number of taxa at any given time or in any given place. Simple diversity is affected not only by climate but also by myriad other factors. Ottens and Nederbragt (1992) were critical of the use of simple diversity for planktonic foraminifera and advocated using a combination of simple diver-

Table 2.2
Foraminiferal assemblages from Late Cretaceous rocks of Israel

Planktonic foraminifera assemblage types

Assemblage Type	Planktonic foraminifers (%)	Dry sediment (abundance per gram)	Diversity (number of species per sample)	Dominance	Environmental interpretation
1	20–80	200–40,000 variable	>20 (total) >10 Globotrun.	Hetero. > Globig.	Normal marine
2	20–60	500–5000 variable	10–20 (total) 5–10 Globotrun.	Hetero. = Globig.	↓
3	10–60	500–5000 variable	5–10 (total) <5 Globotrun.	Globig. > Hetero.	↓
4	0–5	<500 very low	0–5 (total) no Globotrun.	Globig?/Hetero?	Most productive

Globig., *Globigerinelloides;* Globotrun., *Globotruncana;* Hetero., *Heterohelix.*

Benthic foraminifera assemblage types

Assemblage Type	Planktonic foraminifers (%)	Dry sediment (abundance per gram)	Diversity (number of species per sample)	Dominance	Environmental interpretation
1	95–100	<10,000	<10	95–100% buliminids	Most dysaerobic
2	95–100	>20,000	<10	95–100% buliminids	↓
3	40–90	variable	10–20	60–90% buliminids	↓
4	20–80	variable	>20	rotaliids> buliminids	Least dysaerobic

From Almogi-Labin, Bein, and Sass (1993) in *Paleoceanography* 8:671–90, copyright by the American Geophysical Union.

sity, Shannon diversity, and equitability. Shannon diversity is dependent on having good abundance data, as well as presence/absence data, and is calculated as

$$H' = -\sum p_i \ln p_i, \tag{2.1}$$

where p_i is the proportion of each taxon in a fauna. Equitability is a measure of how common each taxon is relative to the other taxa and is calculated

$$E' = eH'/s, \tag{2.2}$$

where s is the number of taxa in a sample. If all species present are in the same proportions, E' is equal to 1.

Ottens and Nederbragt (1992) observed in modern foraminiferal faunas of the North Atlantic and Indian Ocean that the indices—simple and Shannon diversity and equitability—vary differently in different oceanographic conditions. For example, at the mixing boundary between water masses, simple and Shannon diversities are higher than in the water masses themselves, whereas equitability is similar. In variable oceanic environments such as upwelling zones, all three indices are low compared to adjacent water masses. Hypersaline environments will have very low simple diversities, low Shannon diversities, but high-intermediate equitabilities.

Benthic foraminifera. The paleobiogeography of benthic foraminifera has been used to understand oxygenation levels in bottom waters, the character and evolution of bottom-water masses, and changes in productivity at the surface. The first is the use to which Almogi-Labin, Bein, and Sass (1993) put the benthic foraminifera at Zin.

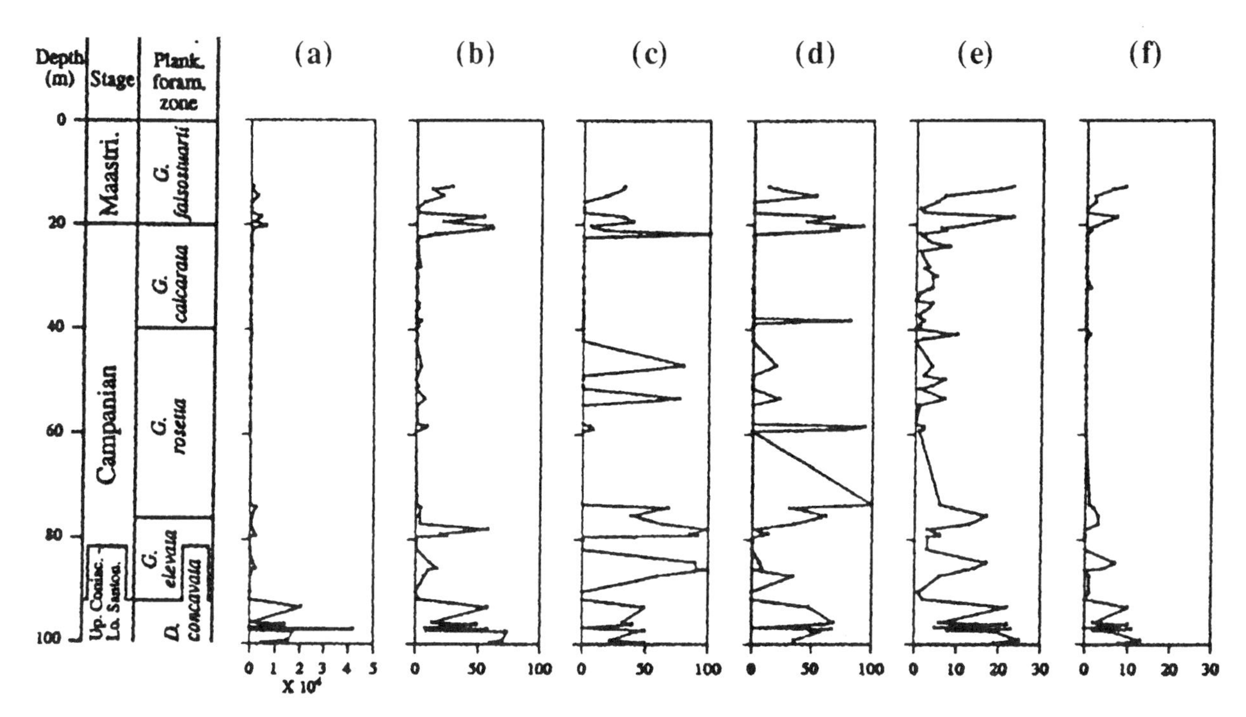

Figure 2.6. Distribution of planktonic foraminifera through time at Zin, Israel. *(a)* Total abundance. *(b)* Relative abundance of planktonic foraminifera. *(c)* Relative abundance of globigerinelloids. *(d)* Relative abundance of heterohelicids. *(e)* Diversity of planktonic foraminifera. *(f)* Diversity of keeled globotruncanids. From Almogi, Bein, and Sass (1993) *Paleoceanography* 8:671–90, copyright by the American Geophysical Union.

They discerned four assemblages (see table 2.2) that they interpreted as indicating decreasing bottom-water oxygen levels in a Late Cretaceous upwelling deposit in Israel. They developed this oxygenation index in conjunction with the productivity index based on planktonic foraminifera, discussed in the previous section. All the assemblages were interpreted to have lived in dysaerobic waters. The assemblage interpreted as living in the least dysaerobic waters occurred in slightly bioturbated sediments; the assemblage interpreted as living in the most dysaerobic waters, dominated by a specialized buliminid fauna, occurred with phosphate. Changes in dominance between the more specialized buliminids and the less specialized buliminids and rotaliids (figure 2.8) were interpreted as indicative of varying oxygenation levels (see table 2.2). The relationship between productivity and bottom-water oxygenation at Zin is shown by the variations among the planktonic and benthic assemblages (see figure 2.7).

A study of Miocene benthic foraminifera in the Pacific Ocean by Woodruff (1985) illustrates the use of factor analysis in biogeography (in this case, principal components analysis), incorporates a unique graphical method, and shows the value of graphically presenting biogeographic data in different ways. The distribution of samples by age and depth are illustrated in figure 2.9. Woodruff (1985) used the age-depth plots to present the results of the factor analysis of biogeography, which has the advantage of showing factor distribution continuously through time.

The first four factors accounted for 27% of the variance. Twenty-seven percent is low; in many factor analytic studies, 80% to >90% of the variance would be explained in four factors. However, Woodruff (1985) concluded that 27% was acceptable, considering the size of the data set, the length of time covered, and the variations in environment, latitude, and depth. Illustration of her interpretations of the four factors is worthwhile to emphasize that factor analytic scores do not inherently provide information about causality and to show that variability in a data set can be influenced by widely differing processes. Her interpretations of the factors are as follows (figure 2.10): Factor 1 is stratigraphic, that is, before and after about 16 to 13 Ma, as seen in the distinct vertical discontinuity in the factor scores. This represents an oceanwide change in the

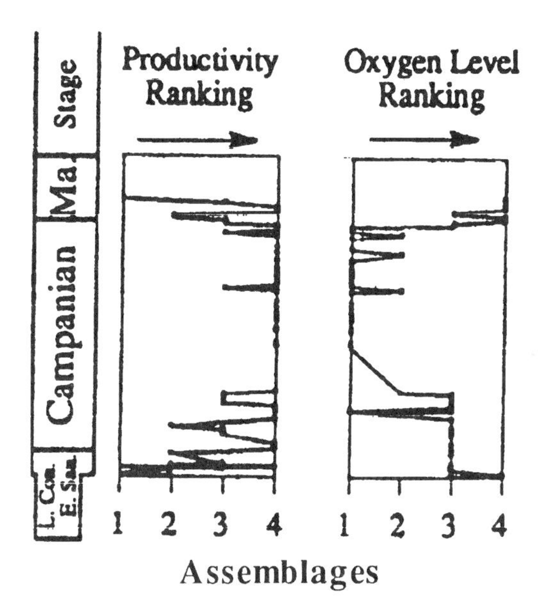

Figure 2.7. Interpretation of productivity and bottom-water oxygen levels from planktonic and benthic foraminiferal assemblages, respectively. The numbered assemblages and their environmental interpretations are described in table 2.2. Note that productivity and bottom-water oxygenation appear to be related: when productivity is higher, bottom-water oxygenation is lower, and vice versa. From Almogi, Bein, and Sass (1993) Paleoceanography 8:671–90, copyright by the American Geophysical Union.

fauna because it occurred at all sites and depths. Factor 2 reflects variations in diversity, either real or related to dissolution, and distinguishes samples with 40 to 60 species from those with 20 to 39. Geographic plots of Factor 2 showed that the samples with factor loadings for high diversity were near the equator in the western and central Pacific and that the sites with those factor loadings changed through time as the Pacific Plate drifted beneath the equator. The highest negative scores (lowest diversity) were in the eastern equatorial Pacific and along western North America, where upwelling and high productivity were likely to have occurred. Factor 3 is concentrated in the early Miocene and represents depth-related assemblages—note the vertical arrangement of factor scores in figure 2.10. Because the stratigraphic division was strong in this factor, Woodruff (1985) divided the data set into early and late subsets (before and after 15 Ma), and the results showed that the depth relationship is similar for both intervals. The intermediate-water assemblages (1.2 to 2.4 km) may be related either to depth or low oxygen, and the top of the bottom-water assemblage might correlate with the lysocline. Woodruff (1985) interpreted Factor 4 as an ocean-basin/ocean-margin factor (see figure 2.10), possibly representing low surface productivity and low sediment accumulation rates (positive factor scores) versus high surface productivity and high sediment accumulation rates (negative factor scores), rather than strictly geography.

Woodruff (1985) plotted Factor 4 scores against time for comparison with time-series plots of lithology and chemistry for Site 289 in the western Pacific, which produced an exceptionally complete and well-studied core (figure 2.11). The high negative scores for Factor 4 between about 16 and 13 Ma (figure 2.11f), which were interpreted to indicate high productivity and high sediment-accumulation rates, are roughly correlative with high carbon isotope values, high sediment-accumulation rates, a high density of laminated sediments, and high planktonic-to-benthic ratios, all of which are regarded to be effects of high biologic productivity.

Woodruff (1985) also used the age-depth plot to illustrate the samples most likely to have been affected by burial diagenesis, which can degrade the fossils and prohibit accurate identification of taxa. This permitted direct comparison of the factor score plots (see figure 2.10) with the dissolution plots (not shown) in order to determine the magnitude of effect dissolution might have had on each factor. The major preservation problem with benthic foraminifera, however, is dissolution at or above the sediment-water interface. One way to assess the effects of dissolution is to determine abundance ratios of planktonic to benthic foraminifera; benthics are buried soon after death and thus are less susceptible to preburial diagenesis. A ratio of 500:1, for example, is consistent with little dissolution and/or high productivity of planktonics, whereas a ratio of 30:1 would indicate significant dissolution and/or low productivity of planktonics (Woodruff 1985). This method is used by many workers to estimate dissolution effects.

Benthic foraminiferal paleobiogeography has also been used extensively to track productivity. High biologic productivity in the oceans occurs at the surface, in upwelling zones, but the production of organic matter at the surface in these regions influences the ecology of the oceans

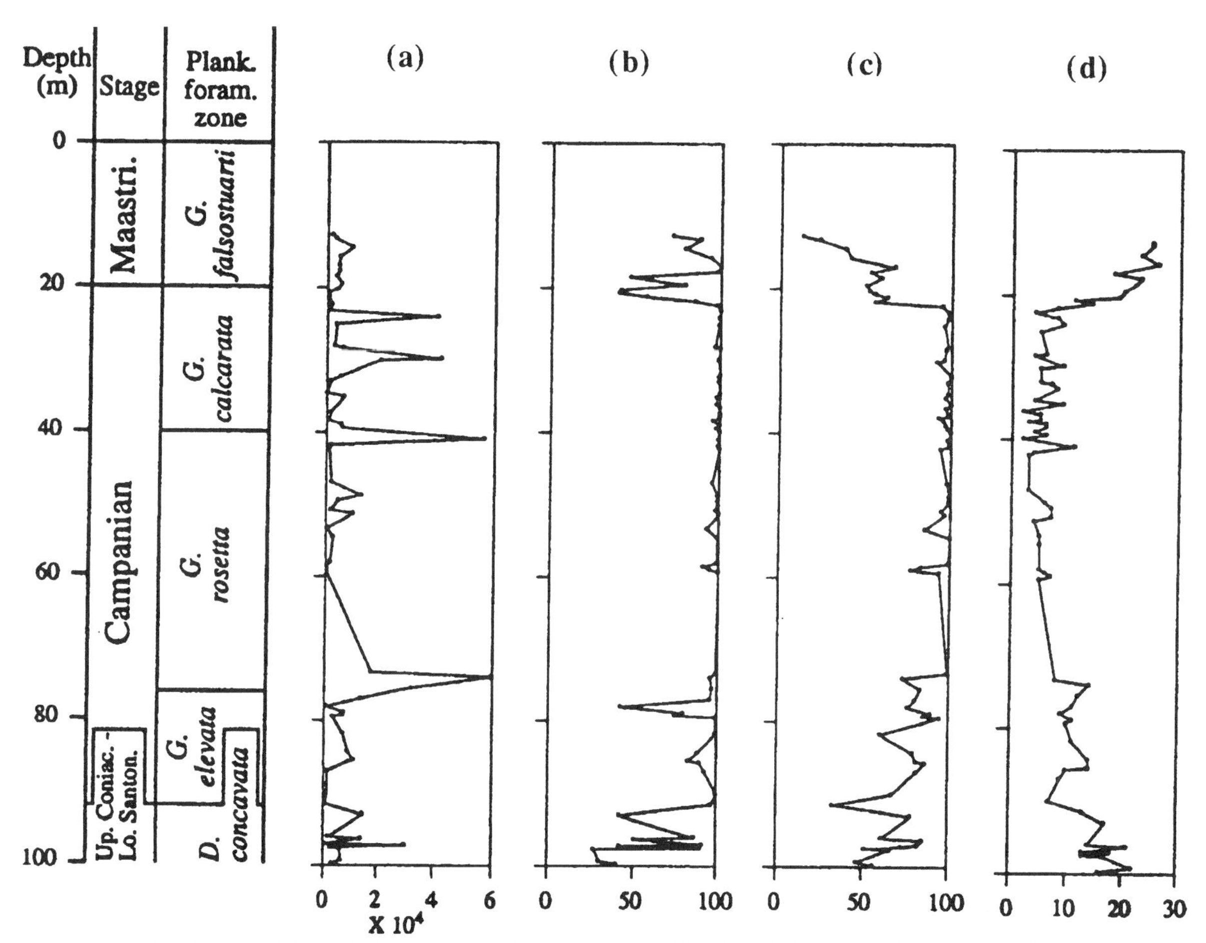

Figure 2.8. Distribution of benthic foraminifera through time at Zin, Israel. *(a)* Total abundance. *(b)* Relative abundance of benthic foraminifera. *(c)* Relative abundance of buliminids. *(d)* Diversity of benthic foraminifera. From Almogi, Bein, and Sass (1993) *Paleoceanography* 8:671–90, copyright by the American Geophysical Union.

throughout the water column down to the sediments. Many lithologic indicators of productivity are not preserved in the geologic record at greater than shelf or slope depths, so the identification of specific biological productivity indicators in deeper waters is even more crucial than in shallow waters.

The distributions of some taxa, such as certain species of the genera *Bolivina* and *Uvigerina*, appear to be related to organic-matter content of the sediments (Sen Gupta and Machain-Castillo 1993; Smart et al. 1994), oxygenation of the water column (Douglas 1979, 1981; Woodruff 1985; Lagoe 1987), or productivity (Pedersen et al. 1988 and references therein). These are long-lived genera commonly used as indicators of low-oxygen bottom waters and/or high-productivity surface waters, although Sen Gupta and Machain-Castillo (1993) argued against such usage; not all species of *Bolivina*, for example, occupy low-oxygen environments (Douglas 1979). The small benthic foraminifera *Epistominella exigua* and related taxa may also be productivity indicators (Smart et al. 1994). All of these taxa are useful at least back through the Miocene.

Strong compositional shifts in modern faunas correspond to sharp productivity gradients at the equator and the southern convergence (Loubere 1991). Measured productivity and productivity estimated from the first four principal components derived from multivariate analysis were highly correlated ($r^2 = 0.87$). Although the elevated production of organic matter is at the surface, more than 3000 m above the sample sites, the living benthic faunas nevertheless faithfully record the lateral productivity gradient in the underlying sediments (Loubere 1991). Thomas and Gooday (1996) recorded a similar effect in Tertiary deep-sea sediments. Coincident with cooling, beginning in the latest Eocene, benthic foramini-

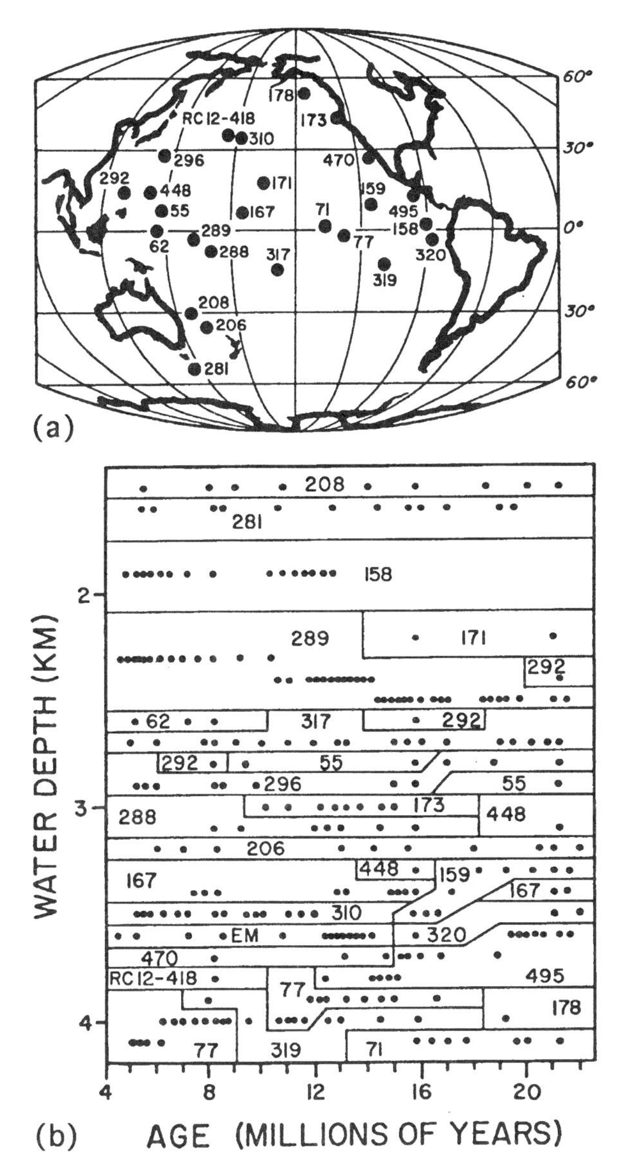

Figure 2.9. *(a)* Distribution of samples of Miocene benthic foraminifera in the Pacific Ocean. *(b)* Distribution by age and water depth at the time of deposition. Numbers on the age-depth plot (b) are site numbers, and each site occupies one or more fields on the plot. Gaps between the fields for each site indicate hiatuses in the data resulting from erosion, nondeposition, or gaps in coring or sampling. Offsets in the data along the depth axis indicate changes in water depth through time, either deepening as a result of subsidence (e.g., Site 77) or shallowing as a result of uplift or basin filling (e.g., Site 289). From Woodruff (1985) in *The Miocene Ocean: Paleoceanography and Biogeography,* J.P. Kennett, ed. Reproduced with permission of the publisher, the Geological Society of America, Boulder, Colorado, USA. Copyright © 1985 Geological Society of America, Inc.

fera developed a strong latitudinal diversity gradient and, at the same time, the diversity of opportunistic taxa, such as *E. exigua*, increased. The lack of correspondence of *E. exigua* to any other factor is support for the species as a productivity indicator (Smart et al. 1994). *E. exigua* occurs in waters ranging from −1°C to 10°C; in other words, there is no correspondence between abundance of the foraminifera and deep-water hydrography. No correlation exists among the abundance peaks of *E. exigua* from different parts of the ocean, and the greatest fluctuations in abundance are where production is greatest, for example, in the Bay of Biscay [Deep Sea Drilling Project (DSDP) Site 400A], consistent with the patchiness of biologic productivity in space and time (e.g., Thunell and Sautter 1992). The same lack of synchroneity in fluctuations of abundance were observed in the deep Pacific Ocean. There, an early Miocene peak in *E. exigua* does not correspond to any isotopic excursion, sea level acme, diatom abundance change, or higher organic content. From this, Smart and Murray (1994) concluded that changes in its abundance must be related to the supply of food from surface waters.

Paleobiogeography of larger foraminifera. The term "larger foraminifera" is purely descriptive, with no taxonomic meaning; it refers to those taxa that are easily discernable to the naked eye. Based on their treatment in the literature, different groups of larger foraminifera have been important during two times in Earth history, Carboniferous-Permian and Cretaceous-Paleogene. The fusulinids, which were fusiform and sometimes extremely elogated, were important during the Carboniferous and Permian Periods. The superfamilies Lituolacea (agglutinating), Miliolacea, Orbitoidacea, and Rotaliacea were important during the Cretaceous and Paleogene. The discocyclinids, which were disk shaped and include the well-known [and problematic (Hottinger 1973)] genus *Nummulites*, were important during the Paleogene.

Notable paleobiogeographic studies of larger foraminifera are those of Ross (1973) for the Carboniferous, Gobbett (1973) for the Permian, Dilley (1973) for the Cretaceous, and Adams (1973) for the Eocene. The biogeography of larger foraminifera has been treated with the biogeographic methods generally applied to marine macrofossils, that is, nonquantitative determinations of global biogeographic units, based on moderate to

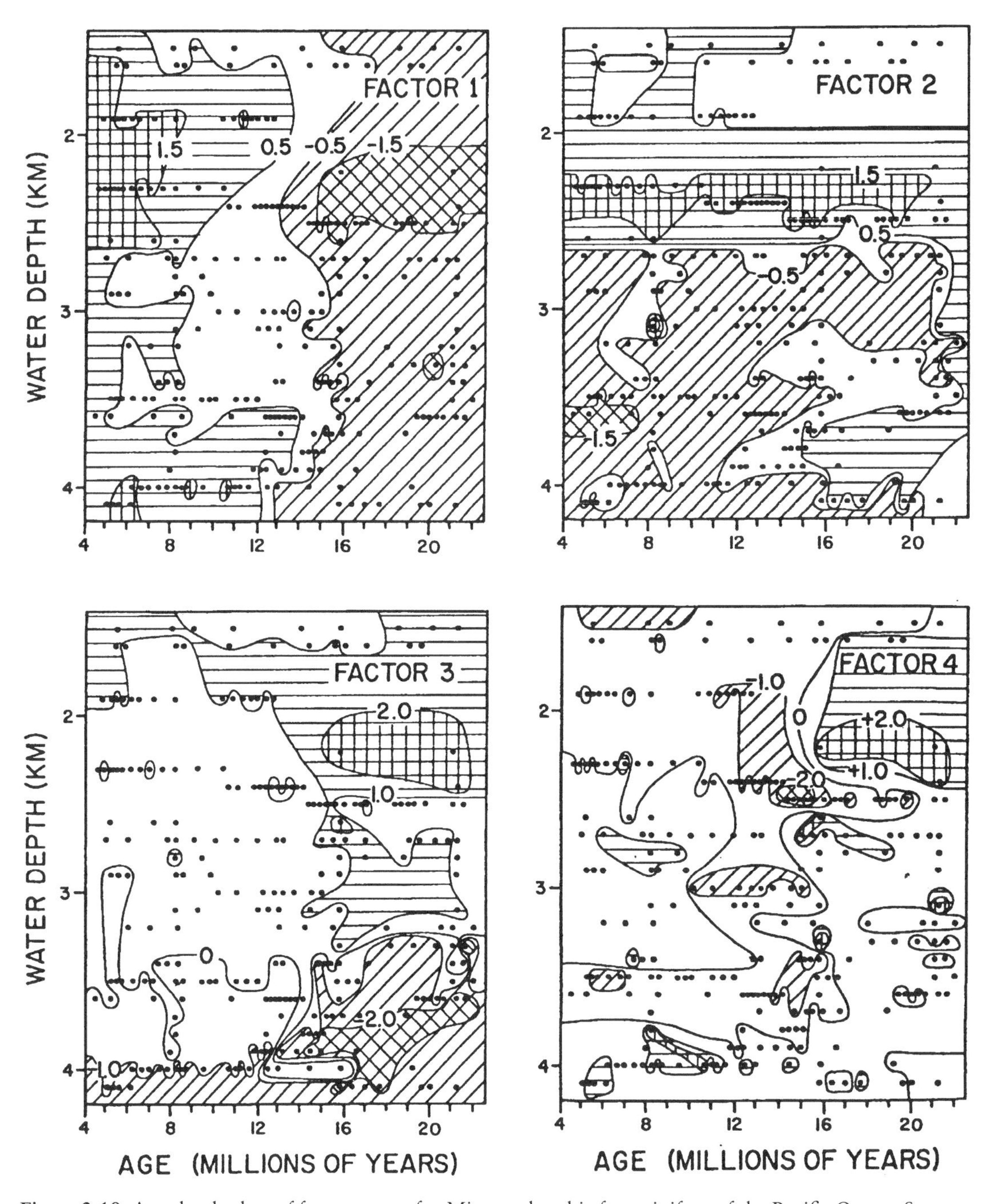

Figure 2.10. Age-depth plots of factor scores for Miocene benthic foraminifera of the Pacific Ocean. See text for interpretation of factor scores. From Woodruff (1985) in *The Miocene Ocean: Paleoceanography and Biogeography*, J.P. Kennett, ed. Reproduced with permission of the publisher, the Geological Society of America, Boulder, Colorado, USA. Copyright © 1985 Geological Society of America, Inc.

high degrees of endemism and relatively few localities, plotted on maps through time. In addition, Tertiary larger foraminifera are assumed to have been subtropical to tropical forms. Although the assumption is rarely stated explicitly, it is based on the distribution of modern larger foraminifera, which are limited to waters ≥ 20°C in the autumn season (Adams, Lee, and Rosen 1990). Latitudinal diversity gradients are vague in this group, decreasing at 45°N and 25°S (modern latitudes; Adams 1973). The discontinuity at 25°S has no climatic significance, as few outcrops of the right age exist in the Southern Hemisphere above that latitude (Adams 1973).

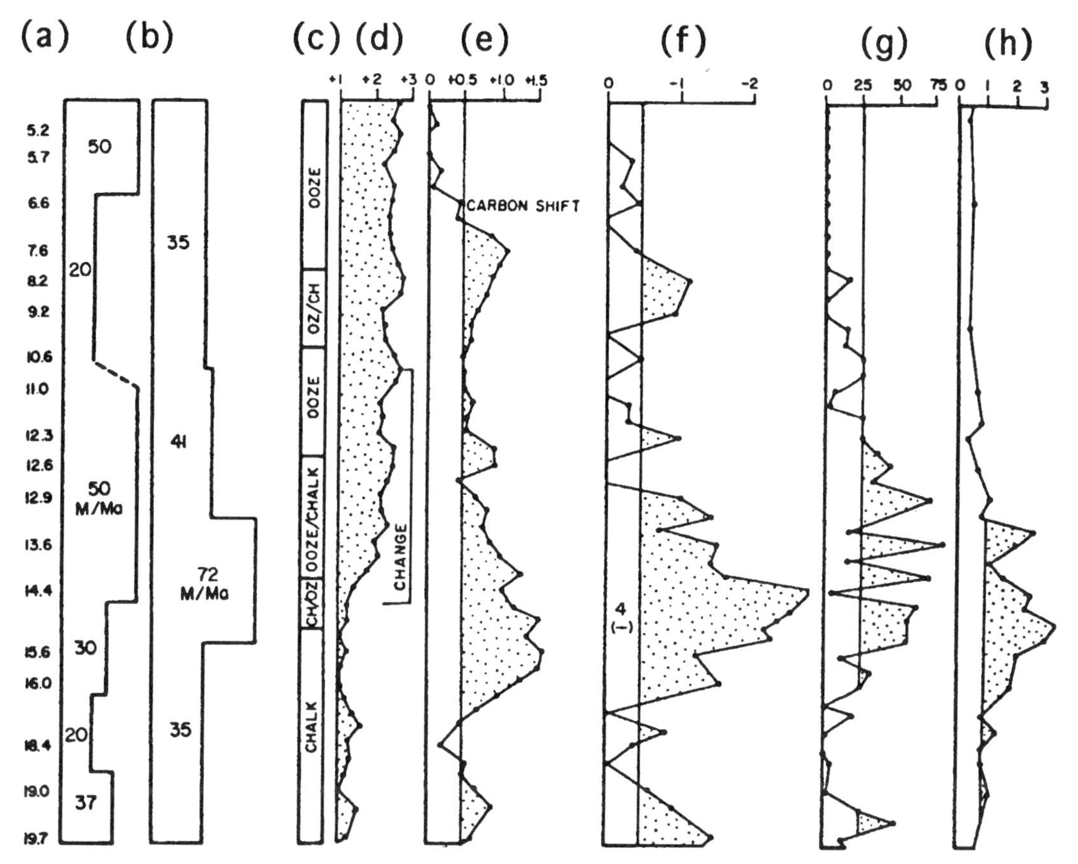

Figure 2.11. Downcore plots of geologic and paleontologic data for Site 289, western Pacific Ocean, compared with time series of Factor 4 scores. *(a)* Age (Ma). *(b)* Two estimates of sediment accumulation rates. *(c)* Lithology: ch, chalk; oz, ooze. *(d)* $\delta^{18}O$. *(e)* $\delta^{13}C$. *(f)* Factor 4 scores. *(g)* Number of centimeters of parallel laminations per core. *(h)* Planktonic/benthic ratios (×1000). From Woodruff (1985) and references therein, in *The Miocene Ocean: Paleoceanography and Biogeography,* J.P. Kennett, ed. Reproduced with permission of the publisher, the Geological Society of America, Boulder, Colorado, USA. Copyright © 1985 Geological Society of America, Inc.

The most comprehensive studies of the paleobiogeography of larger foraminifera are of Tertiary fossils by Adams (1992; Adams, Lee, and Rosen 1990). Adams, Lee, and Rosen (1990) compared the distribution of larger foraminifera with those of mangroves and reef corals, both of which were assumed to have been subtropical to tropical in the past as well as today. All these groups had at least as wide a latitudinal range and as high diversity in the middle Eocene and middle Miocene as today. Adams, Lee, and Rosen (1990) argued that if tropical temperatures were as cool as indicated by isotope data (e.g., Shackleton 1984; but see Shackleton 1986), all these groups would have had to have changed their physiology, especially the physiology of reproduction, which is commonly more narrowly constrained by environmental conditions than is occurrence. This argument is common in paleobiogeography, and it seems to be generally agreed that although some tropical groups might slowly adapt their physiological tolerances, it is highly unlikely that many unrelated groups would do so in synchrony, although warming, as has occurred from the Paleogene forward, would be less stressful than cooling. Adams later (1992) presented an argument that diversity gradients, rather than absolute latitudinal range of occurrences, are more reliable for paleoclimate interpretation, despite the problem he illustrated with discontinuities in distribution of foraminifera-bearing rocks (Adams 1973) and cross-oceanic differences in diversity gradients (see figure 1.7; Adams 1992).

Stable isotopes of oxygen and carbon in foraminifera

Many of the most interesting and controversial issues in global paleoclimatology are addressed with studies of the oxygen and carbon stable isotopic ratios in foraminifera. In most of these studies, the biogeography of the organisms is not explicitly stated as a factor, but it is a factor, if only because the taxa examined are chosen for their distributions, as well as for attributes such as preservability. Problems facing the paleoclimatologist in terms of sampling are not, in these studies, limited to just sample quality for chemical analysis but also include concerns about the sampling of taxa in critical regions or at critical depths.

Stable isotope analysis of foraminifera has been used to address the following paleoclimatologic issues:

1. *Temperature changes in the oceans through time.* Until recently, the most widely reproduced temperature curve based on oxygen isotopes was the one published for the Tertiary by Savin (1977), based on large body of work by Savin himself, R. G. Douglas, J. P. Kennett, N. J. Shackleton, F. Woodruff, and others (see Savin 1977 for a comprehensive review of this work). This curve showed a strong cooling trend in benthic foraminifera coupled with relatively little change in planktonic foraminifera. The interpretation at the time was that bottom-water temperatures reflected high-latitude temperatures, as they do today, and that the cooling trend was the result of cooling at high latitudes, culminating ultimately in glaciation.
2. *Volume of the polar ice caps and temperature versus ice-volume effects.* Another widely reproduced oxygen isotope curve was published by Hays, Imbrie, and Shackleton (1976), who correlated fluctuations in the curve to changes in the Earth's orbital parameters and documented the purported connection between insolation and glaciation. The oxygen isotope curve, which was derived from deep-sea cores in the tropics, was not taken as a temperature curve but, rather, as a recorder of the volume of glacial ice, which sequesters ^{16}O preferentially.

   Although the competing effects of ice volume and temperature changes on oxygen isotopes had been recognized for a long time, much work in recent years has been directed toward teasing apart the ice-volume and temperature signals. In addition, although the presence of ice sheets as far back as the Oligocene was suggested early on (Shackleton and Kennett 1975a; Kennett and Shackleton 1976; Kennett 1977 and references therein), just how much ice occurred and when is still a source of controversy.
3. *Vertical water-mass structure of the oceans.* The work reviewed by Savin (1977) was implicitly a study of the vertical temperature structure of the oceans. Since then, a more detailed picture of the temperature structure has been drawn for some times and some parts of the world, and changes in the vertical temperature profile have been related to global climatic events. The depth zonation of planktonic foraminifera in the upper few hundred meters of the ocean has allowed study of changes in the thermocline. Isotope studies from benthic foraminifera that lived at different depths have begun to draw a picture of the large-scale water-mass structure of the deep ocean.
4. *Abrupt, extreme events in the world ocean.* As the isotope record has been more finely resolved, partly because of technological breakthroughs allowing the analysis of smaller and smaller samples, rapid changes in the record have been revealed that suggest drastic changes in the world ocean. Some but not all of these events are accompanied by mass extinctions. An example is the isotope shift close to the Paleocene-Eocene boundary.
5. *Changes in the carbon budget.* Estimating biologic productivity, global organic carbon burial, and changes in the carbon dioxide content of the atmosphere and oceans is accomplished with the use of carbon isotope records. Although much of this work is done in conjunction with studies of sedimentary organic carbon, some has relied entirely on carbon isotope data from fossils.

In the remainder of this section, examples are presented of work that deals with the aforementioned issues and processes. Comprehensive coverage of this subject would occupy a book in itself, but the cited papers illustrate the broad range of applications of isotopic and paleobiogeographic information.

Temperature changes in the oceans. Few studies of oxygen isotopic ratios in foraminifera do *not* discuss temperature. Most present the isotope data as time series, either for a single locality or as a more generalized, average curve. Generalized isotope time-series curves have the advantage of appearing to summarize regional, if not global, climatic changes. The use of a large number of widely distributed samples in compiling such curves is partly in recognition of the fact that the oceans are not isothermal at the surface, not isohaline, and that lower-latitude surface waters are generally lighter isotopically because they are warmer. However, although the curves can be extremely valuable, they must be used with caution. For example, the strong isotopic differentiation between planktonic and benthic foraminifera during the Cenozoic Era was the key feature in discussions of the temperature structure of the oceans for that time. What few workers have discussed explicitly, and even fewer have taken advantage of, is the fact that benthic foraminifera live at a wide range of depths. The conclusion that the differentiation between planktonic and benthic foraminifera reflects some global paleoclimatic process incorporates the implicit assumption that the oceans were isothermal and isochemical at all depths at which the sampled benthic foraminifera grew. This assumption evidently is drawn from the observation that the modern oceans are essentially isothermal below 500 to 1000 m. In addition, available data, which come from deep-sea drilling, are limited. That the oceans certainly were not isochemical at depth in the past is illustrated by the curves in figure 2.12 (Miller, Fairbanks, and Mountain 1987). Whether the discrepancies in these curves result from different depths of growth or different circulation processes in the two oceans, neither curve by itself shows exclusively global events, despite the large number and wide distribution of the samples. Modern studies that have acknowledged and taken advantage of the heterogeneity of benthic isotopic signatures are discussed in the section on water-mass studies.

If data from different sites show the same isotopic shifts, clearly they are reflecting some nonlocal, possibly global process. Comparing numerous records of time-series data can help reveal those processes. Global processes can also be studied by explicitly examining the geography of such data. For example, Savin et al. (1985) presented synoptic oxygen isotope maps for three time slices, the early early Miocene (22 to 21 Ma), late early Miocene (16 Ma), and middle late Miocene (8 Ma) (figure 2.13). They chose the time slices based on observed stability of benthic foraminiferal isotopic compositions for at least 1 million years at those times. This is important for the construction of synoptic maps because the point of such maps is to show climate throughout the world as it was at one particular time. In practice, of course, it is impossible to have precise correlation from place to place, so Savin et al. (1985) reasoned that if the isotopic signature of the benthic foraminifera was stable for a long period of time everywhere they sampled, then small errors in stratigraphic correlation (that is, less than the specified interval) would not affect the results if the data were plotted geographically. As data have accumulated, particularly from DSDP and the Ocean Drilling Program (ODP), it has been possible to reconstruct the temperature structure of the oceans in even greater detail than shown in figure 2.13. Of particular interest is establishment of latitudinal gradients. Meridional heat-transport mechanisms are obscure, and the revelation of very low temperature gradients (e.g., Huber, Hodell, and

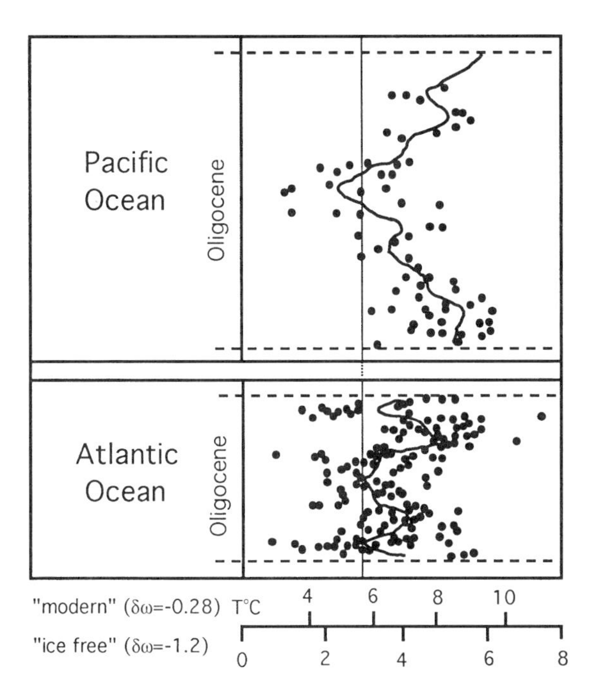

Figure 2.12. Composite oxygen isotope records for Oligocene benthic foraminifera in the Pacific *(top)* and Atlantic *(bottom)* Oceans. Dots are individual data points, which were taken from numerous deep-sea drilling cores. The reference line is at a $\delta^{18}O$ value of +1.8‰, the threshold for indicating presence of significant ice. Modified from Miller, Fairbanks, and Mountain (1987) *Paleoceanography* 2:1–19, copyright by the American Geophysical Union.

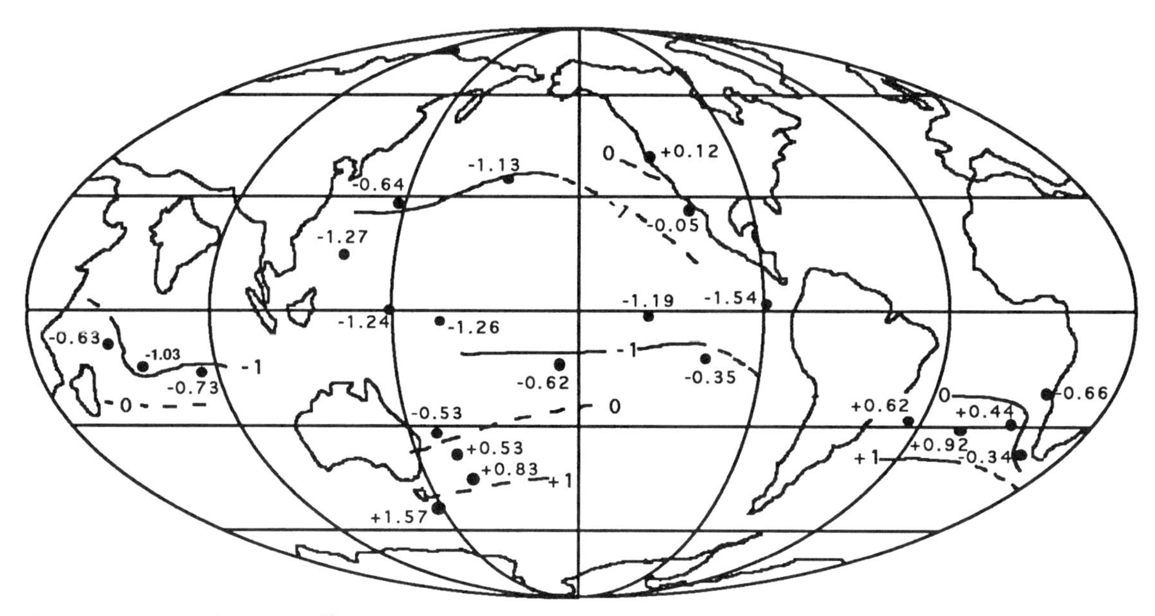

Figure 2.13. Map of average $\delta^{18}O$ values of the planktonic foraminiferal species with the most negative values at each site for middle late Miocene (8 Ma) time. From Savin et al. (1985) in *The Miocene Ocean: Paleoceanography and Biogeography*, J.P. Kennett, ed. Reproduced with permission of the publisher, the Geological Society of America, Boulder, Colorado, USA. Copyright © 1985 Geological Society of America, Inc.

Hamilton 1995) has been of particular interest to climate modelers (e.g., Barron et al. 1995).

Volume of the polar ice caps, and temperature versus ice-volume effects. It has long been recognized that changes in $\delta^{18}O$ in Quaternary marine sediments record primarily changes in ice volume, rather than temperature (Shackleton 1967). Indeed, biogeographical estimates of low-latitude sea-surface temperature are not necessarily in phase with the $\delta^{18}O$ fluctuations during the Quaternary (e.g., Hays, Imbrie, and Shackleton 1976). It has also been long recognized that the formation of ice in the polar regions has affected the pre-Quaternary $\delta^{18}O$ record (e.g., Shackleton and Kennett 1975b), but distinguishing ice-volume and temperature effects has been more difficult.

The heaviest benthic $\delta^{18}O$ values possible in an ice-free world are +1.8‰, because that represents a temperature of 0°C (Miller, Fairbanks, and Mountain 1987). If the highest values are greater than +1.8‰, there must be an ice-volume effect in the isotope signal. Using this reasoning, Miller, Fairbanks, and Mountain (1987) showed that ice probably existed at the end of the dramatic cooling event at the Eocene-Oligocene boundary (figure 2.14) and continued until the early Miocene. In the middle Miocene, $\delta^{18}O$ values dropped

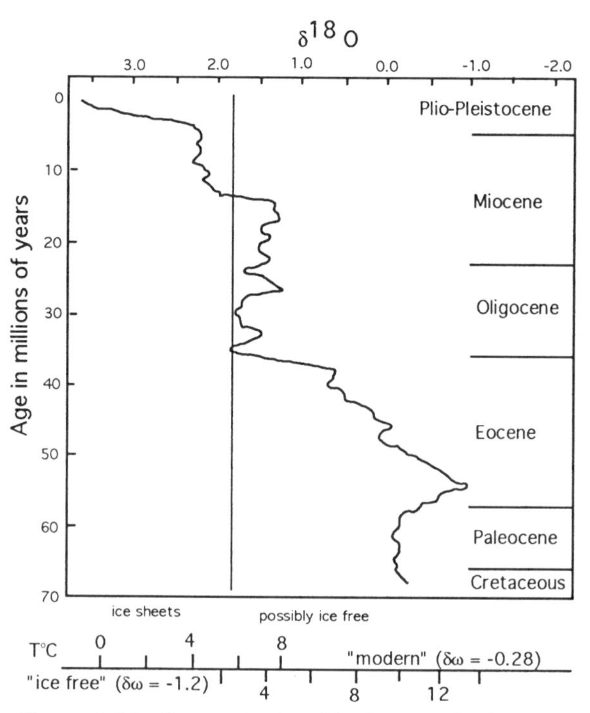

Figure 2.14. Composite benthic foraminiferal oxygen isotope record for the Tertiary Atlantic Ocean. Reference line as in figure 2.12. Modified from Miller, Fairbanks, and Mountain (1987) *Paleoceanography* 2:1–19, copyright by the American Geophysical Union.

below the threshold, where they have been ever since (figure 2.14). Miller, Fairbanks, and Mountain's (1987) synthesis only marginally dealt with temperature. Rather, they analyzed the use of the isotope curve as a proxy for sea level in the Tertiary. They concluded that post-Eocene sea-level changes were glacio-eustatic, whereas earlier changes were related to changes in sea-floor spreading rates (Vail, Mitchum, and Thompson 1977).

Following Miller, Fairbanks, and Mountain (1987), Zachos et al. (1992) interpreted most of the $\delta^{18}O$ shift after the Eocene-Oligocene boundary, sampled in a high-latitude ocean-drilling core, as a result of ice-volume changes, but they also attempted to determine what proportion of the isotopic shift was the result of a change in ice volume as opposed to cooling at high latitudes. To do this, they used regression lines on late Eocene and early Oligocene data to estimate planetary $\delta^{18}O$ gradients (figure 2.15). They argued that if the Oligocene gradient remained the same but was merely offset from the earlier one, the $\delta^{18}O$ shift would be entirely the result of a change in ice volume. Alternatively, if the gradient changed, cooling took place. A comparison between the early Eocene curve in figure 2.15a and the curves in figure 2.15b shows that the slopes are markedly different, with the largest difference at high latitudes, which shifted by about 2‰, versus a shift of >0.5‰ at low latitudes. Therefore the difference between the early and late Eocene curves is primarily the result of cooling.

Zachos et al. (1992) also concluded that the early Oligocene gradient was slightly steeper than the late Eocene gradient (figure 2.15b), with a mean offset between the two curves of 0.3‰ to 0.4‰ [assuming constant tropical sea-surface temperature (SST)], representing the ice-volume effect. The offset at 70° latitude is 0.6‰ (total offset at 70° minus the mean offset), of which 0.2‰ to 0.3‰ would be cooling equal to about 1°C. Zachos et al.'s (1992) study incorporated two important assumptions, that changes in deep-water temperatures were paralleled by changes in high-latitude temperatures, as proposed by Savin (1977) for the Tertiary, and that tropical SSTs were stable. The former assumption has been called into question by workers tracking water-mass characteristics, as discussed in the next section. The latter assumption, although supported by Adams, Lee, and Rosen (1990) and Matthews and Poore (1980), has been questioned on the basis of nannofossil diversity (Wei, Villa, and Wise 1992). In addition, Zachos, Stott, and Lohmann (1994) noted that many of the sites sampled in the late Eocene and early Oligocene could have been affected by upwelling, so changes in the tropical temperatures might indicate variations in upwelling.

Another complication of estimating temperature versus ice-volume effects was brought out by a comparison of sea-level and ice-volume estimates from $\delta^{18}O$ (Chappell and Shackleton 1986). If ice volume were the sole control on sea level and on benthic $\delta^{18}O$, the curves should match, and sea level and $\delta^{18}O$ should correlate. That they do not led Chappell and Shackleton

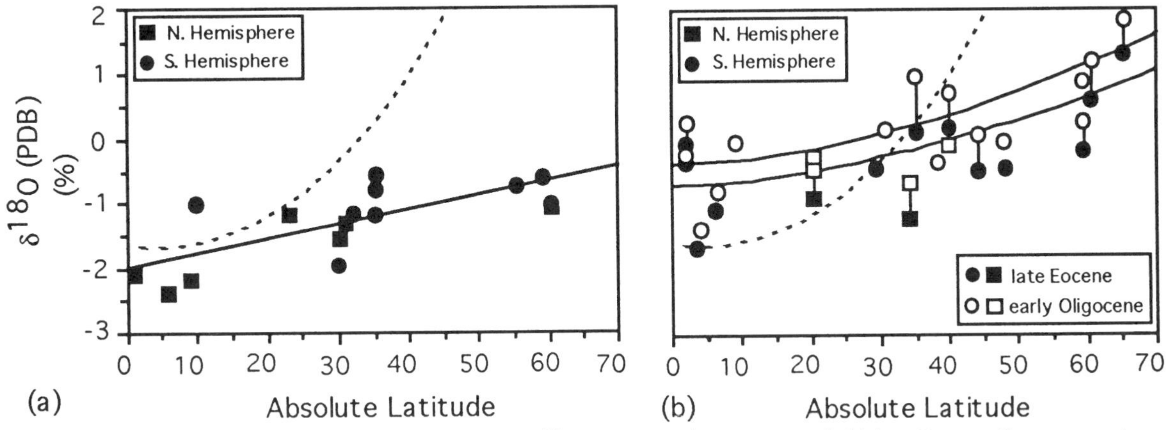

Figure 2.15. Latitudinal gradients in surface-water $\delta^{18}O$ for *(a)* early Eocene and *(b)* late Eocene (*lower curve*) and early Oligocene (*upper curve*); Holocene gradient is shown as *dashed line.* Data points connected by lines in (b) are from the same DSDP/ODP sites. Adapted from Zachos et al. (1992).

(1986) to propose that deep-water temperatures varied independently of ice-volume changes (see also Emiliani and Ericson 1991); a similar conclusion was reached by Prentice and Matthews (1988) for the Tertiary as a whole. Another factor was the gradual change in the isotopic composition of the ice as it accumulated; as an ice sheet expands, the snow that falls in the colder, more central portion becomes lighter (Chappell and Shackleton 1986).

Water-mass studies. In the modern oceans, planktonic foraminifera grow in relatively narrow depth ranges within the photic zone (upper 100 to 200 m) and record the isotopic composition of the waters in which they grow, so that in an area with a steep thermocline, the isotopic composition of the various taxa will vary (Fairbanks et al. 1982). Fossil foraminiferal species also show a range of isotope values, and some taxa maintain the same relative ranking along a $\delta^{18}O$ scale regardless of where they are sampled (figure 2.16). Consistency in ranking within each time slice, across several time slices, and across a range of latitudes in different oceans is strong evidence that the isotope values are controlled by depth and therefore temperature, with the higher values at each latitude indicating deeper-dwelling forms. Note that many of the taxa occur across a broad latitudinal range.

Based on observations such as those in figure 2.16, Keller (1985) related the isotopic composition of foraminifera to their positions in deep-, intermediate-, and shallow-dwelling assemblages. She then was able to determine variations in the distribution of these assemblages and thus approach the problem of variations in the vertical temperature structure of the oceans in space and time. Keller's (1985) reasoning for this, presented schematically in figure 2.17, depends on using both cosmopolitan and endemic taxa. Examples

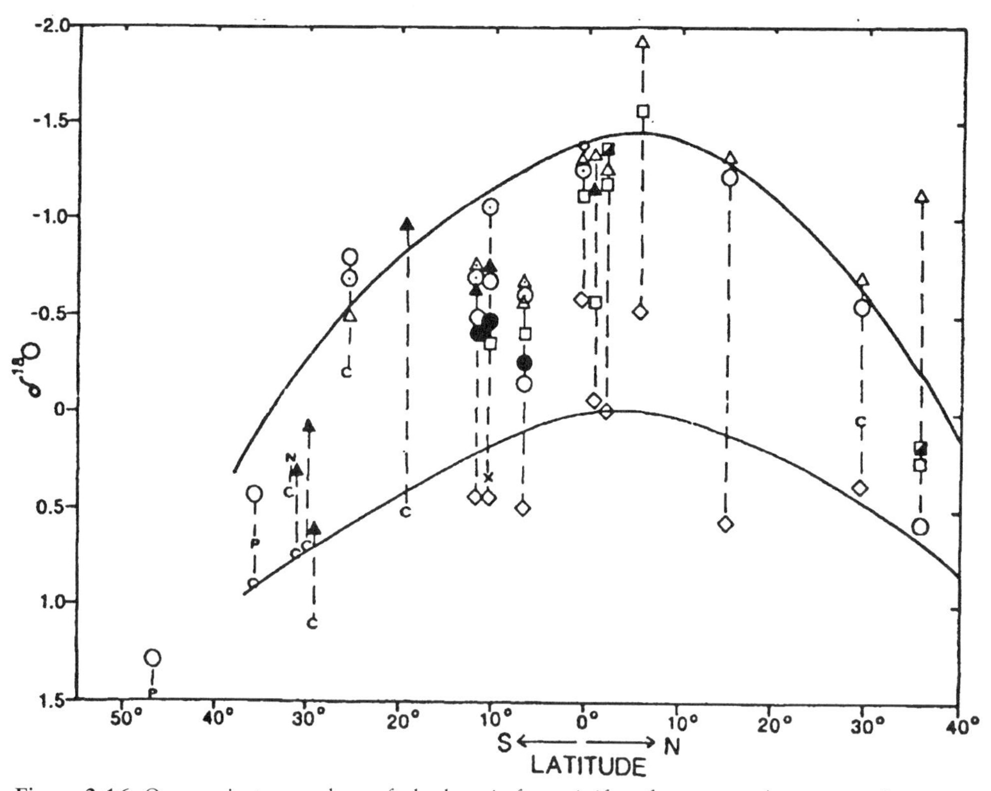

Figure 2.16. Oxygen isotope values of planktonic foraminifera from several oceans and across a range of latitudes. *Upper line* corresponds to surface values and *lower line* to values at the thermocline. Each symbol represents a different species or species group of foraminifera. Note that, although the isotope values for each species change with latitude, the different species tend to maintain the same relative positions on the $\delta^{18}O$ scale. From Keller (1985) in *The Miocene Ocean: Paleoceanography and Biogeography,* J.P. Kennett, ed. Reproduced with permission of the publisher, the Geological Society of America, Boulder, Colorado, USA. Copyright © 1985 Geological Society of America, Inc.

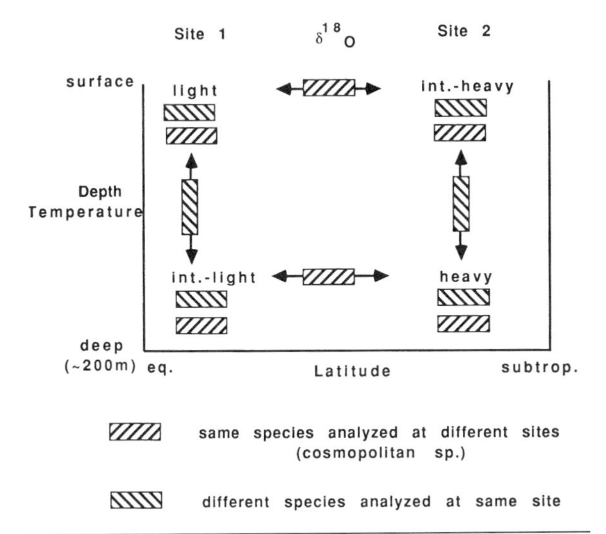

Figure 2.17. Schematic diagram showing the use of foraminiferal biogeography and oxygen isotopic chemistry in the study of water-mass characteristics, shown for two depth categories at two hypothetical sites. Depth categories are determined by analyzing both endemic and cosmopolitan species at each site (see figure 2.16), and latitudinal gradients in each depth category are determined by analyzing cosmopolitan species; this diagram indicates four hypothetical assemblages. The disappearance of one assemblage would indicate a change in the vertical water-mass structure of this schematic ocean.

of variation in depth stratification in different parts of the Pacific Ocean in the late Miocene are shown in figure 2.18. Changes in these profiles through time can be used to track changes in the vertical water-mass structure of the upper 200 meters of the ocean, which commonly includes the thermocline (Keller 1985). This work was combined with factor analysis of the paleobiogeography of the foraminifera (Kennett, Keller, and Srinivasan 1985) to reconstruct changes in the surface circulation.

The factor analysis performed by Kennett, Keller, and Srinivasan (1985) revealed paleobiogeographic differentiation across the equatorial Pacific. Keller (1985) exploited this pattern to construct "faunal climate curves" for the eastern and western Pacific. The curve for the western Pacific was based on the ratio $W/(E + W)$, where E is the abundance of eastern taxa and W is the abundance of western taxa. The faunal climate curve for the eastern Pacific was based on the ratio, $E/(E + W)$. These are similar to climate indices created for other types of fossils. Generally, the $\delta^{18}O$ and faunal climate curves agreed over many fluctuations.

Methods that utilize depth ranks have potential problems and limitations (Poore and Matthews 1984; Kennett, Keller, and Srinivasan 1985; Savin et al. 1985). These problems include the relationship between the upper part of the photic zone and the thermocline, which can control isotopic variation with depth (Fairbanks et al. 1982; Thunell, Curry, and Honjo 1983; Kennett, Keller, and Srinivasan 1985), seasonal bias in calcification (Savin et al. 1985; Spero 1992), disequilibrium precipitation of calcite (Fairbanks, Wiebe, and Bé 1980), calcification related to gametogenesis (Duplessy, Bé , and Blanc 1981), and variability in depth ranking (Poore and Matthews 1984).

Carbon isotopes also are useful in water-mass studies. Water-mass $\delta^{13}C$ is related to biological productivity at the surface and, in the deep sea, to the age of the water mass, that is, the amount of time since the water was at the surface. Under conditions of high biologic productivity, $\delta^{13}C$ in surface-dwelling foraminifera is enriched in ^{13}C because $^{12}CO_2$ is preferentially taken up in the organic matter of the phytoplankton, leaving a ^{13}C-enriched reservoir from which the foraminifera build their tests. Thus surface waters are lighter in less productive regions than in more productive regions. As a surficial water mass sinks in a convergence zone, the effects of productivity reverse themselves, and $^{12}CO_2$, derived from sinking and decaying organic matter, begins to accumulate. The more time since the water mass sank, the more enriched in $^{12}CO_2$ it becomes, resulting in lower $\delta^{13}C$; this effect can be accelerated if the overlying waters are highly productive.

The relationship between $\delta^{18}O$ and temperature, $\delta^{13}C$ and total dissolved CO_2, and expected profiles in high- and low-productivity surface waters are shown in figure 2.19. In modern oceans, waters are productive where the thermocline has been diffused by upwelling and nonproductive where the thermocline is strong. In nonproductive waters, $\delta^{13}C$ is not useful for depth ranking because a gradient is not established by preferential uptake of ^{12}C at the surface. However, $\delta^{18}O$ is useful for depth ranking in such waters because of the strong thermocline (see figure 2.19c). The opposite is true under productive waters (see figure 2.19d; Gasperi and Kennett 1992).

The use of $\delta^{13}C$ in upwelling zones is complicated by the fact that the upwelled water can have low $\delta^{13}C$ values as a result of the transport of iso-

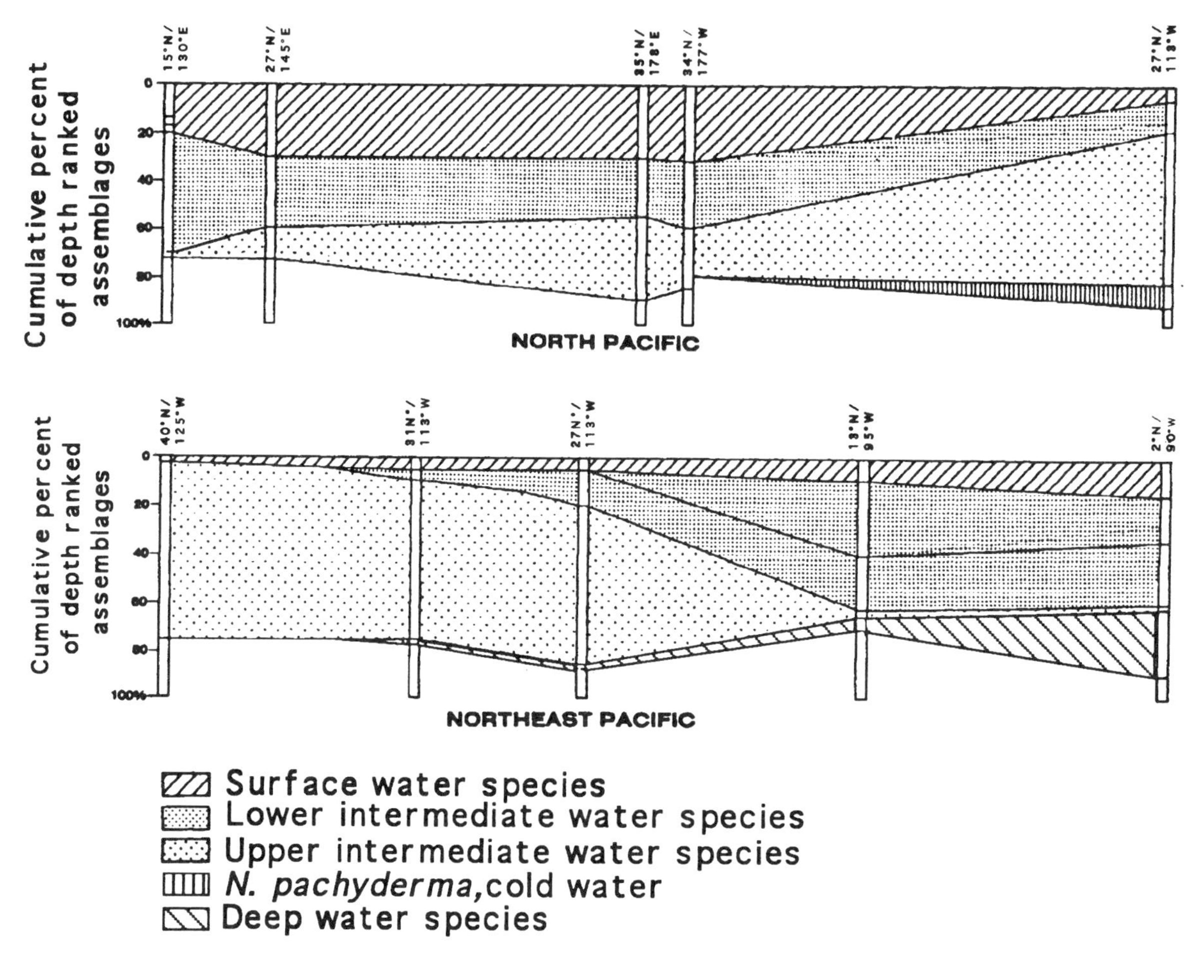

Figure 2.18. Profiles of depth-related foraminiferal assemblages in two transects across the Pacific Ocean in the late Miocene. Vertical axes are cumulative percentages of the various assemblages, but the resulting profiles resemble the thermal profiles from the modern Pacific Ocean, with shallowing of isotherms east to west in the North Pacific and south to north in the Northeast Pacific. Adapted from Keller (1985) in *The Miocene Ocean: Paleoceanography and Biogeography*, J.P. Kennett, ed. Reproduced with permission of the publisher, the Geological Society of America, Boulder, Colorado, USA. Copyright © 1985 Geological Society of America, Inc.

topically light organic matter from the surface. For example, Boersma and Premoli-Silva (1989) noted that biserial heterohelicids in the Paleogene consistently have the lowest $\delta^{13}C$ of all genera and $\delta^{18}O$ signatures indicating an intermediate depth, fitting neither of the two cases in figure 2.19c,d. Boersma and Premoli-Silva (1989) proposed that these genera were associated with the oxygen-minimum zone (OMZ) in high-productivity regions and could be used to map the OMZ in time and/or space.

Recently, high-resolution data sets have permitted very detailed interpretation of water-mass characteristics. The following example shows not only the value of such data sets but also how salinity variations might be interpreted from the record. Kennett and Stott (1990) analyzed $\delta^{18}O$ in specimens of one benthic (*Cibicidoides* sp.) and two planktonic (*Acarinina* sp. and *Subbotina* sp.) genera, densely sampled from two ocean-drilling sites (Sites 689, 690) 60 km apart in the Southern Ocean. These two sites were midwater sites in the Paleogene, Site 689 subsiding from 1200 m to 1790 m between 60 and 25 Ma, and Site 690 subsiding from 2040 m to 2625 m in the same time interval.

Kennett and Stott (1990) made both between-site and within-site comparisons of the $\delta^{18}O$ gradients. The between-site comparisons were done with the benthic foraminifera *Cibicidoides*. $\delta^{18}O$ was higher at Site 689 (figure 2.20), with a few brief reversals, until the late Oligocene, when the between-site gradient reversed. Within-site gradients were determined from the planktonic and

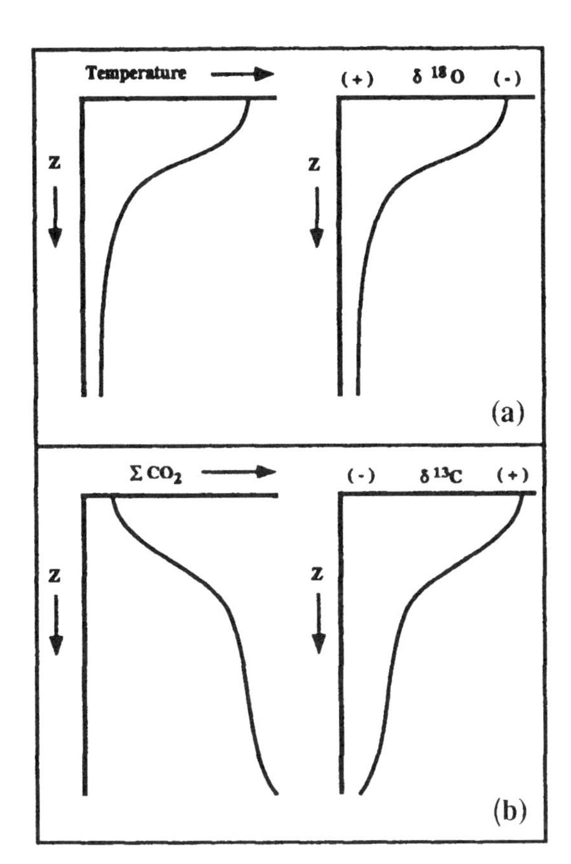

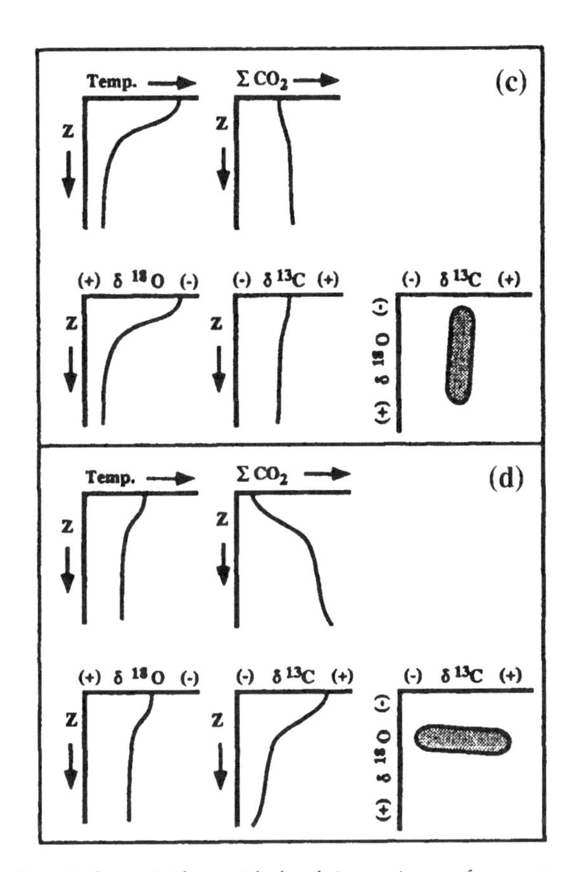

Figure 2.19. Schematic profiles of $\delta^{18}O$ and $\delta^{13}C$ from planktonic foraminifera with depth in marine surface waters. *(a)* Relationship between temperature and $\delta^{18}O$. *(b)* Relationship between total dissolved CO_2 and $\delta^{13}C$. *(c,d)* Profiles of temperature, total dissolved CO_2, $\delta^{18}O$, $\delta^{13}C$, and plots of $\delta^{18}O$ versus $\delta^{13}C$ for a region with a well-developed vertical thermal gradient and low surface-water biological productivity (c) and an upwelling region with high biological productivity (d). Adapted from Gasperi and Kennett (1992) in *Pacific Neogene: Environment, Evolution, and Events,* University of Tokyo Press.

benthic foraminifera. $\delta^{18}O$ at both sites showed a gradient between the surface-dwelling *Acarinina* and the other two genera, but little or no gradient between the planktonic *Subbotina* and *Cibicidoides.*

Kennett and Stott (1990) interpreted these data as indicative of warm, saline, deep water during the Eocene and early Oligocene. The lower $\delta^{18}O$ values for the deeper Site 690 indicate higher temperatures, but the waters could not have been simply warmer than those at Site 689 because that would have led to an unstable water column. The waters at Site 690 must also have been saline. Although higher salinities also imply higher $\delta^{18}O$, the salinity effect is less than that of temperature. Kennett and Stott (1990) considered and eliminated other possible explanations for the differences between the two sites, and it is instructive to reiterate them, as they are the kinds of questions that must be asked of any such data set. These explanations and Kennett and Stott's (1990) reasoning in rejecting them are (1) stratigraphic miscorrelation—the stratigraphy is well constrained by magneto- and biostratigraphy at these sites and the $\delta^{13}C$ changes between the two sites are similar; (2) differential subsidence—the sites are close together and no evidence exists on the seismic lines for any tectonic discontinuity between them; (3) vital effects resulting from biogeographic differences in ecology—the sites are too close together (60 km) for this to be a likely problem; (4) diagenesis—there is no difference in burial depths or lithology; and (5) interlaboratory variability—the various workers ran duplicate samples.

Between-basin differences in isotope profiles can be exploited to trace large-scale water-mass variations in much the same way more-local variations were studied by Kennett and Stott (1990). Using $\delta^{18}O$ and, mainly, $\delta^{13}C$ of the benthic genera *Cibicidoides* spp. and *Planulina wuellerstorfi,* Wright, Miller, and Fairbanks (1992) undertook

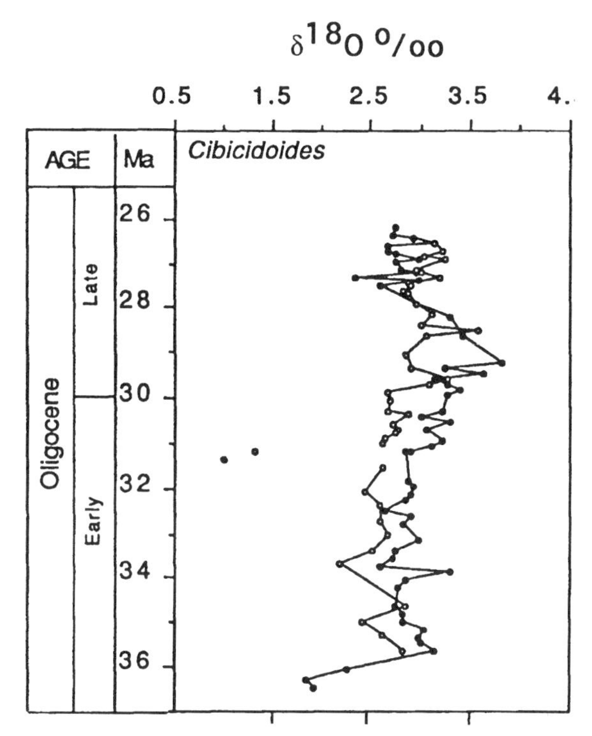

Figure 2.20. Oxygen isotope values from the benthic foraminiferan *Cibicidoides* in Oligocene sediments from ODP Sites 689 (*closed circles*) and 690 (*open circles*). Note where the curves cross at about 28 Ma. From Kennett and Stott (1990).

a global study of deep-water circulation for the Miocene. The basic assumptions were that (1) higher $\delta^{13}C$ indicates younger water, and (2) changes in the same direction and magnitude over all basins indicate global production or burial of organic carbon, but (3) differences among basins are local effects. The records from three representative sites are shown in figure 2.21. A positive excursion at about 19 to 16 Ma in the Southern and Pacific Oceans (the interval is missing from the North Atlantic) indicates a change in the global carbon budget, in this case corresponding to burial of large amounts of ^{12}C-rich organic matter (see Vincent and Berger 1985). After 19 Ma, the North Atlantic (Site 563) and west equatorial Pacific (Site 289) Oceans both show a +1.6‰ change, but the North Atlantic data are consistently higher; data from the Southern Ocean site are intermediate between the two (figure 2.21). The intermediate values for the Southern Ocean suggest mixing of North Atlantic bottom waters with Pacific waters (see also Flower and Kennett 1995).

Abrupt, extreme events. Two events for which isotopes in foraminifera have proved useful are the Cretaceous-Tertiary and near-Paleocene-Eocene boundary events. Both have been the sub-

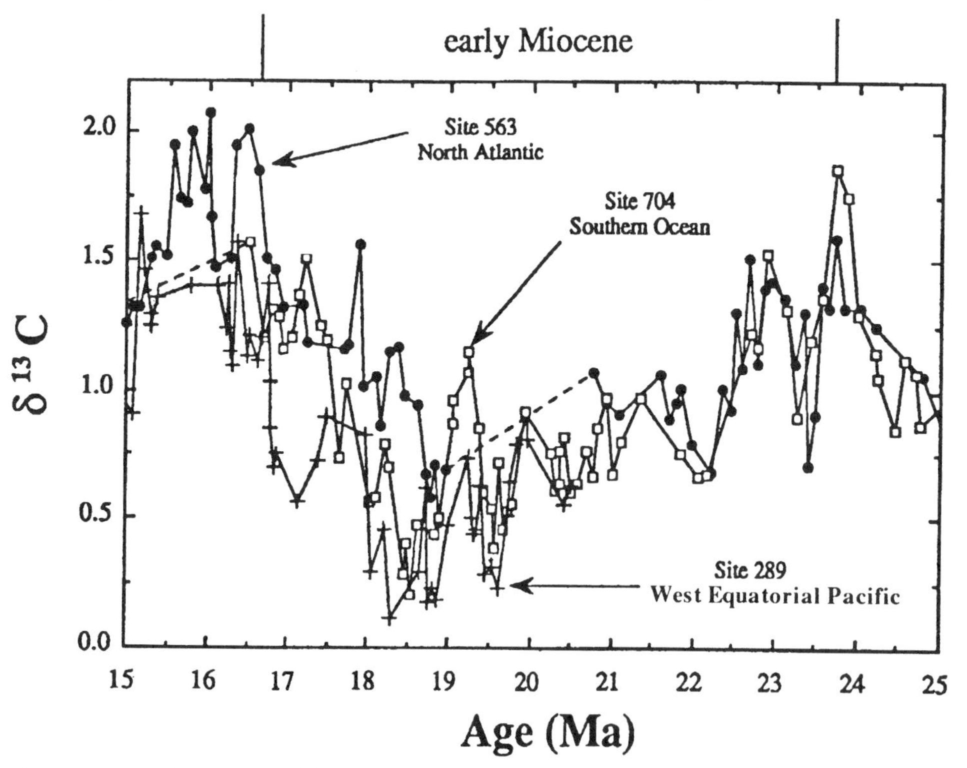

Figure 2.21. Early Miocene $\delta^{13}C$ records for DSDP and ODP Sites 563, 704, and 289. See text for explanation. From Wright, Miller, and Fairbanks (1992) *Paleoceanography* 7:357–89, copyright by the American Geophysical Union.

ject of a vast literature; only two illustrative examples will be presented here.

Zachos, Arthur, and Dean (1989) observed that surface and benthic foraminiferal $\delta^{13}C$ values converge at the Cretaceous-Tertiary boundary. They interpreted this as a reduction of the $\delta^{13}C$ gradient caused by cessation of surface productivity. In addition, the relationships among benthic foraminifera changed, also converging at the boundary. Zachos, Arthur, and Dean (1989) attributed the initial differences among the benthic taxa to their different microhabitats. In situ decay of organic matter decreases the $\delta^{13}C$ of taxa that live within the sediment relative to those that live on top, and the magnitude of the effect is partly a function of the amount of organic matter (Zahn, Winn, and Sarnthein 1986). Thus carbon isotopic convergence between infaunal and epifaunal groups suggests lower organic matter contents and/or accumulation rates resulting from lower productivity. Other evidence for decreased productivity includes a decrease in the accumulation rate of barium and a reduction in the amount of carbonate accompanied by better preservation of microfossils, which shows that the reduction in carbonate is not simply a change in dissolution.

The event near the Paleocene-Eocene boundary extends into the terrestrial realm (e.g., Koch, Zachos, and Gingerich 1992). In the oceans, this event is manifested as a warming of deep waters associated with widespread extinction of benthic foraminifera (e.g., Kennett and Stott 1991; see Bralower et al. 1995 and references therein). A comparison of the isotopic signatures of shallow-dwelling planktonic, deep-dwelling planktonic, and benthic species of foraminifera showed a gradient among the species in $\delta^{18}O$, and between the deep-dwelling planktonic and benthic species in $\delta^{13}C$, in the pre-boundary part of record (Kennett and Stott 1991; figure 2.22). Kennett and Stott (1991) interpreted the $\delta^{13}C$ record to indicate that *Subbotina* sp. inhabited a strong oxygen-minimum zone that disappeared after the event. Gradients in both $\delta^{18}O$ and $\delta^{13}C$ disappeared during the event, coincident with the extinction of benthic foraminifera. Benthic diversity in the larger foraminifera (the >150-μm fraction) decreased coincident with the $\delta^{18}O$ shift and before the $\delta^{13}C$ shift. Abundances also decreased, although small specimens (>150 μm) remained abundant. Kennett and Stott (1991) suggested that the extinction of the benthic foraminifera was caused by the low oxygen content of the much warmer deep waters. They noted that both planktonic foraminifera and calcareous nan-

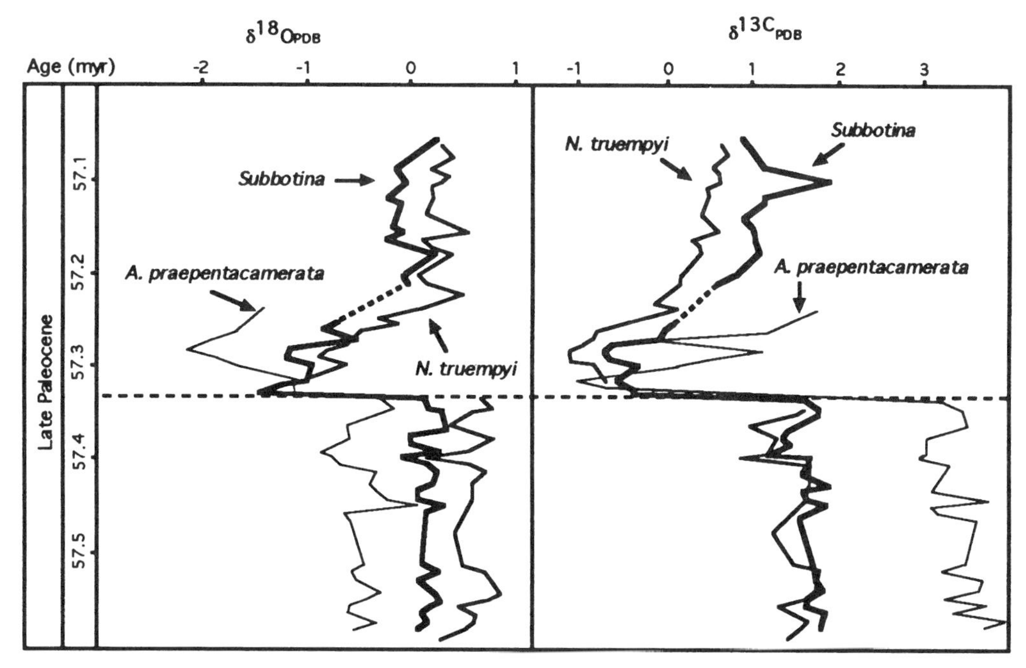

Figure 2.22. Stable isotope curves near the Paleocene-Eocene bounday at ODP Site 690B near Antarctica from shallow-dwelling planktonic (*Acarinina praepentacamerata*), deep-dwelling planktonic (*Subbotina* sp.), and benthic (*Nuttallides truempyi*) foraminifera. From Kennett and Stott (1991). Reprinted with permission from *Nature* 353:225–9. Copyright 1991 Macmillan Magazines Limited.

nofossils became more diverse, which would be expected with warming of the surface waters, and that benthic foraminifera in the more oxygenated shelfal waters were unaffected. The disappearance of a gradient in $\delta^{13}C$ suggests lowering of productivity during the event.

Changes in the carbon budget. Global variations in $\delta^{13}C$, reflected in isotope profiles from anywhere in the world, indicate global events, such as the ones at the Cretaceous-Tertiary boundary and near the Paleocene-Eocene boundary, the Cenomanian-Turonian (Late Cretaceous) event (Scholle and Arthur 1980), and the so-called Monterey event (Vincent and Berger 1985). These global events indicate changes in burial or weathering rates of organic carbon. Increased burial rates are interpreted as either increased preservation, indicative of widespread low-oxygen waters in the oceans, or increased production, usually (though not always) productivity in the oceans.

The Monterey event (Vincent and Berger 1985) has become a benchmark in Neogene isotope studies. It is a 1‰ positive shift in $\delta^{13}C$ between 17.5 and 13.5 Ma that is accompanied by a shift in $\delta^{18}O$ toward higher values during the latter part of the $\delta^{13}C$ shift (figure 2.23). A notable feature of this event is that carbon production and burial have been documented independently of the carbon isotope data. During the same interval, thick, organic-rich, diatomaceous sediments were deposited in upwelling zones around the margins of the North Pacific, including the Monterey Formation in California, for which the event was named (Flower and Kennett 1993, 1994; but see Compton, Snyder, and Hodell 1990). Vincent and Berger (1985) proposed that upwelling and extraction of CO_2 from the atmosphere, initiated by some unspecified mechanism, started a positive feedback loop of cooling (shown in the $\delta^{18}O$ shift) by lowering CO_2, which led to a stronger, ocean-wide thermocline (which permits accumulation of nutrients that are then upwelled at the margins), which led to

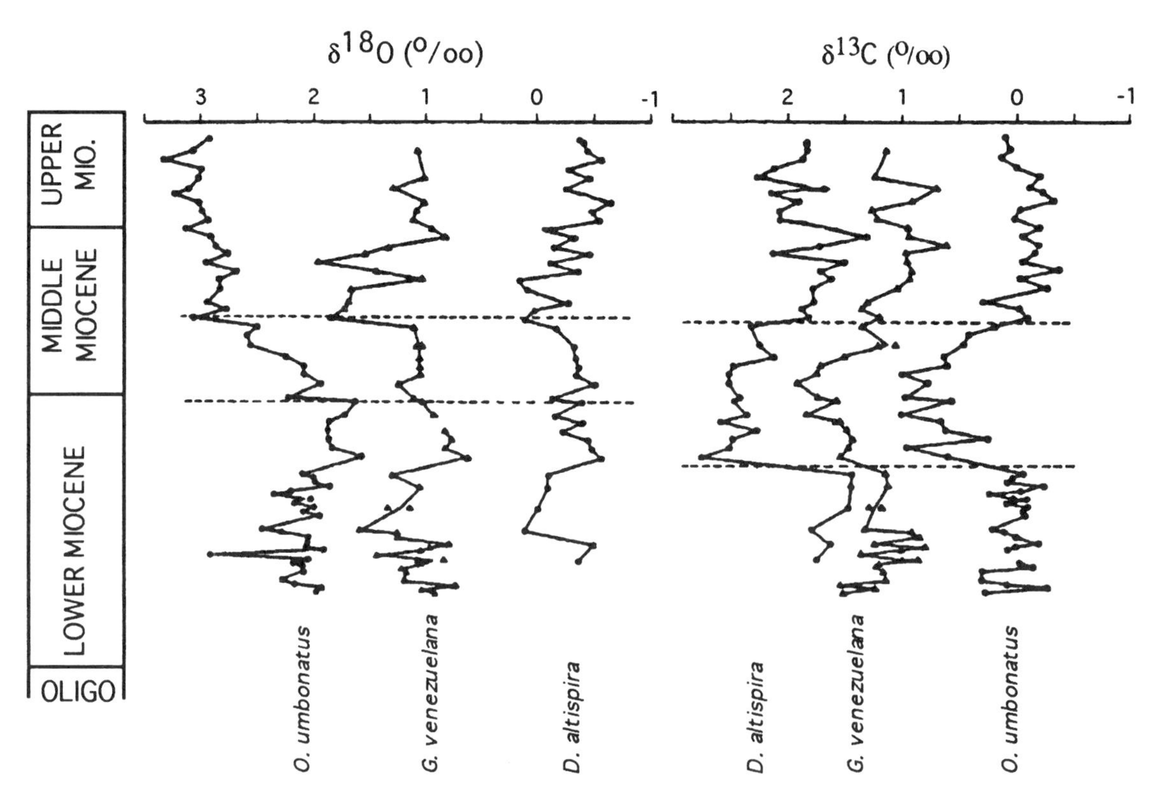

Figure 2.23. Stable isotope data from the Indian Ocean (DSDP Site 216) showing the "Monterey event," which is the shift to higher values between the *dotted lines* on the $\delta^{13}C$ curves. *Dotted lines* on the $\delta^{18}O$ curves bracket a positive shift interpreted to indicate cooling in high latitudes. *Oridorsalis umbonatus*, *Globoquadrina venezuelana*, and *Dentoglobigerina altispira* are benthic, deep-dwelling planktonic, and shallow-dwelling planktonic foraminifera, respectively. Modified from Vincent and Berger (1985) in *The Carbon Cycle and Atmospheric CO_2: Natural Variations Archean to Present,* pp. 455–68. Geophysical Monograph 32, copyright by the American Geophysical Union.

more carbon burial, and so on. The loop was broken as basin configuration and other factors changed.

Perhaps the best-known use of a global positive excursion in $\delta^{13}C$ as an indicator of widespread carbon burial is the Cenomanian-Turonian boundary event first noted by Scholle and Arthur (1980) and later discussed at length by Schlanger et al. (1987), Arthur, Schlanger, and Jenkyns (1987), and many others. The discovery of the global $\delta^{13}C$ shift dovetailed with earlier descriptions of widespread, organic-rich rocks of about that age that were described as representing an "anoxic event" (Schlanger and Jenkyns 1976; the term they used was "oceanic anoxic event," or OAE). In a synthesis of all that was known at the time about the event, Schlanger et al. (1987) and Arthur, Schlanger, and Jenkyns (1987) proposed that the organic-rich rocks were deposited in a widespread oxygen-minimum zone that intruded into epeiric seaways; Arthur, Schlanger, and Jenkyns (1987) also commented that elevated productivity in some locations or even globally could have been responsible for the organic richness. The $\delta^{13}C$ excursion is seen everywhere the boundary is preserved, in both organic-rich rocks and in organic-poor limestones (Scholle and Arthur 1980; Arthur, Schlanger, and Jenkyns 1987; Schlanger et al. 1987).

Subsequent investigations of the Cenomanian-Turonian event, using detailed nannofossil biostratigraphy, revealed slight diachroneity in the the $\delta^{13}C$ shift among several basins (Bralower 1988). Bralower (1988) discussed several possible effects on $\delta^{13}C$, other than carbon burial: dissolution/reprecipitation, especially in different environments of diagenesis; oxidation and sulfate reduction, which can shift $\delta^{13}C$ to lower values; and methanogenesis, which can shift $\delta^{13}C$ to higher values. Another possible problem is the differing signals in carbonate versus organic matter; $\delta^{13}C$ is affected by the type of organic matter, so a shift in organic-carbon $\delta^{13}C$ could be simply a change in the type of organic matter represented in the sample. Bralower (1988) proposed that the overall oceanic environment might have been favorable for the development of anoxia, but that the "event" was actually a series of events in different basins that became anoxic because of local, rather than global, conditions.

Like $\delta^{18}O$, $\delta^{13}C$ can be mapped in space. For example, a map of $\delta^{13}C$ for an interval near the Cenomanian-Turonian boundary in Europe showed an east-west gradient across Europe (Hilbrecht, Hubberten, and Oberhänsli 1992). High values corresponded with upwelling zones in northern Germany and the eastern North Sea predicted from climate models (Parrish and Curtis 1982; Barron 1985a; Scotese and Summerhayes 1986), intermediate values in Great Britain and France likely represented the edge of the upwelling zone, and low values were separated from the upwelling system by the Rhenish-Bohemian Massif in southern Germany.

Other aspects of foraminiferal chemistry and paleoclimate

Cadmium in benthic and planktonic foraminifera has been proposed as a tracer for nutrients in the oceans (Boyle 1988), which is important for understanding the carbon cycle. Phosphorus, which is a key nutrient for biologic productivity, and cadmium, a trace metal, have similar distributions in the oceans (Boyle 1986). Unlike phosphorus, which is taken up by the soft tissue during growth, cadmium is incorporated into the tests of the foraminifera. $\delta^{13}C$ also is closely related to phosphorus (Boyle 1986). Coupled with analyses of $\delta^{13}C$ from the same samples, Cd/Ca ratios should give estimates of the contrast in dissolved CO_2 between surface and deep waters (Boyle 1986). This is potentially a direct measure of nutrient distribution throughout the oceans; modeling studies incorporating nutrient cycles normally make assumptions about nutrient distribution based on the modern ocean (e.g., Sarmiento, Herbert, and Toggweiler 1988a). However, Cd/Ca presents a more complicated picture than initially thought (Boyle 1988, 1993; van Geen, McCorkle, and Klinkhammer 1995) and, for the Miocene oceans, for example (Delaney 1990), is not a good proxy for phosphorus, as it is in the Quaternary oceans (e.g., Lea and Boyle 1990). Discrepancies between the Cd/Ca and $\delta^{13}C$ record also have been noted (e.g., Broecker 1993).

Like Cd, barium substitutes for calcium in foraminiferal tests, and the Ba content of the tests reliably records variations in seawater Ba (Lea and Boyle 1991; Lea and Spero 1994). Ba/Ca ratios in benthic foraminifera have been used to reconstruct nutrient distributions in the late Pleistocene oceans (Lea and Boyle 1990). During the Last Glacial Maximum, Ba contents of deep waters in the Atlantic, Pacific, and Antarctic were similar, whereas Cd was much lower in the Atlantic than in the Pacific. Because Ba also is sedimented as particles, its behavior is different from that of Cd and $\delta^{13}C$. Ba is more sensitive to upwelling, and Lea and Boyle (1990) suggested that upwelling in the Atlantic might have transferred Ba to deep waters, reducing the difference in nutrient status with the Pacific that was reflected in the Cd/Ca ratios. Lea and Boyle (1990) concluded that Ba might be a good tracer for nutrients regenerated in the deep ocean.

Strontium isotopes in seawater have been studied mostly as a tool for geochronology (e.g., DePaolo and Ingram 1985; DePaolo and Finger 1991), but because the Sr-isotopic composition of the oceans is partly dependent on continental weathering rates and riverine influx, and because continental weathering plays a role in the CO_2 budget, Sr isotopes are studied for their paleoclimatic implications as well. Raymo, Ruddiman, and Froelich (1988) and Raymo (1991) argued that the Cenozoic increase in Sr and the Cenozoic cooling since the Eocene could both be explained by uplift and weathering of the Himalaya-Tibet region. Modeling this process, Richter, Rowley, and DePaolo (1992) related variations in the Sr-isotope curve to events during the collision of India and Asia. Their model required a steep increase in erosion rates after about 10 Ma, even though the convergence rate between India and Asia decreased. They suggested that this continued increase in erosion rate could be the result of a complex interaction between the rising topography and climate. Hodell, Mead, and Mueller (1990) argued that, after 8 Ma, uplift and weathering of the Himalaya-Tibet region could not account for rising $^{87}Sr/^{86}Sr$. They proposed a contribution from Precambrian shield areas exposed by continental ice sheets (see also Miller et al. 1991); Quade et al. (1997) found high-Sr metamorphic rocks in the Himalayas. Goddéris and François (1995) argued the problem from another angle. According to their model, CO_2 drawdown by weathering of the Himalaya-Tibet region was not sufficient to cause the Cenozoic cooling; they proposed decreased inputs of CO_2 from mid-ocean ridges and changes in weathering rates elsewhere. Modeling of the geochemical cycles involved in these processes has been carried out by Berner and Rye (1992), François and Walker (1992), Goddéris and François (1995), and others.

Calcareous Nannofossils

Calcareous nannofossils (Haptophyta) are the remains of single-cell algae that secrete several of the hard parts per cell. Oval or circular hard parts are called coccoliths, star-shaped or other nonelliptical forms are called nannoliths (Siesser 1993). They appeared in the Late Triassic (see figure 2.1) and were highly diverse by the Early Cretaceous (Tappan 1980). The paleobiogeography of calcareous nannofossils is used in paleoclimate interpretation. In addition, they are the most important constituent of chalk. Calcareous nannofossils consist of low-Mg calcite and thus are relatively resistant to dissolution, although, as with foraminifera, the variability in their morphology leads to wide differences in dissolution that tend to bias the record toward more robust forms. Dissolution apparently occurs mostly in the water column during settling (Geitzenauer, Roche, and McIntyre 1977).

High percentages and/or diversities of discoasters, an extinct form of nannofossils, are considered indicative of warm temperatures, whereas high percentages of *Coccolithus pelagicus* or *Emiliana huxleyi* are considered indicative of cool temperatures. For example, Pospichal and Wise (1990), in a study of the Weddell Sea, noted that a shift toward more negative $\delta^{18}O$ values in the uppermost Paleocene corresponded with peak discoaster diversity, whereas a shift toward more positive $\delta^{18}O$ values in the middle Eocene corresponded with an absence of discoasters and lower coccolith diversity. In a study in the Kerguelen Plateau, Beaufort and Aubry (1992) used relative abundances of *C. pelagicus* and the cosmopolitan genus *Reticulofenestra* to define paleoceanographic events.

Roth and Krumbach (1986) used factor analysis to define paleobiogeographic patterns in calcareous nannofossils in the Mesozoic oceans. They first closely considered effects of what they called the calcium compensation zone, in reference to the fact that there appeared to be no clear break in diversity that could be used to define a lysocline, sensu Berger (1968), or calcium compensation depth (see figure 2.2). In this zone, species diversity was closely tied to preservation in that the best-preserved samples contained the most-diverse assemblages. They could not ascertain a clear relationship between diversity and depth except that the deepest sites had the most poorly preserved samples. They also noted that sediments from the mid-water zone of poor preservation around 2.5 km have high organic contents. Bearing these limitations in mind, Roth and Krumbach (1986) used only the best-preserved, highest-diversity assemblages for their paleobiogeographic analysis.

Wei, Villa, and Wise (1992) employed changes in nannofossil diversity gradients to examine paleobiogeographic change coeval with a $\delta^{18}O$ shift at the Eocene-Oligocene boundary. They compared changes in nannofossil diversity from the equatorial ODP Site 711 and the high-latitude Site 748 in the Indian Ocean (figure 2.24). Both curves show a decrease in diversity, consistent with cooling, and the narrowing of the gap between the curves is indicative of a reduction in the latitudinal diversity gradient, consistent with a stronger cooling near the equator, at least in the Indian Ocean. This contradicts the assumption by Zachos et al. (1992) that tropical sea-surface temperatures were stable.

At Site 748, shortly after the Eocene-Oligocene boundary and at the same level as an increase in ice-rafted debris noted by Zachos et al. (1992), an abrupt increase in the abundance of cool-water taxa, from about 20% to about 90%, indicates a drop in temperature. Wei, Villa, and Wise (1992) reasoned that this decrease must be about 4°C because the change in the abundance in cool-water taxa over that interval of time is the same magnitude as the difference between the assemblages at 30°S and 65°S latitudes in the late Eocene, and the temperature difference between those latitudes at that time was 4°C, as determined from $\delta^{18}O$. They concluded that temperature change accounted for most of the $\delta^{18}O$ shift and and that the ice-volume effect accounted for very little of it. Wei, Villa, and Wise attributed the discrepancy between their conclusion and that of Zachos et al. (1992) to problems with using regression lines to determine latitudinal gradients. Wei, Villa, and Wise (1992) pointed out that the small number of data points (see figure 2.15) made the regression lines sensitive to the distribution and density of data (acknowledged by Zachos et al. 1992), and emphasized the poor fit of the regression lines ($R^2 \leq 0.65$). Finally, they noted that $\delta^{18}O$ gradients are not necessarily thermal gradients, illustrating their point with an example from the Holocene, when the $\delta^{18}O$ difference between 0° and 60° latitude was about 3.4‰, corresponding to a temperature difference of about 14°C; the actual temperature difference was twice that.

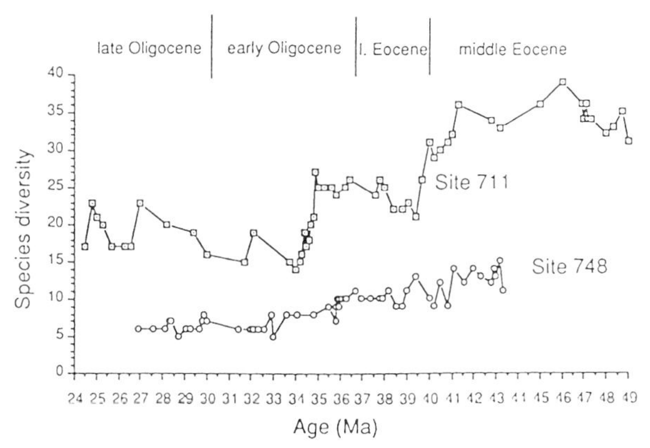

Figure 2.24. Species diversity of calcareous nannofossils in the Indian Ocean, ODP Sites 711 and 748. From Wei, Villa, and Wise (1992).

One of the more enigmatic types of occurrences in the geologic history of the oceans is the sudden, brief appearance of chalk consisting entirely of single species of the nannofossil genus *Braarudosphaera* over vast areas of the oceans; this nannofossil is otherwise a very minor component of cool-water nannofloras in the Tertiary. A completely satisfactory explanation has never been offered for these events. The species might somehow have responded to shifts in the circum-Antarctic current, climatic cooling, and/or higher production rates [presumably as a result of higher nutrient supply (Wise and Hsü 1971; Bralower 1988; Siesser, Bralower, and De Carlo 1992)]. Alternatively, *Braarudosphaera* might have been opportunistic, capable of proliferating when other organisms were under stress (Fischer and Arthur 1977) or the assemblage might simply be an artifact of the exclusion of warmer-water taxa by cold waters and/or dissolution of more delicate forms at the sediment-water interface by cold bottom waters (Wise and Hsü1971).

Marine Ostracodes

Ostracodes are benthic, microscopic crustaceans with bivalved carapaces consisting of chitin-rich calcite. In the modern oceans, there are two depth-related assemblages, the thermospheric assemblage, which lives in shallow water, usually >500 m deep, and the psychrospheric assemblage, which lives in deep water, usually >500 m, with a few exceptions at high latitudes or in upwelling zones (Kennett 1982). Ostracodes were among the earliest marine microfossils (see figure 2.1), and they have been particularly common in marine rocks since the Ordovician Period.

Most of the work on pre-Quaternary marine ostracodes has relied on biogeographical analyses of various kinds, resulting in qualitative and quantitative conclusions about environmental controls on their distribution. Transfer functions are particularly important in marine ostracode studies.

Paleobiogeography of ostracodes

Ostracodes are highly sensitive to oxygenation levels (e.g., Dingle and Giraudeau 1993). Their preferred habitats and feeding habits can be inferred from their morphology. Filter-feeding ostracodes have straight commisures (that is, the line along which the two valves come together, opposite the hinge) and symmetrical valves, whereas deposit feeders tend to have curved commisures and asymmetrical valves. In general, filter feeders are low diversity, but they dominate faunas that live in low-oxygen conditions because they can actively pass water and oxygen over the gas-exchanging tissue. Lethiers and Whatley (1994) used these observations to track oxygenation levels through time in the Paleozoic. The percentage of filter-feeding species and the stratigraphic ranges of the sampled faunas are shown in figure 2.25. From this distribution, Lethiers and Whatley (1994) drew a curve showing the change in mean percentage of filter feeders through time. In the Paleozoic ostracode samples, temperature does not appear to have been a major control, possibly because the samples all came from the low-latitude platforms of what is now western Europe. Lethiers and Whatley (1994) related percentages of filter-feeders to environments of deposition and overall abundance of ostracodes, and they calibrated the oxygenation scale using sedimentological characteristics (Thompson et al. 1985). Using this scale, oxygenation levels would be predicted to have been lower in the Eifelian through early Famennian and highest during the early Tournaisian (see figure 2.25). The period predicted to have low oxygen corresponds in general to a worldwide increase in the burial of organic matter, as indicated by a striking shift in $\delta^{13}C$ (Garrels and Lerman

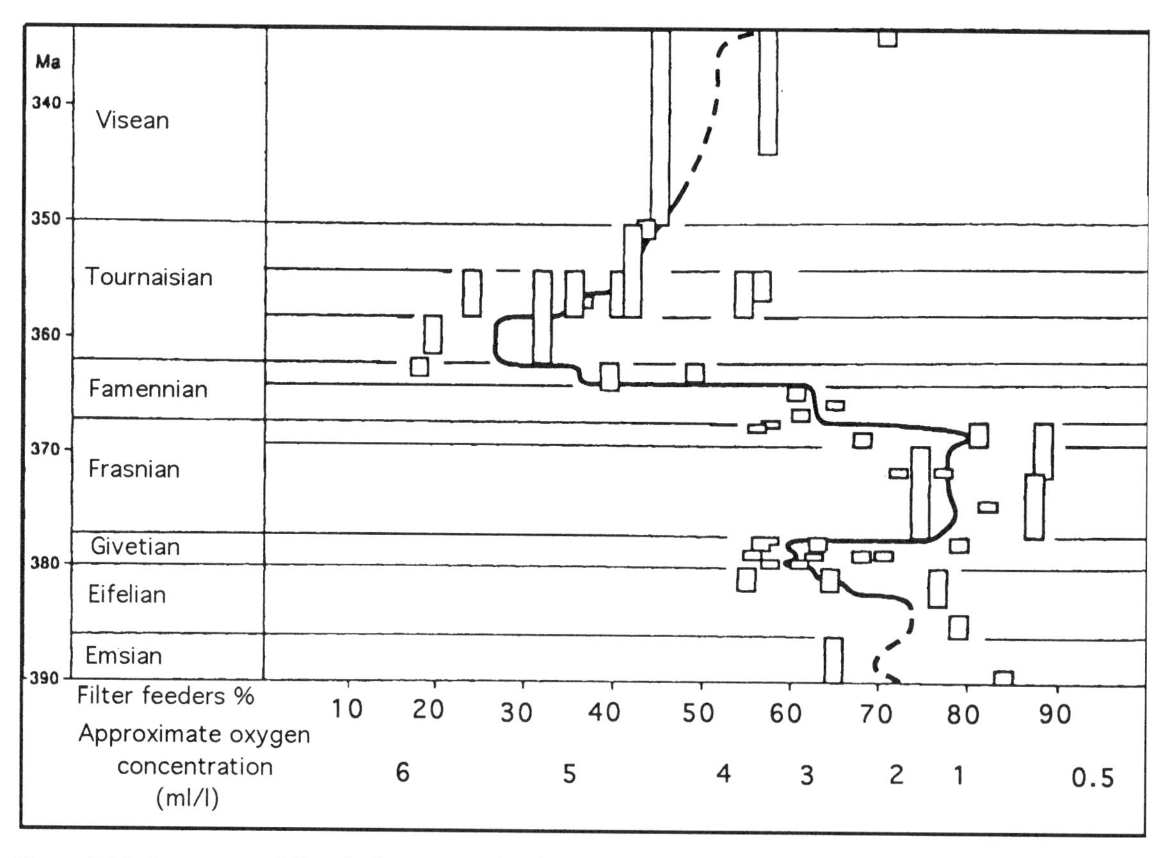

Figure 2.25. Percentage of filter-feeders among benthic ostracodes in Devonian-Carboniferous rocks of western Europe. *Heavy line* is the mean percentage through time; *boxes* represent the stratigraphic ranges of sampled faunas. Note the approximate oxygenation scale. Modified from *Marine Micropaleontology* 24, Lethiers, F. and R. Whatley, "The use of Ostracoda to reconstruct the oxygen levels of Late Palaeozoic oceans," pp. 57–69 (1994) with kind permission of Elsevier Science-NL, Sara Burgerhartstraat 25, 1055 KV Amsterdam, The Netherlands.

1981). Ostracodes have also been the subject of more traditional paleobiogeographic analysis (e.g., Babinot and Colin 1992).

Transfer functions

A graphical transfer function method was devised by Hazel (1988) for ostracodes (figure 2.26). The method involves plotting the present temperature ranges for species found in a geologic unit and noting the greatest degree of overlap. The fauna in figure 2.26 contains species that prefer relatively cool water (species 1 to 5), species that prefer relatively warm water (19 and 20), and a number of species that might be described as intermediate (6 to 18). The temperature ranges of all the species overlap between 12.5° to 15°C and 17.5° to 20.0°C, so Hazel (1988) concluded that the temperature of the waters in which the lower Yorktown Formation was deposited fell in that range.

Cronin and Dowsett (1990) developed mathematical transfer functions for the group for the northwestern Atlantic. Rather than using individual species, they used taxonomic categories, most of which are individual genera, but some of which are groups of species or genera. Part of the rationale for using categories rather than just individual genera was to reduce the size of the data matrix. Ostracode genera are much more species rich than foraminiferal genera, and if each species were treated separately, the database would have been unwieldy. Therefore Cronin and Dowsett (1990) included in each category taxa that are restricted to one climate zone, grouped together. The transfer function model was constructed with factor analysis; an example of the relationship between percent abundance and February bottom temperatures for a modern genus and an example of a factor analytic result are shown in figure 2.27. Seven factors accounted for 80% of the variance. The major limitation of this method is that most of the samples were from waters >100 m in depth, and ostracodes living in such shallow water are highly endemic, so their transfer functions can be applied only to fossil faunas from the same region.

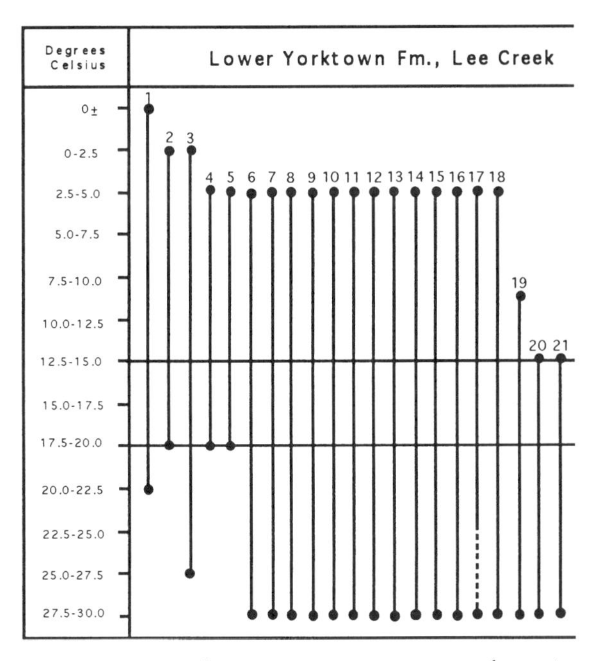

Figure 2.26. Modern temperature ranges of species found in the lower Yorktown Formation (Pliocene; North Carolina, U.S.). From Hazel (1988). The paleotemperature for this unit is defined by the overlap of these ranges between the two *horizontal lines*.

Cronin (1991) used the transfer functions to estimate summer and winter ocean temperatures in the northwestern Atlantic in the late Pleistocene and middle Pliocene. Modern temperatures and the late Pleistocene estimated temperatures were similar and cooler than the middle Pliocene except south of Cape Hatteras. Cronin (1991) attributed these cooler Pliocene temperatures to more vigorous upwelling associated with the Gulf Stream, but it is also possible that the transfer function method does not work there because the fossil faunas were less like the modern ones. Hazel (1988), for example, commented that his data seemed to imply that some taxa occurred farther south (in presumably warmer waters) in the past than they do today. Used in a transfer function, such taxa would tend to underestimate temperatures in the warmer regions. In contrast, Brouwers, Jørgensen, and Cronin (1991) found that ostracodes from the Pliocene Kap København Formation in northern Greenland that are still extant are all Arctic and subarctic, but the congeners of the extinct forms are from much warmer water today. In addition, the extant species occur in different proportions today than in the Kap København Formation. These problems illustrate why transfer functions cannot be extended very far back in time.

Given the limitations of the transfer function method alone, Cronin et al. (1994) employed supplementary methods of analyzing ostracode assemblages for interpretation of climate of the Pliocene Yabuta Formation in Japan. This unit was deposited in a regime of fluctuating sea level. This presented a potential problem for analysis because taxa that live in shallow waters at high

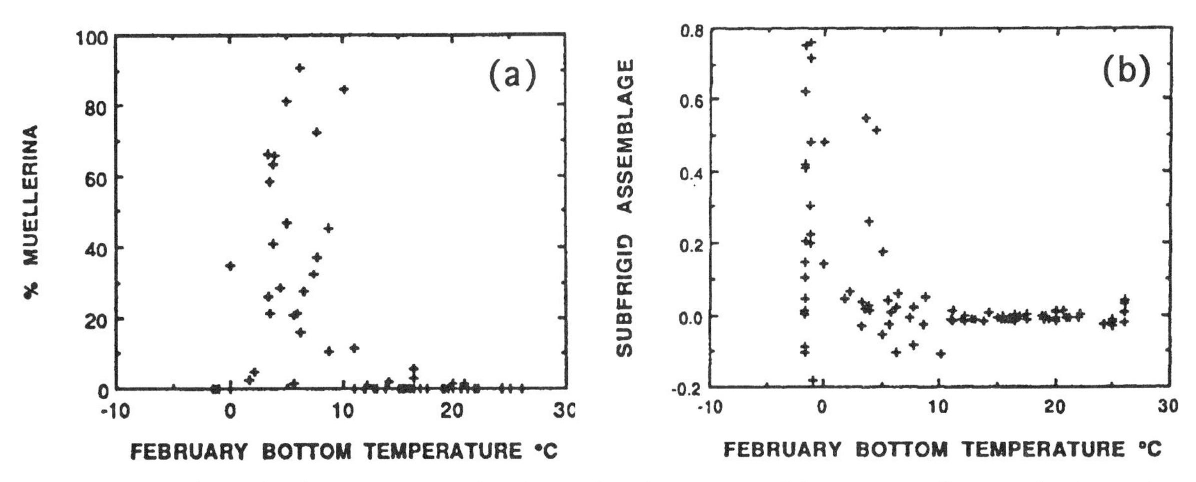

Figure 2.27. *(a)* Relationship between percent abundance of modern species of the genus *Muellerina* and February bottom temperatures. This is one example of the type of data on which the transfer function model of Cronin and Dowsett (1990) was based. *(b)* Example of factor analytic results used to develop the transfer function model, February bottom temperature for the subfrigid assemblage. Reprinted from *Marine Micropaleontology* 16, Cronin, T.M. and H.J. Dowsett, "A quantitative micropaleontologic method for shallow marine paleoclimatology: Application to Pliocene deposits of the western North Atlantic Ocean," pp. 117–47 (1990) with kind permission of Elsevier Science-NL, Sara Burgerhartstraat 25, 1055 KV Amsterdam, The Netherlands.

latitudes can live in deeper waters in low latitudes, so the effects of temperature and sea level change for each horizon must be distinguished in order to arrive at conclusions about climatic change. First, Cronin et al. (1994) used a form of the nearest-living-relative (NLR) method, sorting species into environmentally diagnostic groups based on the environmental tolerances of their extant congeners. Second, they devised a "*Schizocythere* index," that is, the ratio of *S.* cf. *okhotskensis* to *S. kishinouyei*, which are cold- and warm-water species, respectively. In many localities, this ratio changes along with other faunal and oceanographic indicators, for example, the proportions of right- and left-coiling forms of the foraminifera *Neoglobigerina pachyderma.* Third, Cronin et al. (1994) used a version of the transfer function method using dissimilarity coefficients. It is important to note that the NLR and dissimilarity methods are not entirely independent, as they are both based on the ecological ranges of modern taxa. Nevertheless, concordance of several indicators provides strong support for conclusions about climate change in this area.

Marine Diatoms

Diatoms (Chrysophyta) are microscopic algae with boxlike, siliceous shells that occur in all aqueous environments, "practically everywhere that light and moisture occur . . . soils, ice, and attached to rocks and other substrates within spray or splash zones" (Barron 1993), even on the leaves of terrestrial plants (Tappan 1980). In the open ocean, diatoms are planktonic and sometimes form colonies; in water up to about 180 m deep, benthic forms also are common (Burckle 1978; Barron 1993). Diatoms are autotrophs, requiring light, although some facultative and a few obligate heterotrophs also exist in the group (Tappan 1980). Most diatom species are eurythermal, and their ability to thrive in temperature extremes is dependent on nutrient supply. Most are also euryhaline, but diatoms are more sensitive to particular ions than to overall salinity (Tappan 1980). Diatoms appeared in the Jurassic and underwent a major evolutionary radiation in the mid-Cretaceous (Kennett 1982; see figure 2.1).

Marine diatoms are most diverse in higher latitudes. In the Antarctic, there are three latitudinal biogeographic zones, including an ice-edge flora (Gersonde 1986). Ice-edge floras are recognizable in the fossil record (Leventer and Dunbar 1988; Burckle, Gersonde, and Abrams 1990; Lagoe et al. 1993). Diatoms are important for studies of high-latitude environments and of oceanic environments in which deposition is below the carbonate compensation depth because these environments typically lack calcareous fossils. The abundance of diatoms is related to productivity, and high-productivity regions are marked by specific assemblages (Kennett 1982).

Diatom zonation in the Pacific Ocean is similar in general to the latitudinal zonation seen in other marine microfossils, but it has distinctive elements (Jousé, Kozlova, and Muhina 1971), especially the very high latitude units (Sancetta and Silvestri 1986). Some biogeographic units are related specifically to upwelling and high productivity. As with other microfossils, sediment assemblages reflect the living surface biogeographic units but have different taxonomic compositions because of dissolution effects (Jousé, Kozlova, and Muhina 1971). Modern diatoms are highly endemic compared with foraminifera. Diatom floras in north and south high latitudes share only about 50% of genera (Jousé, Kozlova, and Muhina 1971; Hasle 1976).

Diatoms have been studied using a variety of paleobiogeographic approaches and, recently, some workers have experimented with oxygen isotopes in diatoms.

Paleobiogeography of diatoms

Diatoms are well associated with high-productivity waters in modern oceans. The increased silica contents in sediments below oceanic upwelling zones is almost entirely the result of the high content of diatoms (Lisitzin 1972). However, although upwelling is indicated by high silica contents, the reverse is not necessarily true. Under some circumstances, diatom mats that form at the surface and are sedimented are more closely related to oceanic frontal activity than to upwelling (Kemp 1995; Bodén and Backman 1996). Biotic paleotemperature indices have been constructed for diatoms, and they are similar to those developed for foraminifera by Keller (1985) and for ostracodes by Cronin et al. (1994).

Burckle (1978), borrowing from earlier Japanese work (Kanaya and Koizumi 1966), quantified counts of warm- and cold-water species into a method of calculating relative paleotemperatures from Pleistocene floras. Diatom temperature (T_d) is the ratio of the total count of warm-water species to the sum of warm- and cold-water species. In his study, there were no cold-water species, so he substituted cosmopolitan species. He found a loose correlation with several other indicators (e.g., $\delta^{18}O$), suggesting that T_d curves might be useful for paleoclimate interpretation in the absence of other types of information, for example, at high latitudes, where foraminifera are sparse.

Barron (1992) extended Burckle's (1978) method to Pliocene diatom floras in high latitudes (north of the Subarctic Front). He noted that several diatom species are not strictly subtropical, even though they grouped with subtropical species in a cluster analysis of the data. He designated these species "warm transitional" and used the equation

$$Twt = (Xw + 0.5Xt)/(Xc + Xt + Xw), \quad (2.3)$$

where Twt is temperature, Xw the number of warm-water forms, Xc the number of cold-water forms, and Xt the number of transitional forms. Barron (1992) removed two species of *Coscinodiscus* that are known or suspected to be concentrated by dissolution. This resulted in a slightly warmer prediction for earlier intervals. By comparison with Holocene temperatures, Barron (1992) surmised that the northernmost site he studied was as much as 3°C warmer in the summer and 5.5°C warmer in the winter than it is today.

Transfer functions have also been developed for diatoms (e.g., Koç Karpuz and Jansen 1992; Pichon et al. 1992), but so far they have been applied only to Quaternary floras. The usefulness of transfer functions for older records is called into question by the diagenetic and no-analog situations noted by Barron (1992).

Stable isotope studies on diatoms

The temperature equation for $\delta^{18}O$ in diatoms, established empirically by Juillet-Leclerc and Labeyrie (1987), was given in the section on isotopes in chapter 1. Juillet-Leclerc and Schrader (1987) determined oxygen isotopic paleotemperatures in diatoms for the past 3000 y in the Gulf of California and compared the data with changes in the absolute abundance of the silicoflagellate *Dictyocha messanensis*, which they interpreted as an indicator of warm waters and low productivity. Low abundances of *D. messanensis* correspond to high values of $\delta^{18}O$. Juillet-Leclerc and Schrader (1987) therefore concluded that the cold temperatures resulted from upwelling (which resulted in high productivity), and not from intrusion of the cold California Current into the gulf.

In diatoms, $\delta^{18}O$ has also been used to determine ice-volume effects, paleotemperatures, and controls on variation in atmospheric CO_2 (Shemesh, Charles, and Fairbanks 1992; Shemesh et al. 1993; Shemesh, Burckle, and Hays 1995). Shemesh, Charles, and Fairbanks (1992) considered some of the problems with the assumption that high-latitude and bottom-water temperatures are equivalent and with the practice of using bottom-water $\delta^{18}O$ changes as an indicator of changes in ice volume. They proposed that a solution might be found in comparing the $\delta^{18}O$ of diatoms and foraminifera because the temperature relationships to the fractionation of oxygen isotopes are different in silica and carbonate. Using $\delta^{18}O$ data from both diatoms and foraminifera permits the solution of two variables, temperature and $\delta^{18}O_w$, provided the diatoms and foraminifera grew at the same depth and season. Rather than calculating absolute values, Shemesh, Charles, and Fairbanks (1992) calculated change in temperature and $\delta^{18}O_w$ from one interval to the next. The reason for this is that the intercept for the silica paleotemperature equation is uncertain. If diagenesis of the diatom silica can be shown to be minimal (e.g., see Isaacs, Pisciotto, and Garrison 1983), the method should be applicable to older rocks.

Shemesh et al. (1993) also investigated the use of $\delta^{13}C$ and $\delta^{15}N$ in preserved amino acids from diatoms to track changes in productivity through time and, through that, changes in atmospheric CO_2. The rationale for using diatom $\delta^{13}C$ rather than foraminiferal $\delta^{13}C$ was that the most direct connection between the atmosphere and the deep ocean is at high latitudes, where bottom waters are generated, and, during the much of the Tertiary, diatoms have been highly prolific at high latitudes. Oceanic processes are largely responsible for changes in atmospheric CO_2 because the oceanic reservoir is roughly 60 times the size the of the atmospheric reservoir (Berger and Keir 1984), so the connection between the atmospheric and deep oceanic carbon reservoirs is mediated by productivity at high latitudes.

In organic matter in plankton, ^{13}C is controlled by dissolved molecular CO_2, such that higher concentrations result in lower ^{13}C, depending on the photosynthetic carbon demand. The ^{13}C of the organic matter thus depends on the composition of the inorganic carbon used for photosynthesis, total dissolved CO_2, and the demand for carbon, that is, primary production. Shemesh et al. (1993) argued that one can distinguish among these mechanisms using changes in ^{15}N (in proteins in the test), which are dependent solely on primary production. The relationship of nitrogen isotopes to productivity varies depending on the supply of nutrients and the demand for them, such that $\delta^{15}N$ in the organic matter is lower than the $\delta^{15}N$ in the upwelling water where supply is greater than demand, and the two values are equal where nutrients are completely consumed; thus $\delta^{15}N$ increases in the organic matter as the nutrients are consumed (François, Altabet, and Burckle 1992). To determine whether the proteins from the biogenic opal of the diatoms had been altered during diagenesis, they compared amino acid profiles from Holocene samples of the core with amino acid profiles from their older samples. Although reliable and systematic preservation of amino acids is unusual in older rocks, other organic compounds are preserved, so it is possible that a variant of Shemesh et al.'s (1993) method might eventually be developed for older floras.

Germanium/silicon ratios in diatoms

Germanium (Ge) substitutes readily for silicon (Si) in diatom tests. In the open ocean, at least, biogenic opal tends to reflect the original oceanic Ge/Si ratio, which changes with variations in riverine influx and weathering. Thus several workers have proposed using Ge/Si ratios in biogenic opal, mostly from diatoms, as tracers for global fluxes (review in Shemesh, Mortlock, and Froelich 1989).

The value of Ge/Si as an indicator depends on the assumption that fractionation between Ge and Si during biogenic opal production was constant through time. Murnane and Stallard (1988) demonstrated that fractionation of ambient Ge/Si does change, apparently in concert with changes in productivity. They measured opal Ge/Si in diatoms from the laminated Miocene Monterey Formation, California, and found that values were lower in the light laminae. These laminae result from high productivity and intense production of opal, and the low Ge/Si ratios were thus consistent with the interpretation that productivity and opal Ge/Si are linked. This would imply that Ge/Si could be used as a paleoproductivity indicator. However, the range of values of Ge/Si is greater in the Monterey Formation than would be expected from productivity alone (Murnane and Stallard 1988), and, more recently, Kumar et al. (1993) concluded that Ge/Si cannot be used as a paleoproductivity indicator because the elements are not fractionated during the formation of diatom tests.

Radiolarians

Radiolarians are heterotrophs that appeared in the Cambrian (see figure 2.1), and radiolarian cherts occur throughout the geologic record from that time on (e.g., Hein and Parrish, 1987). Radiolarians are entirely marine and stenohaline (Casey 1971a, 1993). Their tests are composed of silica; polycystine radiolarians are by far the most important as fossils. Paleoclimatic studies of radiolarians use their paleobiogeography.

Like planktonic foraminifera, radiolarians are concentrated in or just below the photic zone, in the upper 100 to 200 m (e.g., Casey 1971b, 1977; Kling 1979). Some species are bipolar, and cosmopolitan forms tend to be limited to deeper water [>200 m (Casey 1977)], but in general, faunas from the two hemispheres are distinct. Major provinces are generally latitudinal, but there are also meridional assemblages (Casey 1993). For example, the California and Humboldt Currents, off California and Peru, respectively, are upwelling zones that have their own unique faunas and have been considered separate biogeographic units by some workers (e.g., Casey 1971a). Complex biogeographic patterns have been documented in some areas, for example, the Gulf of Tehuantepec, Mexico (Molina-Cruz and Martinez-López 1994) and the California Current (Boltovskoy and Reidel 1987).

Radiolarian diversity and abundance are highest in the equatorial upwelling zones and usually high in upwelling zones in general (Casey 1971a). However, radiolarians today are commonly out-produced by nannofossils along the equator and by diatoms at high latitudes, so they do not necessarily make up a significant proportion of the sediment, especially above the carbonate compensation depth (Riedel and Sanfilippo 1977). Radiolarian-rich sediments do occur in the very deep ocean under upwelling zones because radiolarian tests are less soluble than carbonate tests and much more robust than diatom tests. Before the evolution of diatoms, radiolarians were the principal component of biogenic cherts (Riedel and Sanfilippo 1977).

Certain modern taxa are indicative of specific oceanographic conditions (Molina-Cruz 1984; Molina-Cruz and Martinez-López 1994), and one of those, *Cycladophora davisiana*, extends back into the Pliocene. *C. davisiana* is indicative of frontal boundaries between water masses, commonly near upwelling zones (Molina-Cruz 1984; Molina-Cruz and Martinez-López 1994) and, in the North Atlantic, tended to occur in greatest abundances in glacial stages during the Pliocene and Pleistocene (Ciesielski and Bjørklund 1995).

Nigrini (1970) formulated a radiolarian temperature equation similar to those proposed for other groups:

$$Tr = [Xw/(Xt + Xc) + Xw] \times 100, \quad (2.4)$$

where Tr is the radiolarian temperature, Xw the number of warm-water species, Xc the number of cold-water species, and Xt the total number of species, and the warm- and cold-water species were defined by statistical analysis of the biogeographic distribution of the species. Temperature curves determined by this method resemble those obtained by other methods.

Factor analysis has been used on radiolarian biogeography to formulate transfer functions for the study of the Quaternary record (Lozano and Hays 1976). Hays, Pisias, and Roelofs (1989) subsequently applied the same methods to modern, Pleistocene, and Pliocene radiolarians from the equatorial Pacific to determine geographic and temporal variations in seasonality, a rather unusual but interesting approach to paleoclimatic change in time and space. Seasonality and the variation in seasonality through time were much higher for the eastern site than for the central equatorial site (figure 2.28). The seasonality gradient, that is, the change in seasonality with longitude, was much steeper in the Pliocene than now (figure 2.28b).

Comparison of factor analytic results for both modern and late Miocene radiolarian faunas allowed Romine (1985) to formulate a kind of qualitative transfer function to study late Miocene (8 Ma) radiolarians in the Pacific Ocean. She defined modern and late Miocene faunas by factor analysis, using taxa or lineages that were common to both modern and late Miocene faunas. She then calculated communalities among the resulting factors and used these communalities as a measure of how good an analog the modern faunas are for the late Miocene ones (figure 2.29). Values of 1.0 would represent perfect analogy, that is, a factor model based on the modern fauna would perfectly describe the late Miocene one. Most of the values were quite high, with the highest communalities where the modern assemblages represent cold waters at high

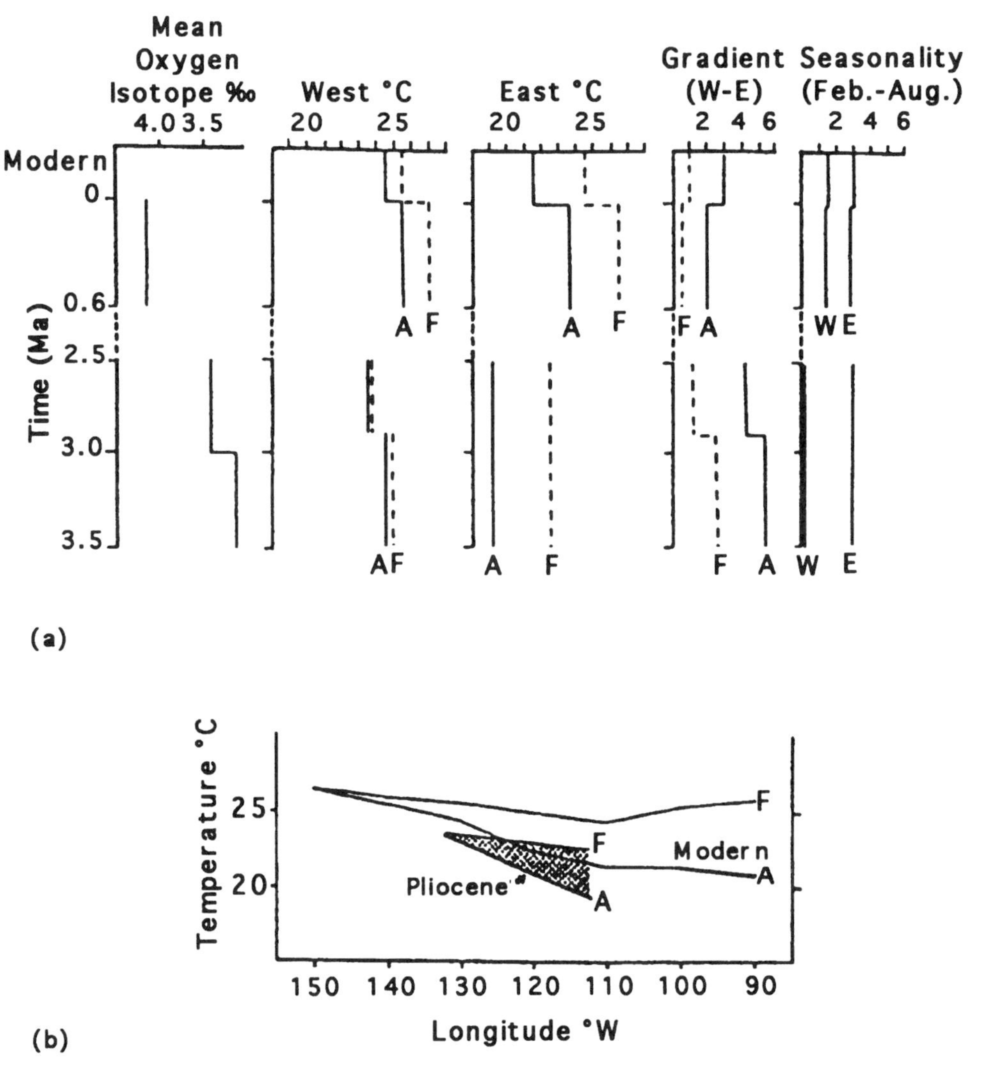

Figure 2.28. Paleoclimate interpretation from radiolarian transfer functions. *(a)* Seasonal temperature estimates for the central and eastern Pacific, east-west temperature gradients, and seasonality gradients through time, compared with mean oxygen isotope data from benthic foraminifera. *(b)* Comparison of the seasonality gradients in the modern and Pliocene equatorial Pacific Ocean. The *lines* are February and August mean temperatures. Both intervals show a decrease in seasonality toward the central Pacific but a much steeper gradient in the Pliocene. From Hays, Pisias, and Roelofs (1989). *Paleoceanography* 4:57–73, copyright by the American Geophysical Union.

latitudes or in the eastern boundary currents (figure 2.29), and Romine (1985) assumed that high communalities indicate similar environmental conditions as well as taxonomic similarity. A factor-by-factor comparison of present and late Miocene radiolarian faunas showed that only two of the factors from modern faunas were well expressed in the late Miocene because of the vagaries of sampling and the taxonomic differences between modern and late Miocene faunas in low latitudes. An example is the subpolar assemblage; its distributions in the Pacific Ocean today and in the late Miocene are very similar. The species associated with that factor at both times are very similar and, presumably, the water masses were also very similar.

Dinoflagellates

Dinoflagellates are organic-walled, motile, photosynthesizing, single-celled organisms. They are classified with algae (Pyrrhophyta), but some are

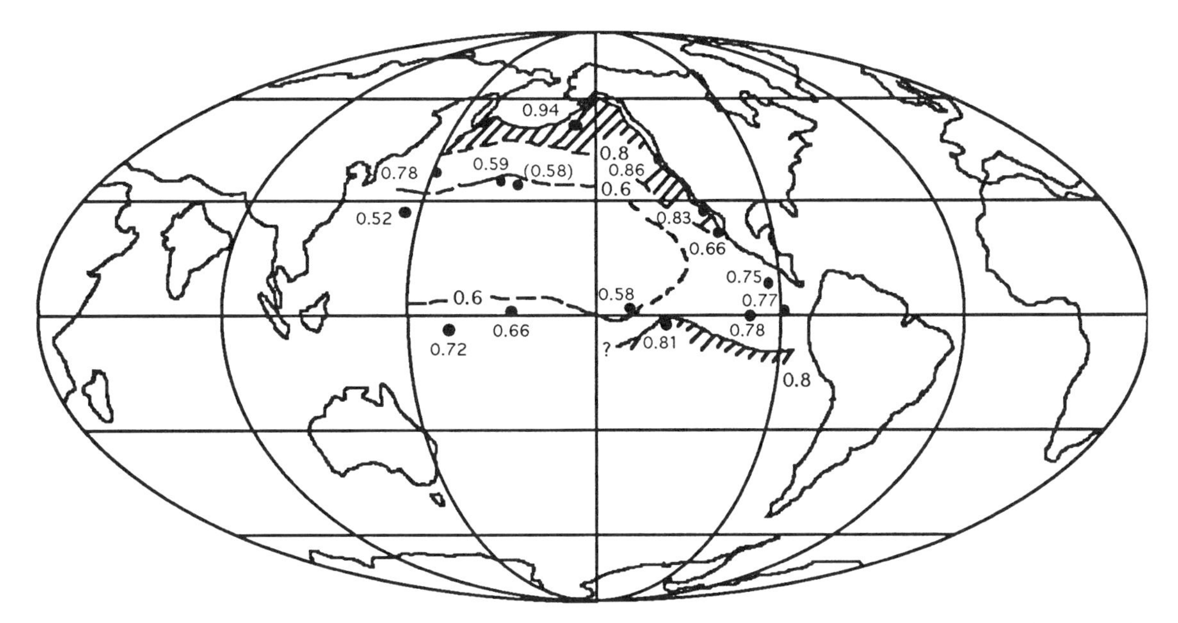

Figure 2.29. Communalities between modern and late Miocene radiolarian samples in the equatorial and northern Pacific Ocean. From Romine (1985) in *The Miocene Ocean: Paleoceanography and Biogeography,* J.P. Kennett, ed. Reproduced with permission of the publisher, the Geological Society of America, Boulder, Colorado, USA. Copyright © 1985 Geological Society of America, Inc.

heterotrophic (Williams 1977). The life cycle of dinoflagellates includes vegetative and dormant, encysted stages, and it is the cysts, called dinocysts, that are preserved in the fossil record. The group's geologic history is controversial. Unquestioned dinocysts occur from the Silurian onward (see figure 2.1). If acritarchs are dinocysts, then the group ranges into the Precambrian; Tappan (1980) presented evidence that acritarchs could be from a number of groups, and she preferred to simply assign them to the algae (see also Mendelson 1993). Dinocysts were diverse in the Late Jurassic, Cretaceous, and early Tertiary (Williams 1977). Dinocyst paleobiogeography is used in paleoclimate studies.

Temperature and onshore-offshore environmental gradients appear to be the primary controls on dinocyst biogeography, which is very complex (Williams 1977; Wall et al. 1977). Dinocysts have been used principally for biostratigraphy and for distinguishing marine from nonmarine environments. Tappan (1980) made an intriguing statement that, to this author's knowledge, has not been followed up: "The less widespread extinctions of dinoflagellates than of coccolithophores at the end of the Cretaceous may result from their lower nutrient requirement: Modern dinoflagellates are most abundant in the nutrient-poor warmer waters, and their annual blooms commonly follow the spring diatom blooms as the nutrients in the water are depleted." Although she compared two types of events of very different temporal scales, this aspect of the ecology of dinoflagellates might have the potential to contribute to ocean-chemistry studies.

Like many organisms, dinoflagellates exhibit a latitudinal diversity gradient. This was used explicitly as a indicator of relative temperature by Mohr (1990) in studies of Paleogene ocean sediments near Antarctica. Extremely low diversities (two species) were taken to indicate cold water, whereas relatively high diversities (12 to 15 species) indicated "at least temperate" conditions.

A strong floral differentiation between neritic and slope and deeper environments exists in modern floras (Wall et al. 1977), which Brinkhuis and Biffi (1993) used to indicate sea-level changes. Near the Eocene-Oligocene boundary in Italy, marginal-marine dinoflagellates and high-latitude (presumably cool-water) dinoflagellates occur together, indicating a link between cool temperatures and low sea level there (Brinkhuis and Biffi 1993).

Perhaps the most comprehensive attempt to use dinoflagellates as paleoclimatic indicators in the pre-Quaternary record was a study of North Atlantic Pliocene floras by Edwards, Mudie, and de Vernal (1991). The study is also instructive in that they compared four different quantitative methods that were developed for estimation of winter (WSST) and/or summer (SSST) sea surface temperatures, all based on dinoflagellates, similar in concept to the ostracode-based indicators used by Cronin et al. (1994). The four dinoflagellate methods are the *Impagidinium* index, transfer functions, actualistic method, and gonyaulacoid-to-protoperidinioid ratio.

The *Impagidinium* index is based on the observation that subtropical and tropical assemblages have significant numbers of species of *Impagidinium* as well as other warm-water taxa. The method is sensitive to low numbers of specimens and to transported specimens. The transfer functions for winter and summer paleotemperatures were calculated using the methods of Imbrie and Kipp (1971) developed for foraminifera, that is, principal components analysis on the flora and then multiple regression on the assemblage factor loadings versus modern oceanographic parameters. The actualistic method is similar to the graphical transfer function method developed for ostracodes by Hazel (1988; figure 2.26), except that Edwards, Mudie, and de Vernal (1991) incorporated information on the temperatures of maximum abundance, as well as presence/absence. As with ostracodes (Hazel 1988), problems of interpretation arose when the fossil sample did not match any of the diagnostic modern assemblages. This method is highly sensitive to the end points of the distributions and to broad abundance groupings. More problems arise in the decisions about how to handle extinct taxa, for which substitutes in the modern flora must be chosen.

The only method used in older rocks, the gonyaulacoid-to-protoperidinioid ratio (sometimes reported as the inverse), is controlled partly by temperature and is based on the numbers of specimens of each of the two taxonomic groups. The principal problem with this method is that the ratio can be affected by parameters other than temperature. For example, protoperidinioids favor the cool waters of upwelling currents as well as high latitudes, and gonyaulacoids favor estuarine environments as well as warm temperatures. Without some independent evidence, distinguishing these environmental influences from the latitudinal temperature gradient is impossible.

In Cretaceous rocks of Israel, peridinioid-to-gonyaulacoid (P/G) specimen ratios were used to interpret productivity, rather than temperature, with high ratios indicating high productivity (Eshet, Almogi-Labin, and Bein 1994). Peridinioids are characteristic of high-productivity systems as well as cold water; they are facultatively heterotrophic and feed on diatoms (Edwards, Mudie, and de Vernal 1991). In contrast, gonyaulacoids are principally autotrophic. Eshet, Almogi-Labin, and Bein (1994) cross-checked the P/G ratio with the foraminiferal paleoproductivity indicator proposed by Almogi-Labin, Bein, and Sass (1993; see figure 2.7) and with geochemical measures of total organic matter (figure 2.30). The curves for the P-type foraminifera and P/G are remarkably similar. An advantage of the P/G method is that it can be used in rocks where foraminifera are not preserved.

Conodonts

All the microfossils discussed up to this point are the hard parts of single-celled organisms and have close living relatives. In contrast, conodonts are the microscopic hard parts of a larger, extinct organism. A number of possibilities for the affinities of the "conodont animal" have been proposed, including chaetognath worms (Briggs, Clarkson,

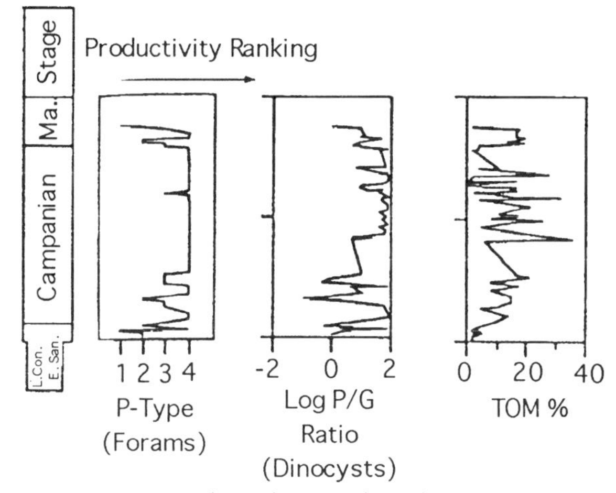

Figure 2.30. Peridinioid/gonyaulacoid specimen ratios (middle column) through the Campanian and Maastrichtian in the Zin Basin, Israel, compared with the distribution of P (productivity)-type foraminifera (*left*, from Almogi-Labin, Bein, and Sass 1993; see table 2.2 and figure 2.7) and percent total organic matter (TOM; *right*). P-type foraminiferal assemblages range from 1, normal marine, to 4, most productive. Reprinted from *Marine Micropaleontology* 23, Eshet, Y., A. Almogi-Labin and A. Bein, "Dinoflagellate cysts, paleoproductivity and upwelling systems: A Late Cretaceous example from Israel," pp. 231–40 (1994) with kind permission of Elsevier Science-NL, Sara Burgerhartstraat 25, 1055 KV Amsterdam, The Netherlands.

and Aldridge 1983) and vertebrates (Gabbott, Aldridge, and Theron 1995; Janvier 1993).

Paleobiogeography of conodonts

Conodont provincialism was similar in some respects to that of other Paleozoic marine organisms, supporting the notion that the conodont animal was a shallow-water dweller. However, unlike brachiopods, for example (e.g., Ziegler 1965; Ziegler, Cocks, and Bambach 1968), conodont distribution is not tied to lithofacies distribution (Bergström 1973), and thus the animals are interpreted to have been pelagic (Sweet 1988). Nevertheless, a strong onshore-offshore paleobiogeographic differentiation existed that is unlikely to be related to climate but rather to hydrographic differences between the two settings (e.g., Sandberg 1973; Weddige and Ziegler 1987) or, possibly, depth zonation of the faunas.

The most comprehensive analyses of global conodont paleobiogeography were by Charpentier (1984) and Bergström (1990). Earlier attempts were frustrated not only by the lack of good paleogeographic reconstructions for the Paleozoic but also by the changeover in the 1970s from single-element to multi-element taxonomy of conodonts, which resulted from the recognition that several conodont elements of very different morphologies existed in the same organism.

Charpentier (1984), who continued to use form genera (Bergström 1990), used similarity nets to identify three times of strong provincialism, probably controlled by temperature, but not necessarily related to a latitudinal temperature gradient. These times were the Ordovician, Early Devonian, and Pennsylvanian-Permian, which had two provinces. Otherwise, conodonts were cosmopolitan, or the samples were so few that paleobiogeographic differentiation was impossible. The later work by Bergström (1990), based on more data, showed provinciality in the Late Cambrian and more provinces in the Ordovician than found by Charpentier (1984).

An example of a similarity net, for the Early Ordovician, is shown in figure 2.31. Note that grouping of sample sites is done in relation to the continental plates, with the exception of Gondwana. Had Charpentier (1984) grouped all the Gondwana sites together, the result would have been high levels of similarity between that continent and all the others because the continent was so large, close to several other continents that were widely separated from each other, and spanned many latitudinal zones. The Ordovician provinces were probably related to the latitudinal temperature gradient. The high degree of endemism coupled with relatively low similarities support placement of Baltica into a separate

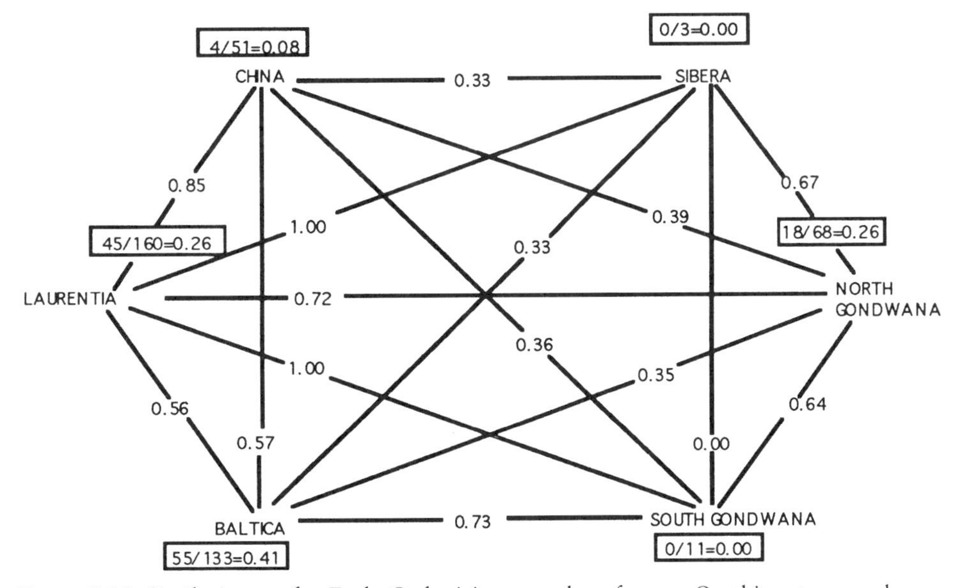

Figure 2.31. Similarity net for Early Ordovician conodont faunas. On this net, a number on the line connecting two sites is the similarity between the sites. The numbers in the rectangles at each site are the number of endemic taxa, the total number of taxa, and the percent endemic taxa, respectively. From Charpentier (1984) in *Conodont Biofacies and Provincialism*, D.L. Clark, ed. Reproduced with permission of the publisher, the Geological Society of America, Boulder, Colorado, USA. Copyright © 1984 Geological Society of America, Inc.

province that was probably partly controlled by temperature and partly controlled by geographic separation from the other continents. The low diversity of the South Gondwana province supports the notion that it was a high-latitude province. The paleoclimatic significance of Charpentier's (1984) work was simply to indicate that the biogeography of conodonts was consistent with existence of a latitudinal temperature gradient, except perhaps in the Silurian.

Interpretations of conodont biogeography are complicated by the onshore-offshore biogeographic differentiation of the faunas (Charpentier 1984). The onshore and offshore faunas might be quite distinct, but they would nevertheless be considered biofacies within a single province. Sweet and Bergström (1984) observed that in the Late Ordovician, the endemics were largely nearshore, whereas the deeper-water species tended to be cosmopolitan (Klapper and Johnson 1980); this is likely to have been true throughout the Paleozoic (and was also true for trilobites). If this is true for conodonts throughout their history, as is likely, drawing boundaries among provinces would be even more difficult.

Bergström (1990) overcame the problem of onshore-offshore variation by limiting his analysis to conodonts from similar depositional environments. He used the coefficient of similarity and constructed similarity nets directly on paleogeographic maps; he also used multi-element taxonomy. Bergström (1990) interpreted a strong latitudinal temperature control for conodont biogeography. He noted that many provinces in the Ordovician are limited to individual continents, but in his view, a steep latitudinal temperature gradient isolates the continents more than distance alone. This would explain why, after the latest Ordovician glaciation, provinciality disappeared even though the continents were only slightly closer together.

Isotope studies on conodonts

Conodonts are the only Paleozoic microfossils that are common enough and well enough preserved to be considered for isotope studies, and they have the further advantage of being composed of phosphate, which may be slightly more stable than carbonate. Conodonts consist of francolite, which has the approximate chemical composition $Ca_5Na_{0.14}(PO_4)_{3.01}(CO_3)_{0.16}F_{0.73}(H_2O)_{0.85}$ (Luz, Kolodny, and Kovach 1984). All errors considered, the precision is estimated to be about 0.5‰ (Luz, Kolodny, and Kovach 1984).

Luz, Kolodny, and Kovach (1984) studied a series of samples from Ordovician and mid-Devonian to Carboniferous rocks from North America (figure 2.32). They checked their results with subsamples of conodonts or fish remains, which are also phosphatic, from the same beds. The trend shown in figure 2.32 is similar to the trend in Paleozoic cherts (Knauth and Epstein 1976). To calculate temperature, Luz, Kolodny, and Kovach (1984) used the apatite equation, and they assumed that the Earth was unglaciated and that seawater was −1‰ on the SMOW scale. The resulting estimated temperatures for the Paleozoic are mostly above 30°C, some >40°C (figure 2.32), and the conodont values are on the light side of the envelope.

Data from Geitgey and Carr (1987) from North America are also shown in figure 2.32, but Geitgey and Carr (1987) did not attempt to assign absolute temperatures to the $\delta^{18}O$ values because of the problems of calibration in extinct organisms and lack of knowledge of the isotopic composition of the Paleozoic oceans.

Sr isotopes can also be measured from conodonts (Bertram et al. 1992). Sr determinations are reliable only in the 0.2 M HNO_3 dissolvable fraction, which excludes all fish and some conodonts, and thermally altered conodonts are entirely unreliable (Bertram et al. 1992). From reliable Sr data, Bertram et al. (1992) tentatively estimated a 10% increase in the flux of Sr to the oceans resulting from continental erosion during the Silurian. This increase postdated the end-Ordovician glaciation. Sea level rose during the Silurian, and climate is generally considered to have become warmer, mostly based on the expansion of reefs and the cosmopolitan nature of the global fauna (e.g., Ziegler et al. 1981).

Silicoflagellates

Silicoflagellates are classified with the Chrysophyta; they appeared during the Early Cretaceous and were diverse by the Late Cretaceous (McCartney 1993). Living species of silicoflagellates probably number fewer than six (Tappan 1980; McCartney 1987). Ebridians are similar organisms, but Tappan (1980) placed them with the Pyrrhophyta rather than the Chrysophyta; silicoflagellates are primarily autotrophic, whereas the ebridians are facultatively heterotrophic (Haq 1978). Their use in paleoclimatology is in their paleobiogeography and, to a lesser extent, variations in morphology.

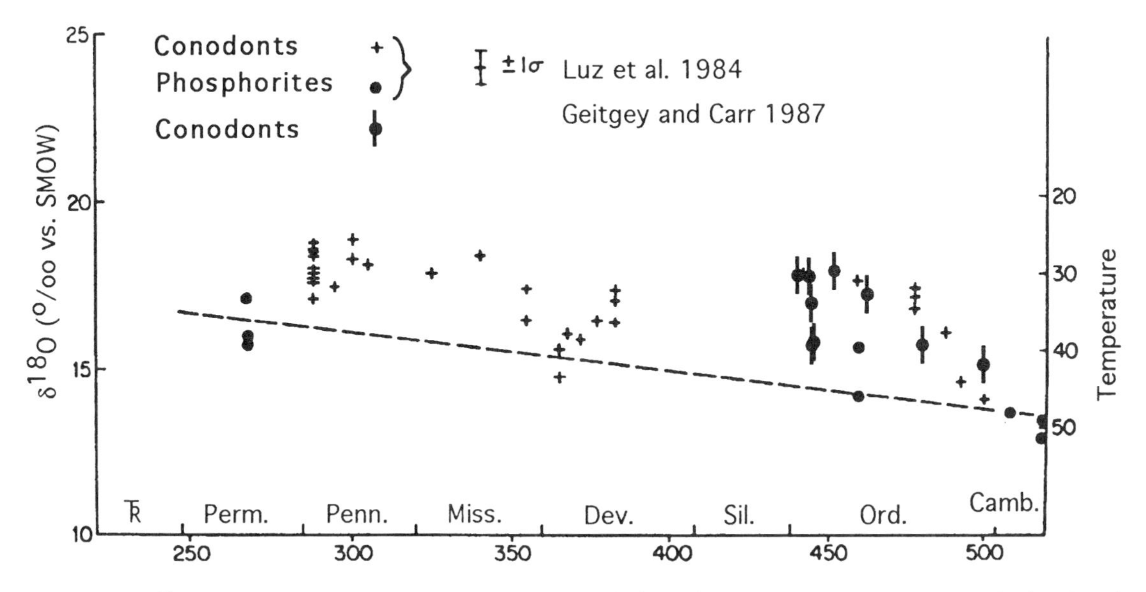

Figure 2.32. $\delta^{18}O$ from phosphorite and conodonts through the Paleozoic. Temperatures were calculated with the assumption that seawater was −1‰ $\delta^{18}O$. *Dashed line* is phosphorite $\delta^{18}O$ trend from Shemesh, Kolodny, and Luz (1983). Reprinted from *Earth and Planetary Science Letters* 69, Luz, B., Y. Kolodny and J. Kovach, "Oxygen isotope variations in phosphate of biogenic apatites, III. Conodonts," pp. 255–62 (1984) with kind permission of Elsevier Science-NL, Sara Burgerhartstraat 25, 1055 KV Amsterdam, The Netherlands, with additional data from Geitgey and Carr (1987).

Because silicoflagellate taxa are widespread (Martini 1977; Haq 1978), abundance ratios among taxa, rather than presence/absence, have been the methods used for paleoclimate interpretation; ratios between the genera *Dictyocha* and *Distephanus* appear to have been most useful (Jendrzejewski and Zarillo 1972). *Dictyocha* is a warm-water form and *Distephanus* a temperate form. Bukry (1981) formalized the method with an equation similar to those used for diatoms and foraminifera. Some assemblages, the definitions of which might include different morphologies within single genera (Ciesielski and Case 1989; McCartney and Wise 1990), are also related to upwelling, e.g., *Bachmannocena*.

Other Microfossils

Two other microfossils are worth mentioning in this book because they have specific, if limited, paleoclimatic significance. These are the organic-walled tasmanitids and *Bolboforma*, a calcareous microfossil of uncertain affinities. Tasmanitids, which Tappan (1980) classified with the prasinophycean green algae, are commonly noted in discussions of organic-rich or black shales. They can be abundant enough in marine rocks to give the rock the appearance of coal, hence the term "marine" coals (Williams 1978). So-called *Tasmanites* shales are especially common in the Devonian, but they occur throughout the geologic record and are commonly taken as indicative of high biologic productivity. Williams (1978) suggested that they represent algal blooms and surmised that they were cysts of green algae. *Bolboforma* is a middle Tertiary genus found in continental-margin and deep-sea sediments (Poag and Karowe 1986). Its paleobiogeographic distribution and relationship to oceanographic events defined by foraminifera, oxygen isotope studies, and other methods suggests that the genus is specifically indicative of temperate waters. For example, it disappeared from subantarctic waters in the earliest Oligocene with the onset of cooling (Poag and Karowe 1986; Kennett and Kennett 1990).

MARINE MACROFOSSILS

Most information about paleoclimate derived from marine macrofossils comes from their biogeography, although a growing number of studies utilizes stable isotopes of oxygen and carbon to interpret paleoclimate and related paleowater chemistry. For the Jurassic and earlier, marine macrofossils provide most of the paleontological information about marine climates. Marine macrofossils are most useful where marine microfossils are generally least useful—on the continental shelves, where clastic input commonly dilutes the microfossil record. In addition, the shelf faunas provide an important link between terrestrial and oceanic climates. Resolution of paleoclimatic events and features seems poorer in marine macrofossils than in marine microfossils, but this is partly because marine macrofossil studies are most common in the older record, where both temporal and spatial resolution are poorer. The use of marine macrofossils as paleoclimatic indicators had its heyday in the 1970s and early 1980s (e.g., papers in Hallam 1973; Stevens 1980; Ziegler et al. 1981), when plate tectonics was a relatively new theory and biogeographic patterns could finally be studied in their plate tectonic contexts.

A feature of paleobiogeographic studies is that they tend to concentrate on one or two geologic periods and, at the most, one or two groups of organisms, generally at taxonomic levels lower than phylum. For example, the volume on Paleozoic biogeography edited by McKerrow and Scotese (1990) contained 20 papers on global-scale biogeography of marine invertebrates. Of these, nine papers dealt with one or part of one geologic period and five with two geologic periods. Only four papers (Bambach 1990; Boucot 1990; Cocks and Fortey 1990; Ross and Ross 1990) attempted to integrate the biogeographic information from more than one group of organisms. The result is that paleoclimate interpretations from period to period are inconsistent, and perceptions of the climate of a particular time tend to be driven by the most biogeographically differentiated or the best-studied group for each period (see also Hallam 1994). One of the few papers that attempted to overcome these limitations was that by Ziegler et al. (1981). The geologic stages they studied and the groups of organisms they used are listed in table 2.3; it should be noted that they integrated continental phytogeography into this study as well. This and other integrated studies are discussed at the end of the section on marine invertebrates.

This part of the chapter is divided into two major sections, marine invertebrates and marine vertebrates. Only the major groups whose biogeography and/or chemistry have been used explicitly for paleoclimate studies are discussed here. This is in contrast to the very widespread use of marine macrofossil biogeography to interpret paleogeography, for example, continental positions and the open-ocean connections of epeiric seaways. Paleogeography is important for understanding paleoclimatology, so a few such paleogeographically oriented studies are cited here, but as a complete discussion of the use of organisms in constraining paleogeographic reconstructions is outside the scope of this book, much of this work will not appear in the following sections.

Table 2.3.
Marine invertebrate macrofossils used for biogeographic/paleoclimatic analysis of the Paleozoic by Ziegler et al. (1981)

Stage	Marine macrofossils
Franconian (Upper Cambrian)	Trilobites
Llandeilo-Caradoc (Middle Ordovician)	Brachiopods*
	Trilobites
Llandovery (Middle Silurian)	Brachiopods*
	Corals
	Bivalves
Emsian (Lower Devonian)	Brachiopods*
	Trilobites
	Corals
Visean (Lower Carboniferous)	Corals
Westphalian (Upper Carboniferous)	Fusulinids*
	Corals
Kazanian (Upper Permian)	Brachiopods*
	Bryozoans*
	Fusulinids
	Corals

*These groups were the primary basis for biogeographic divisions where more than one group was used.

Paleontologists studying the biogeographic distributions of Paleozoic faunas, in particular, still commonly do so at least partly, if not entirely, in the context of testing paleogeographic models rather than as a way of interpreting paleoclimate (e.g., many papers in McKerrow and Scotese 1990, at the behest of the editors). This introduces an element of circularity in paleobiogeographic and paleoclimate studies that has been almost impossible to avoid. Nowhere is this problem more evident than in the study of the distributions of Paleozoic marine invertebrates. In many studies, available Paleozoic global paleogeographic reconstructions are taken as given, but those reconstructions were often at least partly based on biogeographic information. In most of the studies, certain assumptions about climate—generally, that it was zonal at all times—

are used uncritically. Thus, studies (1) conclude that the paleogeographic reconstructions are wrong, (2) conclude that climatic and oceanographic patterns were not zonal, (3) conclude that the distributions of the organisms were not controlled by climate, or (4) result in internally consistent biogeographic and climate models that might be wrong despite their internal consistency. Internal consistency is considered the ideal, so resistance is great to the notion that an internally consistent construct is incorrect. With regard to the several possible negative conclusions, (1) is the most common, (3) is the next most common, and (2) is by far the least common. The control on paleobiogeography most commonly cited, other than climate, is continental configuration.

It is commonly not clear whether organisms and/or their biogeographic patterns, especially in the Paleozoic, are considered indicative of climate because of where they are found or because of some independent quality. It can rarely be stated that a fauna or organism is used to indicate a particular type of climate. Rather, internal consistency is stressed. As a final point, it should be noted that internal consistency can be fluid. An internally consistent construct of paleogeography, predicted climatic patterns, and biogeography for one geologic period and one group of organisms may not appear so tidy in the context of the biogeographies of other groups of organisms at that time or even the biogeography of the same group over several time intervals.

Despite all the aforementioned problems, it should be stressed that marine macrofossil biogeography is sometimes the best available information for interpretation of the climate of a particular time, particularly in the Paleozoic, because marine fossils are so widespread and well studied. We have learned a great deal about the climate of certain times by studying marine macrofossil biogeography. Despite the use of these data for both paleoclimatic and paleogeographic reconstructions, a degree of objectivity is possible (Rosen 1992), and much useful information can be extracted from the fossils.

Marine Invertebrates

Brachiopods

Both the biogeography and the stable isotopic chemistry of brachiopod shells have been used for paleoclimate studies. In addition, Li/Ca ratios have shown some promise as paleotemperature indicators (Delaney et al. 1989). However, the Li/Ca method has not come into widespread use, so the reader is referred to the original paper.

Biogeography. Brachiopod biogeography has figured prominently in paleoclimate interpretations for the Paleozoic (see table 2.3). Important papers are those of Waterhouse (1969), Cocks (1972), Copper (1977), Ziegler et al. (1977, 1981), Boucot and Gray (1982), Raymond, Kelley, and Lutken (1989), Kelley, Raymond, and Lutken (1990), and numerous papers in McKerrow and Scotese (1990). The papers by Waterhouse (1969), Cocks (1972), and Copper (1977) discuss high-latitude brachiopods, whereas the other papers are all global in scope. Most of the papers took a qualitative approach to the definition of biogeographic units.

The mid-Paleozoic Malvinokaffric Realm is defined on the basis of brachiopod biogeography, most notably the distribution and abundance of the genus *Clarkeia* in the Silurian and a low-diversity fauna in the Devonian. A striking characteristic of these brachiopod assemblages is the "overwhelming numbers" (Cocks 1972) of *Clarkeia antisiensis* (d'Orbigny); assemblages are commonly monospecific. Cocks (1972) likened the *Clarkeia* assemblages to the low-latitude *Eocoelia* community of Ziegler, Cocks, and Bambach (1968), which also formed high-dominance assemblages in shallow water. Thus Cocks (1972) attributed the *Clarkeia* fauna to sedimentary facies rather than climate, but subsequent workers (Ziegler et al. 1981; Van der Voo 1988) have treated it as an indicator of high paleolatitudes and, by implication, cool or cold climate.

Boucot (1975, 1990) explicitly identified the Malvinokaffric Realm as indicative of cool or cold water on the basis of low faunal diversity and the absence of limestones, reefs, and nonmarine redbeds, which were widespread elsewhere. He also noted that the clastic sediments contain abundant unweathered white mica and biotite, and he regarded that as suggestive of cool climates. The Devonian Malvinokaffric Realm is also low diversity, but dominated by bivalves, and also explicitly described as indicative of cool or cold climates (Copper 1977). Like Boucot (1975, 1990), Copper (1977) used absence of generally warm-water elements such as sponges, corals, and calcareous algae, as well as the presence of an endemic trilobite fauna, the presence of purported Devonian glacial sediments in South America, and the proximity of the southern paleomagnetic pole to support his interpretation. High-latitude, low-diversity faunas also have been documented in the Ordovician and Permian (Waterhouse 1969; Ziegler et al. 1981; Boucot 1990).

In the Silurian, when faunas in general were cosmopolitan, brachiopods also were cosmopolitan; the principal biogeographic differentiation was along sedimentary-facies, rather than climate, gradients (Ziegler 1965; Ziegler, Cocks, and Bambach 1968). Only the *Clarkeia* fauna stood out as a separate biogeographic unit. Cocks and Fortey (1990) also showed a mid-southern-latitude diverse fauna distinct from the low-latitude cosmopolitan fauna, but they argued that this fauna was at least partly facies dependent (see also Cocks and Scotese 1991). Ziegler et al. (1981) took the paleogeographic reconstructions of Scotese et al. (1979) as given, whereas Cocks and Fortey (1990) approached the biogeography as a test of the reconstructions (of Scotese and McKerrow 1990). The results were comparable with regard to paleolatitudinal and, at least implicitly, paleoclimate interpretation of the faunas. Nevertheless, Boucot (1990) opined that the Silurian reconstructions were arbitrary.

A common approach to interpreting paleoclimatic change from paleobiogeographic change is to document changes in the latitudinal positions of paleobiogeographic boundaries. For example, Waterhouse and Bonham-Carter (1975) regarded changes in the latitudinal distributions of high-, mid-, and low-latitude groups as proxies for climatic change in the Permian. So long as it can be determined that the latitudinal change is real and not just a change in the distribution of available data for each successive time period, this approach is valid. Commonly, however, scatter in the data makes this determination impossible. Thus in a series of papers examining possible relationships between brachiopod extinctions and climate in parts of the Carboniferous, Raymond, Kelley, and Lutken (1989) and Kelley, Raymond, and Lutken (1990) took a slightly different approach, assuming a relationship between faunal diversity and latitude and tracking change in diversity in fixed paleolatitudinal belts through time. They took currently available paleogeographic reconstructions as given, and although reconstructions of Carboniferous continental positions changed during the course of their studies, their conclusions did not. A feature of their work is that the methods were quantitative but flexible enough to overcome some of the problems of uneven sampling.

Raymond, Kelley, and Lutken (1989) and Kelley, Raymond, and Lutken (1990) grouped sample localities of equatorial and Northern Hemisphere brachiopods into three latitudinal zones. They then divided the Tournaisian-Namurian interval into nine stages and separately plotted the percentage of genera whose northern and southern ranges changed during each stage, that is, whether their occurrence in each zone changed (figure 2.33). Statistically significant numbers of range changes occurred between the middle and late Visean, when 42% of the genera present in both stages moved northward, and between the Namurian A and B, when 33% of the genera present in both stages moved their northern ranges south and 13% moved their southern ranges north. Raymond, Kelley, and Lutken (1989) and Kelley, Raymond, and Lutken (1990) interpreted these results to indicate high-latitude warming in the Visean and high-latitude cooling accompanied by equatorial warming in the Namurian.

Raymond, Kelley, and Lutken (1989) and Kelley, Raymond, and Lutken (1990) also determined latitudinal diversity gradients, separating each latitudinal zone into Tethyan and Cordilleran sectors in recognition of the possibility of meridional differences in diversity gradients. Assuming that the latitudinal diversity gradients are related to temperature, changes in the Carboniferous diversity gradients are consistent with the range boundary changes, that is, with an increase in nonequatorial diversity in the late Visean and migration of forms into mid-latitudes from both directions during the Namurian B. Raymond, Kelley, and Lutken (1989) and Kelley, Raymond, and Lutken (1990) attributed these two major changes to continental collision and onset of glaciation, respectively.

Stable isotopes in brachiopods. Brachiopods have been used extensively for oxygen isotope studies in Paleozoic rocks, partly because their shells consist of low-Mg calcite, the most stable form (Hudson and Anderson 1989). Popp, Anderson, and Sandberg (1986b) argued that brachiopods are the best candidates for Paleozoic isotope reference standards because

1. they are abundant and commonly used as stratigraphic index fossils;
2. they were apparently relatively stenohaline (but see Carpenter and Lohmann 1995);

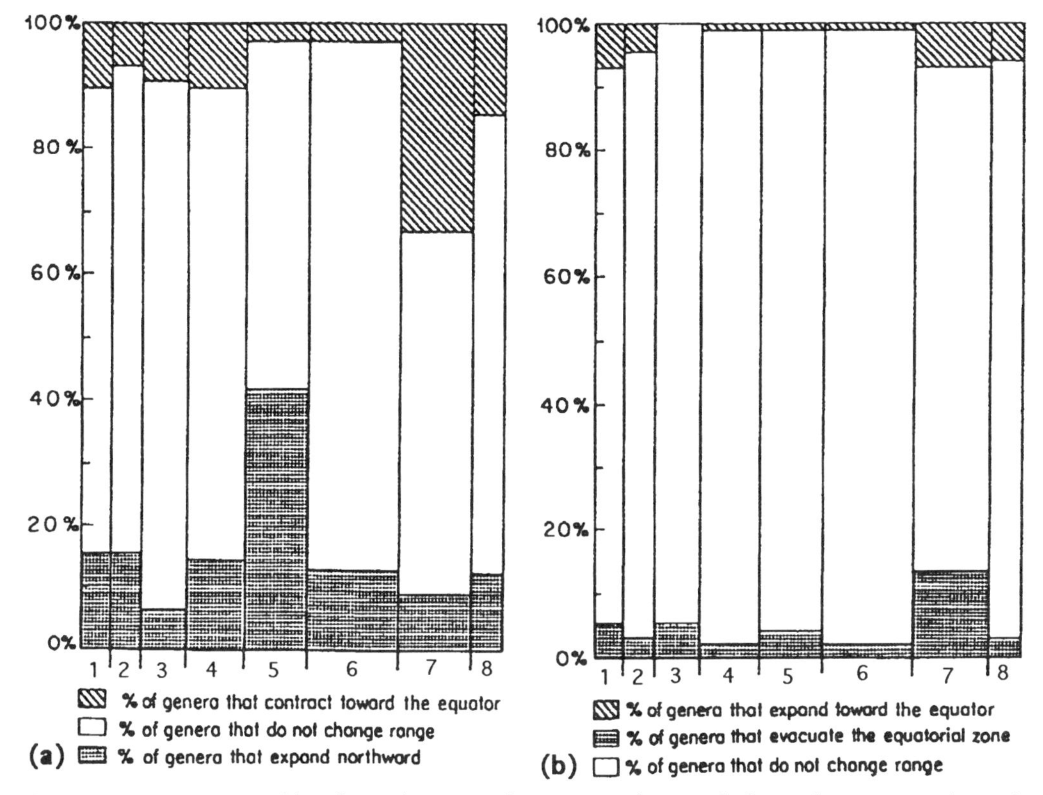

Figure 2.33. Percentage of brachiopod genera whose *(a)* northern and *(b)* southern ranges changed during each of nine intervals during the Dinantian-Namurian (Carboniferous). Numbers represent the intervals between which the faunal changes occurred: 1, early-middle Tournaisian; 2, middle-late Tournaisian; 3, late Tournaisian-early Visean; 4, early-middle Visean; 5, middle-late Visean; 6, late Visean-Namurian A; 7, Namurian A-B; 8, Namurian B-C. From Raymond, Kelley, and Lutken (1989) in *Geology*. Reproduced with permission of the publisher, the Geological Society of America, Boulder, Colorado, USA. Copyright © 1989 Geological Society of America, Inc.

3. modern forms precipitate their shells in approximate isotopic equilibrium with ambient waters, and it might be assumed that extinct forms did so as well (see also Lowenstam 1961; Carpenter and Lohmann 1995; James, Bone, and Kyser 1997);
4. they precipitate stable, low-Mg calcite;
5. shell microstructure is relatively well known, permitting microscopic recognition of recrystallization; and
6. they are large enough to provide splits for different kinds of microscopic and chemical analyses.

A Paleozoic $\delta^{18}O$ curve derived partly from brachiopod data is shown in figure 2.34.

Popp, Anderson, and Sandberg (1986b) compared the isotopic compositions of whole brachiopods, unaltered portions of brachiopods, and other whole fossils from the same intervals, and estimates of the original isotopic compositions of coeval marine cements. Unaltered portions of brachiopods were determined by cathodoluminescence, staining, and trace element analysis. Strontium, ^{18}O, and ^{13}C are depleted and iron and manganese enriched in diagenetic carbonates. Popp, Anderson, and Sandberg (1986b) used staining techniques to eliminate samples enriched in iron, cathodoluminescence to eliminate those enriched in manganese, and lower contents of Sr, ^{18}O, and ^{13}C to indicate those most likely to have been altered by diagenesis. However, Popp, Anderson, and Sandberg (1986b) observed that Sr and Mg concentrations show a strong vital effect in brachiopods, leaving Fe and Mn as the most reliable trace element indicators of diagenesis in these organisms.

In general, Popp, Anderson, and Sandberg's (1986b) data were internally consistent, with the nonluminescent portions of the brachiopods giving narrower ranges of values than whole fossils, and consistent with other types of data. However, variation in the $\delta^{18}O$ values was great and the nonluminescent brachiopod samples were

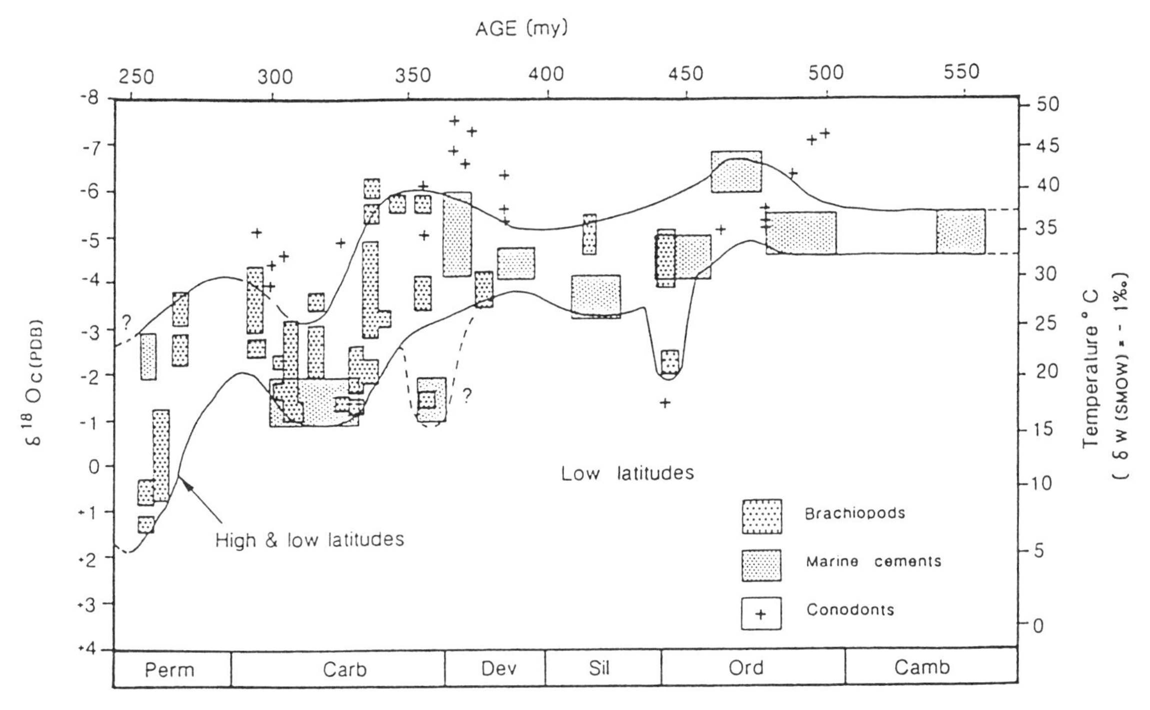

Figure 2.34. Oxygen isotope data from brachiopods, conodonts, and marine cements in the Paleozoic. Conodont data are the same as those shown in figure 2.32. From Hudson and Anderson (1989). Reproduced by permission of the Royal Society of Edinburgh and J.D. Hudson from *Transactions of the Royal Society of Edinburgh:Earth Sciences,* volume 80, parts 3 & 4 (1989), pp. 183–92.

not consistently heavier than the other types of data, which is what would be expected if nonluminescence were an indicator of lack of alteration. $\delta^{13}C$ was also variable, and, in addition, it revealed a possible vital effect among different species that could explain the relatively wide variation in samples from the Carboniferous. This finding tempers the assumption that extinct brachiopods precipitated their shells in isotopic equilibrium with ambient waters, an assumption that has recently been questioned by Carpenter and Lohmann (1995), who found that some parts of modern brachiopods shells, for example, the hinge area, are clearly precipitated out of equilibrium with seawater, and the other parts of the shells are precipitated only approximately in equilibrium.

As benthic organisms, brachiopods would be expected to have lived at different depths, and their relationship to sedimentary regime bears this out (e.g., Ziegler 1965). Thus brachiopods could be useful for studying vertical temperature and salinity profiles in shallow waters on the continental shelves, just as foraminifera are for the open ocean (Adlis et al. 1988; Railsback et al. 1990; Grossman, Zhang, and Yancey 1991). In one study, brachiopods were collected from one closely sampled section in Late Carboniferous rocks in Texas (Adlis et al. 1988; Grossman, Zhang, and Yancey 1991). The data set was consistent taxonomically in different parts of the section, and Adlis et al. (1988) defined depth on the basis of sedimentology and paleontology (table 2.4). The fossils and sediments in each zone in table 2.4 occur in regular, Waltherian successions over several cycles. Grossman, Zhang, and Yancey (1991) and Grossman, Nii, and Yancey (1993) limited their sampling to shales, whose relative impermeability makes diagenetic alteration of shells by meteoric fluids less likely. In general, higher $\delta^{18}O$ values occur at greater water depths. The simplest explanation is that the water column was temperature stratified. However, $\delta^{18}O$ values are dependent on both temperature and ice volume because the changes in depth zones indicate changes in sea level, probably glacio-eustatic, and the two effects would work in opposite directions. Transgression, bringing deeper, colder (higher $\delta^{18}O$) waters over the site, was the result of ice-volume decrease, which would lower the $\delta^{18}O$ of seawater. Grossman, Zhang, and Yancey (1991) estimated a minimum depth change of 50 to 100 m from the $\delta^{18}O$ change and from a reasonable tempera-

Table 2.4.
Depth zonation of Late Pennsylvanian rocks of central Texas and paleontological and sedimtentological characteristics

	Zone	Description
Increasing depth ↓	Myalinid zone	Sand/shale Thick-shelled species
	Fusulinid zone	Limestone or shale Fusulinids common High diversity biota Phylloid algae Few ammonoids
	Ammonoid zone	Shale Ammonoids common Moderate to high diversity biota Phosphate nodules
	Gondolellid zone	Shale Gondolellid conodonts Low diversity biota Few ammonoids Few phosphate nodules

From Grossman, Zhang, and Yancey (1991, fig. 3).

ture-depth profile, which can aid in estimates of ice volume for that time. Similarly, Brenchley et al. (1994) combined information on sea-level changes with isotope data from brachiopods to describe the Late Ordovician glacial episode (figure 2.35). Except in the U.S. Midwest, where part of the relevant section may be missing, data from a number of localities show a pronounced shift in both ^{18}O and ^{13}C at or near the Ordovician-Silurian boundary, coinciding with evidence for a sea-level fall just below the boundary. This event also was accompanied by a profound extinction event among the brachiopods (Sheehan and Coorough 1990).

Grossman, Nii, and Yancey (1993) took the first steps toward delineation and interpretation of the types of $\delta^{13}C$ shifts in the global record that have proved so interesting in the younger geologic record. Previous work prohibited all but the most general conclusions. They used their own data and data from nonluminescent brachiopods reported by other workers to show that for the Late Carboniferous, $\delta^{13}C$ data fall into two groups at about the Lower-Upper Carboniferous boundary. In particular, data from Russia and Spain increased 1‰ or more than data from North America. This shift was coincident with the closure of the seaway between Laurussia and Gondwana and the consequent isolation of paleo-Tethys from eastern Panthalassa. $\delta^{18}O$ showed no comparable effect, and for that as well as other reasons, Grossman, Nii, and Yancey (1993) regarded the $\delta^{13}C$ shift as real. They attributed the shift to changes in ocean circulation resulting from the collision, which would have altered vertical circulation on both coasts of Pangea (Grossman 1994). Their work is a hopeful sign that, despite the lack of sea-floor fossils comparable to those available for the Cretaceous and Tertiary, detailed paleoceanographic studies might be possible in the older record.

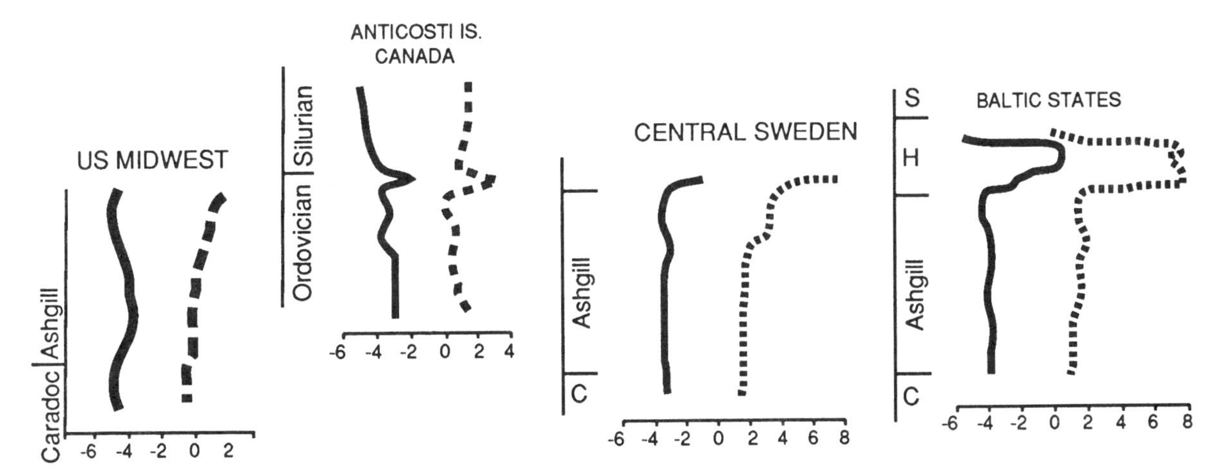

Figure 2.35. Stable isotopic compositions of brachiopods from composite stratigraphic sections in North America and Scandinavia, Late Ordovician to earliest Silurian. *Solid lines,* $\delta^{18}O$; *dotted lines,* $\delta^{13}C$. From Brenchley et al. (1994) in *Geology.* Reproduced with permission of the publisher, the Geological Society of America, Boulder, Colorado, USA. Copyright © 1994 Geological Society of America, Inc.

Mollusks

Mollusks are among the most abundant marine invertebrates in the record. Their shells consist of calcite alone, aragonite alone, or a mixture of the two minerals in different shell layers. Molluskan biogeography has proved less amenable to paleoclimatic analysis than that of other groups. However, the chemistry of molluskan shells has proved useful; indeed, mollusks were among the first invertebrates subjected to isotopic analysis.

"Indicator" taxa and morphology. Bivalves in low latitudes have larger and thicker shells than do those in high latitudes, and gastropods in low latitudes are more elaborately ornamented than those in high latitudes (Vermeij 1978). Bivalves in low-productivity faunas in the West Pacific have inflated shells, whereas those in higher-productivity systems are flatter (Vermeij 1990). Such characteristics are occasionally cited as supporting information in paleoclimate interpretation. For example, Hallam (1977) noted that 17 of the 27 genera of Jurassic bivalves that occurred between 30°N and 30°S (reconstructions of Smith, Briden, and Drewry 1973) had thick shells.

Single taxa of different types of organisms are occasionally adopted as paleoclimatic indicators. The bivalve *Eurydesma* is one such taxon (Runnegar 1979). *Eurydesma* is widespread in high-latitude Permian sequences of Gondwana (Veevers and Powell 1987; Parrish et al. 1996). It is associated with bryozoans and spiriferoid and productoid brachiopods (Waterhouse 1969), occurred in deep, presumably cold, water, and is commonly associated with glaciomarine sediments in shallow water (Runnegar 1979). Rudistid bivalves formed reefs in the late Mesozoic.

Although the morphology of bivalves and gastropods alone probably cannot be used for paleoclimate reconstructions, such information is potentially helpful in combination with paleobiogeographic analysis. For example, Blodgett, Rohr, and Boucot (1990) studied Early and Middle Devonian gastropod biogeography and determined that the biogeographic units were similar to those for other organisms at the same time, with a high-diversity Old World Realm, a moderately diverse Eastern Americas Realm, and a low-diversity Malvinokaffric Realm. As had other workers, Blodgett, Rohr, and Boucot (1990) interpreted these as arrayed along a temperature gradient. This interpretation was bolstered by variations in the degree of ornamentation in the gastropods, which was greatest in the high-diversity fauna and least in the low-diversity fauna.

Biogeography. Molluskan biogeography has been most important in studies of the climate of the Mesozoic, usually at global scale, and Cenozoic, usually at regional scale. Studies have been conducted by numerous workers (e.g., Addicott 1970; Kauffman 1973; Scott 1975; Imlay 1980; Westermann 1981; Tozer 1982; Taylor et al. 1984; Newton 1988; Doyle 1992; Dhondt 1992; and many others).

It is not clear that molluskan biogeography, particularly that of the cephalopods, is consistently related to paleoclimate on a global scale (Hallam 1983, 1984b, 1994; Westermann 1985; Crick 1990). Rather, continental configuration appears to play a large role. The following statement by Hallam (1984b), in reference to the Tethyan-Boreal division in the Jurassic, might apply to molluskan paleobiogeography in general: ". . . it seems evident that the provinciality gives us no information on climate that is not better deduced on other grounds." Nevertheless, molluskan biogeographic units are generally arrayed in latitude-parallel bands (e.g., Kauffman 1973; Stevens 1980; Hallam 1984b) and latitudinal diversity gradients have been documented for some regions and times (e.g., Addicott 1970).

Triassic faunas are loosely divided into Tethyan and what might be termed non-Tethyan realms, characterized in the latter case by a lack of endemic forms. Probably the most confusing aspect of molluskan biogeography in the Triassic came from the geographic juxtaposition of Tethyan and non-Tethyan faunas in western North America (see review in Tozer 1982). Once it was recognized that the Tethyan faunas occur on so-called suspect terranes (Davis, Monger, and Burchfield 1978; Coney, Jones, and Monger 1980; Howell et al. 1983), paleobiogeographers and paleomagnetists seized on the faunas as evidence that the terranes had moved significant distances, and the question became, how far did they move? Because Tethyan faunas were concentrated in Tethys, on the other side of the paleo-Pacific Ocean, some workers proposed that the terranes drifted the entire width of the ocean; the implied model was the modern Indo-Pacific Realm, which is a large realm in the warm-water Indian and western Pacific Oceans. Newton (1988), however, noted that the modern Indo-Pacific Realm is actually pan-tropical, with a few outliers in the warmer parts of the eastern

Pacific, and proposed a similar model for Tethyan bivalves in the western North American suspect terranes.

Jurassic biogeography is commonly based on cephalopods or cephalopods plus bivalves because those groups are the best studied. The style of biogeographic differentiation seen in the Triassic continues into the Jurassic, although by then some endemic forms are known outside the Tethyan Realm, and Jurassic biogeography became more differentiated through time. The Boreal and Triassic-Early Jurassic Maorian (Paleoaustral) Realms occurred at higher paleolatitudes than the Tethyan Realm, and it is commonly assumed that they represent cool-water faunas (e.g., Stevens 1980). However, latitudinal diversity gradients are commonly lacking or even reversed (Hallam 1975, 1984b) and the Maorian Realm was either very short-lived or broader than originally defined; for most of the Jurassic, biogeographic differentiation in the Southern Hemisphere was slight or nonexistent (Stevens 1980; von Hillebrandt et al. 1992; Hallam 1994). In addition, some evidence exists that the faunal realms might have been controlled by substrate and/or salinity rather than climate (Hallam 1975).

On a regional scale, variations in diversity and composition of molluskan faunas have sometimes been used as indicators of changes in water temperature, salinity, and/or oxygenation, environmental conditions that might have as much to do with paleogeography as climate (e.g., Scott 1975). So long as changes in other controlling factors such as sedimentary regime or sea level can be constrained, these faunal changes can be informative for paleoclimate.

As with other organisms, mollusks in younger rocks can provide more detailed information about paleoclimate because they can be closely related to modern forms. For example, Addicott (1969) analyzed the distribution of bivalve and gastropod genera in Tertiary (Oligocene to Pliocene) faunas of western North America (figure 2.36), limiting his analysis to genera that are still extant and whose northernmost limits today are in warm-water provinces that now lie to the south of the Tertiary sites. The data showed, first, that many warm-water genera lived farther north during much of the study interval than they do today. Only in the northernmost sites and only in the Pliocene were warm-water genera lacking entirely (figure 2.37d,e). The most striking aspects of the data, however, are the in-

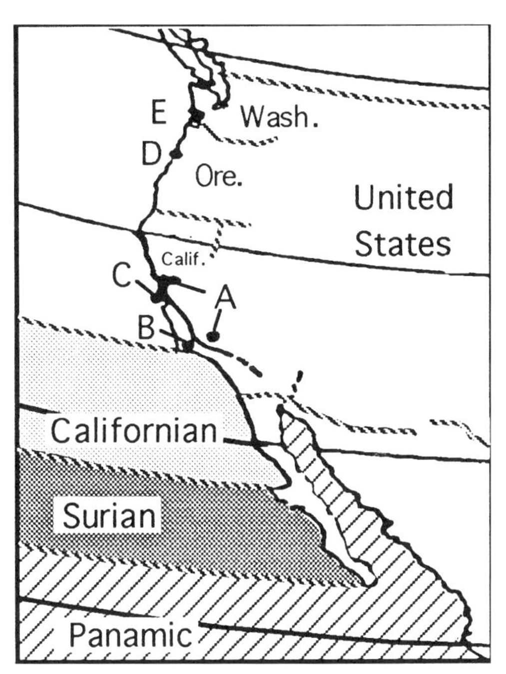

Figure 2.36. Modern shallow-, warm-water molluskan provinces and sample locations for Tertiary mollusks. *Letters* refer to the plots in figure 2.38. Note that all sites studied are north of the northernmost limit of the modern provinces. From Addicott (1969).

crease in percent of warm-water genera at all sites during the middle part of the Miocene and the consistent latitudinal gradient in percent warm-water genera through time (figure 2.37). Addicott (1969) noted that these data are consistent with changes in diversity of foraminifera. The interpretation was that climate in the region studied became warmer, allowing northward extension of the ranges of warm-water taxa. The consistency of the pattern in the sampled region indicates that the control was unlikely to have been a change in sedimentary regime, and a change in sea level would not account for latitudinal change in this setting.

Mollusks in low-oxygen and high-productivity environments. Bivalves have been particularly important in the interpretation of low-oxygen environments. They are the hard-shelled organisms most tolerant of low-oxygen conditions and commonly the only fossils preserved in organic-rich units. The assemblages are very low diversity or even monotypic, but the bivalves in them are typically extremely abundant. In addition, the bivalves tend to have similar morphologies, that is, thin, flat valves. Examples of taxa that form such assemblages are "*Lucina*" *miniscula* Blake in the

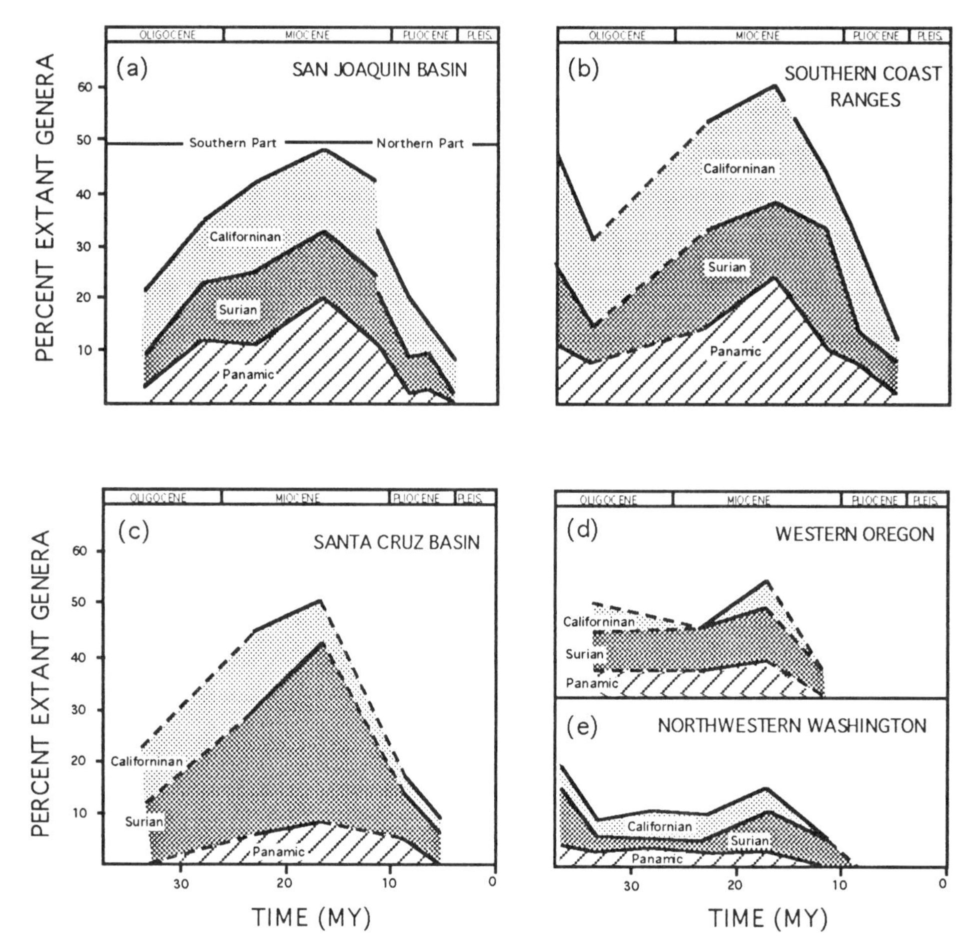

Figure 2.37. Cumulative percentages of warm-water genera in Tertiary faunas of western North America; see figure 2.36 for locations. From Addicott (1969).

Jurassic Kimmeridge Clay, northwestern Europe (Oschmann 1988); *Halobia* in the Triassic Shublik Formation, Alaska (Parrish 1987a); and *Bositra* in the Jurassic Posidonienschiefer, Germany (Kauffman 1978, 1981).

Quite different models have been proposed to explain the assemblages. Oschmann (1988) regarded the bivalves in the low-oxygen facies of the Kimmeridge Clay as infaunal suspension feeders tolerant of low oxygen conditions. He noted that the thin shells and flat shape of these bivalves would maximize oxygen exchange in low-oxygen environments, and he proposed that some groups were adapted for consuming H_2S-oxidizing bacteria. The animals were gradually driven to the sediment surface by a shallowing of the O_2-H_2S interface within the sediments, and death and formation of the shell-rich layers occurred when the interface rose above the sediment-water interface. Tozer (1982) and Parrish (1987a and unpublished data) suggested that the thin- and flat-shelled *Halobia* was a pelagic or epipelagic suspension feeder and that the shell assemblages represented death assemblages of a high-productivity system in which the bivalves proliferated in the water column. Kauffman (1978, 1981) regarded *Bositra* as an epifaunal suspension feeder that was simply extremely tolerant of low-oxygen conditions. Whether these models simply reflect the paleoecologies of the different taxa or whether one model could explain all "flat-clam" assemblages (Savrda and Bottjer 1988) is still debated. Nearly everyone agrees, however, that no exact ecological equivalents to these assemblages are found in the modern oceans.

A key issue is whether the low-oxygen conditions were the result of partial stagnation or high productivity. A benthic habit is not ruled out in high-productivity systems; for example, Oschmann (1988) proposed that the Kimmeridge Clay was also an upwelling deposit. However, a pelagic or nektonic habit, coupled with high productivity, better explains the taphonomic features of the flat-clam assemblages. If it can be demonstrated that some or all of these bivalves lived in the water column, their concentrated assemblages would be excellent upwelling indicators.

Jeffries and Minton (1965) and earlier workers (see references in Jeffries and Minton 1965) emphasized the extremely high densities of shells on bedding planes, commonly on every parting over many meters of rock. Jeffries and Minton (1965) argued that *Bositra buchi* and, probably, *"Posidonia" radiata* were nektoplanktonic on distributional (see also Hayami 1969) and morphological grounds, that is, lack of a byssus, morphological evidence of swimming ability, and experimental evidence showing that active swimming was possible.

The principal challenge to this interpretation of flat-clam ecology came from Kauffman (1978, 1981) and later from Wignall and Ruffell (1990) and Wignall (1993), all of whom rejected with no explanation some of the observations made by the earlier workers. The challenge view has subsequently been incorporated into biotic models for low-oxygen environments (e.g., Sageman, Wignall, and Kauffman 1991). Kauffman's (1978, 1981) arguments for a benthic habit and against a planktonic or pseudoplanktonic (i.e., attached to wood or seaweed) habit, and some of the countering observations and arguments, were the following:

1. *Bositra* is not found with logs or ammonites (Wignall and Simms 1990).
2. The musculature resembles that of benthic bivalves. Jeffries and Minton (1965) had argued that the musculature, as well as the shell morphology, was structured in such a way as to make a swimming habit not only possible but efficient. However, Kauffman (1978, 1981) regarded the similarity of the musculature to that of benthic bivalves as more compelling.
3. Other benthic organisms are present in association with the flat clams. Kauffman (1978) was not specific about the facies in which these other benthic organisms occur and in at least one case—stemmed crinoids—his own illustrations (in Kauffman 1981) show a possible pseudoplanktonic habit for those organisms. Jeffries and Minton (1965) acknowledged the presence of crabs but noted that they are very rare and that they also were very mobile in life. Later, Kauffman (1981) acknowledged that all the benthics other than the clams were "rare" or "sparse."
4. The clams occur only in the laminated, dysoxic or anoxic facies (also Wignall and Ruffell 1990; Wignall 1993). However, *Bositra* occurs in a variety of facies and is only most concentrated in the black, laminated facies (Jeffries and Minton 1965). Many of the genera are global in distribution, which is typical of pelagic organisms. Both observations are also true for *Halobia* (Parrish 1987a).
5. The true thickness of the shells is not preserved and, in life, they were too thick and heavy for a pelagic habit. However, well-preserved shells of *Bositra buchi* contain both the laminar and prismatic layers (Jeffries and Minton 1965), and the largest forms of *Bositra* never occur with the presumed pelagic forms, which are typically >10 mm long and have shells 10 to 20 μm thick (Oschmann 1993).

An important line of evidence for a pelagic habit is the taphonomy of the assemblages. High-density assemblages over great thicknesses of section can be explained only by very slow sedimentation rates or by rapid influx of organisms. Slow sedimentation rates are improbable because anoxic waters are corrosive to shells and they would be only poorly preserved, even as impressions, once they were finally buried; preservation is usually excellent. If the bivalves were benthic, the densities indicate that they would smother each other, especially if oxygen were already at a premium. Although such assemblages show low dominance-diversity values considered diagnostic of low-oxygen environments (Wignall and Myers 1988), such values are also typical of high-productivity (eutrophic) faunas. Therefore Parrish (1987a and unpublished data) concluded that the high densities indicate rapid influxes of shells to the sediment from the water column, where the individuals could grow unhampered by crowding.

Another key observation that supports a pelagic habit is the occurrence of the clams in facies other than the black shale facies in which they are most common. This suggests that they preferentially concentrated in one environment but occasionally moved to other environments. This pattern is common in pelagic organisms. Conditions in the water column itself, such as sites of high productivity, could limit the distribution of the organisms at the surface. Those same areas would tend to be underlain by soft sediments (Wignall 1993), thus giving the appearance of substrate preference once the clams died and fell to the bottom. In the same vein, Kauffman's (1978, 1981) idea that they were benthic and preferred low-oxygen environments, coupled with their abundance in those environments, suggests narrow ecological tolerances, which would then have precluded the clams from being global in distribution. Finally, there is the argument that organic-rich sediments are themselves indicative of high productivity in the water column, not stagnation (Parrish 1995).

The role of high biologic productivity in producing extremely dense accumulations of organisms showing high dominance was also addressed by Allmon (1993; Allmon et al. 1996) in studies of the "Pinecrest Sand" (upper Tamiami Formation, Pliocene; Florida, U.S.), which is a highly fossiliferous unit dominated by bivalves and gastropods. Evidence for influx of nutrients and cooling associated with upwelling includes beds consisting almost entirely of the gastropod *Turritella*, which today dominates molluskan faunas where waters are cool and nutrient rich (Allmon 1988); the presence of cool- and/or deep-water taxa among the warm-water taxa, along with taxa characteristic of tropical upwelling zones (Allmon et al. 1996); the absence of calcareous algae and presence of solitary and cool-water corals (Allmon et al. 1996); oceanic microfossils suggestive of upwelling and instability (Allmon et al. 1996, citing Keller, Zenker, and Stone 1989); and a bed consisting primarily of bones of cormorants and other birds and of marine mammals, including whales (Allmon 1993; Allmon et al. 1996). Cormorants, in particular, are strongly associated with upwelling today, as are many marine mammals, especially whales. Finally, isotope evidence supports the hypothesis of upwelling (Allmon et al. 1996).

No evidence exists that the "Pinecrest Sand" mollusks lived in a low-oxygen environment. Several explanations can be offered for the absence of such a facies: (1) the bottom-water currents were such that the produced organic matter did not reach the bottom, where its accumulation would consume oxygen in decay; (2) plankton productivity was high enough to boost mollusk productivity but not enough to drive the system to anoxia; or (3) the mollusk productivity represents an edge effect (Mullins et al. 1985) of a well-developed upwelling zone, and the organic-rich sediments from that system are farther offshore. The latter is suggested by the presence of a rich phosphate deposit, an indicator of upwelling, just below the Tamiami Formation (Allmon et al. 1996). If the Tamiami Formation was deposited when sea level was lower and if the upwelling zone that formed the phosphate still existed, the Tamiami would represent the inner edge of that upwelling zone.

Stable isotopes in mollusks—oxygen and carbon. Pre-Quaternary mollusks are generally not preferred for stable isotope studies because the shells of most forms are composed largely or completely of aragonite, which is unstable. Exceptions are the oysters, belemnites, and rudistids, which are composed mostly or entirely of low-Mg calcite, and inoceramids, some of which have thick calcitic layers (Hudson and Anderson 1989; Steuber 1996). On the other hand, where the aragonite is preserved, it can be an excellent recorder of the original isotope signal (Hudson and Anderson 1989; Wilson and Opdyke 1996). Despite these limitations, mollusks have been important in isotope studies. The reasons for this are that they are abundant in the geologic record, they commonly construct large shells, and they are commonly the only calcareous fossils that are well preserved. Important early papers utilizing mollusks were by Lowenstam and Epstein (1954, 1959), Bowen (1961), Lowenstam (1964), Tourtelot and Rye (1969), Spaeth, Hoefs, and Vetter (1971), Stevens (1971), Stevens and Clayton (1971), and Dodd and Stanton (1975).

Isotope studies of mollusks are almost all in Mesozoic, mostly Cretaceous, faunas. This is because the groups of mollusks that secrete largely low-Mg calcite shells did not evolve until the Mesozoic and because foraminifera are better suited than mollusks for Tertiary studies (Emiliani 1955). With the advent of deep-sea drilling, isotope studies of mollusks became less common, as attention turned to the deep-sea record, although some recent studies have been important for interpreting paleoclimate in microfossil-poor rocks in the deep sea, high latitudes, and epicontinental seaways (e.g., Saltzman and Barron 1982; Barron, Saltzman, and Price 1984; Wright 1987; Pirrie and Marshall 1990b; Ditchfield 1997; Price and Sellwood 1997). Sediments from the Jurassic and Triassic are rare or absent in the deep-sea record. On the other hand, mollusks are very common in rocks of that age, and thus they provide the best potential isotope database. Recently, there has been renewed interest in Jurassic mollusks (e.g., Ditchfield 1997; Price and Sellwood 1997), but studies of Triassic mollusks are still unusual.

Mollusks have been used for paleotemperature determinations in epeiric seaways, which gives some information about continental climates. Excellent examples that, though several decades old, still stand as a standard for mollusk isotope studies, come from the work done by H. Lowenstam and S. Epstein. Lowenstam and Epstein (1954, 1959) compiled a large amount of isotope data from Cretaceous mollusks of Europe and North America. They were concerned with latitudinal temperature gradients for intervals during the Cretaceous (figure 2.38). An advantage of shallow-water mollusks from epeiric seaways or continental shelves that applies even to the Cenozoic is that they record seasonal temperatures (e.g., Jones and Quitmyer 1996). A 14°C seasonal variation in Eocene snails, for example, led Andreasson and Schmitz (1996) to question the notion of equable climates for that time.

Mollusks have been used in studies of the vertical structure of the waters in epeiric seaways (Tourtelot and Rye 1969; Wright 1987). They also have potential for tying together the deep- and epeiric-sea isotope records. This would throw light, for example, on hypotheses that link the two environments, such as extension of the oxygen-minimum from the open ocean into epeiric seaways (e.g., Arthur, Schlanger, and Jenkyns 1987). Despite possible problems with a vital effect (Tourtelot and Rye 1969; but see Pirrie and Marshall 1990a and MacLeod and Hoppe 1992), *Inoceramus* has the advantages of being common in Cretaceous, fine-grained marine sediments and of occurring across a broad range of depths, from shallow-water in epicontinental seaways to abyssal depths sampled by deep-sea drilling cores. The thick shells of this clam are composed of coarse calcite crystals, making the organisms ideal for isotope studies. If the vital effect is linear with temperature, the isotope variations would still reflect real variations in temperature.

Isotope analyses of *Inoceramus* were used to confirm the presence of warm water at depth in the opening South Atlantic (Saltzman and Barron 1982; Barron, Saltzman, and Price 1984), which had been predicted in the warm, saline bottom-water hypothesis of Brass, Southam, and Peterson (1982). Saltzman and Barron (1982) and Barron, Saltzman, and Price (1984) also found differences between the Angola and Brazil Basins, which they attributed to paleogeographic isolation of the deepest portions of the Angola Basin from the world ocean and to generation of warm, saline bottom water in the adjacent epicontinental seaways on Africa.

Isotopic signatures of belemnites, the aragonitic and calcitic internal skeletons of extinct cephalopods, have also been used extensively in paleoclimate studies (Lowenstam and Epstein 1959; Bowen 1961; Lowenstam 1964; Spaeth,

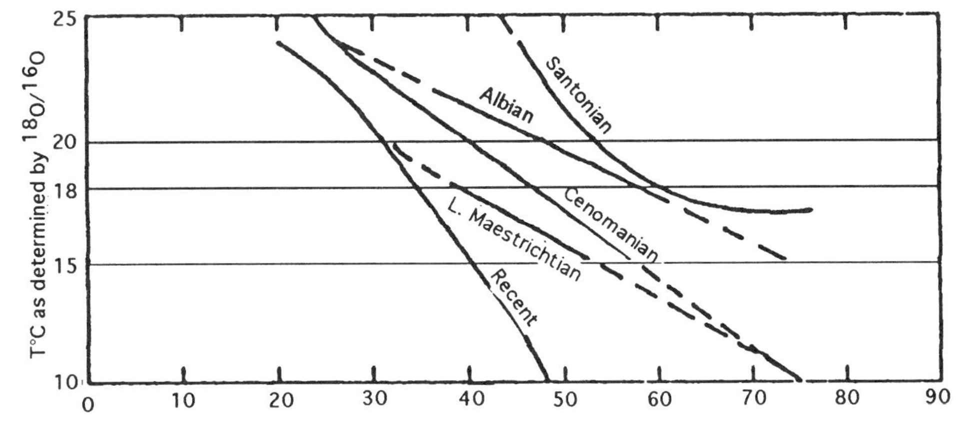

Figure 2.38. Latitudinal paleotemperature gradients for several stages of the Cretaceous determined from oxygen isotope data in mollusks, mostly belemnites, compared with the modern Northern Hemisphere gradient. From Lowenstam (1964) in *Problems in Palaeoclimatology,* A.E.M. Nairn, ed. Copyright John Wiley and Sons, Ltd. Reproduced with permission.

Hoefs, and Vetter 1971; Anderson et al. 1994; Sælen, Doyle, and Talbot 1996). Work has tended to concentrate on the calcitic rostrum, which is more stable isotopically than the aragonitic phragmocone. Spaeth, Hoefs, and Vetter (1971) noted that single specimens of belemnites could give 5‰ variations in $\delta^{18}O$ and 14‰ variations in $\delta^{13}C$. For this reason, they concluded that true temperature records in belemnites cannot be distinguished from diagenetically altered records. Later studies, however, have successfully used stable isotope results from belemnites, suggesting that Spaeth, Hoefs, and Vetter's (1971) results were influenced by diagenesis of the rostra; the methods for using trace elements to detect diagenesis had not yet been developed when Spaeth, Hoefs, and Vetter (1971) conducted their study.

Using belemnites and other organisms, Anderson et al. (1994) conducted a detailed study of the Peterborough Member of the Oxford Clay Formation (Jurassic; United Kingdom). Their sampling strategy included animals from nektonic and benthic habits and both invertebrates and vertebrates; the results from the vertebrates are discussed later in this chapter. Invertebrates sampled, in addition to belemnites, were ammonites and calcitic and aragonitic bivalves. The results are shown in figure 2.39. No systematic trend is visible from older to younger beds, meaning that the variation in isotope results reflects environmental variation on relatively short time scales within an otherwise stable system.

To calculate paleotemperatures (figure 2.40a), Anderson et al. (1994) corrected for the relative fractionation of aragonite and calcite and as-

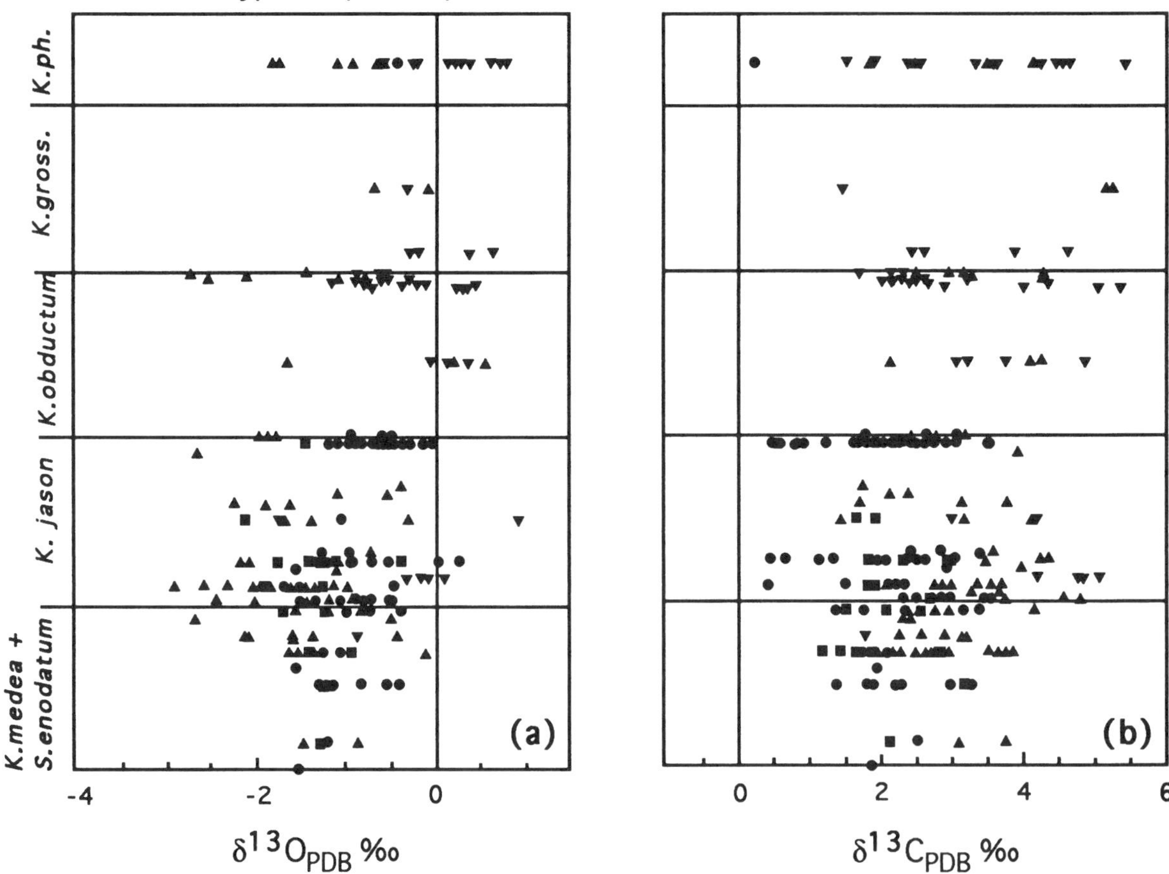

Figure 2.39. Stable isotopic compositions of mollusks in the Peterborough Member of the Oxford Clay (Jurassic; United Kingdom) presented in stratigraphic order. *(a)* Oxygen. *(b)* Carbon. From Anderson et al. (1994). *Journal of the Geological Society* of London, © 1994 The Geological Society.

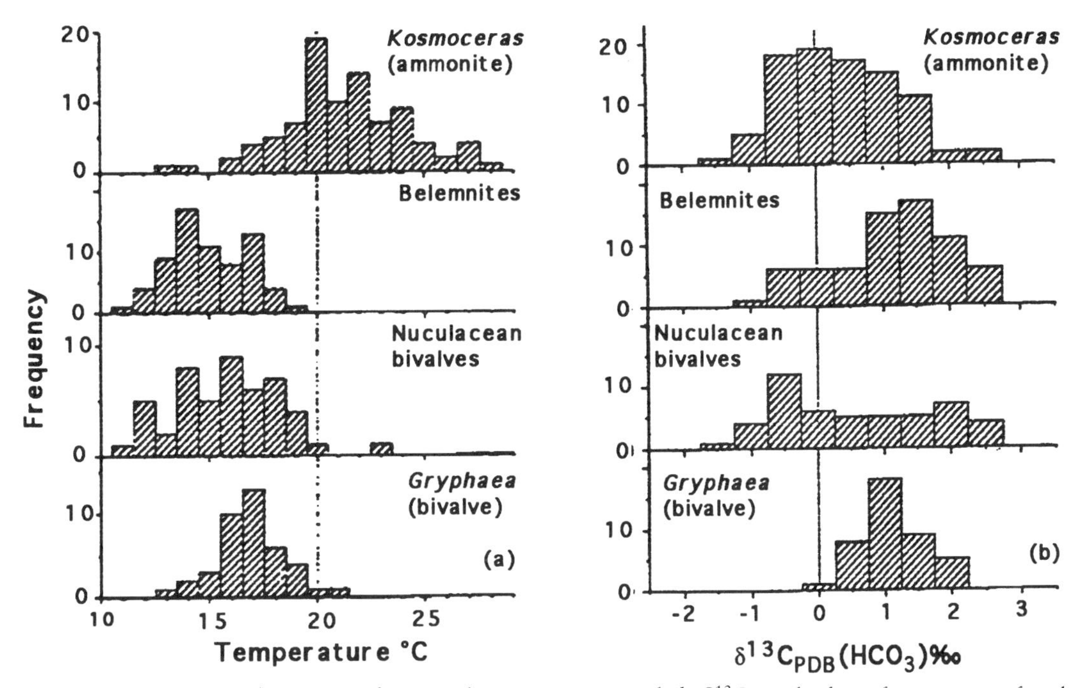

Figure 2.40. Frequency histograms for *(a)* paleotemperature and *(b)* $\delta^{13}C$ results from four groups of mollusks in the Peterborough Member of the Oxford Clay (Jurassic; United Kingdom). The means for paleotemperature from *Gryphaea* (17°C), nuculacean bivalves (16°C), and belemnites (15°C) are not significantly different; the mean paleotemperature from *Kosmoceras* is 21°C, significantly higher. From Anderson et al. (1994). *Journal of the Geological Society* of London, © 1994 The Geological Society.

sumed average seawater for the modern, ice-free Earth (−1.0‰ SMOW). The lowest temperatures were for the benthic bivalves and the belemnites and the highest temperatures for the nektonic ammonite *Kosmoceras*, but all groups exhibited variability over a wide temperature range. All groups showed overlap in $\delta^{13}C$ values as well (figure 2.40b), with *Kosmoceras* slightly lower, a result that is surprising because benthic organisms would be expected to have lighter values from ambient accumulated organic carbon. Anderson et al. (1994) concluded that aragonite-calcite fractionation differences would not have significantly affected the relative $\delta^{13}C$ signatures of the different groups, but they raised the possibility that *Kosmoceras* did not precipitate aragonite in isotopic equilibrium with respect to $\delta^{18}O$, citing work that suggested that the estimated temperatures could be 3° to 4°C too high. If this were the case, both the $\delta^{13}C$ and $\delta^{18}O$ data indicate a well-mixed water column. Alternatively, the $\delta^{18}O$ results could indicate that *Kosmoceras* lived at shallower depths than belemnites (figure 2.40a; see also Martill et al. 1994), and this was the explanation preferred by Anderson et al. (1994).

Anderson et al. (1994) did not find a strong effect of salinity on $\delta^{18}O$ during deposition of the Oxford Clay. In contrast, Sælen, Doyle, and Talbot (1996) argued for a substantial salinity effect in stable isotope results from belemnites in certain parts of the Whitby Mudstone Formation (Jurassic; United Kingdom). Assuming normal salinities and seawater composition of −1‰ SMOW (ice-free Earth), paleotemperatures from belemnites were 13.4° to 36.9°C, with most falling between 15° and 30°C. By analogy with modern temperature distributions, the Whitby Mudstone belemnite temperatures are consistent with sea-surface temperatures (SSTs) at the latitude at which the Whitby Mudstone was deposited, that is, about 30° to 40°N.

Despite these positive results, Sælen, Doyle, and Talbot (1996) extended their study to try to account for the organic-rich beds and low-diversity biota, which other authors had attributed to stagnation by stable stratification of the water column. Sælen, Doyle, and Talbot (1996) argued that belemnites probably recorded the temperatures of a mixture of surface and deeper water, either because they migrated vertically through the water column or because they migrated horizontally in and out of upwelling waters (cf. Anderson et al. 1994). Therefore the indicated temperatures were cooler than the actual SSTs. Assuming that the vertical water-column temperature gradient was 2°C, Sælen, Doyle, and Talbot (1996) adjusted the SSTs accordingly. Then they argued that adjusted SSTs higher than 31°C must indicate a salinity effect. Assuming a $\delta^{18}O$ composition of freshwater runoff of –7‰ SMOW, freshwater dissolved inorganic carbon $\delta^{13}C$ of –9‰ to –6‰ PDB, and a range of productivities, Sælen, Doyle, and Talbot (1996) argued that parts of the Whitby Mudstone Formation had surface salinities as low as 29‰ (versus normal marine salinities of 34‰), and that therefore the organic richness and low-diversity biota were, indeed, the result of water-column stratification.

It is important to emphasize the complex stacked assumptions in Sælen, Doyle, and Talbot's (1996) work. The first was the a priori assumption that the organic richness and low diversity were the result of oxygen depletion in a stratified water column, even though this was also presented as a result. Other assumptions were the 2°C vertical temperature gradient, the ecology of the belemnites (which was addressed more directly in the comparative study by Anderson et al. 1994), the maximum SST of 31°C, and the isotopic composition of the freshwater, among other assumptions about productivity, and so on. Complex stacked assumptions are common in paleoclimatology and emphasized throughout this book, and fruitful research is performed examining each one, but one must be particularly circumspect in interpreting those that lead to numerical results.

In the previous section, the study by Allmon et al. (1996) was cited in which they proposed that dense accumulations of mollusk shells in the Pliocene Tamiami Formation (Florida, U.S.), along with other evidence, were indicative of upwelling. Part of this evidence came from carbon and isotopic compositions of some of the mollusks. Their prediction for the isotopic composition of the shells and surrounding waters in an upwelling system is shown in figure 2.41. The key feature is the relatively low $\delta^{13}C$ and high $\delta^{18}O$ values expected in the Pinecrest fauna. Seasonal variations in upwelling would show changes in $\delta^{13}C$ and $\delta^{18}O$ in opposite directions. The isotope results are shown in figure 2.42 and indicate variations in isotopic composition consistent with variation in the strength of the upwelling zone illustrated in figure 2.41, with $\delta^{18}O$ and $\delta^{13}C$ trending in opposite directions. An exception is at sample 20 in *Turritella apicalis*. That sample shows both $\delta^{13}C$ and $\delta^{18}O$ decreasing, which is consistent with an influx of freshwater, which has lower $\delta^{18}O$ than marine water, and terrestrial organic matter, which has lower $\delta^{13}C$ than marine organic matter.

Stable isotopes in mollusks—strontium. Recently, workers have begun to examine mollusks as sample material for strontium isotopes (Ingram and DePaolo 1993; Jones et al. 1994; Jones, Jenkyns, and Hesselbo 1994). Ingram and DePaolo (1993) used variations in the $^{87}Sr/^{86}Sr$ ratio in the shells of bivalves to estimate salinity over the past 4300 years in the San Francisco Bay estuary. The $^{87}Sr/^{86}Sr$ ratio correlates with salinity because of the large difference between marine waters entering the bay ($^{87}Sr/^{86}Sr$ = 0.7092) and the average freshwaters (0.7065). Ingram and DePaolo (1993) went on to estimate variations in freshwater influx during the study interval.

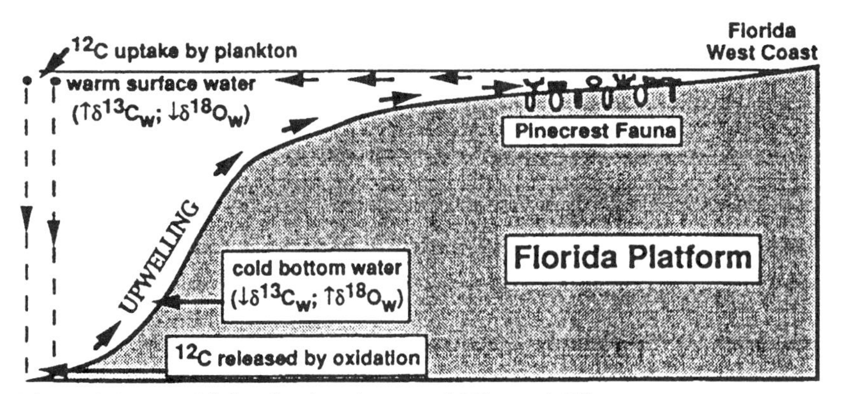

Figure 2.41. Model for the distribution of $\delta^{13}C$ and $\delta^{18}O$ variations in an upwelling system. The Pinecrest fauna (Pliocene; southeastern U.S.) is predicted to have grown in relatively cool, nutrient-rich water. From Allmon et al. (1996) in *Journal of Geology*, © 1996 by The University of Chicago.

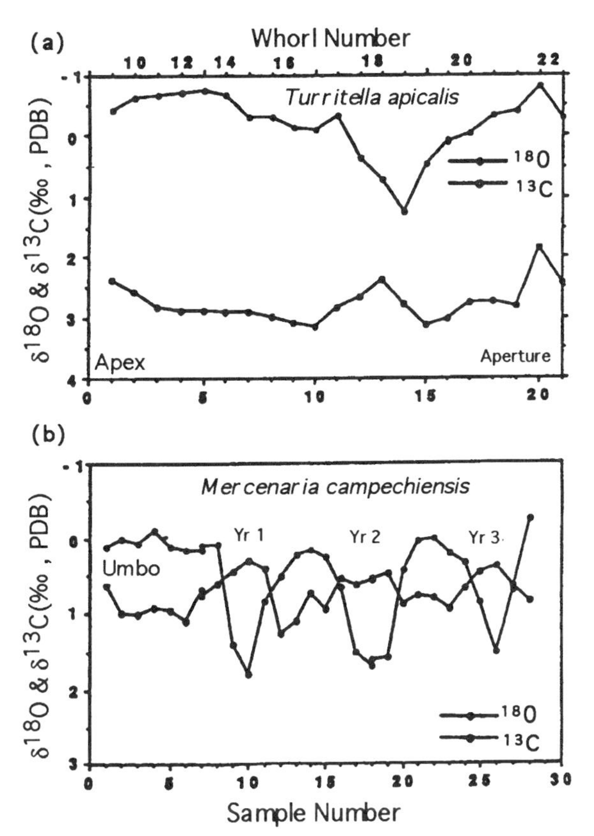

Figure 2.42. Stable isotope profiles across shells of the gastropod, *Turritella apicalis (a)*, and the bivalve, *Mercenaria campechiensis (b)*. See text for explanation. From Allmon et al. (1996) in *Journal of Geology*, © 1996 by The University of Chicago.

As with all isotope studies, there are several unknowns, in this case, the isotopic compositions of the seawater and freshwater. For such a short and recent record, Ingram and DePaolo (1993) could reasonably assume that the isotopic compositions of the waters were the same as they are today and thus estimate freshwater influx. For the older geologic record, seawater $^{87}Sr/^{86}Sr$ could be determined from shells of organisms from fully marine environments, leaving the freshwater influx and the isotopic composition of the fresh waters as unknowns. By measuring $^{87}Sr/^{86}Sr$ in fully freshwater organisms, estimates could be made for freshwater influx in ancient estuarine systems. However, in the older record, diagenesis becomes a major consideration, as it does for the isotopes of oxygen and carbon. Indeed, as noted previously, variation in Sr content is sometimes used as an indicator of diagenetic alteration (Popp, Anderson, and Sandberg 1986b). Therefore much current work is focused more on producing good data from mollusks in the older geologic record (Jones, Jenkyns, and Hesselbo 1994) than on using those data for paleoenvironmental interpretations like those of Ingram and DePaolo (1993).

Variations in marine $^{87}Sr/^{86}Sr$ are also used to interpret global weathering. Interpretations from the older record of mollusks are conflicting, however. For example, Jones et al. (1994) concluded that the relatively narrow range within which the $^{87}Sr/^{86}Sr$ curve varies for the Triassic, Jurassic, and Early Cretaceous and the time scale on which the changes occur are consistent with the scale of variations resulting from changes in hydrothermal activity and do not require changes in riverine flux. They then suggested that the curve might be used to track hydrothermal activity associated with the breakup of Pangea. The interesting aspect of this work for this book is the converse of their conclusion, that weathering and riverine input might not have played an important role during much of the Mesozoic. This has important implications for climatic processes at that time, especially hypotheses about monsoonal climate during the Pangean interval (Parrish 1993). However, Berner and Rye (1992) argued that changes in the composition of the source rocks drive Sr variation, rather than weathering flux or hydrothermal influx. Their reasoning, from results with a geochemical model, is that changed spreading rates per se are not related to changes in the Sr curve. Higher spreading rates also result in greater production of volcanics on the continents, which are then weathered.

Corals

Biogeography. Coral reefs, which define the belt of warm water that lies in the tropics and subtropics (Veron 1995), are biogenic structures, and their use as paleoclimatic indicators transcends the taxa that they comprise. In contrast, coral taxa per se have not been used extensively in paleoclimatology. Not all corals are reef-builders, and the biogeography of corals has been used in paleoclimatology independent of their particular habits. However, few such studies have been performed. Corals have been used mostly as paleogeographic, rather than paleoclimatic, indicators.

In one of the few studies that used coral biogeography to indicate climate variation, Beauvais (1992) postulated warm and cold currents to account for occurrences of certain Early Cretaceous coral genera outside of the latitudinal belts in which they generally occurred. For example, she suggested that the reestablishment of Boreal Realm in the Cenomanian (after its disappearance in the Aptian) indicated the formation of cold currents across North America. She also postulated factors other than temperature to explain the scarcity of corals at certain times. For example, she noted that belemnites seemed to have a normal distribution during the Valanginian, whereas corals are uncommon, and suggested that some other factor, perhaps turbidity, limited the corals (Hallam 1975; Ziegler et al. 1984). Provinciality of the corals, however, appeared to have little to do with paleoclimate but rather was dependent on paleogeography. In contrast, in the Silurian, coral biogeography was apparently independent of paleogeography and dependent on latitude (Hill 1959; Ziegler et al. 1981).

Stable isotopes in corals. Modern corals are known to deposit their aragonite skeletons out of equilibrium with the isotopic composition of seawater (Weber and Woodhead 1972; Hudson and Anderson 1989). However, for any coral genus, the isotopic offset is constant at any given temperature in the rapid-growth portions of the skeleton (e.g., McConnaughey 1989), so relative changes are meaningful. Very few studies of the isotopic composition of corals have been done on pre-Quaternary corals, and arguments about the reasonableness of the results rely on other types of information (e.g., Scherer 1977). An exception is a study by Roulier and Quinn (1995), who conducted a very detailed study of a sample of coral to examine seasonality and other short-term climatic changes in the Pliocene of Florida. The average annual isotopic range in the coral is 0.7‰ to 2.27‰ $\delta^{18}O$, 0.31‰ to 2.21‰ $\delta^{13}C$. Isotope mass-balance calculations on data from modern corals indicate a partitioning of the $\delta^{18}O$ signal into 8% to 12% salinity signal and 92% to 88% temperature signal. If Roulier and Quinn's (1995) data are interpreted as 100% temperature signal, the seasonal variation for the Pliocene coral was 7° ± 1.7°C, which is 3.5°C less than today. An interesting feature of their data was that the peak $\delta^{18}O$ values led peak $\delta^{13}C$ values by about 2 months. In the same corals today, $\delta^{13}C$ is correlated with the number of sunshine hours per day. Thus in the Pliocene coral, the minimum sea-surface-temperature values were followed 2 months later by maximum sunshine, as occurs in southwestern Florida today, where the summers are warm and cloudy and the winters cool and dry. Their results are intriguing because of the indirect connection of their data to cloudiness, but $\delta^{13}C$ signatures in corals in general are so complex (e.g., Weber and Woodhead 1970; Land, Lang, and Barnes 1975; and many others) that similar investigations elsewhere have yet not been encouraging for obtaining a paleocloud indicator (T. M. Quinn, personal communication).

Trilobites

Trilobite biogeographic studies have been most important for the early Paleozoic and Devonian (e.g., Ziegler et al. 1981; see table 2.3). These studies have tended to take a traditional approach, and few have gone beyond simply noting a relationship between biogeography and some simplified model of climate. For example, it was noted by Cocks and Fortey (1990) that trilobite distribution in the Early Ordovician was strongly related to latitude and, presumably, temperature, especially the pelagic faunas, which would have crossed ocean barriers with relative ease. However, paleogeography was clearly also a major control on trilobite distributions, especially the shallow-water faunas.

Trilobite faunas in deep-water facies, determined paleogeographically and sedimentologically, are taxonomically different from the pelagic and shallow-water faunas (Taylor and Forester 1979; Cocks and Fortey 1990) and were much more widespread than shallow-water taxa, having lived on widely separated continental plates. Taylor and Forester (1979) drew an analogy with modern marine isopods, which are morphologically and, apparently, ecologically similar to some trilobite taxa. Isopods show the highest generic similarities within warm-water biofacies, which are all in shallow water, and within cold-water biofacies, regardless of depth; there is little similarity between the warm- and cold-water biofacies. Taylor and Forester (1979) argued that Cambrian trilobite biogeography indicated the same pattern, dividing the trilobite faunas into two realms, the psychrospheric (cold-water) and thermospheric (warm-water) realms. Worldwide dispersal of the psychrospheric fauna could occur at depth, whereas the shallow-water, thermospheric provinces would evolve in isolation from one another. Explicitly with respect to climate, the pattern seen in the Cambrian indicates a thermally stratified ocean, in which cold bottom waters were generated in high latitudes (Taylor and Forester 1979).

As a result of this work, Taylor and Forester (1979) cautioned against the uncritical use of biogeographic information for continental plate reconstruction. If deep- and shallow-water faunas are treated together, biogeographic differentiation among continents is much less. This point was emphasized by Cocks and Fortey (1990), who found a similar shallow- and deep-water differentiation in the trilobite faunas in the Ordovician.

Graptolites

Graptolites are extinct organisms with neither living relatives nor close modern analogs. They are thought to have been pelagic, partly on the basis of their morphology and partly because individual taxa are relatively widely distributed. Little consensus exists about either the patterns of graptolite biogeography during the Ordovician and Silurian or the causes of biogeographic differentiation. For example, Finney and Chen (1990) described the change from two graptolite provinces in the Early Ordovician to a cosmopolitan fauna in the Middle Ordovician as the result of climatic cooling, which would have narrowed the zone of warm waters suitable for graptolites so much that provinciality could not be sustained because of a lack of space. However, their hypothesis was inconsistent with recent paleogeographic maps (Finney and Chen 1990; Scotese and McKerrow 1990), and their attempt to reconstruct paleocurrents on the basis of graptolite biogeography was unsuccessful (see also Rickards, Rigby, and Harris 1990). Finney and Chen (1990) identified the problem as flawed continental reconstructions and recommended altering them.

In contrast, Berry and Wilde (1990) identified the same provinces as Finney and Chen (1990) for the Early Ordovician but showed very different distributions for them. In addition, they indicated that essentially the same two provinces persisted through the Middle Ordovician and into the Silurian. In their reconstructions, the provinces are nearly parallel to latitude, and the longitudinal variations are consistent with modeled current patterns of Ziegler et al. (1981; see also Rickards, Rigby, and Harris 1990). Neither Finney and Chen (1990) nor Berry and Wilde (1990) considered the suggestion that the two provinces represented shallow- and deep-water faunas (Fortey and Cocks 1986).

Rickards, Rigby, and Harris (1990) strongly advocated controls in addition to latitude (i.e., temperature) for graptolite distribution, such as upwelling and depth (Berry and Wilde 1978, 1990; Wilde and Berry 1982). So close is the association of graptolites and black shales that Berry (1974) and others have called such deposits "graptolite facies." Berry and Wilde (1978) suggested that the graptolites flourished at the surface on the edges of upwelling zones, where nutrient levels would have been high, and they interpreted the dense accumulations of graptolites as mass-kill deposits resulting from upwelling of anoxic waters and suffocation. In 1990, Berry and Wilde proposed an alternative model that had the graptolites living along the oxic-anoxic interface, either in upwelling zones or in an ocean-wide oxygen-minimum zone (see also Berry, Wilde, and Quinby-Hunt 1987). Many of the upwelling zones that they specifically mentioned had been predicted from climate models (Parrish 1982; Moore, Hayashida, and Ross 1993).

Integrated biogeographic studies on marine invertebrates

Paleobiogeographic studies that are based on single groups of organisms tend to result in different interpretations depending on the vagaries of the data available for each of the groups. If the observed paleobiogeographic patterns for all groups truly reflected the relationship of their distributions to aspects of their physical environment such as paleogeography or paleoclimate, this would not be a problem. However, how paleobiogeographic units are defined can be affected by completely extrinsic factors such as monographic effect and the distribution of localities. In addition, different groups might, by their very nature, record change with different levels of resolution, as illustrated in table 2.5. During the late Paleozoic, different groups had different levels of endemism, cosmopolitanism, and what Bambach (1990) called regional ligation. Paleobiogeographic differentiation will be harder to interpret in groups with high levels of endemism (e.g., echinoderms) and cosmopolitanism (e.g., bivalves in the Paleozoic), and easier to interpret in groups whose members mostly occur in more than one place but less than all places sampled (Bambach 1990). The limitations of using one

Table 2.5.
Percentage of taxa from each taxonomic group in each of three biogeographic categories

	Cosmopolitan	Regionally ligated	Endemic
Bivalvia	53.3	20.5	26.2
Pedunculate Articulata	44.5	16.1	39.4
Strophomenida	28.3	27.5	44.2
Tabulata	23.2	29.3	47.5
Bryozoa	19.5	31.2	49.3
Rugosa	9.4	41.6	49
Average	28.1	28.7	43.2

From Bambach (1990) in *Palaeozoic Palaeogeography and Biogeography*, W. S. McKerrow and C. R. Scotese, eds., © 1990 The Geological Society.

group of organisms can be at least partially overcome by combining paleobiogeographic information from many groups. This approach has been used occasionally in studies of single periods (e.g., Hallam 1975 for the Jurassic Period) because extrinsic influences, such as monographic effect, tend to be less severe at that level, but the approach has been used much less commonly over longer time periods.

The integrated approach also has the advantage of providing greater continuity through time, facilitating the understanding of changes in paleoclimates and/or paleogeography. For example, an examination of table 2.3 shows that the principal groups of organisms for each time period are different. This is largely a monographic effect, dependent on how well various groups have been studied. It is not difficult to imagine the concerns that might arise from a study that attempted to track paleoclimatic or paleogeographic change through the Paleozoic by examining only trilobites in the Cambrian, brachiopods in the Ordovician through Devonian, corals in the Early Carboniferous, fusulinids in the Late Carboniferous, and, again, brachiopods in the Permian. However, by integrating biogeographic information from many groups, some continuity through time is achieved—all the time periods have groups in common with the adjacent time periods. An additional advantage is expansion of the total database (Bambach 1990). This approach allowed Ziegler et al. (1981) to draw some general conclusions about the evolution of Paleozoic climates that had not previously been recognized and that would not otherwise have been possible. For example, they were able to analyze the relationship between the predicted warm and cold currents and the biogeographic units. They found that many groups showed a relationship between paleocurrents and high-diversity, warm-water faunas. Although showing these types of relationships might have been possible for individual time periods or groups of organisms, the integrated approach allowed Ziegler et al. (1981) to conduct a more systematic study.

Bambach's (1990) study of late Paleozoic provinciality also used an integrated approach. He first examined biogeographic patterns in each group individually (partly summarized in table 2.5) and then combined information from all the groups (figure 2.43). Several realms are defined for each time interval, and each realm has at least two provinces. Bambach (1990) defined provinces as areas that had at least 30% endemic genera, more than one endemic genus, and a paleogeography that indicates likely barriers to dispersal. He noted that the numbers of provinces are comparable to the number occurring today. In contrast, studies of single groups of Paleozoic organisms generally resulted in identification of only a few provinces. A striking outcome of Bambach's (1990) study, which had been only hinted at in studies of individual groups of organisms, was the restriction of high-diversity faunas to very low latitudes in the Late Carboniferous (figure 2.44). Even without other evidence of large-scale glaciation at that time, such a conspicuous pattern would be taken as evidence of cooling in higher latitudes.

Marine Vertebrates

Biogeography

The biogeography of marine fishes has not received a great deal of attention in the literature, possibly because they are relatively poorly preserved compared with freshwater fishes. The biogeography of marine mammals, reptiles, and birds, however, has provided some intriguing information about paleoclimate at some times during Earth history. Marine reptiles were a prominent component of marine communities during the Mesozoic, and marine mammals are common in the geologic record from the Eocene on. These organisms are high in, or at the top of, the food chain. Thus their distribution appears to have been controlled more directly by the distribution

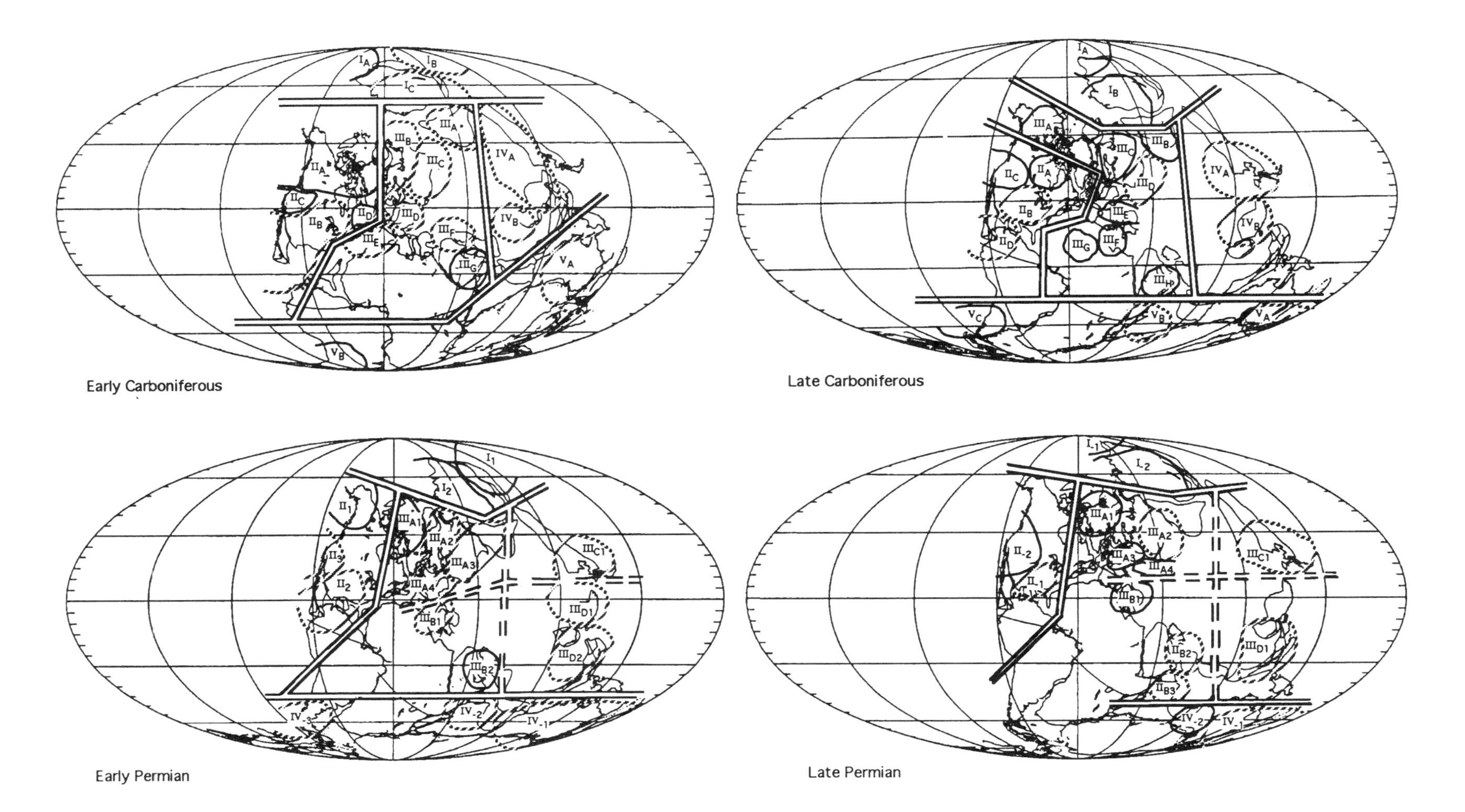

Figure 2.43. Distribution of biogeographic regions based on bivalves, pedunculate articulate brachiopods, strophomenid brachiopods, tabulate and rugose corals, bryozoans, crinoids, and ammonoids. Ellipsoids are drawn around the areas from which major faunas were sampled. *Solid-line ellipsoids* are provinces with at least 30% endemic genera, more than one endemic genus, and a paleogeography that indicates likely barriers to dispersal. *Dashed-line ellipsoids* indicate faunas that are not diverse enough or do not have sufficient levels of endemism to meet the criteria for provinces; Bambach (1990) called these potential provinces. *Solid double lines* separate realms, *dashed double lines* separate regions within realms. From Bambach (1990) in *Palaeozoic Palaeogeography and Biogeography,* W.S. McKerrow and C.R. Scotese, eds., © 1990 The Geological Society.

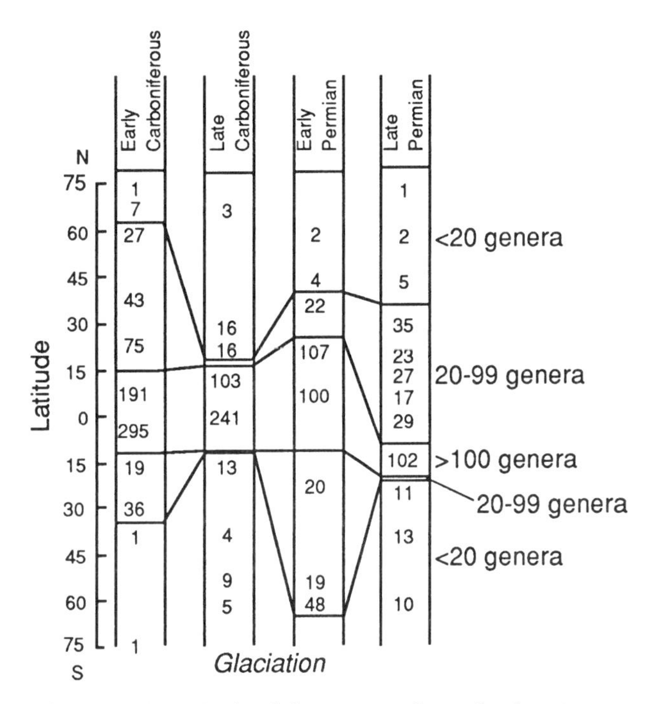

Figure 2.44. Latitudinal diversity gradients for four intervals in the Paleozoic. Note the constriction of high-diversity faunas to low latitudes in the Late Carboniferous and the asymmetry of the gradient, particularly in the Late Permian. From Bambach (1990) in *Palaeozoic Palaeogeography and Biogeography,* W.S. McKerrow and C.R. Scotese, eds., © 1990 The Geological Society.

of their food, and by climate to the extent that the distribution of food has been controlled by climate. The most abundant food resources are along the continental margins and, especially, in upwelling zones, where biologic productivity is highest. Thus it is not surprising that accumulations of marine vertebrates are commonly used as indicators of high biologic productivity, whether on a local or global scale.

Studies of the potential paleoclimatic aspects of marine reptiles are few, and the subject has not been fully treated in the literature as yet. In an abstract, Parrish and Parrish (1983) described a strong correlation between the distribution of marine reptiles and the distribution of predicted upwelling zones (Parrish and Curtis 1982) and suggested that the animals were the Mesozoic ecological equivalents of whales, whose feeding ranges are strongly related to upwelling in the present oceans. Their hypothesis received some support from an ecological analysis of the feeding apparatus of some marine reptiles by Massare (1987), but this work has yet to be pursued further. Recently, however, Martill et al. (1994) analyzed the trophic structure of the biota, including marine reptiles and large fish, in a Jurassic unit in the United Kingdom, the Peterborough Member of the Oxford Clay Formation. They showed that in this unit, at least, the marine reptiles were in fact part of a high-productivity ecosystem. Martill et al. (1994) concluded that the elaborate trophic structure of the fauna indicates high biologic productivity in the Peterborough sea. Other features of the Peterborough ecosystem are the relatively short paths from primary producers to top carnivores and the presence of giant planktivores, which also are typical of high-productivity ecosystems (Lipps and Mitchell 1976).

The most complete treatment of marine vertebrate paleobiogeography in relation to climate was by Fordyce (1982), who discussed Australian and New Zealand marine reptiles, birds, and mammals, concentrating on cetaceans (whales, dolphins, and porpoises). He emphasized that many taxa are so poorly known that it is difficult to discern whether biogeographic patterns are real or artifacts of poor systematics. The record of penguins extends from the Eocene onward. Fordyce (1982) noted that these birds overlap in size and ecology with small cetaceans and that the evolutionary history of the group is similar. A strict uniformitarian approach to the interpretation of penguin distribution requires that they be interpreted as having lived in cold water. If they are taken as cold-water indicators, their presence in Australia and New Zealand supports a higher-latitude position in the Tertiary than today and/or a marked cooling of the Southern Ocean during the Oligocene.

Cetaceans also are known from the Eocene on. The group underwent a striking evolutionary radiation, from four species in the Early Oligocene to 45 by the end of the Oligocene (Fordyce 1980, 1982). Most of this diversification occurred in the Southern Hemisphere and involved the evolution of odontocetes (toothed whales that use echolocation to locate their prey) and mysticetes (baleen whales) in the early Oligocene. In explaining this diversification, Fordyce (1980) stressed the ecology of whales, whose feeding ranges are concentrated in upwelling zones. The largest such zone today is in the circum-Antarctic region, and Fordyce (1980) proposed that cooling associated with an increase in ice volume (see figure 2.14) established "localized high-productivity areas" in the Southern Ocean that triggered odontocete and mysticete evolution. The increase in numbers and distribution of cetaceans through the rest of the Tertiary paralleled "modernization" of the structure of the oceans (Fordyce 1980). Although the evolution of the forms cannot be considered indicative of paleoclimate, since paleoclimatic change was proposed as an explanation, Fordyce's (1980) work does emphasize that these vertebrates are indicative of high biologic productivity.

Oxygen isotopes in marine vertebrates

The use of oxygen isotopes to interpret paleotemperature from the phosphate of vertebrate skeletons was pioneered by Y. Kolodny and his colleagues (Kolodny, Luz, and Navon 1983; Luz, Kolodny, and Horowitz 1984; Luz and Kolodny 1985; Kolodny and Raab 1988; Kolodny and Luz 1991). The 1988 study by Kolodny and Raab was the first to make use of oxygen isotopes from marine fishes and reptiles to estimate paleotemperatures. They sampled fish, sharks, and a few reptiles from Cretaceous to Eocene rocks in the Mediterranean region, mostly Israel. Their results are shown in figure 2.45. The data suggest a cooling trend through the Late Cretaceous and warming by the early Eocene. The trends were consistent with results of isotope analyses from other types of organisms, but the estimated paleotemperatures were higher. Kolodny and Raab (1988) explained the difference as a result of the water in which the vertebrates lived. The region was close to the tropics at the time and all the samples in their study came from shallow seas—probably the warmest marine environment on the Earth at that time. Kolodny and Raab (1988) also emphasized that depth habitat of vertebrates is important for understanding the isotope results, as deep-water dwelling forms are likely to give cooler temperatures than surface dwellers. However, in their estimation, the waters did not exceed several tens of meters depth. Thus temperature stratification was likely not to have been a source of error. High salinities would have given lower-than-true temperatures, so the high estimated temperatures would seem to indicate that salinity variations also were not a great source of error.

A more detailed study of fish debris (mostly shark and ray teeth) determined paleotemperatures from both benthic and nektonic taxa across the Cretaceous-Tertiary boundary in northern Morocco (Lécuyer et al. 1993). These data also showed a cooling during the Late Cretaceous followed by warming in the early Tertiary. This study was also able to show that the water column, which Lécuyer et al. (1993) estimated to be no deeper than 200 m, became thermally stratified during times of rapid temperature change and was isothermal during times of stable temperatures.

Many of the vertebrates from the Peterborough

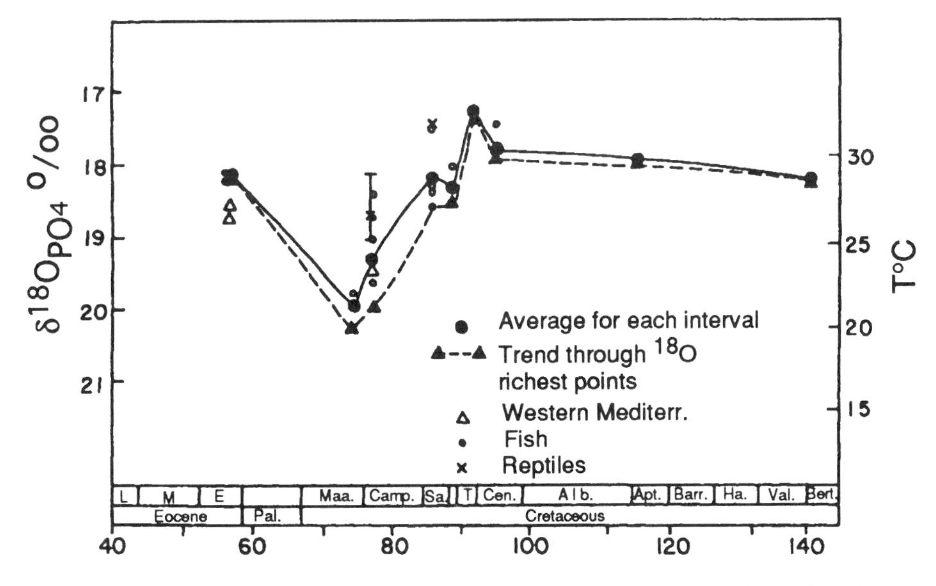

Figure 2.45. Oxygen isotope data and interpreted paleotemperatures from phosphate in the bones of Cretaceous and Eocene marine vertebrates in the Mediterranean region. *Solid line* connects average data from fish for each interval; samples from Israel. Reprinted from *Palaeogeography, Palaeoclimatology, Palaeoecology* 64, Kolodny, Y. and M. Raab, "Oxygen isotopes in phosphatic fish remains from Israel: Paleothermometry of tropical Cretaceous and Tertiary shelf waters," pp. 59–67 (1988) with kind permission of Elsevier Science-NL, Sara Burgerhartstraat 25, 1055 KV Amsterdam, The Netherlands.

Member of the Oxford Clay Formation (Jurassic; United Kingdom) studied by Martill et al. (1994) were analyzed for oxygen isotopes by Anderson et al. (1994), in conjunction with the study on Peterborough mollusks discussed previously. The results are shown in figure 2.46, where they are compared with paleotemperature estimates from the nektonic invertebrates (cephalopods). The values overall are consistent with those of Mesozoic marine fish and reptiles reported by Kolodny and Luz (1991). The estimated paleotemperatures for presumed deep- and shallow-dwelling forms are alike and overlap with those of the ammonite *Kosmoceras*, which gave the warmest estimated paleotemperatures among the mollusks (see figure 2.39; Anderson et al. 1994).

Anderson et al. (1994) rejected the possibility of vital effects to explain the similarity of deep- and shallow-dwelling forms because the results were similar across several taxa. They also rejected diagenetic alteration because such alteration occurs near the sediment-water interface in phosphatic material, but the isotopic compositions, if diagenetic, are consistent with diagenesis well after burial. They thus concluded that the paleoecological interpretations for depth-habitat preference were wrong and/or obscured by the mobility of the organisms.

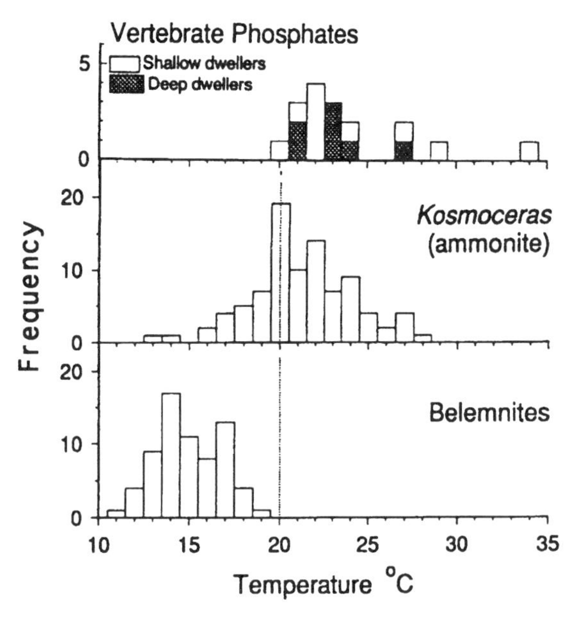

Figure 2.46. Estimated paleotemperatures from $\delta^{18}O$ in vertebrate phosphate in the Peterborough Member of the Oxford Clay Formation (Jurassic; United Kingdom) and comparison with results from nektonic invertebrates in the same unit. Note different frequency scales. From Anderson et al. (1994). *Journal of the Geological Society of London*, © 1994 The Geological Society.

The potential of the isotopic composition of cetacean bones to provide information about water temperature has also been investigated (Barrick et al. 1992). Whales are warm-blooded and thus present problems for the isotope geochemist that are different from those presented by cold-blooded animals because of the influence of body temperature. Barrick et al. (1992) started with the assumption that because whales have constant body temperatures, the isotopic composition of their bones should vary with seawater isotopic composition (rather than seawater temperature directly), even if fractionation occurred; bone $\delta^{18}O$ would simply be offset from seawater $\delta^{18}O$.

Fossil whale bone $\delta^{18}O$ was consistently higher than modern ones by 6‰. This offset is too large to be an ice-volume effect, so Barrick et al. (1992) attributed it to physiological differences and accepted the relative trends in bone $\delta^{18}O$ as real (figure 2.47). If the lowest values indicate no-ice conditions, the highest values are consistent with an increase in ice volume resulting in a 250 m sea-level drop, comparable to Pleistocene fluctuations. Barrick et al. (1992) acknowledged that this result does not fit current ideas about Miocene ice-volume fluctuations and thus this method is as yet not conclusive.

Marine Trace Fossils

Marine trace fossils are commonly used to interpret depth and general environment (e.g., Seilacher 1967; Ekdale and Berger 1978; Ekdale, Bromley, and Pemberton 1984; Droser and Bottjer 1988), but they are also useful for interpreting paleo-oxygenation levels in bottom waters

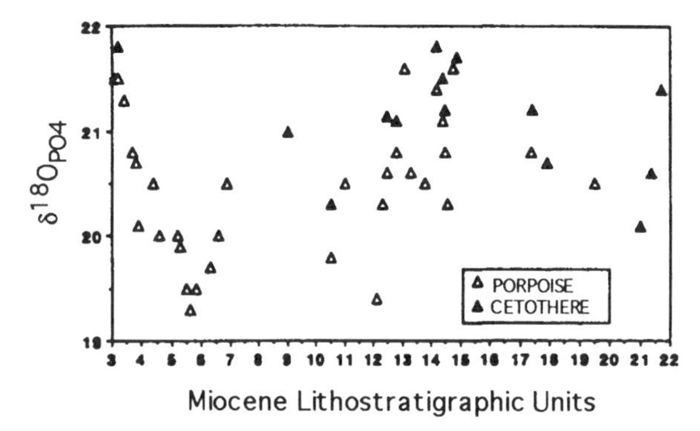

Figure 2.47. Variations in $\delta^{18}O$ of Miocene cetacean bone through a stratigraphic section of the Chesapeake Group (Maryland, U.S.). From Barrick et al. (1992).

(Rhoads and Morse 1971; Savrda, Bottjer, and Gorsline 1984; Savrda and Bottjer 1989). Savrda and Bottjer's (1989) model included not only the types of trace fossils occurring at each level, but also how the trace fossils are tiered, that is, the types of cross-cutting relationships that can be expected as sediments are deposited under changing oxygenation conditions over varying lengths of time and varying sedimentation rates. They also proposed a method for using trace fossils to reconstruct oxygenation levels (figure 2.48). Savrda and Bottjer (1989) applied this method to part of the Smoky Hill Member of the Niobrara Formation (Cretaceous; Western Interior seaway). In that section, the trace-fossil relationships appear to be linked directly to variations in organic carbon, although this was not true for other parts of the unit. Savrda and Bottjer (1989) accepted the interpretation that the chalk-marl cycles in this unit were controlled by variations in freshwater influx and water-column stratification (Pratt 1984; Pratt et al. 1993), but the trace fossils only indicate oxygenation levels and do not rule out the alternative explanation that anoxia and higher carbon accumulation were driven by elevated productivity (e.g., Van Os et al. 1994). Although they discussed a number of limitations to the method, for example, destruction of all burrows by complete bioturbation during prolonged periods of oxygenation, Savrda and Bottjer's (1989) is clearly the best method currently available for semiquantitatively assessing paleo-oxygenation levels.

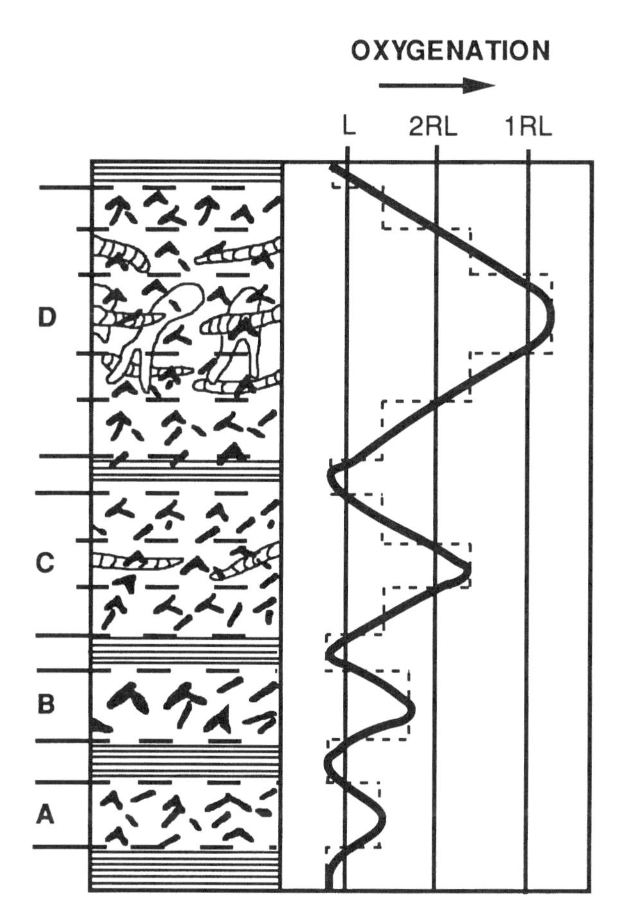

Figure 2.48. Method of reconstructing oxygenation levels from trace-fossil relationships. The geologic column on the left contains bioturbated beds separated by laminated beds representing anoxic bottom waters. The column on the right indicates interpreted oxygenation. L, threshold oxygenation level below which laminations are preserved; 2RL and 1RL, the threshold oxygenation levels above which type 2 and type 1 fossils are found. Reprinted from *Palaeogeography, Palaeoclimatology, Palaeoecology* 74, Savrda, C.E. and D.J. Bottjer, "Trace-fossil model for reconstructing oxygenation histories of ancient marine bottom waters: application to Upper Cretaceous Niobrara Formation, Colorado," pp. 49–74 (1989) with kind permission of Elsevier Science-NL, Sara Burgerhartstraat 25, 1055 KV Amsterdam, The Netherlands.

3

Marine Lithologic Indicators of Paleoclimate

The usefulness of marine plankton for paleoclimate interpretation does not end with the biogeography and chemistry of the organisms. Where biologic productivity is relatively high in the surface waters of the oceans and clastic sediment accumulation low, marine plankton can dominate marine sediments, resulting in characteristic lithologies. The distribution of these rocks, quite apart from the particular fossils that they contain, is useful for interpreting paleoclimate. For example, deep-sea biogenic sediments have been used to interpret nutrient fluxes in and between ocean basins and changes in CO_2 levels, and their distribution on the sea floor can help track zones of high biologic productivity. On the continental shelves, biogenic sediments are most commonly used to interpret temporal and spatial variations in biologic productivity. Finally, the geochemical signatures of some of these rocks, analyzed in whole-rock samples rather than individual fossils, are used in some studies.

Certain minerals in marine rocks also are useful for paleoclimate interpretation. These include manganese, clay minerals, barite, and a spectrum of iron-bearing minerals, which are useful for assessing redox (reduction-oxidation) states in marine waters. Certain sedimentological features of shallow-water clastic sediments provide clues about paleoclimate.

This chapter is divided into two sections, the first on biogenic sediments and the second on nonbiogenic sediments. A recurrent theme in the section on biogenic sediments is biologic productivity at the sea surface, which exerts a major control on sedimentation in the oceans (Summerhayes et al. 1995). The relationship between marine biologic productivity and climate is both direct and indirect. Spatial patterns of high biologic productivity have been predicted using atmospheric paleoclimate models (Parrish 1982; Parrish and Curtis 1982; Parrish 1983; Scotese and Summerhayes 1986; Kruijs and Barron 1990), so understanding how to recognize high-productivity deposits is important in evaluating the efficacy of the models. The distribution of organic carbon in marine sediments is also at least partly a paleoclimate problem, in both the supply (biologic productivity) and the preservation (oxygenation, which is partly related to temperature) of organic matter. Finally, the cycling of organic carbon, including production and burial, plays a major role in determining atmospheric levels of the greenhouse gas, CO_2. Nonbiogenic sediments are usually more direct as indicators of paleoclimate, although the information they contain is commonly distorted by other processes.

BIOGENIC SEDIMENTS

Biogenic sedimentary rocks include biogenic siliceous rock, chalk, phosphorite, organic-rich rock, and reefs and other shallow-water carbonate deposits. Biogenic siliceous rock and chalk are found on the continental shelves and in the deep sea. Phosphorites and reefs, on the other hand, are shelfal deposits. Today, marine organic-rich sediments are limited to the continental shelves and upper slopes, but in the past they were also deposited in the deep sea. Phosphorite and organic-rich rock commonly occur with either biogenic siliceous rock or chalk in continental-shelf upwelling-zone deposits.

Phosphorites and bedded biogenic siliceous rock are regarded as particularly diagnostic of high biologic productivity in the oceans. Some workers also regard chalk and organic-rich rock to be indicative of high productivity. Therefore a discussion is in order at this point of the causes of high biologic productivity and its relationship to climate.

Biologic Productivity in the Oceans

"Productivity" is a general term used by ecologists to describe the amount of biomass created per unit of time. Plant (including autotrophic protist) growth is primary productivity and is important for understanding the distribution of biogenic marine rocks. Changes in biologic productivity in the oceans result from changes in the supply of nutrients to the surface waters, where plants can utilize them. However, for productivity to be recorded in the geologic record as a productivity-related deposit, the nutrient supply must be high and must be constant or seasonally recurring over long periods of time, at least many tens or hundreds of years. In addition, the nutrients must be imported from outside the productive system. The reason for this is that biogenic sediments, especially phosphorite, remove nutrients from the system when they are deposited. If the deposits are to be preserved in the geologic record, that removal is continuous, requiring the importation of nutrients from elsewhere.

The deep ocean is a huge reservoir of dissolved nutrients, the most essential of which are nitrogen and phosphorus and, to a lesser extent, silicon and iron. Nutrients below the photic zone, that is, the zone of penetration of sunlight, cannot be used directly by animals and, of course, cannot be used by plants, either, because of the lack of solar energy. The highest-productivity regions in the oceans are where waters from below the photic zone are brought to the surface in vertical currents called upwelling currents or upwelling zones, and upwelling zones account for 67% of total global new production (Chavez and Toggweiler 1995). Upwelling zones occur in the open oceans and in certain settings on the continental shelves. Productivity is higher in general on the continental shelves than in the open ocean because the nutrients are easily recycled from the sediments by microbial regeneration and physical reworking and because the shelves are closest to nutrient sources on the continents. In upwelling zones, however, the products of high biologic productivity are exported in higher proportion to the sediments, requiring a constant input of nutrients from waters off the continental shelf.

Over geologic time scales, nutrient variability in the world ocean is controlled by river and eolian influx, and major events on land can influence the nutrient status in the oceans (e.g., Filippelli 1997). Some workers have argued that riverine influx of nutrients can be important at regional scales for the formation of high-productivity deposits. There are two problems with this hypothesis. First, rivers themselves have plants that utilize the nutrients as the waters flow downstream. Although rivers account for input of nutrients to the oceans over long time periods, it is questionable whether the influx of nutrients is great enough at river mouths over the time scales required to form high-productivity deposits. The second problem is that modern rivers do, in fact, export excess nutrients to the sea, but it has not been demonstrated that modern rivers are good analogues for rivers in the geologic past. All rivers today are very heavily influenced by human activity, including input of fertilizers, so it might be impossible to assess the actual influence of rivers in the past (Ruttenberg 1993; Ortner and Dagg 1995). Moreover, even the highly nutrient-rich waters from modern rivers affect productivity on a relatively local scale, at the river mouth and delta, so this mechanism is unlikely to account for large deposits observed in the geologic record. The large river that is least affected by humans, the Amazon, exports most of its phosphorus as particulate inorganic phosphorus, which is not immediately accessible for biological uptake (Richey and Victoria 1993). For these reasons, it is assumed for this book that river-borne nutrients were not directly responsible for the formation of local- or regional-scale high-productivity deposits in the geologic record, although the importance of rivers as the ultimate source of nutrients is not questioned.

Modeled global primary productivity is shown in figure 3.1. This pattern closely follows that of dissolved phosphate in surface waters (e.g., Raymont 1963). Primary productivity, which in this case is expressed in grams of dry matter added to

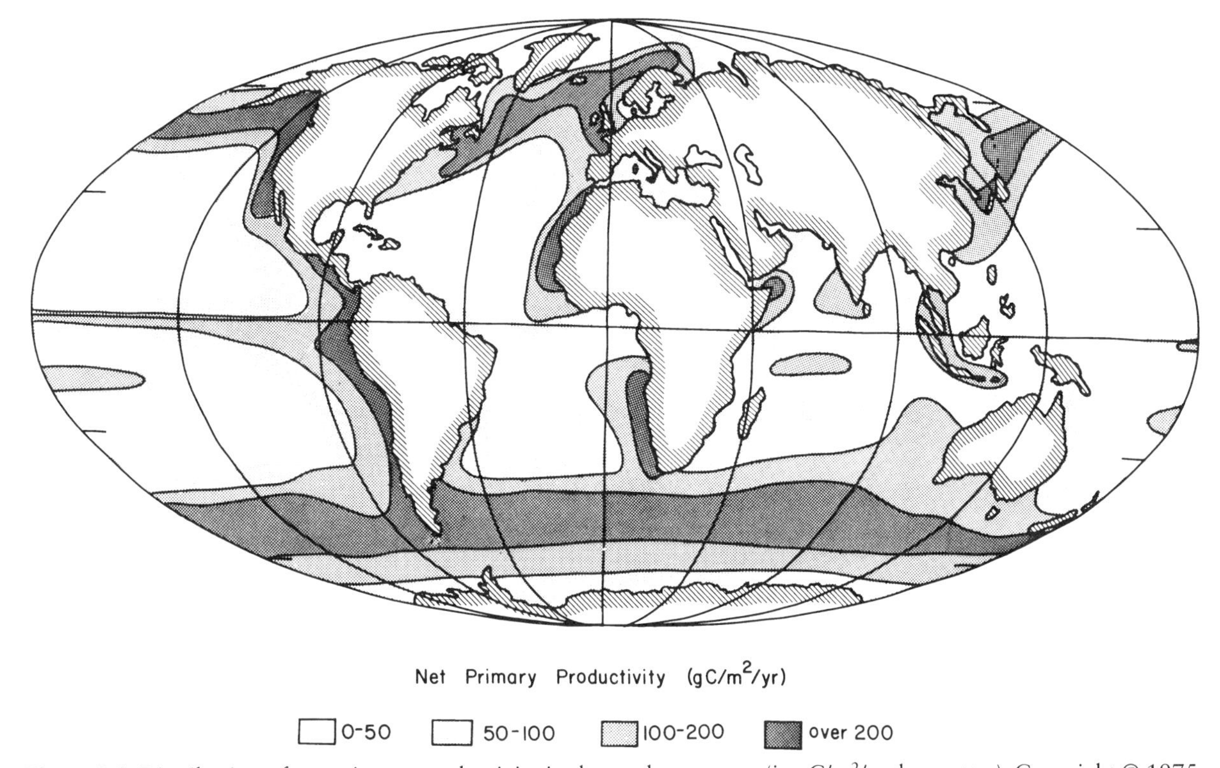

Figure 3.1. Distribution of net primary productivity in the modern oceans (in gC/m²/yr dry matter). Copyright © 1975. Modified by permission of Springer-Verlag New York from *Primary Productivity of the Biosphere,* H. Lieth and R.H. Whittaker, eds.

biomass per square meter per year, is very low over most of the oceans, as is the amount of dissolved phosphate, whereas productivity and plankton biomass are high where the amount of dissolved phosphate is also high. The areas of high biologic productivity are in three types of settings, coastal divergences, high-latitude divergences, and low-latitude divergences. These are all upwelling zones; the term "divergence" refers to the dynamics of the surface waters that allow upward-moving vertical currents.

To understand how divergences and upwelling are related, it is first necessary to briefly discuss the effect of wind on the surface of the ocean. Wind movement over the sea surface creates frictional drag on the water, which is the principal cause of surface waves. However, water molecules, like air molecules, are free to move in any direction and, because of the rotation of the Earth, do not move parallel to the wind. The deflection of the water from a path parallel to the wind is called the Coriolis effect. For the purpose of understanding its effects on water circulation, imagine that the upper few tens of meters are divided into several layers. The upper layer is set into motion by friction with the wind, friction from the upper layer sets into motion the layer below it, and that layer sets into motion the layer below it, and so on. With each incremental increase in depth, more frictional energy is lost, so that, at some depth, water motion can no longer be traced back to friction with the wind at the surface. As each layer is set into motion, the water in that layer is deflected by the Coriolis effect from the direction of the water in the layer above it. The vectors of motion, then, become increasingly deflected from the wind direction and also become shorter. This spiral of vectors is called the Ekman spiral. Adding up the vectors of motion attributable to wind friction results in a net transport of water (Ekman transport) perpendicular to the wind direction, to the right (facing downwind) in the Northern Hemisphere and to the left in the Southern Hemisphere. The depth of Ekman transport varies from as little as a few meters to as much as 1000 m, depending on the geographic conditions and the strength and steadiness of the wind, but it generally involves the upper 50 to 100 m of water.

Upwelling occurs under atmospheric low-pressure systems and in certain settings along the coastlines. The examples shown in figure 3.2

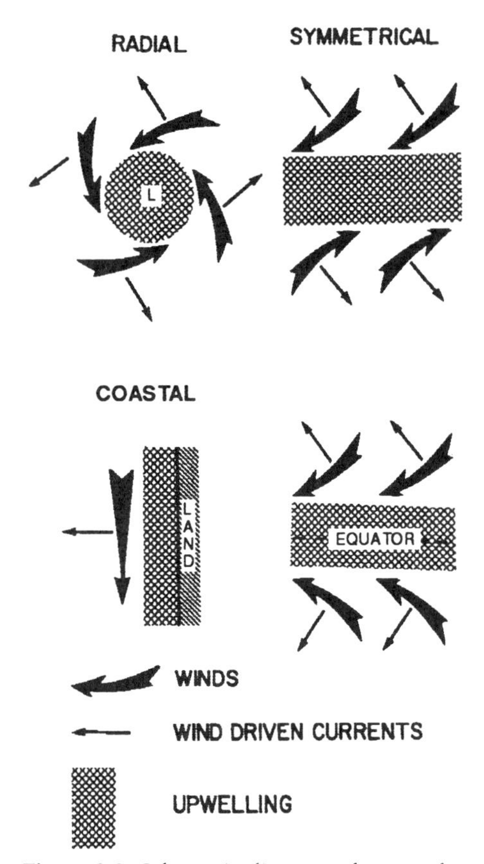

Figure 3.2. Schematic diagram of types of upwelling zones in the Northern Hemisphere and along the equator. Radial, symmetrical, and equatorial upwelling occur under permanent atmospheric low-pressure systems. From J. T. Parrish © 1982, reprinted by permission of the American Association of Petroleum Geologists.

apply to the Northern Hemisphere, where the Ekman transport is to the right of the wind (facing downwind) and to the equator. Because the Coriolis effect changes sign across the equator, in the Southern Hemisphere, all but the equatorial divergence would be mirror images of the ones illustrated in figure 3.2. Because the winds on both sides of the equator are easterlies, the equatorial divergence owes its existence to the change in sign of the Coriolis effect. Upwelling and elevated productivity occur even under storm-scale low-pressure cells, for example, hurricanes. However, the high productivity is transitory and not directly recorded in the geologic record. Upwelling that is likely to be recorded in the geologic record occurs under permanent low-pressure systems or in coastal regions. Today, permanent low-pressure systems are along the equator, in the northern Pacific and Atlantic (radial upwelling), and in the Southern Ocean (symmetrical upwelling; see figure 3.2). The effects of these systems on productivity in modern oceans can be seen by comparing figure 3.1 with figure 1.2.

Coastal upwelling that is persistent enough to affect the geologic record has three requirements. First, the winds must blow parallel to the coastline and in the right direction, with the coast on the left (facing downwind) in the Northern Hemisphere and the coast on the right in the Southern Hemisphere. Second, the winds must be relatively steady and strong for several days at a time for at least part of the year (Lentz 1992; Smith 1995). This is accomplished on the eastern limbs of the subtropical high-pressure cells, in the zonal winds, and in monsoonal regions (see chapter 1). Most of the coastal upwelling zones today are meridional upwelling zones on the west coasts of the major continents, driven by winds associated with the subtropical high-pressure cells (figure 3.3; see figure 1.2). However, both zonal and monsoonal upwelling zones also exist today. Figure 3.3 is extremely idealized, and the actual locations of the upwelling zones will depend on the vagaries of continental positions

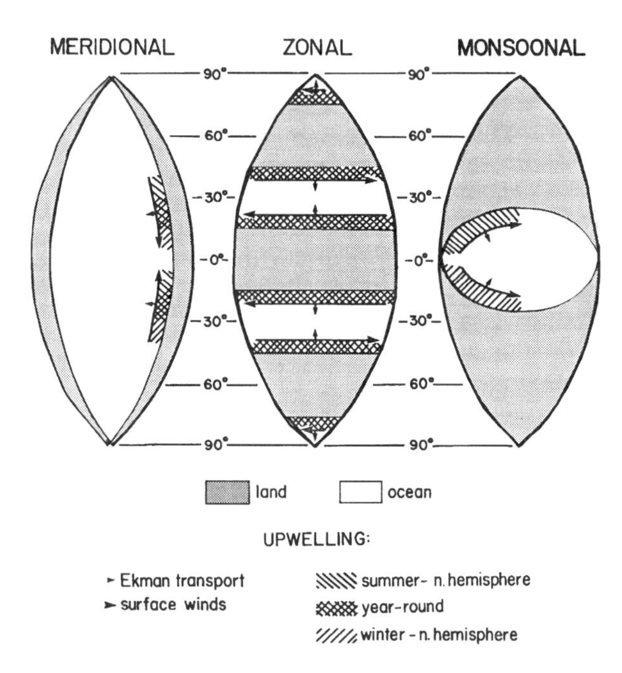

Figure 3.3. Schematic diagram of settings for coastal upwelling. Modern examples of meridional upwelling are California, U.S. and Peru; zonal upwelling, Venezuela; monsoonal upwelling, Somalia. Compare with figure 3.1. From From J. T. Parrish © 1982, reprinted by permission of the American Association of Petroleum Geologists.

and their effects on atmospheric circulation patterns (Parrish 1982).

The locations of ancient upwelling zones are predicted from atmospheric paleoclimate models using the wind predictions. For conceptual paleoclimate models, wind direction and information from present climates on where winds are likely to be steady are used to predict upwelling (Parrish 1982; Scotese and Summerhayes 1986). For numerical climate models, wind direction and calculated wind stress (a combination of strength and steadiness) are used to predict upwelling (Kruijs and Barron 1990).

Upwelling occurs at divergences because as water moves away from the center of the low pressure cell or from the coastline at the surface, it is replaced from below. If the depth from which this replacement water moves is below the photic zone, all other things being equal, that water will be much richer in nutrients than the water it replaces. If the flow is sustained, high biologic productivity will be sustained. For upwelling to occur, a separation between the outgoing flow at the surface and the incoming flow at depth must be maintained. This is not a problem in the open ocean, but it is relevant to upwelling on the continental shelves and, for the geologic record, in epeiric seaways. The oceanographic inner shelf, where the outgoing surface and incoming bottom waters interact, is where the ratio of the water depth to the depth of the vertical boundary layer is 0.4 to 4 (Lentz 1994; Smith 1995). The vertical boundary layer in mid-latitudes is typically 40 m (Smith 1995), so water depths for effective upwelling (Parrish 1982) are at least 16 m to as much as 160 m or more. A 16 m deep upwelling zone would probably not result in high productivity because 16 m is within the photic zone in many waters. However, that is a minimum, and at typical depths, particularly given the turbidity of water in high-productivity regions, upwelling can be effective for transporting nutrients into the photic zone even in relatively shallow waters.

It is important to emphasize that upwelling does not bring water from the deep sea; it is, and is regarded by oceanographers as, a surface-mixing process. In addition, upwelling is distinguished from the large-scale vertical circulation of ocean water that results in general oceanic overturn (see, for example, discussion in Reverdin 1995); upwelling is a much faster process. The two processes are connected in their effects on productivity, as will be seen in the section on biogenic siliceous sediments, but changes in regional-scale upwelling do not necessarily denote changes in oceanic overturn.

Another requirement for upwelling zones to have an effect on the geologic record is that the nutrient content of the upwelled water be high enough to sustain high biologic productivity. Seawater below the photic zone is richer in nutrients than sea-surface water, but the nutrient distribution below the photic zone is not uniform. Deep water accumulates nutrients as organisms in the photic zone are consumed or die and are transported to the deep sea in fecal pellets or by settling. The longer a water mass is below the photic zone, the richer it will be in nutrients. If water is upwelled not long after it is downwelled elsewhere, it will not be very productive. This is the case in the region off the coast of northwestern Africa (see figure 3.1), which is less productive than the upwelling zone off Namibia, for example. The upwelled water off northwestern Africa is low in nutrients because it is derived from waters downwelled not far away in the North Atlantic; that water does not spend enough time below the photic zone to accumulate significant concentrations of nutrients.

Although the largest upwelling zones in the modern oceans are wind driven, a few areas have upwelling zones that are not. These are dynamic upwelling zones and upwelling zones related to bathymetry. Dynamic upwelling zones are associated with strong western boundary currents, such as the Gulf Stream (Yentsch 1974; Pietrafesa 1990; Lee, Yoder, and Atkinson 1991), or with permanent oceanic gyres that flow counterclockwise in the Northern Hemisphere and clockwise in the Southern Hemisphere. These types of upwelling zones today are small and occur over the deep ocean, but they might have been important in the past. For example, the Miocene phosphorites of the southeastern United States have been attributed to dynamic upwelling, when the Gulf Stream flowed over the continental shelf at a highstand in sea level (Riggs 1984).

The effects of high biologic productivity on marine sediments are dependent on depth (e.g., Shimmield and Jahnke 1995). This is illustrated in figure 3.4, which is a map of percent total organic carbon (TOC) and modern phosphorites in surface sediments in the world ocean. Where productivity is high over the continental shelves (see figure 3.1), TOC tends to be high. Where

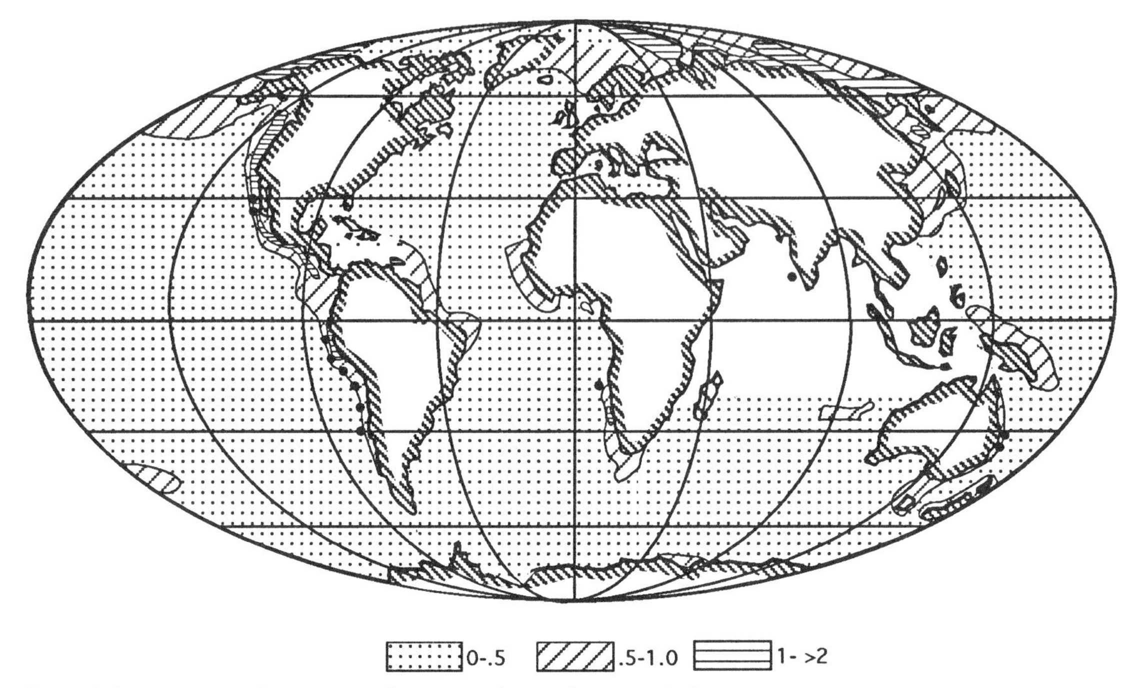

Figure 3.4. Percent total organic carbon in surface sediments of the oceans (*patterns*) and distribution of modern phosphorites (*dots*). Blank areas are no data. Compare with figure 3.1. From Lisitzin (1972) with additional data from Bremner (1983); modern phosphorites from Glenn et al. (1994a).

productivity is highest over continental shelves—off Namibia, southwestern North America, western India, and Peru—phosphorites are forming. The most productive upwelling zone is in the Southern Ocean, yet phosphorites are absent and organic carbon contents of the sediments low. This is because the organic tissue from plankton, which carries most of the organic carbon and phosphorus, is consumed and recycled in the water column before it can reach the sediments. Deep-sea biogenic siliceous and carbonate deposits, on the other hand, are composed of relatively stable material, and they may form in regions of moderate or even low productivity, depending on the depth of the water and, in the case of carbonate, the carbonate compensation depth (CCD). However, the accumulation rates in the deep sea still depend on productivity at the surface.

Biogenic siliceous sediments

Biogenic siliceous sediments are composed principally of diatoms, radiolarians, or, much less commonly, siliceous sponge spicules. Over time, most biogenic siliceous rocks undergo diagenesis to chert, and the original skeletal morphologies become obscured. The major chert-forming organisms have been important at different times in Earth history. Spiculites are the most common type in shelf and slope deposits from the Silurian through the Cretaceous, although these cherts tend to be nodular rather than bedded (Maliva, Knoll, and Siever 1989). Radiolarites and spiculites are common in deep-sea suites deposited during that interval. In contrast, from the Cretaceous onward, and especially from the Eocene, diatomites are the dominant type (Maliva, Knoll, and Siever 1989).

Although chert is common in sedimentary rocks of all types, bedded chert is limited to marine or lacustrine settings. In marine settings, bedded cherts, particularly those younger than Ordovician (Maliva, Knoll, and Siever 1989), are regarded as indicators of high biologic productivity (Jenkyns and Winterer 1982; Hein and Parrish 1987). The bedding itself may be diagenetic, and classic, evenly bedded "ribbon" cherts may be most typical of radiolarites (Murray, Jones, and Buchholtz ten Brink 1992). The presence of chert or siliceous oozes in deep-sea cores is also considered diagnostic of high productivity at the surface because, in modern deep-sea sediments, the distribution of siliceous oozes tracks the distribution of upwelling zones.

Although silica itself is rarely a limiting nutrient, biogenic siliceous sedimentation is dependent on productivity (Lisitzin 1972; Leinen 1979, figure 3.5). Therefore tracking the deposition of silica in the deep sea provides information about variations in nutrient fluxes within and among ocean basins (Berger 1970; Brewster 1980; Miskell, Brass, and Harrison 1985). Shifts in nutrient states among the oceans occur by a process called basin fractionation (Berger 1970). Waters flow from the source basin to the sink basin over long distances, accumulating nutrients on the way[1]. When waters in the sink basin are upwelled into the photic zone, productivity is high, resulting in accumulation of biogenic sediments and depletion of the surface waters, which are then returned to the source basin (Berger 1970). This process is illustrated schematically for silica in figure 3.6. Note the accumulation of silica in the sink basin and note that the concentration of nutrients just below the photic zone in the sink basin is higher than in the source basin. Today, the Atlantic Ocean is a source basin and the Pacific Ocean a sink basin, not just for silica but for other nutrients as well.

The basin fractionation signal is difficult to interpret from the geologic record. First, as little as 1% to 25% of the biogenic silica produced at the surface is deposited and preserved; the rest is dissolved before it reaches the sediment-water interface (Calvert 1974; Archer et al. 1993; Thunell et al. 1994). Second, biogenic silica is diluted by other sediments in many parts of the ocean, ei-

1. Some workers refer to sink basins as lagoonal, and source basins as estuarine, after similar shallow-water processes.

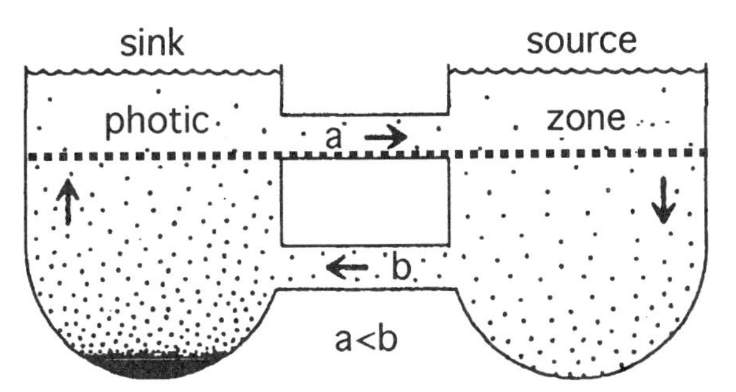

Figure 3.6. Schematic illustration of basin fractionation of silica. Density of *stippling* indicates gradients in silica distribution, and the schematic levels of exchange (a, b) indicate relative depths of exchange. The amount of silica supplied to the sink basin from the source is greater than that returned to the source basin, resulting in a net accumulation of silica in the sink basin. Note that both basins have about the same concentrations of silica in the photic zone. Modified from Berger (1970) in *Geological Society of America Bulletin*. Reproduced with permission of the publisher, the Geological Society of America, Boulder, Colorado, USA. Copyright © 1970 Geological Society of America, Inc.

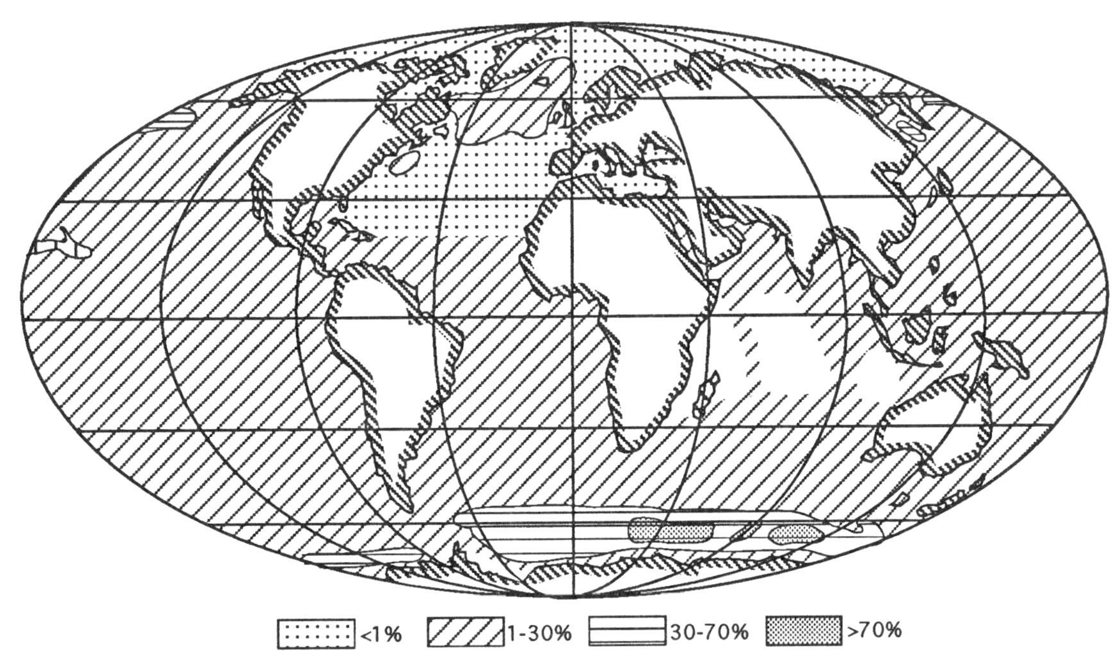

Figure 3.5. Map of the distribution of biogenic silica in surface sediments in the World Ocean as percent dry sediment. Blank areas are no data. Compare with figure 3.1. Simplified from Lisitzin (1972).

ther by carbonate above the carbonate compensation depth or by terrigenous sediment in the circumcontinental regions (Lisitzin 1972). Third, estimates of biogenic silica mass accumulation rates are difficult to make. The biogenic silica must be distinguished not only from nonsiliceous sediment but from nonbiogenic siliceous components as well, including detrital quartz, clay minerals, and volcanic ash (Leinen 1979; Brewster 1980; Miskell, Brass, and Harrison 1985). Finally, estimates for ocean basins through time must take into account the vagaries of sampling brought about by the changing area of total sea floor, of sea floor above the CCD, and the movement of the sea floor under the high-productivity zones (Ramsay 1973; Leinen 1979; Miskell, Brass, and Harrison 1985).

Several workers have addressed shifts in silica deposition through time (Leinen 1979; Brewster 1980; Keller and Barron 1983; Miskell 1983; Miskell, Brass, and Harrison 1985; Woodruff and Savin 1989), the most recent being papers by Barron and Baldauf (1989) and Baldauf and Barron (1990), who documented shifts in silica deposition during the Tertiary. In the early Paleogene and Eocene, latitudinal temperature gradients were low, and biogenic silica accumulation was concentrated in a broad band in the equatorial region. Equatorial biogenic silica accumulation rates were particularly high in the middle Eocene; silica accumulation also increased in the North Atlantic (the silica "burp" of McGowran 1989). Barron and Baldauf (1989) concluded that this was likely an artifact of preservation, as diatom and radiolarian test walls were much thicker at that time, but later they suggested that cooling of the bottom waters led to more vigorous equatorial circulation and high productivity associated with establishment of a Gulf Stream–type current (Baldauf and Barron 1990). Increasingly, biogenic silica accumulation began to shift to high latitudes, although it was still very widespread through the Oligocene and Miocene. In the Pliocene, the modern pattern of relatively geographically restricted upwelling regions was established. Overall, silica deposition increased through the Cenozoic (Miskell, Brass, and Harrison 1985), which Barron and Baldauf (1989) attributed to cooling at high latitudes.

Barron and Baldauf (1989) and Baldauf and Barron (1990) used presence-absence data, whereas other workers (Leinen 1979; Brewster 1980; Miskell, Brass, and Harrison 1985) attempted to calculate accumulation rates. The principal problem with presence-absence data is that changes in basin fractionation may be detectable only in the most extreme cases, where silica production shuts down almost completely in one region and starts up from virtually none in another. On the other hand, problems exist with accumulation rate data as well. Some are the analytical problems described above; a major problem is the location of sampling sites, which might or might not happen to be in the regions of highest biogenic silica production. Probably as a consequence of these problems, Miskell, Brass, and Harrison (1985) and Baldauf and Barron (1990) arrived at contradictory conclusions about the timing of the onset of North Atlantic bottom water production. Wright, Miller, and Fairbanks (1992) argued that neither presence-absence nor biogenic silica mass accumulation rates were reliable, proposing instead that changes in the preservation of delicate diatoms are the most appropriate indicator of changes in deep-water nutrient fluxes, such that enhanced preservation indicates increased nutrient fluxes. They noted that diatom preservation in the North Atlantic improved above the early-middle Miocene boundary, coincident with stable isotope indications of the onset in bottom-water production in the North Atlantic.

For the time being, trying to directly relate production of biogenic silica and abundance of biogenic silica in the geologic record, particularly in deep-sea sediments, may be futile (Archer et al. 1993), and it is instructive to review the reasons why. Archer et al. (1993) modeled preservation of biogenic silica, taking into consideration burial rates, flux rates through the water column (i.e., "rain" rates), pore-water concentrations, and sediment concentrations. A key element of their model was the observation that, in tropical oceans, most silica is dissolved before it reaches the sediment-water interface, so that burial is small compared to production. Thus the accumulation of biogenic silica in tropical deep-sea sediments is driven not just by the production and rain rates but by other factors that have nothing to do with silica production. One common method of determining variations in biogenic silica is to calculate the percent of biogenic silica in the sediments on a calcite-free basis. The purpose of this method is to distinguish changes in biogenic silica production from those of carbonate production, which is also high in tropical waters. The problem with this method is illustrated in figure 3.7. Calculated variations in biogenic silica across the carbonate lysocline show that in sediments deposited above the lysocline, the importance of silica is exaggerated if it is calculated on a calcite-free basis (see figure 3.7b), whereas a more accurate picture of total biogenic silica is obtained at all depths if percent biogenic silica of the total sediment is used. Archer et al. (1993) called this the calcite effect. Using mass accumulation rates (MAR) of biogenic silica creates problems for similar reasons. When the total sediment is composed of significant amounts of material other than biogenic silica, the silica MAR is correlated with total sediment accumulation rate, rather than with silica production, because most of the silica produced is dissolved before it reaches the sediment (see figure 3.7c). This contrasts with the situation in high-latitude oceans today, where most of the rain is biogenic silica and a much higher fraction of the produced silica is buried. Figure 3.7 implies that the best way of measuring biogenic silica production variations in tropical oceans is to calculate percent biogenic silica as a fraction of the total sediment (%$opal_{tot}$; see figure 3.7a). However, no systematic correlation exists between biogenic silica rain rates and %$opal_{tot}$. Archer et al. (1993) concluded that %$opal_{tot}$ is related to an as-yet unknown mechanism for systematic variation in the solubility of the biogenic silica with production, and that interpreting the geologic record of biogenic silica is at this point impossible. This gloomy conclusion refers to interpretation of the details of the record, that is, whether primary productivity can be calculated from biogenic siliceous sediments, and it does not negate the qualitative observation that abundant biogenic silica is related to high biologic productivity. However, the converse, that a lack of abundant biogenic silica means that production was low, is not true.

Chalk

Chalk is the name for the generally poorly lithified rock derived from calcareous oozes composed of coccoliths and/or foraminifera. Chalk is rare in the pre-Jurassic geologic record because these organisms did not become abundant in the oceans until then.

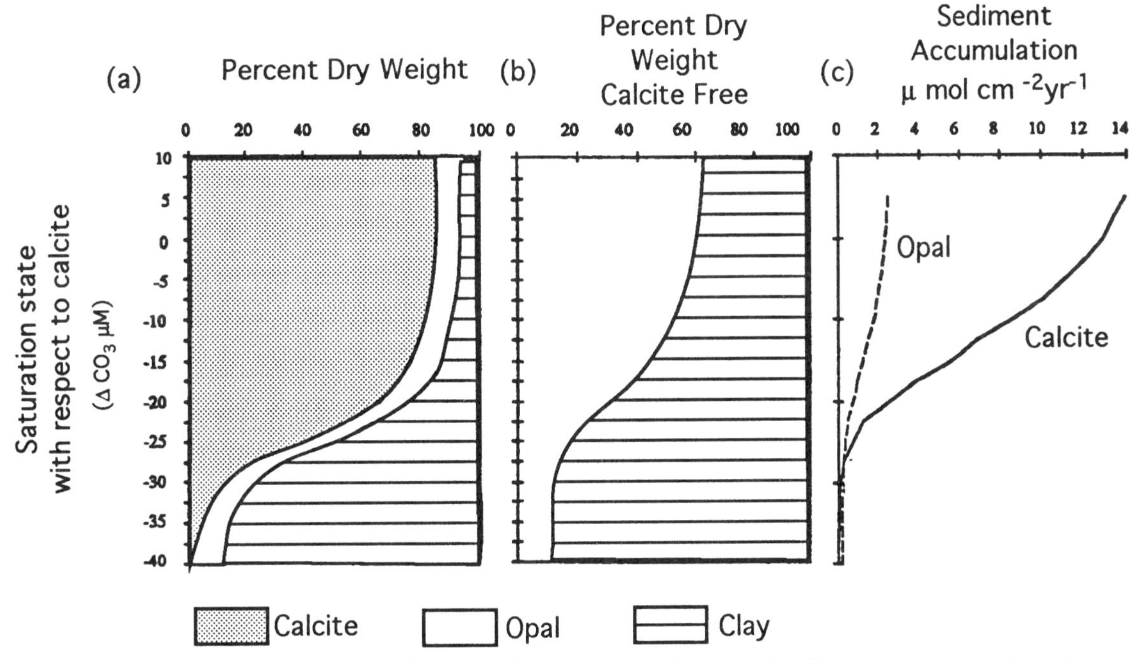

Figure 3.7. Model of the behavior of biogenic silica as part of the overall sediment rain across the carbonate lysocline. Rain rates and dissolution kinetics are constant except for the saturation state of calcite, which changes with depth. As part of the total sediment rain, biogenic silica maintains a relatively constant percentage, even as calcite disappears from the rain *(a)*. However, if calcite is eliminated from the calculations, biogenic silica appears to be a much larger component of the initial sediment rain *(b)*. In modern tropical oceans, where most of the sediment rain is calcite, biogenic silica accumulation rates are correlated with total sediment accumulation rates *(c)*. From Archer et al. (1993) *Paleoceanography* 8:7–21, copyright by the American Geophysical Union.

The distribution of calcium carbonate in modern deep-sea sediments is shown in figure 3.8. Calcium carbonate is high in low latitudes, between about 50°S and 40°N. This is where productivity of the calcareous plankton matches or exceeds that of siliceous plankton, and the pattern is related to temperature and/or slightly lower nutrient contents, although it is important to note that the distributions of carbonate sediments and siliceous sediments (see figure 3.5) are not strictly antithetical. Calcium carbonate concentrations are also high in relatively shallow parts of the oceans. The most striking examples are along the Mid-Atlantic Ridge and among the islands of the southeastern Pacific Ocean; calcium carbonate there is clearly related to bathymetry. The sea floor in these regions is above the CCD (Berger 1968). Although chalk is commonly referred to as a high-productivity deposit, the limitations of the CCD are the reason chalk is not studied as much as biogenic siliceous deposits for variations in biologic productivity. However, it is the relationship to the CCD that makes studying the distribution of chalk useful for understanding other types of paleoceanographic processes, particularly changes in ocean chemistry related to the carbon cycle.

Ocean chemistry and the CCD. The depth of the CCD through time is determined from ocean drilling records. The procedure includes determination of the age of sediments that lack calcite and calculation of the depth and location of the sea floor at each locality through time. The location of each sample site is established by backtracking crustal plate motions. Depth is determined by backtracking subsidence rates using the relationship described by Sclater, Hellinger, and Tapscott (1977) and other workers, wherein the depth of the sea floor is related to $1/t^{-2}$, where t is the age of the sea floor determined from paleomagnetic and/or biostratigraphic information.

Conceptually, the relationship between the CCD and oceanic and atmospheric chemistry is relatively simple. The depth of the CCD depends on the amount of dissolved CO_2 in the bottom waters because dissolved CO_2, in combination with water, creates a slightly acidic environment: $CO_2 + H_2O \leftrightarrow H^+ + HCO_3^-$. Waters at the surface are relatively low in total dissolved CO_2 (tCO_2). As the water sinks and ages in the deep ocean, it gains CO_2 from oxidation of organic matter; this combines with the carbonate ion (CO_3^{2-}) to form bicarbonate, HCO_3^- (Berger and Spitzy 1988). Put another way, as CO_2 from the oxidation of organic matter increases with depth and age of the water, CO_3^{2-} ions, which are re-

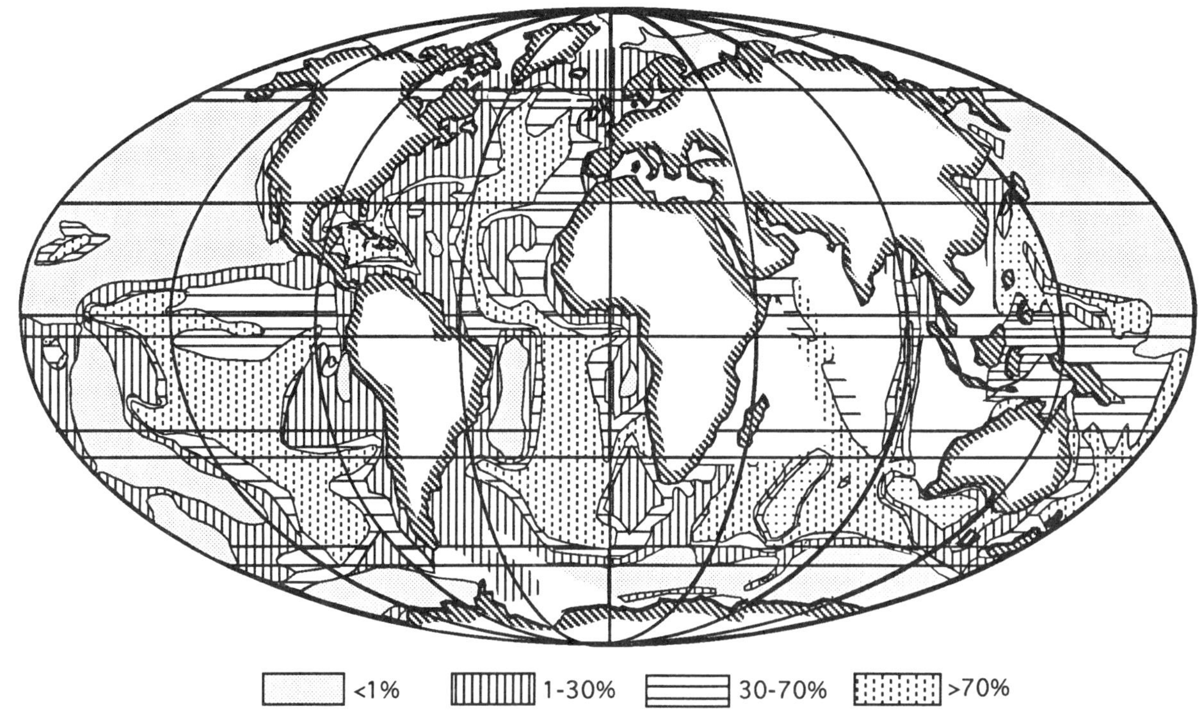

Figure 3.8. Map of the distribution of calcium carbonate in surface sediments in the World Ocean as percent dry sediment. Blank areas are no data. Simplified from Lisitzin (1971).

leased by dissolution of calcite ($CaCO_3$), decrease because they react with the CO_2 (Broecker and Peng 1982). This causes more dissolution of calcite, so that as tCO_2 increases, the CCD rises. This is part of the reason why the CCD is at different depths in different ocean basins. In some basins (the Atlantic), bottom waters are relatively young, so they are not so corrosive. In others (the Pacific), bottom waters are older, and the CCD higher. It is this observation that led workers (e.g., Berger 1970) to propose the notion of basin-basin fractionation (discussed in the section on biogenic silica) and of basin-shelf fractionation (van Andel 1975) to explain overall changes in the level of the CCD.

The partial pressure of CO_2 (pCO_2) in the atmosphere is partly dependent on tCO_2 in the oceans, and variations in atmospheric CO_2 through time are of great interest to paleoclimatologists. Not surprisingly, the distribution of organic carbon and carbonate are sometimes antithetical (e.g., Thierstein 1979). However, those two constituents are not controlled by the same factors. Increases in organic carbon indicate higher productivity, higher terrestrial supply, and/or lower dilution by other constituents (Thierstein 1979), whereas increases in carbonate result from lower dissolution or higher production of carbonate. The partial pressure of CO_2 in seawater, which is the relevant parameter for exchange with the atmosphere, is dependent on tCO_2, total alkalinity, temperature, salinity, and pressure (Berger and Spitzy 1988; Keir 1988; Heinze, Maier-Reimer, and Winn 1991; Berger et al. 1993; Heinze and Broecker 1995).

Chalk cycles. Chalk deposits commonly are cyclic. These cycles have been of particular interest to geologists as possible evidence for Milankovitch cyclicity in the pre-Quaternary record (e.g., Barron, Arthur, and Kauffman 1985; Schwarzacher and Fischer 1982; Fischer 1986, 1993; Research on Cretaceous Cycles Group 1986; and many others). The cycles can occur in a number of styles (Einsele 1982; Research on Cretaceous Cycles Group 1986). Cycle styles include productivity cycles (e.g., Van Os et al. 1994), redox cycles (coupled with clastic dilution at time of salinity cap; e.g., Pratt 1984), and scour cycles, resulting from sea level changes or storms (Research on Cretaceous Cycles Group 1986). Limestone-marl cycles have also been attributed to diagenesis (e.g., Eder 1982; Walther 1982) and to productivity/dissolution cycles related to changes in the CCD (e.g., Weber, Wiedicke, and Riech 1995).

Figure 3.9 shows a very simple model of a stratigraphic sequence grading from marl to limestone. In this case, the frequency of the clastic input (what might be called clastic influx events) remains the same, but the magnitude of clay input (accumulation rate) varies through time. The possible combinations of beds if the frequencies and amplitudes of the environmental factors vary and the cycles are not regular is staggering (figure 3.10). Simply varying one factor at a time can result in a wide variety of bedding types and thicknesses. A common pattern in such

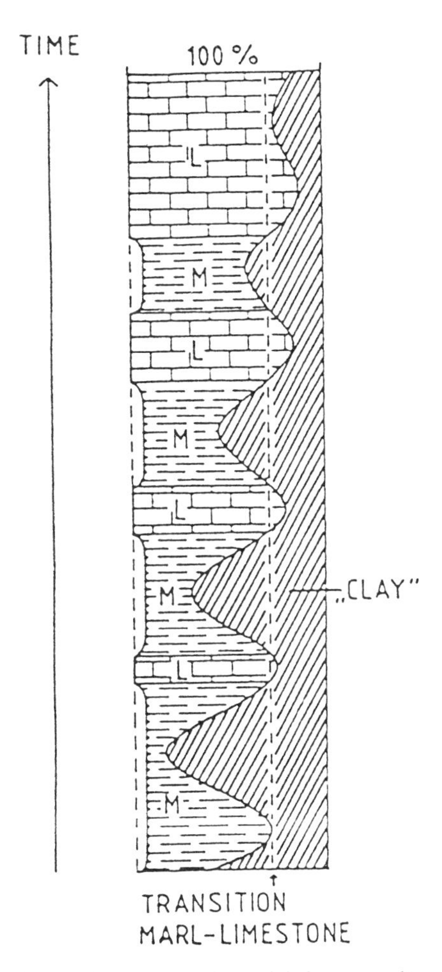

Figure 3.9. General model for a cyclic transition from a pure shale section to a pure limestone section with decreasing clay input. Note that the frequency of the cycles does not change, only the amplitude. Copyright © 1982. Reprinted by permission of Springer-Verlag New York from *Cyclic and Event Stratification,* G. Einsele and A. Seilacher, eds.

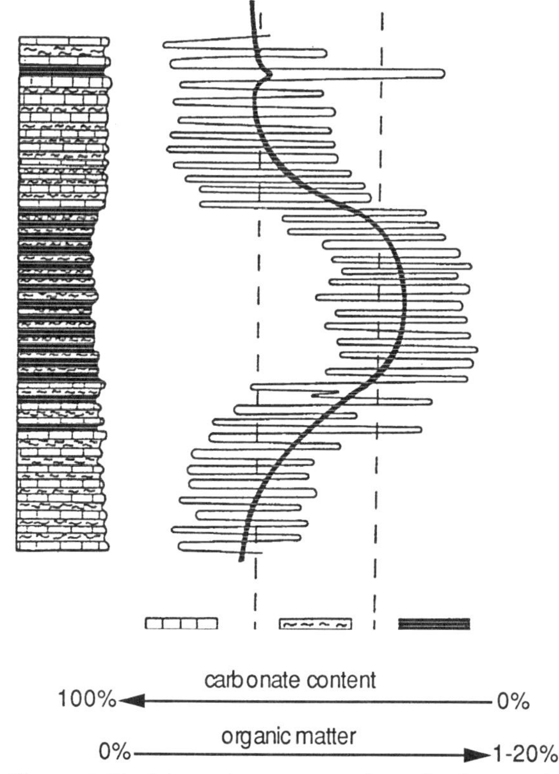

Figure 3.10. Schematic representation of pelagic sediments showing the result of superposition of two slightly irregular cycles. Copyright © 1982. Reprinted by permission of Springer-Verlag New York from *Cyclic and Event Stratification*, G. Einsele and A. Seilacher, eds.

sequences is called bundling, in which, for example, the limestone beds become thicker and then thinner again in a more or less regular cyclic fashion (figure 3.10).

The marls in limestone-marl cycles are commonly organic rich, and the literature is divided on how the chalk and organic-rich layers are related. One interpretation is that the cycles are redox cycles. For example, Pratt (1984) interpreted chalk-marl cycles in the Western Interior seaway (Cretaceous; United States) as periodic freshwater influx leading to stable stratification and deoxygenation of the bottom waters for the marl beds, alternating with overturn and high productivity of calcareous plankton for the chalk beds. A major line of evidence was the oxygen isotope data, which showed heaviest values in the chalk and lightest values in the marls. This pattern is opposite to that seen in most chalk-marl sequences (e.g., Ditchfield and Marshall 1989), and Pratt (1984; Pratt et al. 1993) attributed the light $\delta^{18}O$ values in the marls to the postulated low-salinity cap. The elevated total organic carbon (TOC) contents in the marls were attributed to low levels of oxygen in the bottom waters. Cycles in isotope records are discussed further in the section on stable isotope records in chalk.

In contrast, Van Os et al. (1994) concluded, for Pliocene sediments in the eastern Mediterranean, that the chalk beds were the result of moderate productivity, and the alternating black, organic-rich beds were the result of high productivity of nontestate plankton that out-produced the calcareous plankton (see Tribovillard et al. 1994a for more on the importance of organic-walled plankton). This conclusion was supported by the high TOCs (which they regarded as a priori evidence for high biologic productivity), high barium contents, and great abundance of siliceous plankton and of the opportunistic foraminiferan *Globorotalia puncticulata* in the black beds.

An inverse relationship between TOC and productivity indicators appeared in the data of Williams and Bralower (1995) in Early Cretaceous chalk cycles in the North Sea; data were whole-rock isotopes, abundances of certain taxa (notably the coccolith *Biscutum constans*), total abundances of nannofossils, and contents of $CaCO_3$ and TOC (figure 3.11). Increases in total nannofossil abundance (measured as number of specimens per gram of sediment) and in the abundance of *B. constans* were interpreted as indicative of higher productivity. This was supported by higher $\delta^{13}C$ of the whole-rock carbonates and lower $\delta^{18}O$, which suggests that the productive waters were cooler, as would be expected in an upwelling system. Williams and Bralower (1995) did not comment on the inverse relationship of these high-productivity indicators to TOC, but if Van Os et al.'s (1994) concept applies, the high-productivity indicators described by Williams and Bralower (1995) are relevant only to the coccolith-dominated beds, and the coccoliths were outproduced by a nontestate organism when the high-TOC beds were deposited.

Carbonate accumulation, although indicative of elevated productivity, is not dominant in the highest-productivity environments (e.g., Van Os et al. 1994; Verardo and McIntyre 1994; Weber, Wiedicke, and Riech 1995). However, studies that explicitly compare the deposition of biogenic siliceous sediments and chalk are lacking. This is surprising because both biogenic siliceous rock and chalk are regarded as products of high bio-

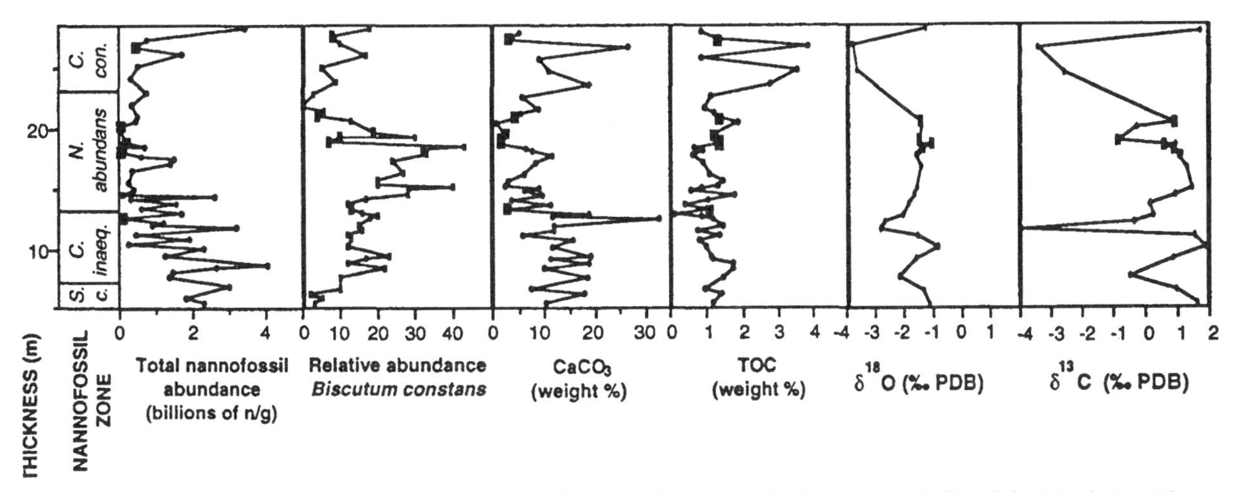

Figure 3.11. Nannofossil assemblage, geochemical, and isotope data for Early Cretaceous chalks of the North Sea. The nannofossil *Biscutum constans* was interpreted as an indicator of high biologic productivity. From Williams and Bralower (1995) in *Paleoceanography* 10:815–39, copyright by the American Geophysical Union.

logic productivity, but such studies would be hampered by the problems raised by Archer et al. (1993). The picture that emerges from the literature is that calcite-secreting organisms are indicative of moderate to high biologic productivity and, possibly, warm water (Verardo and McIntyre 1994), whereas silica-secreting organisms will outproduce calcite-secreting organisms where nutrient levels are very high and/or waters are cold.

Stable isotope records in chalk. Despite its fine grain size, chalk is resistant to diagenetic processes that can alter the isotope record, especially that of $\delta^{13}C$ (Jarvis et al. 1988b; Pratt et al. 1993; Jenkyns, Gale, and Corfield 1994). Because chalk is generally made up of many taxa of coccoliths, one can assume that vital effects of various taxa are averaged out (Jenkyns, Gale, and Corfield 1994; but see Williams and Bralower 1995). In addition, if a chalk is still chalk, that is, if it has not undergone enough cementation and diagenesis to alter the chalky texture, it has not undergone enough diagenesis to significantly alter the isotopic composition (Hudson 1977). Whole-rock carbonate analyses are common on chalk, and some interesting paleoclimate information has come from them. Perhaps one of the most intensely studied events, revealed in $\delta^{13}C$ analyses from chalk worldwide, is a significant shift in $\delta^{13}C$ of 2‰ to 4‰ at the Cenomanian-Turonian (Late Cretaceous) boundary. First described by Scholle and Arthur (1980), that event has been the object of many geochemical, paleontological, and stratigraphic studies since (Pomerol 1983; Leckie 1985; Brumsack and Thurow 1986; Arthur, Schlanger, and Jenkyns 1987; Elder 1987; Schlanger et al. 1987; Arthur, Dean, and Pratt 1988; Bralower 1988; Jarvis et al. 1988a,b; Corfield, Hall, and Brasier 1990; Thurow et al. 1992; Gale et al. 1993; Glancy et al. 1993; Pratt et al. 1993; Hasegawa 1997; and many others).

One of the most detailed studies of isotope data from that interval was by Jenkyns, Gale, and Corfield (1994). They discussed not only the $\delta^{13}C$ shift, but also the $\delta^{18}O$ curves from the same localities. A striking feature of their high-resolution $\delta^{13}C$ curves from East Kent, England, and Gubbio, Italy, is how closely the curves resemble each other (figure 3.12), good evidence that the events controlling the $\delta^{13}C$ signal were at least continent wide. The Cenomanian-Turonian event has been interpreted as a time of widespread production and burial of organic carbon in the oceans, which would preferentially remove ^{12}C from the carbon cycle (e.g., Arthur, Schlanger, and Jenkyns 1987; Schlanger et al. 1987; Jarvis et al. 1988b). Burial of carbon in the sediments would lead to a drawdown of CO_2 in the atmosphere and global cooling, which would appear as increasingly heavy $\delta^{18}O$ values. Jenkyns, Gale, and Corfield's (1994) data show this response clearly; $\delta^{18}O$ values are at a minimum at the Cenomanian-Turonian boundary and then increase after the $\delta^{13}C$ event. This implies a warm pulse associated with the event and cooling following the event, which might be expected if CO_2 levels were lowered by carbon burial.

Positive $\delta^{13}C$ shifts have been observed in other

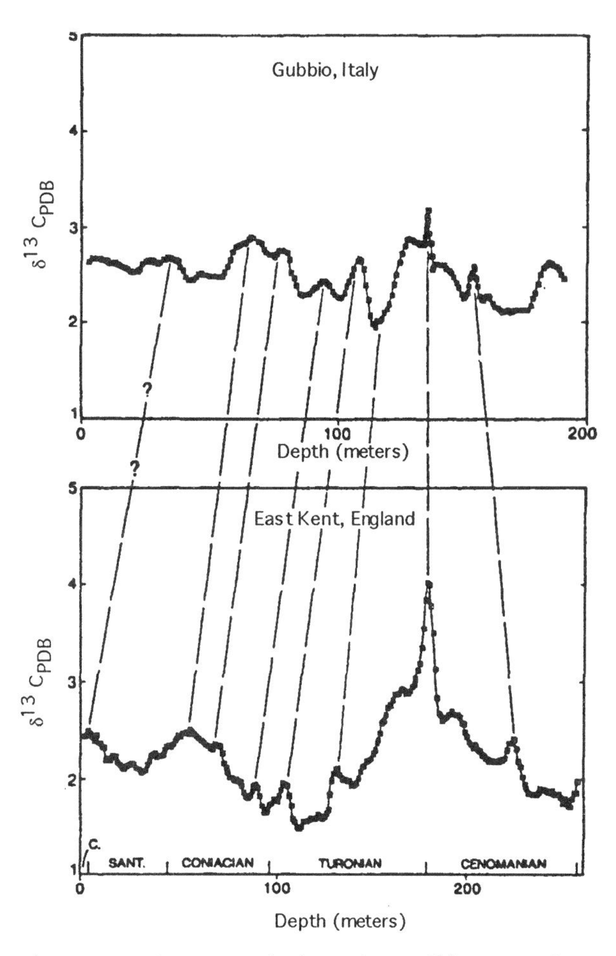

Figure 3.12. Composite, high-resolution $\delta^{13}C$ curves from Gubbio, Italy, and East Kent, England. Curves smoothed with five-point running average. Stages based on macrofossils from the English Chalk, and correlation with Gubbio was accomplished with the inflections in the $\delta^{13}C$ curves. From Jenkyns, Gale, and Corfield (1994) in *Geological Magazine,* Copyright © 1994 Cambridge University Press. Reprinted with the permission of Cambridge University Press.

parts of the record known for widespread organic-carbon burial, for example, Miocene (Vincent and Berger 1985) and Toarcian, Jurassic (Jenkyns and Clayton 1986). A shift in $\delta^{13}C$ toward lower values, opposite to that at the Cenomanian-Turonian boundary, Toarcian, and Miocene was identified at the Cretaceous-Tertiary boundary (Keller and Lindinger 1989; Zachos, Arthur, and Dean 1989). This event also has been recorded worldwide and has been attributed to a decrease in biologic productivity, so that less ^{12}C was buried than normal. An alternative explanation for this event would be weathering of a large reservoir of isotopically light organic carbon. However, the Cretaceous-Tertiary boundary isotopic shift is accompanied by decreases in barium and an increase in $\delta^{13}C$ in benthic foraminifera, both of which support decreasing productivity. Weathering of organic carbon reserves would not decrease barium input to the sea floor. In addition, benthic foraminiferan $\delta^{13}C$ signatures would be expected to parallel $\delta^{13}C$ of the chalk, which represents surface-water isotopic composition, if an organic carbon reservoir were weathered. Instead, the increase in $\delta^{13}C$ indicates reduced productivity and consequent lack of transport of ^{12}C to the sea floor. Although this event was widespread, the collapse of productivity did not occur everywhere. The production of biogenic silica—diatoms and radiolarians, especially spumellarians—increased dramatically in what is now the North Island of New Zealand, starting at the Cretaceous-Tertiary boundary (Hollis, Rodgers, and Parker 1995).

Phosphorites

Phosphorite is rock that contains a significant amount of phosphate, usually as carbonate fluorapatite, a complex and variable mineral ($(Ca_{10-a-b-c}Na_aMg_b(PO_4)_{6-x}(CO_3{\cdot}F)_y(SO_4)F_2$, where $x = y + a + 2c$ and c is the number of Ca vacancies; Nathan 1984), but sometimes as other forms (e.g., vivianite, $Fe_3(PO_4)_2{\cdot}8H_2O$; Berner 1990). Carbonate fluorapatite is usually cryptocrystalline and takes many forms in phosphorite: peloids and ooids; nodules of various irregular shapes and sizes; crusts; beds; and cements (Parrish 1987; Bremner and Rogers 1990; Glenn 1990). Today, phosphorite is forming in upwelling zones (see figure 3.4), and it is generally regarded as a good indicator of upwelling.

Plants and animals concentrate phosphorus because it is vital to metabolic processes, and organic matter is rich in phosphorus. When organic matter is sedimented and begins to break down, phosphate ions are released. If the bottom waters are oxygen depleted, phosphorite will precipitate from the phosphate-rich waters in regions of organic sedimentation (Ingall and Jahnke 1994). Thus phosphorites are commonly associated with organic-rich rocks, biogenic siliceous rock, or, less commonly, chalk. The association of organic matter, phosphorite, and biogenic siliceous rock is considered diagnostic of high biologic productivity (e.g., Cook 1976; Parrish, Ziegler, and Humphreville 1983).

In this section, it is important to understand that there are two scales of deposition of phosphate in sediments. Disseminated removal of phosphate plays by far the largest role in the phosphorus cycle (Ruttenberg 1993; Ruttenberg and Berner 1993). This is why most areas are at most moderately productive and why continental shelves are, in general, more productive than the open ocean. Phosphate is also removed at high rates in certain localized environments. Removal of phosphate at this scale, as phosphatic grains and beds, requires a continuous supply of phosphorus from outside the local area. This is why upwelling is especially compelling as an explanation for these rocks.

Phosphorites have been deposited in a variety of settings—on seamounts and carbonate islands, on continental margins, and in epeiric seas—and the consensus is that nearly all are related to upwelling in some fashion (Glenn et al. 1994a; Föllmi 1996). The most familiar and common settings are continental-margin and, in the past, epeiric-sea upwelling zones (Parrish 1990). Continental-margin upwelling zones where phosphorites are forming today are off Namibia, Peru, and Baja California, Mexico (see figure 3.4). These deposits, particularly those in Namibia and Peru, have been the subject of numerous studies (Veeh, Calvert, and Price 1974; Burnett 1977; Burnett and Veeh 1977; Soutar, Johnson, and Baumgartner 1981; Jahnke et al. 1983; Thomson et al. 1984; Bremner and Rogers 1990; Burnett 1990; Glenn 1990; Glenn et al. 1994b). Sedimentation rates are typically high in the area of persistent, strong upwelling; Burnett (1990) reported several estimates ranging from 0.028 to 1.3 cm/y in the Peru upwelling zone. Much lower sedimentation rates are calculated for the sediments that contain phosphorite. Burnett (1990) attributed this difference to sediment mixing and winnowing in the phosphate zone (see section on high-productivity facies associations), which concentrates the phosphatic grains (see also Compton et al. 1993). Phosphatic crusts in upwelling zones can grow as fast as 1.3 cm/1000 y (Burnett 1990). A few continental-margin phosphorites have been disputed as upwelling deposits, most notably the iron-rich phosphorites from the East Australian shelf, although seasonal upwelling occurs there as well (O'Brien and Veeh 1983; Heggie et al. 1990; O'Brien et al. 1990).

Insular phosphorites result from phosphatization of limestone, and the prevailing view is that the mechanism is leaching of accumulations of bird and bat droppings (guano) by meteoric waters (e.g., Stoddart and Scottin 1983; but see Bourrouilh-Le Jan 1980, cited in Glenn et al. 1994a). These deposits are all Quaternary or younger in age and in a setting that is unlikely to be preserved, particularly in the older record. Seamount and guyot phosphorites and phosphate-rich manganese crusts in the Pacific are as old as Eocene (Glenn and Kronen 1993; Hein et al. 1993; Glenn et al. 1994a). Their formation has been linked to the oxygen minimum zone associated with equatorial upwelling.

Variations in the distribution of phosphorites in time and space. The distribution of phosphorites through time is shown in figure 3.13, expressed in tonnes and as number of deposits. The data are plotted in both ways to illustrate one of the more striking features of the history of phosphorite deposition—the dominance of one deposit in the Permian, the Phosphoria Formation (McKelvey, Swanson, and Sheldon 1952), which accounts for almost all the estimated tonnage of phosphate at that time. The data represent economically significant deposits, a bias that has been pervasive in studies that attempt to understand the distribution of phosphate (indeed, many chemical sediments) in space and time. This bias is important because it does not account for the deposition of smaller amounts of phosphorite or disseminated phosphate (Föllmi 1993, 1995).

Although few dispute the importance of phosphorites for indicating upwelling, which is related to winds, many workers have argued that the variation through time (see figure 3.13) also indicates other processes, some of which are paleoclimatic. For example, Arthur and Jenkyns (1981) considered changes in the carbonate compensation depth, carbon mass accumulation rates, silica production, depth, fluctuating redox conditions, and sea level, none of which was sufficient to explain all aspects of the pattern, although in the interval they emphasized (the Mesozoic), phosphogenesis occurred during warm climates. In contrast, Sheldon (1980) called on sea-level change and large-scale glaciation to explain the episodicity of phosphogenesis through the Phanerozoic. Parrish (1983, 1990) concluded that the distribution of phosphorites is mostly simply dependent on the distribution of upwelling zones over continental shelves. The distribution of phosphorites is statistically correlated with wind-driven upwelling zones predicted from climate models for much of the Phanerozoic and for all the major phosphogenic episodes except the Ordovician and Neogene. The lack of statistical correspondence in the Neogene was the result of the deposits in the southeastern United States, which were deposited in a dynamic upwelling

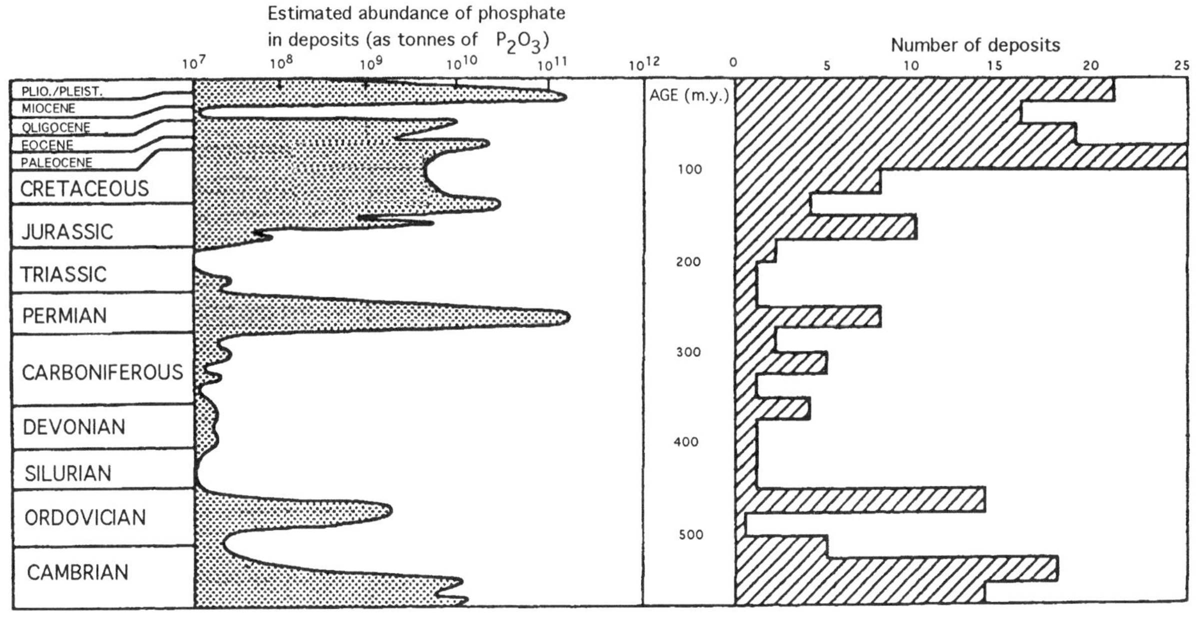

Figure 3.13. Distribution of phosphorites through time, in 25 Ma increments, expressed as metric tons P_2O_5. These are economically significant deposits and do not necessarily reflect fluctuations in total phosphate deposition through time (see figure 3.15; Föllmi 1993). From Cook and McElhinny (1979), with modification for mid-Cretaceous from Föllmi (1993). Reproduced from *Economic Geology,* 1979, Vol. 74, p. 316.

zone (Riggs 1984). Parrish (1990) also found that phosphogenesis during the three glacial phosphogenic periods, the Cambrian, Permian, and Neogene, was largely in meridional upwelling zones (see also Sheldon 1980) but was unable to relate glaciation to variations in meridional wind strength.

Variations in sea level have also been a recurring theme in attempts to model phosphogenesis (Sheldon 1980; Arthur and Jenkyns 1981; O'Brien et al. 1990; Föllmi et al. 1994), but not in any sense directly related to climate (i.e., glacio-eustatic changes). Because most phosphorites form at shelfal depths (<500 m deep), the general argument is that phosphorites are more likely to be deposited when sea level is high because there is more room for them to form (Glenn et al. 1994a), an argument that applies to any sediment that forms in shallow water. Cook and Shergold (1986) and Parrish (1990) argued that no single factor or combination of factors can explain why phosphorites are more abundant at some times than at others. However, it is worth repeating that these and similar studies concentrated on economically significant deposits, and that this skews the arguments about variations in phosphorus accumulation through time.

Global phosphorus accumulation rates. In examining the distribution of economically significant phosphorites, Glenn et al. (1994a) concluded that modern phosphorites are good analogues for ancient ones. They referred to the episodicity of phosphorite deposition noted earlier (see figure 3.13) and commented that today appears to be a phosphogenic episode. Filippelli and Delaney (1992) also concluded that accumulation rates in modern and ancient phosphorites were similar (figure 3.14). In the last few years, however, much more attention has turned toward phosphorus accumulation rates in the deep sea, partly in an attempt to avoid the bias that occurs from concentration on economically significant deposits. The rationale is that much more phosphorus is deposited in nonphosphorite sediments than in phosphorites. For example, Froelich et al. (1982) estimated that 10% or less of global phosphorus ends up in phosphorites, and Yanshin and Zharkov (1986, cited in Glenn et al. 1994a) estimated that the total phosphate in all Phanerozoic phosphorites is a tiny fraction of that buried in disseminated forms. Phosphorus accumulation in the deep sea presumably tracks more faithfully the overall behavior of the phosphorus budget.

In deep-sea cores from the equatorial Pacific, site-specific patterns included a strong relationship between phosphorus accumulation rates

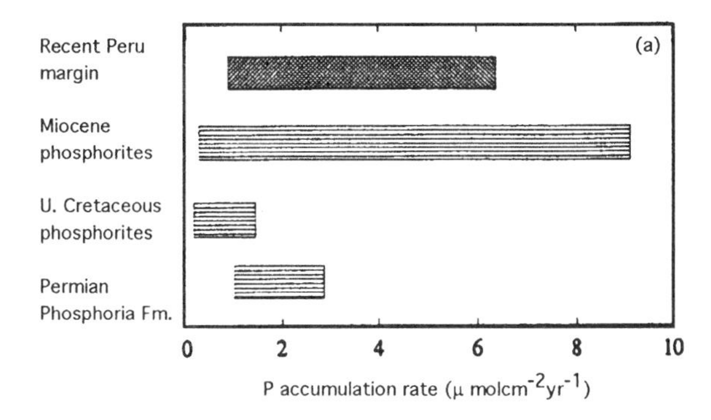

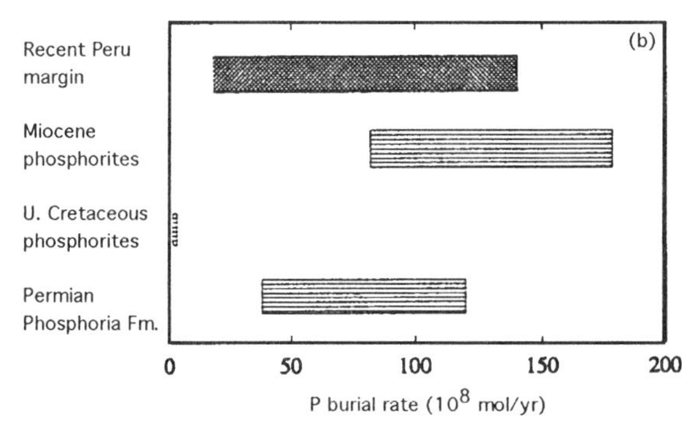

Figure 3.14. Phosphorus accumulation *(a)* and burial *(b)* rates for modern and ancient phosphorites. The Miocene phosphorites include the Monterey Formation (California, U.S.) and Hawthorne and Pungo River Formations (Florida and North Carolina, U.S.). The Upper Cretaceous phosphorites include several deposits in the Middle East. From Filippelli and Delaney (1992) in *Geology.* Reproduced with permission of the publisher, the Geological Society of America, Boulder, Colorado, USA. Copyright © 1992 Geological Society of America, Inc.

(PAR) and sediment mass accumulation rates, and between PAR and surface productivity patterns, with PAR increasing as surface productivity increased (Delaney and Filippelli 1994; Filippelli and Delaney 1994). As expected, depth was also a major influence on PAR, with higher rates in shallower sites. General patterns dominate, for example, a productivity maximum in the late Miocene (6 to 5 Ma).

Removal of disseminated phosphate is balanced by riverine and eolian influx over geologic time, but Delaney and Filippelli (1994) found no relationship between PAR and weathering-flux indicators ($^{87}Sr/^{86}Sr$, Ge/Si, $^{187}Os/^{186}Os$). The weathering-flux indicators indicate increased riverine flux through time, which should have resulted in an increase in phosphorus flux. They concluded that phosphorus cycling must somehow be decoupled from that of the other elements, including carbon. Föllmi (1995), using a much larger data set, also found little correlation with $^{87}Sr/^{86}Sr$, nor did he find a correlation with $\delta^{18}O$. There was a weak correlation with $\delta^{13}C$ and a stronger, though complex, correlation with sea level. The relationship with sea level was as follows: rapid increases in PAR were followed by rises in sea level—at about 160 Ma, 140 Ma, 100 Ma, and 65 Ma—until 32 Ma (figure 3.15), when the pattern was reversed. He linked the early pattern to changes in sea-floor spreading rates, reasoning that increased spreading rates would result in an increase in CO_2 and climatic warming, which in turn would lead to greater weathering rates and influx of phosphorus. The increased PAR would precede the rise in sea level because climate would respond quickly to increased volcanogenic CO_2, but it takes much longer for enough new ocean crust to form to raise sea level. The pattern from 32 to 0 Ma recorded cooling and increased influx of phosphorus resulting from mechanical and chemical weathering by glaciers (Föllmi 1995). Thus in Föllmi's (1996) scheme, phosphorus influx can increase with either climatic warming or cooling, just through different pathways. Despite the uncertain aspects of the work on PAR (Delaney and Filippelli 1994; Filippelli and Delaney 1994; Föllmi 1995), testable hypotheses about climate and the geochemical connection between the continents and oceans have been formulated.

Stable isotopes in phosphorites. Shemesh, Kolodny, and Luz (1983) investigated the possibility of obtaining paleotemperatures from phosphorites, which should be recorded in the phosphate-ion oxygen. They argued that phosphorites are less subject to diagenesis than carbonates and thus should provide more reliable results. A comparison of $\delta^{18}O$ determinations from chert, limestone, and phosphorite in a Cretaceous unit in Israel supported this contention. Data from Paleozoic rocks were generally consistent with $\delta^{18}O$ values obtained from conodonts (see figure 2.32), and trends were similar to those in $\delta^{18}O$ from cherts and carbonates.

More recently, Shemesh, Kolodny, and Luz (1988) and McArthur and Herczeg (1990) showed that the phosphorus-oxygen bond is weaker than Shemesh, Kolodny, and Luz (1983) had thought, although phosphorites are no more subject to diagenesis than carbonates. McArthur and Herczeg (1990) suggested a method of obtaining valid paleotemperatures from even altered phosphorites by defining alteration trends

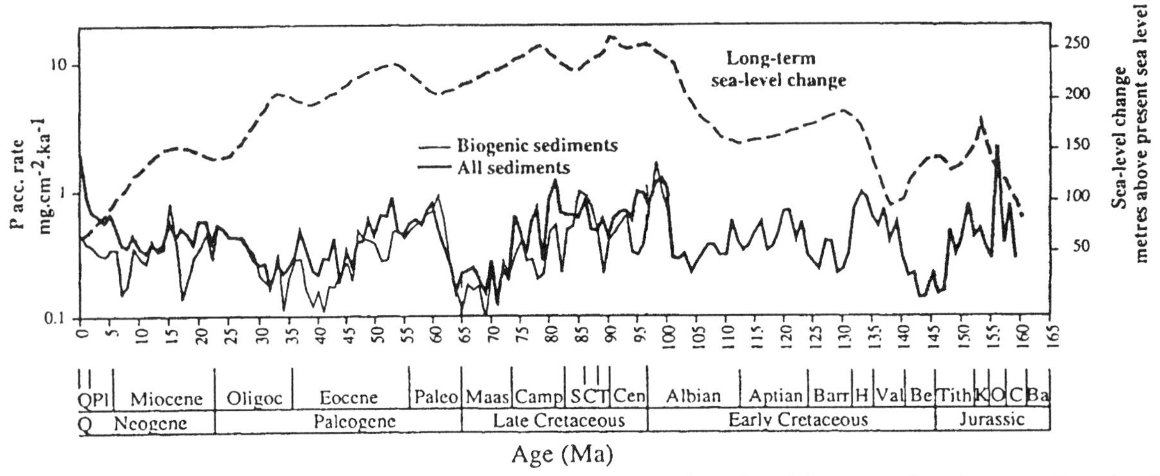

Figure 3.15. Geometrical mean phosphorus accumulation rates of total and biogenic phosphorus and sea-level change for the last 165 Ma. From Föllmi (1995) in *Geology.* Reproduced with permission of the publisher, the Geological Society of America, Boulder, Colorado, USA. Copyright © 1995 Geological Society of America, Inc.

and extrapolating the trends to the point of zero alteration, using Sr/Ca characteristic of minimally altered, offshore phosphate. McArthur and Herczeg (1990) also pointed out that the effects of diagenetic alteration would be greater at higher latitudes because the difference between $\delta^{18}O$ in seawater and $\delta^{18}O$ in meteoric waters increases with latitude.

Organic matter and organic-rich rocks

The presence and distribution of marine organic-rich rocks in the geologic record have been subjects of intense scrutiny, not least because they are major source rocks for petroleum. Understanding the controls on the distribution of source rocks has been important because of the potential of that information to help lead to new petroleum reserves. In a now-classic paper, Demaison and Moore (1980) described several environments in which source rocks might be found (table 3.1). The common element among all of them was the presence of anoxic bottom waters, but the authors' ideas on the generation of anoxia included paleoclimatic processes, including upwelling, influx of freshwater to enclosed seas, and so-called oceanic anoxic events (OAEs), of which the Cenomanian-Turonian boundary event, discussed earlier in this chapter, was only one.

Since Demaison and Moore's (1980) paper appeared, the literature has been dominated by references to anoxia in connection with the formation of organic-rich rocks, to the point that the origin of the organic matter has almost been

Table 3.1.
Marine anoxic basins, their paleogeographic and paleoclimatic settings, and their stratigraphic distributions

Anoxic basin type	Paleogeographic setting	Stratigraphic distribution of anoxic sediments
Anoxic silled basins	*Temperate to warm, rainy.* Intracratonic pockets on shelves.	*Variable.* Tends to be richest at bottom of basin or pocket.
Anoxic layers with upwellings	Oceanic shelves at *low latitudes.* West side of continents.	*Often narrow trends.* But can be widespread. Phosphorites. Diatomites.
Anoxic open ocean	Best developed at times of global warm-ups and major transgression.	Very widespread with little variation. Often synchronous worldwide.

Modified from G. J. Demaison and F. T. Moore © 1982, reprinted by permission of the American Association of Petroleum Geologists.

overlooked. Since about 1990, however, a number of workers have questioned the role of anoxia. This has resulted in controversy and two end-member positions, (1) that organic-rich rocks are a priori evidence for elevated carbon influx, or (2) that organic-rich rocks are a priori evidence of elevated carbon preservation under anoxic waters. This "production versus preservation" controversy (Parrish 1995) has focused particularly on rocks rich in marine organic matter, that is, rocks that are both organic and hydrogen rich (Demaison 1991). The paleoclimatic implications of these models are quite different. Elevated influx of hydrogen-rich organic matter in marine systems implies elevated productivity, which has implications for wind directions and strength and, possibly, weathering on the adjacent continents. In contrast, anoxia independent of elevated productivity implies stagnation of bottom waters, requiring a stably stratified water column, which has implications for freshwater influx, oceanic overturn, and/or temperature-controlled oxygenation levels. In this model, preservation of organic carbon is enhanced to the point where carbon influx rates are unimportant.

The "type" modern anoxic basin is the Black Sea (e.g., Glenn and Arthur 1985). Organic-rich sediments are found in cores below anoxic bottom waters. Higher C_{org} accumulation rates in unit B of figure 3.16a were attributed to a rise in sea level, spillover of saline Mediterranean waters into the freshwater Black Sea, increased turnover rates and elevated productivity, and onset of anoxia (Glenn and Arthur 1985). The conclusion of past (as well as present) anoxia was apparently based solely on the presence of the organic carbon and on sulfur data that Glenn and Arthur (1985) acknowledged was likely faulty and that has since been discredited (Calvert and Karlin 1991; Lyons and Berner 1992). Although many developments in geochemistry have transpired since Glenn and Arthur's (1985) paper appeared (e.g., see Arthur et al. 1994), it was an important paper and is still cited by those adhering to the "anoxia-only" school of thought.

More recent work indicates that the water column in the Black Sea was oxygenated during deposition of the most organic-rich layers: Calvert (1990) used Mn/Al and I/Br ratios, high values of which indicate oxygenated conditions (manganese, for example, is soluble in anoxic waters). Both ratios are higher in the organic-rich unit C

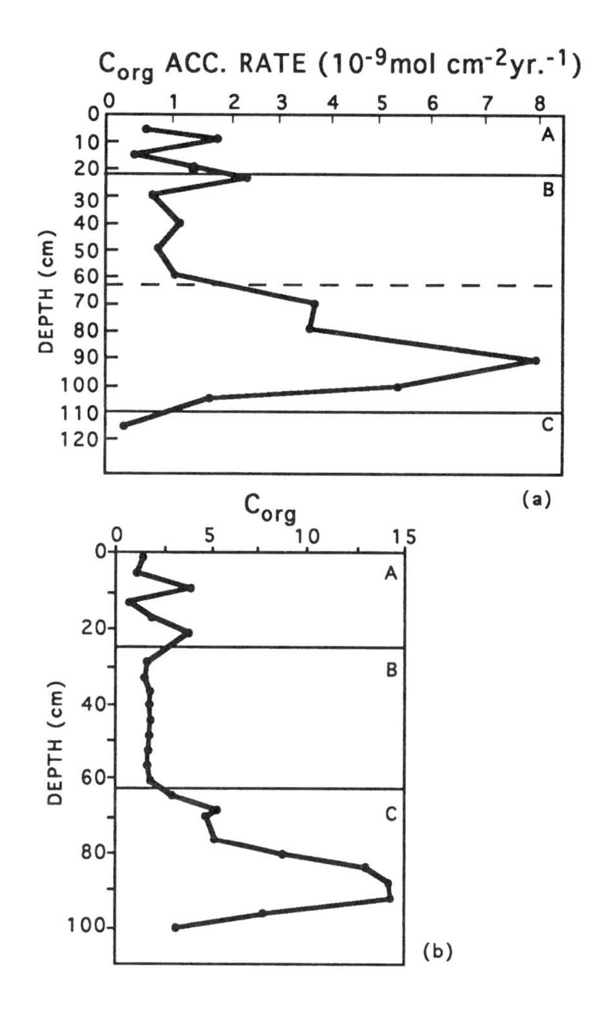

Figure 3.16. Organic carbon accumulation rates with depth *(a)* (from Glenn and Arthur 1985) and percent organic carbon with depth *(b)* (from Calvert 1990) in sediment from core 1432 in the Black Sea abyssal plain, within the anoxic zone. Note that Glenn and Arthur (1985) and Calvert (1990) differed in their definitions of units B and C; the *dashed line* on (a) marks the boundary between units B and C as defined by Calvert (1990). Reprinted from *Chemical Geology* 48, Glenn, C.R. and M.A. Arthur, "Sedimentary and geochemical indicators of productivity and oxygen contents in modern and ancient basins: The Holocene Black Sea as the "type" anoxic basin," pp. 325–54 (1985) with kind permission of Elsevier Science-NL, Sara Burgerhartstraat 25, 1055 KV Amsterdam, The Netherlands (a). Copyright © 1990, Reprinted by permission of Springer-Verlag New York from *Facets of Modern Biogeochemistry,* V. Ittekkot, S. Kempe, W. Michaelis and A. Spitzy, eds. (b).

(see figure 3.16b), indicating oxygenated conditions. Calvert (1990) therefore concluded that the higher organic accumulation rates were in oxygenated waters. He also noted that surface sediments in the Black Sea are organic rich largely by virtue of input of terrestrial organic matter, not by elevated preservation of organic matter produced in the basin (Calvert et al. 1991; see also Arthur et al. 1994). Using radiocarbon ages on the organic matter, Calvert et al. (1991) showed that the ratio of carbon production to carbon accumulation is linearly related to sedimentation rate (Müller and Suess 1979), and that only 0.7% to 2.1% of produced organic carbon in the Black Sea was preserved, not a particularly high preservation rate, and lower than the 2% to 4% estimate of Glenn and Arthur (1985). Calvert (1990) explicitly concluded that the elevated carbon accumulation rates were the result solely of elevated productivity, not onset of anoxia (see also Arthur et al. 1994).

The argument for enhanced preservation of organic matter under anoxic waters has been extended to upwelling zones. For example, Dean, Gardner, and Anderson (1994) argued that the higher TOC and hydrogen-rich organic matter (high $S_1 + S_2$) in laminated sediments from northern California, U.S. (figure 3.17) was evidence for enhanced preservation under anoxic conditions in an overall highly productive system[2]. One possible explanation for the change in the $S_1 + S_2$ curve was simply a change in the type of organic matter, but $\delta^{13}C$ shows that was not the case (the variations in $\delta^{13}C$ in figure 3.17 are insignificant in this regard). They also based the conclusion of enhanced preservation partly on the assumption that "the laminated and bioturbated sediments represent conditions of lower and higher dissolved oxygen, respectively," despite the fact, which they noted, that anoxia does not inevitably lead to laminated sediments. They also acknowledged that the higher levels of biogenic silica (figure 3.17) indicate an increased influx of diatoms. The increases in biogenic silica are direct evidence that the laminated intervals represent higher productivity than the adjacent beds. This alone could account for both the anoxia and the higher values of hydrogen-rich organic matter. They (Gardner, Dean, and Dartnell 1997) later concluded that anoxia was not the controlling factor in organic sedimentation.

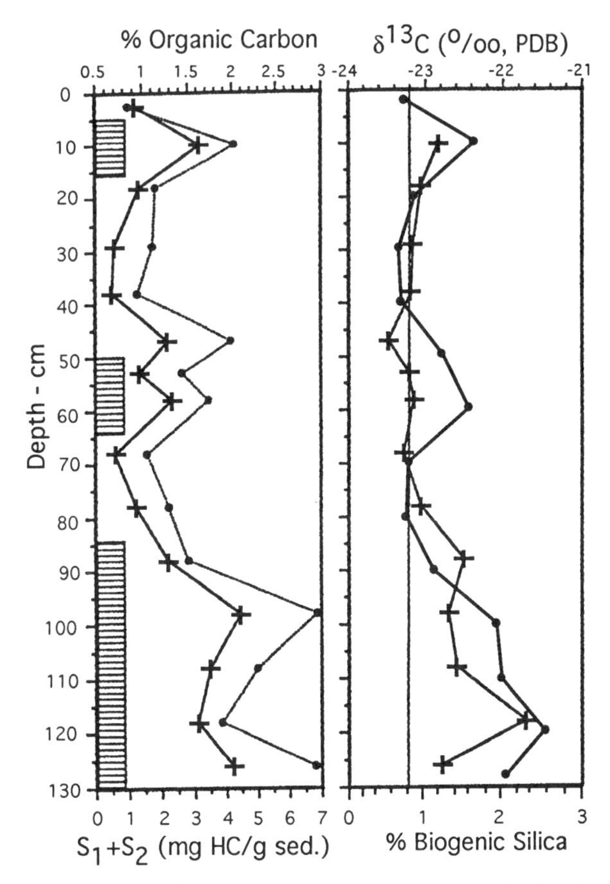

Figure 3.17. Distribution of laminations (*left, hatchures*), total organic carbon (*left, crosses*), hydrogen-rich organic matter ($S_1 + S_2$; *left, dots*), $\delta^{13}C$ of the carbon (*right, crosses*), and biogenic silica (*right, dots*) in a core from the northern California shelf. From Dean, Gardner, and Anderson (1994) in *Paleoceanography* 9:47–61, copyright by the American Geophysical Union.

Parrish (1995) reviewed much of the evidence for and against the production and preservation models and reanalyzed data on the distribution of organic-rich rocks in light of current ideas about the evidence for anoxia, in particular the presence of hydrogen-rich organic matter. The reanalysis showed a very high correlation (93%) between the distribution of upwelling zones predicted from climate models and hydrogen-rich, organic-rich rocks. She concluded, based on her

2. One of the ways petroleum geologists characterize organic matter is by the results of Rock-Eval, a method that combusts organic matter and measures, among other things, the following parameters: S_1, the amount of hydrocarbons already present; S_2, hydrocarbons generated from organic matter in the rock; and the ratios of hydrogen and oxygen to carbon (H/C and O/C, also called the hydrogen and oxygen indices, HI and OI). How organic matter falls on an HI/OI plot is used to determine the type of organic matter and provides clues about its condition. For example, organic matter with very low HI and OI values is considered to be completely degraded, either by biotic or thermal processes. Minimally degraded marine organic matter, derived from phytoplankton, typically has moderate to high HI and moderate to low OI and is considered hydrogen rich (see, for example, Tissot and Welte 1984).

reanalysis, based on the studies reanalyzing Black Sea data cited in this section, and based on studies showing no relationship between organic-carbon content and anoxic bottom waters (e.g., Pedersen and Calvert 1990; Calvert and Pedersen 1992, 1993; Calvert, Bustin, and Pedersen 1992; Calvert et al. 1995), that the weight of the evidence was toward productivity as the principal control on the formation of marine, organic-rich, hydrogen-rich rocks. Additional work on modern sediments has supported this conclusion, especially for upwelling zones (Jahnke and Shimmield 1995).

At the beginning of this section, it was noted that organic-rich rocks are taken by the "preservation" school as a priori evidence for anoxia. However, no evidence exists for the role of anoxia in organic sedimentation independent of the presence of laminated, organic-rich rocks in the geologic record. Although laminated rocks usually (but not always) indicate anoxia, they are not evidence that anoxia was the control on organic sedimentation.

Fine-grained, laminated rocks. The least-disputed evidence for anoxia or dysoxia in the geologic record is the presence of fine-grained, finely laminated rocks. These rocks are regarded as evidence for anoxia based on the argument that the laminations could not survive the bioturbating action of a benthic fauna and that exclusion of the benthic fauna is most easily explained by anoxic bottom waters (Rhoads and Morse 1971; Savrda, Bottjer, and Gorsline 1984; Sageman, Wignall, and Kauffman 1991). Quite apart from the issue of the relationship of anoxia and organic accumulation, it is useful in paleoclimate studies to be able to determine paleo-oxygenation levels because they can indicate climate-controlled stable stratification, high salinities, or warm temperatures.

An important consideration is whether the laminations are primary, detrital sedimentary structures. Examples of laminated rocks in which the lamination is not primary are those in which the laminations consist of compacted fecal pellets (Cuomo and Bartholomew 1991; Parrish and Gautier 1993) or thin, abundant shells of flat clams (Parrish 1987; see figure 4 in Littke et al. 1991). In addition, not all laminated, fine-grained mudstones were deposited under anoxic conditions. For example, if sediment influx is rapid, the resultant turbidity and soft substrates can prevent colonization by benthic organisms (Leithold 1993; Wignall 1993). Laminations can result from binding by algal mats (wavy laminations; O'Brien 1990) or by gentle current reworking of fine-grained sediments (thick laminations; O'Brien 1990). If these effects can be ruled out, however, the presence of laminations is good evidence for anoxic bottom waters. It must also be borne in mind that if laminations indicate anoxia, the reverse is not true. Not all anoxic bottom waters, particularly in upwelling zones, are underlain by laminated sediments (Arthur, Dean, and Stow 1984; Savrda, Bottjer, and Gorsline 1984; Savrda and Bottjer 1988; Calvert, Bustin, and Pedersen 1992; Dean, Gardner, and Anderson 1994).

Widespread organic-rich rocks. Earth history is marked by times of particularly widespread organic-rich rocks. These were termed "oceanic anoxic events" by Schlanger and Jenkyns (1976) and subsequently the subject of many studies on Mesozoic and Paleozoic rocks (Wilde and Berry 1982; Arthur, Schlanger, and Jenkyns 1987; Hallam 1987; Schlanger et al. 1987; Thickpenny and Leggett 1987; Corfield, Hall, and Brasier 1990; and many others). Models for the formation of these deposits typically include at least local upwelling and high productivity as an element (e.g., Berry and Wilde 1978; Arthur, Schlanger, and Jenkyns 1987). A great deal of effort has been expended proposing mechanisms for the development and maintenance of anoxia over large areas, and relatively little on determining mechanisms for elevating productivity. No evidence contradicts the hypothesis that high productivity was solely responsible for marine organic accumulation; Parrish (1995) suggested that oceanic anoxic events might have been oceanic productivity events (see also Hallock and Schlager 1986).

A mechanism for just such widespread productivity was suggested by Ingall, Bustin, and Van Cappellen (1993). In comparisons of laminated and bioturbated shales in several organic-rich units, they found that organic phosphorus occurred in higher concentrations in the bioturbated shales. In the Devonian New Albany Shale (northeastern United States), the ratio $C_{org}:P_{org}$ in the bioturbated shales is 150:1, close to the Redfield ratio. In contrast, the ratio is 3900:1 in the laminated shales. Noting that phosphate is regenerated from sedimentary organic matter under anoxic conditions, Ingall, Bustin, and Van Cappellen (1993) concluded that organic carbon can be sequestered under such conditions without reducing the phosphorus supply. Ingall and Jahnke (1994) proposed a positive feedback of marine productivity, oceanic anoxia, and enhanced phosphorus regeneration. High productivity could provide the organic influx to drive the system to anoxia, especially when the oceans were warm (Arthur, Dean, and Stow 1984). With the establishment of anoxia, the regeneration of phosphorus from the sediments would be much more efficient. The key is a mechanism to get the phosphorus to the photic zone. Once the feedback is initiated, perhaps in local upwelling, vertical circulation could be reduced through time without affecting productivity. This hypothesis has an underlying assumption, that on geologic time scales, phosphorus, not nitrogen, is the limiting nutrient. Ingall and Jahnke (1994) argued for the validity of the assumption on the basis of the ability of some marine organisms to fix nitrogen from the atmosphere. They also argued that the feedback mechanism would not lead to runaway anoxia but just shift all the involved processes to higher rates. An important consequence of this hypothesis is that the formation and burial of phosphorites cannot coincide in time with oceanic anoxic events, and Arthur and Jenkyns (1981) had, in fact, emphasized that in the Mesozoic they do not. However, Föllmi (1996) argued that the relationship is not so clear, even though his data (see figure 3.15) show slight decreases coincident with OAEs. Van Cappellen and Ingall (1994) suggested changes in thermohaline circulation as one way to enhance anoxia. This was an element of Arthur, Schlanger, and Jenkyns' (1987) model for the Cenomanian-Turonian boundary OAE and was also modeled by Wilde and Berry (1982).

Calculations of biologic productivity. The production-versus-preservation controversy about the formation of organic-rich deposits might be solved if we could calculate biologic productivity. Much effort has gone into trying to develop methods for doing this. However, none of the methods developed to date can be used with confidence for rocks of any age (e.g., see discussion in Parrish and Gautier 1993). A detailed discussion of these methods would be beyond the focus on paleoclimatic indicators in this book, so only the salient points are mentioned here.

In modern marine sediments, organic richness is controlled by organic supply, sedimentation rate (or, more properly, sediment accumulation rate; Tyson 1987), and grain size (Müller and Suess 1979; Arthur, Dean, and Stow 1984; Bralower and Thierstein 1984, 1987; Calvert 1987; Tyson 1987; Sarnthein et al. 1988, 1992; but see Pelet 1987). Organic supply depends on productivity (Müller and Suess 1979; Arthur, Dean, and Stow 1984; Lyle 1988), transport of terrestrial organic matter from the continents (Tissot et al. 1980), and transport of organic matter in turbidites from shallower parts of the basin (Dean, Arthur, and Stow 1984; Stow and Dean 1984). Up to a point, a higher sedimentation rate results in greater organic richness for the same input of organic matter to the sediment. Water depth, on which sedimentation rate is partly dependent, also exerts some control on organic richness. Greater depths increase the residence time of organic matter in the water column and thus the opportunity for degradation, so that a lower proportion of organic matter arrives at the sea floor. These observations have been well documented by Müller and Suess (1979), and by many other workers, for shallow water and varying oxygenation states or deep, oxygenated water.

Müller and Suess (1979) and Sarnthein et al. (1988, 1992) empirically derived paleoproductivity equations based on data from modern sediments and productivity of the overlying waters. These equations have in common carbon accumulation rate and sedimentation rate; Sarnthein et al. (1988, 1992) added depth as an explicit variable. In contrast, Bralower and Thierstein (1984, 1987) determined paleoproductivity directly from organic carbon accumulation rates using a standard preservation factor of 2% of produced organic matter; preservation factor is inherent in the paleoproductivity equations. The two methods give very different results for ancient deposits (Parrish and Gautier 1993).

One of the greatest problems for calculating paleoproductivity from the older geologic record is in the estimates of sedimentation rates. Calculated sedimentation rates in organic-rich units are commonly low (e.g., Hudson and Martill 1991). This is part of the reason such units are commonly called condensed or starved (Loutit et al. 1988). As pointed out by Hudson and Martill (1991), many of the features of such deposits are inconsistent with long exposure times at the sediment-water interface, suggesting that the deposits should be regarded as products of active systems in which sediment was constantly deposited and removed (see also Baird and Brett 1991). For example, Hudson and Martill (1991) concluded that sedimentation in the Lower Oxford Clay (Jurassic, United Kingdom) was highly episodic and that the instantaneous sedimentation rates were high because fossils in many of the beds are too well preserved to have sat at the sediment-water interface for long periods of time and lack encrusting biota, and some shell beds show evidence of concentration by winnowing. This is consistent with the great fluctuations in productivity in young sediments calculated by many workers (e.g., Abrantes 1992; Sarnthein et al. 1992; Schrader 1992) and with results of modern, high-resolution sediment-trap work (e.g., Thunell and Sautter 1992).

Biomarkers and organic-matter isotopes. Biomarkers are geologically stable organic molecules that can be traced to particular groups of organisms (Meyers 1994). They have been used in paleoclimate, paleothermometry, and paleoecologic studies in rocks as old as Proterozoic. Examples of biomarkers used in paleoclimate studies and some of their properties are listed in table 3.2. Although some biomarkers are derived from specific precursor molecules, most are simply members of classes of compounds known to be synthesized by specific groups of organisms. The biomarker composition of a sample can be used to interpret paleoclimate, or the carbon isotopic composition of biomarkers can be used, giving more precise isotope information than bulk total organic carbon (Hayes et al. 1989). Using specific biomarkers for carbon isotope studies has the advantage of avoiding a diagenetic signal—if diagenesis occurs the molecule is either destroyed or altered to another compound.

One of the most promising applications of biomarkers has been the use of long-chain ketones, known as alkenones, for paleotemperature and paleo-CO_2 studies. The only known source of alkenones (table 3.2) is prymnesiophycean algae, which include the coccolithophorids (Brassell et al. 1986). Alkenones are known in rocks as old as Cretaceous (Jasper and Hayes 1990), so the methods utilizing them have the potential to be applied at least that far back. Paleotemperature is determined from the alkenone unsaturation index, U^k_{37} (Brassell et al. 1986). The alkenone unsaturation index is calculated from concentrations of di- and triunsaturated C_{37} alkenones, using the equation, $U^k_{37} = [37{:}2]/[37{:}3 + 37{:}2]$, where 37:2 indicates the diunsaturated alkenones and 37:3 the triunsaturated alkenones. Sea-surface temperatures are derived from U^k_{37} using the equation, SST (°C) = $(U^k_{37} - 0.039)/0.034$ (Prahl, Muehlhausen, and Zahnle 1988); 0.01 U^k_{37} unit is approximately equivalent to 0.3°C. U^k_{37} reflects mean sea-surface temperature (Lyle, Prahl, and Sparrow 1992; Schneider, Müller, and Ruhland 1995).

Table 3.2.
Examples of biomarkers and of organic geochemical analyses performed on pre-Quaternary rocks used in environmental interpretation

Biomarker	Notes
Alkenones	Specific to algae such as coccolithophorids
Geoporphyrins	Derived from chlorophyll, indicative of lacustrine (in lakes) or marine (in oceans) algae; chlorophyll from terrestrial plants does not survive transport and depositional processes
Hydrogen index (HI)	Higher when OM is marine and immature
Oxygen index (HI)	Higher when OM is terrestrial and/or very mature and/or oxidized
Carbon preference index (CPI)	Higher when more terrestrial OM is present, used for long-chain hydrocarbons
C_{29} steranes	Higher when more terrestrial OM is present
Phytane/n-C_{18}	Amount of oil that has been generated
Hopanes	Biomarkers, bacteria

Derived from information in Hayes et al. (1989), Jasper and Hayes (1990), Jasper and Gagosian (1990), Littke et al. (1991), Schoell et al. (1994), and Kenig et al. (1994).

OM, organic matter; "mature" and "immature" indicate the degree of catalysis.

Some recent studies have shown that source-specific biomarkers may be stable enough to provide information in the older geologic record, as predicted by Jasper and Hayes (1990; see review in Beerling 1997). For example, Schoell et al. (1994) proposed a method of interpreting paleotemperatures from biomarkers in the Miocene Monterey Formation (California, U.S.). They used variations in $\delta^{13}C$ in C_{27} steranes, which are derived from shallow-water organisms, and C_{35} hopanes, which are from bacteria living at the base of the photic zone. The $\delta^{13}C$ signals of the two classes of compounds are decoupled (figure

3.18), which Schoell et al. (1994) attributed to the effects of different habitats of the source organisms. If the source organisms and proposed habitats for them are correct, the difference in $\delta^{13}C$ between the steranes and hopanes should approximate the difference from the top to the bottom of the photic zone. A difference in $\delta^{13}C$ would result either from differences in the isotopic composition of dissolved inorganic carbon (DIC) or from biologic fractionation of varying concentrations of dissolved CO_2. The difference of 6‰ to 7‰ at the top of the section (figure 3.18) is too great to be explained by differences in DIC, according to Schoell et al. (1994), so they concluded that varying concentrations of dissolved CO_2, which is related to temperature, must account for the differences in the isotopic composition of steranes and hopanes. The small difference in the lower part of the section indicated a well-mixed water column within the photic zone and a concomitant small difference in temperature. The increasing difference in the isotopic composition of steranes and hopanes indicated increasing temperature stratification within the photic zone, which might be expected as a result of the cooling recorded by $\delta^{18}O$ in benthic foraminifera from the same rocks.

Carbon isotopic compositions of selected biomarkers have also been used to study variations in atmospheric CO_2 (Jasper and Hayes 1990; Freeman and Hayes 1992; Hayes 1993). Photo-

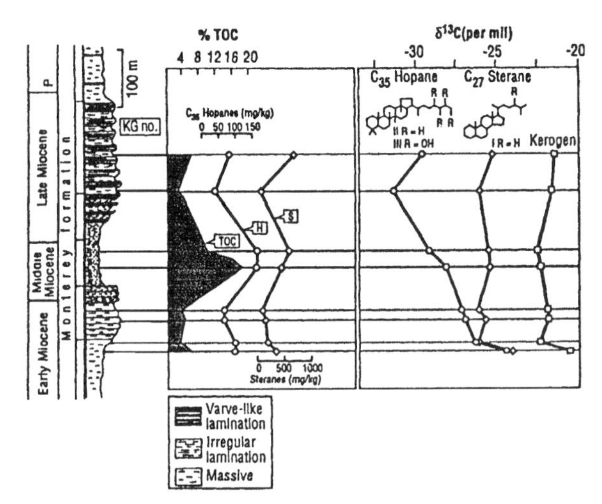

Figure 3.18. Stratigraphy of the Miocene Monterey Formation at Naples Beach, California, U.S. (*left*); abundance of total organic carbon, steranes (S), and hopanes (H; *center*); and $\delta^{13}C$ of the steranes, hopanes, and bulk kerogen (*right*). From Schoell et al. (1994), © American Association for the Advancement of Science.

synthetic processes in marine algae are such that if CO_2 increases, the ^{13}C difference between dissolved inorganic carbon and organic carbon should increase. This relationship is expressed by the empirical equation

$$\varepsilon_p = a \log c + b,$$

where ε_p is the isotope effect associated with photosynthetic fixation, c is concentration of dissolved CO_2, and a and b are constants. To determine ε_p, the isotopic composition of dissolved CO_2 and of the phytoplankton must be reconstructed. Dissolved CO_2 today is depleted relative to total DIC by 8.8‰ (at T = 24°C, pH = 8.2), and foraminiferan tests are depleted relative to DIC by 0.5‰. Taking advantage of this relationship, Jasper and Hayes (1990) used foraminiferan $\delta^{13}C$ as a proxy for dissolved CO_2 and $\delta^{13}C$ of the alkenones as a proxy for the phytoplankton (depleted by 3.8‰ from total algal biomass). Their results compared favorably with determinations of pCO_2 from the Vostok ice cores (Barnola et al. 1987); discrepancies might be the result of the considerable problems of correlating ice cores and the marine sediment record.

Popp et al. (1989) used a method similar to Jasper and Hayes' (1990) to track relative variations in CO_2 in rocks from Late Triassic to Pleistocene age. They tackled the observation that the difference in $\delta^{13}C$ of carbonates and marine organic matter was 5‰ to 6‰ greater in the Mesozoic and early Cenozoic than it is now, but that the isotopic difference between carbonates and terrestrial organic matter has not changed. They chose to examine the $\delta^{13}C$ of geoporphyrins, which are derived from chlorophyll, because those molecules in marine sediments are derived specifically from marine organic matter. They also analyzed lignin, which is unique to terrestrial plants, from Cretaceous rocks. Using an equation similar to that used by Jasper and Hayes (1990), Popp et al. (1989) calculated the fractionation effect, ε_p, and offered an explanation for why marine organic matter changed but terrestrial organic matter did not. In marine phytoplankton, the rate of diffusion of CO_2 into the cell is enhanced by high concentrations of CO_2 such that high levels of CO_2 "decrease the extent to which the rate of fixation is limited by mass transport." In other words, the more CO_2 that is available, the more cells can discriminate against ^{13}C. In terrestrial plants, diffusion is controlled by the stomata, and the number of stomata in some modern plants is reduced under higher CO_2 levels, so Popp et al. (1989) speculated that older terrestrial plants likewise kept diffusion rates, and therefore isotopic fractionation, constant.

Bulk-sediment nitrogen isotopes. Bulk-sediment nitrogen isotopes, which are part of organic matter, can be an independent check on carbon isotope data on surface productivity fluctuations, and they are not complicated by varying transport and utilization with depth as seen in $\delta^{13}C$. Fractionation of $^{15}N/^{14}N$ depends on the supply of nutrients. Where nutrients are abundant, as in upwelling zones, organisms preferentially use ^{14}N and the organic matter has low $\delta^{15}N$. When nutrients are in short supply, organisms are not as selective, and organic matter $\delta^{15}N$ is higher (discussion and references in Altabet and François 1994). Nitrogen isotopes have been applied extensively in Quaternary and Holocene sediments (e.g., Calvert, Nielsen, and Fontugne 1992; François, Altabet, and Burckle 1992; Farrell et al. 1995) but rarely in the pre-Quaternary record. An example of application of nitrogen isotopes to pre-Quaternary rocks is a study by Rau, Arthur, and Dean (1987). Extremely low $\delta^{15}N$ values were noted in Cretaceous organic-rich shales from DSDP Sites 367 and 530. Rau, Arthur, and Dean (1987) argued that increased denitrification, which removes nitrogen, and consequent reduced euphotic zone nitrate availability might have resulted in phytoplankton assemblages that were dominated by plants, such as blue-green algae, that could fix nitrogen from the atmosphere. These plants have very low $\delta^{15}N$ signatures. Note that if this interpretation is correct, reduced nutrient availability, in this case nitrate, resulted in lower $\delta^{15}N$, rather than in higher values as described earlier.

High-productivity lithofacies associations in continental-shelf upwelling zones

As noted earlier, the association of biogenic siliceous rock, organic-rich rock, and phosphorite is taken as a strong indicator of high paleoproductivity; chalk might replace biogenic siliceous rock in certain systems. Organic-rich rock and phosphorite, along with sediments containing the reduced-iron clay mineral, glauconite, commonly occur in a specific facies array in upwelling zones. Understanding these associations can be particularly helpful for identifying ancient continental-shelf upwelling zones.

The best-developed upwelling zones on the modern continental shelves are those off Namibia and Peru. The sediments accumulating in these areas consist of diatomaceous and clayey sediment containing phosphorite, glauconitic sands, and abundant organic matter (Soutar, Johnson, and Baumgartner 1981; Baturin 1983; Bremner 1983; Burnett, Roe, and Piper 1983; Calvert and Price 1983; Glenn et al. 1994a; Parrish 1995). These are arranged in a particular pattern related to the average pattern of productivity at the surface (figure 3.19). The most organic-rich sediments underlie the zone of highest biologic productivity. This zone is commonly overlain by bottom waters driven to anoxia by the enormous influx of organic matter. Surrounding the organic-rich sediments are phosphatic sediments that are also organic rich. The phosphorite precipitates from phosphate ions released from the organic matter in the presence of low levels of oxygen and can take on a variety of forms depending on conditions in the sediments and bottom-water currents (Glenn et al. 1994a). This zone is surrounded, in turn, by a zone of glauconitic sediments. Glauconite is an indicator not of high biologic productivity but of low oxygen levels (in this case, in the water column, but see the second part of this chapter, on nonbiogenic sediments). The sediments overall are diatomaceous, so silica does not form a distinct geographic facies, although it commonly is a distinct facies in ancient high-productivity deposits. The zonation can be complex, depending on bottom currents, and it varies along strike (see

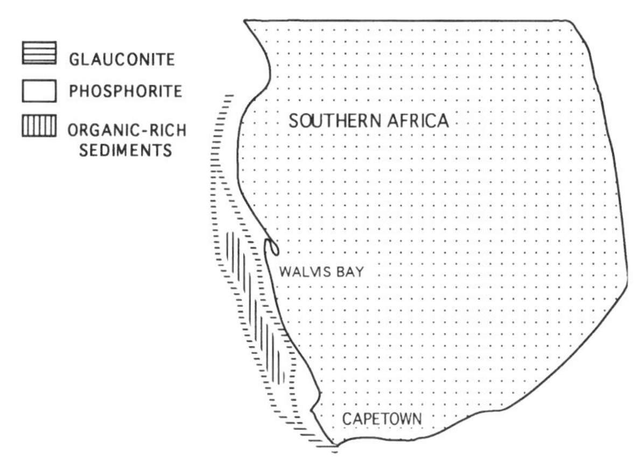

Figure 3.19. Schematic distribution of glauconite, phosphorite, and organic-rich sediments in a well-developed, modern upwelling zone, southwestern Africa. Surrounding sediments are clastic- and/or diatom-rich. From information in Bremner and Rogers (1981).

figure 3.19), but the general facies relationship holds.

An example of a well-mapped unit that shows the distinct facies array characteristic of continental shelf upwelling zones is the Shublik Formation (Triassic; northern Alaska, U.S.; figure 3.20). The Triassic shoreline was to the north (present coordinates) and, in core and outcrop, the Shublik is glauconitic closest to the shoreline, phosphatic farther out, and highly organic rich closest to what is now the Brooks Range front (Dingus 1984; Parrish 1987; Kupecz 1995). A siliceous facies containing bedded radiolarian chert, which might have formed farthest offshore, constitutes the partly correlative Otuk Formation (Mull et al. 1982; Bodnar 1984).

Shallow-Water Carbonates

Taken as a whole, shallow-water carbonates tend to be deposited in low latitudes, mostly <40°, even most of those that are considered to have formed in cool water. Carbonate, especially aragonite, is much more soluble in cold water. For these reasons, shallow-water carbonates tend to be regarded as warm-water indicators (Briden 1968). However, this really applies only to reef and platform carbonates because coquinas and mixed clastics/bioclastic carbonates can occur at any latitude (Nelson 1988).

Reefs and carbonate platforms

Modern reefs are built primarily by scleractinian corals, but this has not been the case for most of Earth history (Newell 1972; Fagerstrom 1987). Reef communities have been divided into a number of ecological "guilds," of which the "constructor" guild is the most important. The organisms in that guild form the rigid framework of the reef (Fagerstrom 1987); they are also called framework builders and have comprised different taxa at different times in Earth history. Reefs have varied in importance through time; a major gap, during which no known reefs were formed, was in the Early Triassic (Fagerstrom 1987). Such fluctuations have been regarded as important for understanding the role of carbonate sedimentation and basin-shelf fractionation in the history of changes in atmospheric CO_2 (Berger 1982; Walker and Opdyke 1995). Reefs have

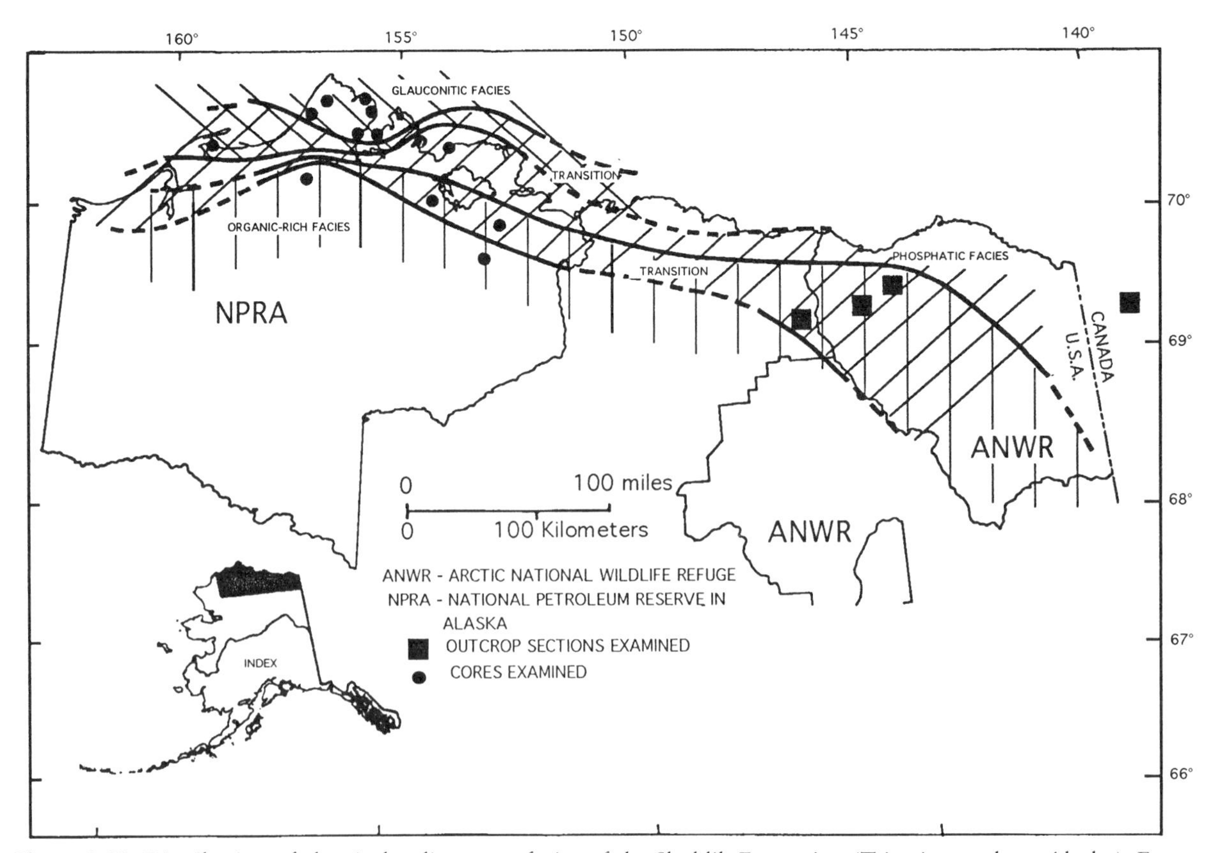

Figure 3.20. Distribution of chemical sedimentary facies of the Shublik Formation (Triassic; northern Alaska). From Parrish (1987a).

formed either as isolated structures in deeper-water sediments or as laterally extensive structures that form broad shelves. These latter features are called carbonate platforms and, if they include reefs, are climatically equivalent to reefs. For the purposes of interpreting paleoclimate, it is important to distinguish between photic-zone and non-photic-zone reefs. The latter have much broader environmental tolerances than do photic-zone reefs (Fagerstrom 1987). Only a few such reefs have been identified in the fossil record (Cairns and Stanley 1987, cited in Fagerstrom 1987). They are distinguished by relatively low diversities and, except for those in deep water off the southeastern United States, have an antithetical geographic distribution to photic-zone reefs. The terms "hermatypic" and "ahermatypic" are sometimes used to distinguish between photic- and non-photic-zone reefs, respectively, but Fagerstrom (1987) pointed out that this distinction is not appropriate and preferred the terms "zooxanthellate" and "nonzooxanthellate," referring to the symbiotic algae that characterize modern reef-building corals and that dictate the requirement for light. No real evidence exists to contradict the hypothesis that ancient and modern carbonate reefs are ecologically and paleoclimatically equivalent, despite the very different faunas, so it is assumed that the environmental tolerances of ancient reefs are the same as those of modern ones.

Distribution of reefs. Modern coral reefs appear to be limited by the 16°C isotherms in the Pacific and the 18°C isotherms in the Atlantic; the latitudinal limits are 37°N to 35°S and 37°N to 21°S, respectively (Rosen 1984). There is a positive correlation between growth rates and temperature in corals (Fagerstrom 1987), although not between reef accretion and temperature (Hallock 1988). An argument in support of temperature as a limiting factor is the general observation that reefs tended to occur farther poleward during times that are considered to have been warm on other grounds. For example, reefs in the Cretaceous occurred at paleolatitudes as high as 55°, regardless of their taxonomic makeup (Flügel and Flügel-Kahler 1992; Scott 1995; Johnson et al. 1996; but see Gili, Masse, and Skelton 1995). On the other hand, the data demonstrating expansion and contraction of reef zones are rather poor (Ziegler et al. 1984; Fulthorpe and Schlanger 1989; Johnson et al. 1996), and no obvious contraction of reef range can be demonstrated between the Cretaceous and Tertiary despite the strong global cooling interpreted from other data. Fagerstrom (1987) emphasized that the Pleistocene glaciations had a limited effect on reefs and noted how similar the Pleistocene and Holocene reefs are.

A small percentage of fossil reefs occurred north or south of 35° paleolatitude in the past. Flügel and Flügel-Kahler (1992) compiled an extensive database on reefs through time; plotted on the newer maps of Scotese and Golonka

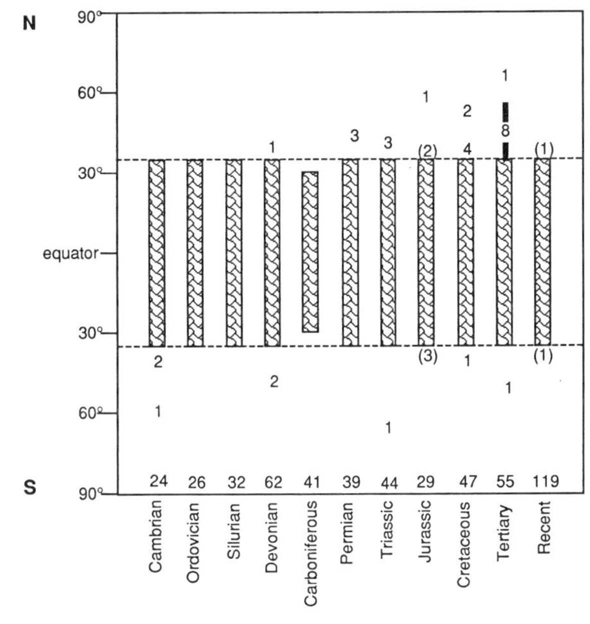

Figure 3.21. Paleolatitudinal distribution of reefs through time, based on data from Flügel and Flügel-Kahler (1992) for fossil reefs, and Rosen (1984) for modern reefs. The *dashed horizontal lines* are the 35° parallels, and most reefs occurred between them. Except for the Carboniferous, when reefs had a significantly narrower paleolatitudinal range within the reef zone, the bars simply extend through the reef zone and do not imply that reefs extended to 35° in both hemispheres. The numbers above and below the bars and their paleolatitudinal positions indicate the number of reefs at each latitude outside the general reef zone. Numbers in parentheses occur between 35° and 37° paleolatitude. The numbers at the bottom of the graph are the total number of reefs for each time. At least one of the eight reefs north of the reef zone in the Tertiary was a non-photic-zone reef (see text). The Scotese and Golonka (1992) maps used for each interval are the following and were based on the times when the largest number of reefs for each period occurred and on the map intervals chosen by Flügel and Flügel-Kahler (1992): Early Cambrian, Late Ordovician, Middle Silurian, Middle Devonian, Late Carboniferous, Late Permian, Late Triassic, Late Jurassic, middle Cretaceous (94 Ma), middle Miocene.

(1992), they yield the distributions in figure 3.21. Excluding one of the higher-latitude reefs in the Tertiary, which was a non-photic-zone reef, Flügel and Flügel-Kahler (1992) tabulated 398 fossil reefs; 35 of these (9%) occur above 35° paleolatitude on the Scotese and Golonka (1992) reconstructions. The narrowest latitudinal ranges for reefs, with all reefs at 35° paleolatitude or less, were in the Ordovician, Silurian, and Carboniferous. The range for Carboniferous reefs was particularly narrow, and only two reefs today fall outside the 35°-latitude limit, at <37° latitude. The widest latitudinal ranges were in the Cambrian, Triassic, and Tertiary, with significant numbers of reefs outside the "reef zone" in the Cretaceous; overall, reefs in the Mesozoic were more widespread than at other times. These data show that if global temperature and reef distribution are related, the relationship is certainly not linear. The most striking feature of figure 3.21 is the latitudinal constriction of reefs in the Carboniferous, when, at least during the latter part of the period, significant glaciation occurred in the Southern Hemisphere.

Ziegler et al. (1984) proposed that light is the principal limiter to reef growth, partly on the basis of the importance of the zooxanthellae. Ziegler et al. (1984) noted that the amount of light penetrating the water column is much lower above 30° latitude and that the number of reefs drops off dramatically there. In addition, nutrient-rich waters are more turbid than low-latitude, oligotrophic ocean waters, so that light is blocked that way as well. Ziegler et al.'s (1984) hypothesis is not easily testable for ancient reefs because no direct evidence exists that ancient reef builders, for example, rudistid clams, were dependent on zooxanthellae.

Modern coral reefs are also inhibited by excess nutrients (Hallock and Schlager 1986; Hallock et al. 1988) because the normal reef-dwelling and -building organisms are out-competed by opportunistic taxa able to take advantage of the additional resources (Hallock 1988). The fact that changes in the communities do not involve just die-off of framework builders but also proliferation of other forms supports the hypothesis that it is the nutrients themselves, and not the turbidity and light reduction they bring about, that are inimical to the reef builders.

Hallock and Schlager (1986) invoked changes in nutrient levels to explain drowned reefs and carbonate platforms in the fossil record, which are paradoxical (Schlager 1981). Drowning of reefs and platforms occurs when submergence rate exceeds growth rate. The paradox is that growth rates of modern corals and reefs exceed by one or two orders of magnitude all the physical processes, such as tectonic subsidence or eustatic sea-level rise, that could result in the removal of reef organisms from the photic zone (Schlager 1981). Drowned carbonate platforms should therefore be rare but, in fact, are common. Hallock and Schlager (1986) proposed that drowned platforms are an indication of increased nutrient fluxes, which would have two effects: (1) increased turbidity of the water column with concomitant shallowing of the euphotic zone, and (2) replacement of fast-growing, reef-building organisms by slower-growing forms and an increase in bioeroders. They proposed three mechanisms for increased nutrient influx: (1) influx of terrestrial nutrients and reduced production and accumulation by reef-building organisms before excess nutrients can be exported; (2) where exposed platforms become submerged, supply of excess nutrients from soils, so that the reef builders can neither become established nor thrive; and (3) changes in local or regional upwelling or global oceanic overturn. Although the first two mechanisms might explain the demise of individual platforms, Hallock and Schlager (1986) favored the third for widespread drowning of reefs and carbonate platforms. Such events occurred in the Devonian, Jurassic, mid-Cretaceous, Eocene, and Miocene (Schlager 1981; Bourrouilh-Le Jan and Hottinger 1988). Hallock and Schlager (1986) suggested that oceanic anoxic events were actually oceanic productivity events, which makes a connection between OAEs and reef drowning.

Reefs that are close to drowning are dominated by benthic animals. Stratigraphically, drowned reefs are commonly separated from overlying deep-water deposits by bored hardgrounds, ferromanganese oxide crusts, phosphates, and/or glauconite, which Hallock (1988) attributed to increased nutrient availability. For example, Cretaceous Tethyan platforms change from healthy platforms consisting of corals, large benthic foraminifera, oolite, and little biogenic silica to platforms lacking framework builders and oolite and dominated by crinoids, bryozoans, epibenthic bivalves, and terebratulid and rhynconellid brachiopods, along with abundant biogenic silica. Above these later mesotrophic platform communities were phosphatized drowning surfaces (Föllmi et al. 1994).

In summary, reefs and carbonate platforms are generally taken as indicative of warm water, >18°C, although the stability of their latitudinal distribution through time suggests that they might respond to some other control as well, perhaps light. In addition, changes in reef or carbonate platform formation through time, especially if accompanied by chemical and biotic changes, are indicative of variations in local to global nutrient status of the oceans.

Reefs as current indicators. Isolated reefs, as opposed to those that are parts of carbonate platforms, are potential wind indicators (Schlager and Philip 1990). For example, Masse and Philip (1981) noted that isolated Cretaceous reefs in France all have the same orientation of facies, which they attributed to the direction of the dominant swell (figure 3.22). In waters shallow enough for reef growth, the dominant swell would be wind driven and, indeed, the current direction suggested by these reefs is consistent with winds predicted for that region (Parrish and Curtis 1982).

Stable isotope studies. Carbonate platforms are commonly the source of shells on which carbon isotope analyses are performed. A large carbonate platform creates its own chemical environment by virtue of its shallow depth and restricted exchange of water between the interior of the platform and the open ocean. With this in mind, Patterson and Walter (1994) sampled three different types of platforms: Florida Bay, which has significant freshwater influx and evaporation (Lloyd 1964); the Great and Little Bahama Banks, which have significant evaporation; and the Atlantic reef tract on the ocean side of the Florida Keys. They found that seawater on the Bahama Banks is depleted as much as 4‰ $\delta^{13}C$ relative to surrounding seawater and relative to the Atlantic reef tract. The depletion occurred at both higher and lower than normal salinities. Two important points are taken from these results. First, extreme salinity variations are not required to significantly affect the $\delta^{13}C$ values. Second, care must be taken to make sure that shallowing-upward sequences that show ^{13}C depletion and ^{18}O enrichment are not related to restriction of surface waters and evaporative concentration, rather than cooling associated with glaciation. Patterson and Walter (1994) suggested that the most reliable data for regional or global paleoclimate interpretations would come from oolites and organisms within the reefs themselves. However, the very characteristics that make the interiors of carbonate platforms unsuitable for regional- and global-scale studies make them ideal for local paleoclimate studies because the interiors are so sensitive. For example, Mutti and Weissert (1995) suggested that very depleted $\delta^{18}O$ values in Triassic carbonate platforms of central Europe were indicative of high freshwater influx associated with the Pangean monsoon.

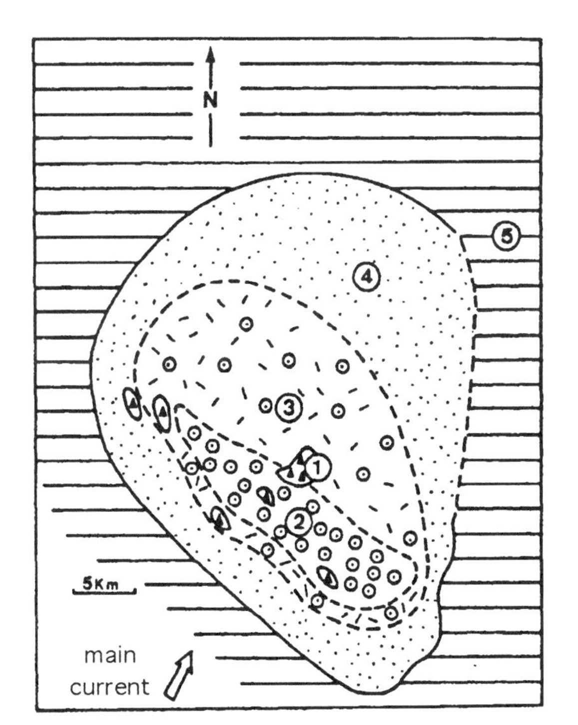

Figure 3.22. Facies in an isolated reef (Cretaceous; France) and predominant current direction. In waters shallow enough for reef growth, such currents would be wind driven, making these types of reefs good wind-direction indicators. Facies: 1, coral patches; 2, oolite; 3, oolite/bioclastic sand; 4, bioclastic, micritic sand; 5, fine-grained terrigenous limestone. From Masse and Philip (1981).

Nontropical limestones

Changes in the biota of reefs and carbonate platforms with increasing nutrient availability, for example, increase in bryozoans, are similar in some ways to changes in the biota with increasing latitude and decreasing temperature (Schlanger 1981; Fulthorpe and Schlanger 1989; Adams, Lee, and Rosen 1990; Brachert et al. 1993). As it is not always possible to discern the framework and structure of a reef in the fossil record, or the preserved part of a platform may not include the reef facies, a set of criteria has been developed to distinguish carbonates within the "reef zone," which are assumed to be warm-water carbonates, from those outside. Warm-water carbonates fall into what Lees and Buller (1972) called the chlorazoan facies, and Schlanger (1981), the coral-algal facies. Later, Lees (1975) divided the chlorazoan facies into the chlorazoan and chloralgal facies. The modern chlorazoan facies is dominated by corals, whereas the chloralgal facies has a higher component of green, calcareous algae; ancient equivalents of the chlorazoan facies would be dominated by the main framework builder. The mineralogy in both facies is dominantly aragonite, and nonskeletal grains such as ooids, grapestone, and peloids are common; carbonate mud is more common in the chloralgal facies.

Lees and Buller's (1972) term for cool-water carbonates was "foramol," and such carbonates today consist predominately of bryozoans, brachiopods, mollusks, foraminifera, barnacles, and calcareous red algae. The mineralogy is dominantly calcite (Nelson 1978; Leonard et al. 1981), and carbonate-dominated gravels and sands, up to 97% carbonate, are known from as far poleward as southern Alaska and the Antarctic shelf (Leonard et al. 1981; Domack 1988). Rao and Nelson (1992) adopted the term "bryomol" to describe cool-water assemblages dominated by bryozoans, mollusks, and foraminifera, similar to the highest-latitude facies described by Schlanger (1981). They compared the $\delta^{18}O$ and $\delta^{13}C$ signatures of cool-water carbonates of different ages in Australia and Tasmania with those of warm-water carbonates and deep-sea carbonates (Rao and Jayawardane 1994). The cool-water carbonates overlapped substantially with each other and with deep-sea carbonates but very little with warm-water carbonates. As with reef limestones, the taxonomic composition of cool-water limestones has varied through time.

Hallock and Schlager (1986) were critical of Lees's (1975) classification because it concentrated on salinity and temperature, and the variations he was trying to explain also can be explained by nutrient variability; they pointed out that the failure of Lees's (1975) model to correctly predict carbonate facies in Florida Bay was the result of this deficiency. Carannante et al. (1988) thus divided the foramol facies (Lees and Buller 1972; Lees 1975) into two facies: rhodalgal, consisting of encrusting coralline algae and bryozoans with lesser benthic foraminifera and barnacles, and molechfor, consisting of mollusks, arenaceous benthic foraminifera, echinoids, and barnacles, with lesser bryozoans and serpulid worms. They retained the chlorazoan and chloralgal facies in their scheme and incorporated a facies they called bryo-algal as the transition between the rhodalgal and molechfor. Their model implicitly includes the considerations of nutrient availability (Hallock and Schlager 1986).

On the Antarctic shelf, in the Miocene of southeastern Alaska, and in the Permian of Australia, carbonate sediments are intimately associated with glacially derived sediments and faceted dropstones (Rao 1981a,b; Domack 1988; Eyles and Lagoe 1989); these workers independently used very similar models for the formation of the deposits. The deposits consist of diamictite (very poorly sorted, glacially derived sediments) interbedded with limestones consisting of either planar or lenticular coquinas. These very high-latitude carbonate occurrences are dominated by mollusks. Both Rao (1981a) and Domack (1988) explicitly discussed upwelling or high plankton productivity as a factor in the formation of these assemblages; the abundant food permits more rapid reproduction of shell-forming organisms.

Many of the cool-water limestones are specifically the product of winnowing by storms or associated with storm deposits (Eyles and Lagoe 1989). Carbonate storm deposits include flat-pebble conglomerates (Sepkoski 1982).

NONBIOGENIC SEDIMENTS

Nonbiogenic sediments and minerals in the marine realm provide information about physical and chemical conditions of sedimentation and of marine waters that can be related to climatic processes. Shallow-water clastic sediments can record changing climate through changes in influx of clastics to the sea or degrees of storminess. Clay minerals in deep-sea sediments provide information about adjacent continental climates. Various iron-bearing minerals and manganese record different chemical states of the ocean waters. Other trace and major elements in marine sediments can, under certain circumstances, provide climatically related information about ocean chemistry. Finally, the mineral ikaite is considered indicative of particular climatic conditions. Clay minerals, storm deposits, certain types of iron occurrences, and ikaite are the most direct indicators of climate. All the other indicators discussed in this section are supplementary indicators.

Several types of indicators found in marine sediments are not included here because they are more indicative of adjacent terrestrial climates than of marine climates; these indicators are discussed in chapter 5. Eolian sediments in the deep sea provide information about winds and climatic conditions on distant continents, as does debris borne by ice. Although oolitic ironstones (chamosite and verdine) are commonly marine and closely related mineralogically to glauconite, the paleoclimatic controls on their formation are distinct and closely related to climate on land, whereas the distribution of glauconite is controlled by marine climates. Finally, most, but not all, evaporites are derived from marine waters, but even the marine evaporites are more indicative of terrestrial than marine climates. Clay minerals and oolitic ironstones also reflect climatic conditions on the continents. However, for clay minerals, the signal is commonly clearer in the adjacent oceans than in terrestrial deposits, so they are discussed in this chapter.

Clay Minerals in Marine Sediments

The use of clay minerals in deep-sea sediments as paleoclimatic indicators of adjacent continental climates has been developed and promoted principally by French geologists, notably G. Millot, C. Robert, H. Chamley, and G. S. Odin. The basis for the use of clays comes from the weathering sequence of crystalline rocks in various environments (Paquet 1970):

rock → illite (from feldspars) + chlorite (from Fe-Mg minerals) → montmorillonite → kaolinite (in regions of high temperature, good drainage).

The weathering reactions for albite illustrate the variability of weathering and its relationship to climate (Hay 1992). Note particularly the role of water, which implies poorly drained soils:

$2NaAlSi_3O_8$ (albite) + $2CO_2$ + $11H_2O$ → $Al_2Si_2O_5(OH)_4$ (kaolinite) + $2Na^+$ + $2HCO_3^-$ + $4H_4SiO_4$, and

$2NaAlSi_3O_8$ + $2CO_2$ + $6H_2O$ → $Al_2Si_4O_{10}(OH)_2$ (montmorillonite[3]) + $2Na^+$ + $2HCO_3^-$ + $2H_4SiO_4$.

Kaolinite, indicative of warm climates, and vermiculite, indicative of cool climates, are probably the most reliable indicators. In addition, three relatively rare clays—sepiolite, palygorskite, and attapulgite—are diagnostic of warm, arid climates (e.g., Paquet 1970; McFadden 1988). Clay assemblages dominated by illite and montmorillonite can be taken as indicative of temperate climate, but these clays are also the most common anywhere, a fact that itself might simply be indicative of the preponderance of temperate continental climates given the present global geography and climate. Chlorite is a product of mechanical weathering, so chlorite abundance is only indirectly useful as a paleoclimatic indicator. Smectite forms in warm, humid climates with seasonal rainfall, but it is also a weathering product of volcanic ash (Singer 1984). The important clay indicator might not be the most dominant mineral or even the dominant clay mineral in a given deposit.

Complicating the use of clays as paleoclimatic indicators is the fact that the source terrane for the clays may influence clay composition (Griffin, Windom, and Goldberg 1968). For example, kaolinite in the Arctic Ocean is derived from weathered, previously kaolinitized rocks (Singer 1984). Tectonic activity and varying erosion rates also can affect the types of clays transported to the oceans. Volcanic ash contributes material for authigenic clays (Griffin, Windom, and Goldberg 1968).

The distribution of several clay minerals in surface sediments in the oceans was compiled by Griffin, Windom, and Goldberg (1968), and this can be compared to a generalized map of weathering and weathering products (e.g., kaolinite; figure 3.23). From their compilation, Griffin, Windom, and Goldberg (1968) drew the following conclusions. Chlorite is representative of high latitudes with less intense chemical weathering and more intense mechanical weathering (Singer 1984). Its distribution is reciprocal to that of kaolinite (figure 3.23) and gibbsite. Chlorite appears to be mostly derived from glacial sediments except near Hawaii, where gibbsite is undergoing diagenesis to chlorite, and near Australia, where the chlorite is apparently volcanic. Montmorillonite is mostly derived from volcanic ash, especially where it is found with volcanic glass shards, for example, in the Indian Ocean and the tropical Pacific volcanic province. Illite is more common in the Northern Hemisphere because of the higher detrital input; in the Pacific, a band of abundant illite co-occurs with eolian quartz. Illite also might

3. This structure is highly simplified for the montmorillonite group of clays, which can have variable amounts of Ca, Na, Mg, and Fe.

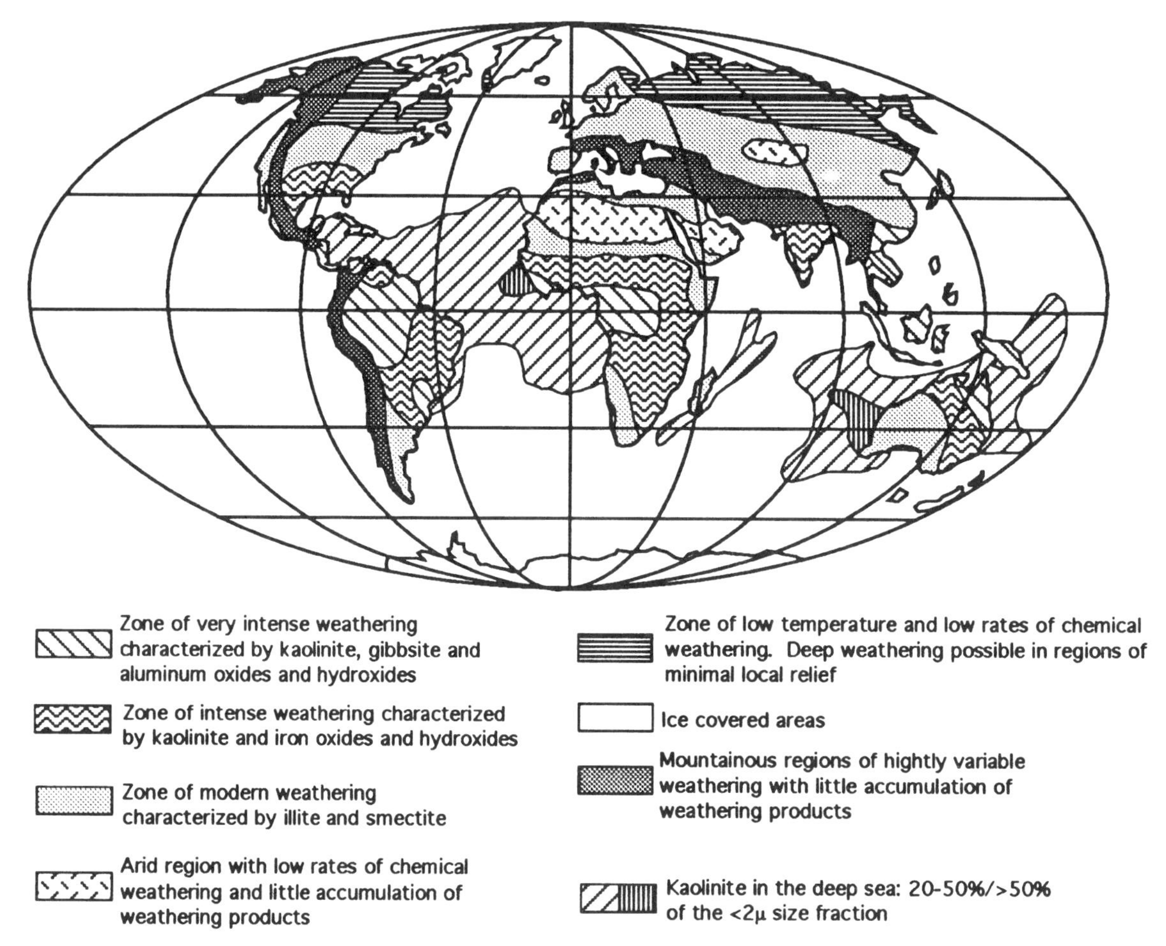

Figure 3.23. Kaolinite concentrations in the clay-size fraction of sediments in the World Ocean and pattern of weathering products on the continents. Adapted from *Deep-Sea Research* 15, Griffin, J.J., H. Windom, and E.D. Goldberg, "The distribution of clay minerals in the World Ocean," pp. 433–59 1968, with permission from Elsevier Science Ltd, The Boulevard, Langford Lane, Kidlington OX5 1GB, UK (oceanic kaolinite); and *Journal of the Geological Society of London* 147, Curtis, C.D., © 1990 The Geological Society (continental weathering).

be influenced by glaciation, as off southern South America and Alaska. In general, illite is indicative of low-intensity weathering (Singer 1984), both mechanical and chemical. Kaolinite (see figure 3.23) is a low-latitude mineral, derived from rocks that are intensely weathered.

The connection between land and sea is complicated by factors such as the locations of river drainages and volcanic input (Curtis 1990). In addition, the clay-mineral assemblages in the deep sea can be obscured by selectivity of erosion (that is, different physicochemical properties of different clay minerals), extent and intensity of erosion, transport, water currents, ice, wind (which can transport large amounts of clay minerals), size sorting, and diagenesis (Singer 1984). Moreover, differentiating authigenic and detrital clays can be difficult (Singer 1984). Nevertheless, the relationship between weathering products on land and the clay mineral assemblage in the oceans is strong.

Clay minerals in the deep sea through time

The use of clays in deep-sea sediments to interpret climate change was pioneered by Robert and Chamley (1987, 1991), who traced the history of clay mineralogy in the deep sea through the Cenozoic in most of the World Ocean, excluding only the East Pacific. More recently, Robert and Kennett (1992, 1994) related changes in clay mineralogy off Antarctica to water-mass studies using oxygen isotopes in planktonic foraminifera in early Tertiary sediments of the South Atlantic Ocean. Smectite was abundant throughout the region (70% to 100%), which Robert and Kennett (1992) interpreted to indicate warm and seasonal climates everywhere. They interpreted the smectite as pedogenic rather than volcanogenic in these sediments because the abundance of smectite and the abundance of volcaniclastics were not related; the smectite was also more poorly crystalline than either diagenetic or volcanic smectite.

In contrast to the widespread abundance of smectite, kaolinite occurred only near Antarctica in the early Tertiary, which Robert and Kennett (1992, 1997) interpreted to indicate high rainfall associated with warm temperature on the continent. Kaolinite was especially abundant in the latest Paleocene and early Eocene, when $\delta^{18}O$ indicated maximum warmth in surface and intermediate waters. Illite increased dramatically in the Oligocene, suggesting onset of active mechanical erosion or erosion of poorly developed soils. Oxygen isotopic ratios indicated that by the Oligocene, significant glaciation was underway (Shackleton and Kennett 1975; Miller, Fairbanks, and Mountain 1987; Robert and Kennett 1997).

In a subsequent paper, Robert and Kennett (1994) concentrated on the details of $\delta^{18}O$ changes and clay mineralogy at the Paleocene-Eocene boundary (figure 3.24). Kaolinite increased dramatically, indicating a major increase in temperature and/or precipitation (see also Robert and Maillot 1990). At low latitudes, the Paleocene-Eocene boundary was marked by an increase in palygorskite, indicative of arid conditions (Robert and Chamley 1991). Near Antarctica, the increase in kaolinite was followed by an increase in the importance of smectite, which Robert and Kennett (1994) interpreted as a brief increase in annual precipitation and seasonality.

Although most data on clay mineralogy (and other indicators in the deep sea) are presented as time series, such graphs are of necessity limited to one site or are very generalized. To overcome these limitations, and noting that the magnitudes of change were less important than the directions of change, Robert and Chamley (1987) plotted changes in clay mineralogy as different symbols a global map (figure 3.25). This unusual graphical method makes it very easy to see the distribution of changes, and it might be considered for other types of data where magnitude of change is less important than the sign of change.

Clay minerals in epeiric seas

The Silurian marine Keefer Sandstone and Rose Hill Formation of the eastern United States contain oolitic hematite, which indicates strong weathering and humid climate on the adjacent continent. The same rocks contain abundant kaolinite (Meyer, Textoris, and Dennison 1992),

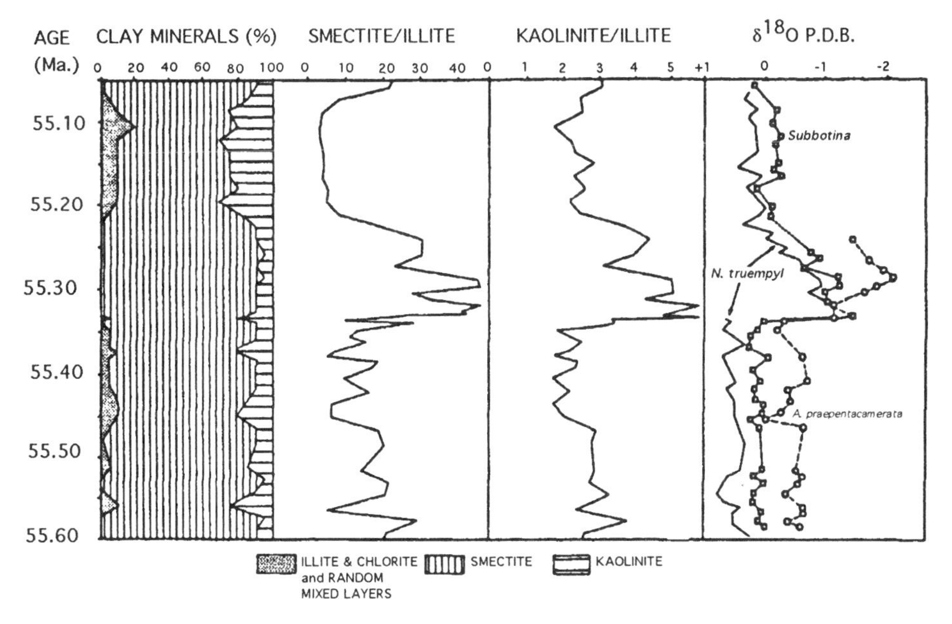

Figure 3.24. Variations in the concentrations of clay minerals, clay mineral ratios, and $\delta^{18}O$ at ocean-drilling Site 690B, Maud Rise, near Antarctica. The terminal Paleocene event is between 55.33 Ma and 55.21 Ma. *Acarinina praepentacamerata* and *Subbotina* are planktonic foraminifera; *Nuttalides truempyi* is a benthic foraminiferan. From Robert and Kennett (1994) in *Geology.* Reproduced with permission of the publisher, the Geological Society of America, Boulder, Colorado, USA. Copyright © 1994 Geological Society of America, Inc.

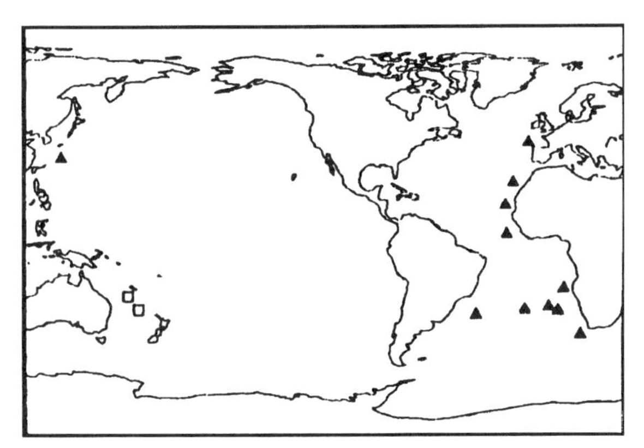

Figure 3.25. Changes in kaolinite content in deep-sea sediments from the late Oligocene to the early Miocene. *Filled triangles*, kaolinite change; *open squares*, no change. Reprinted from *Palaeogeography, Palaeoclimatology, Palaeoecology* 60, Robert, C. and H. Chamley, "Cenozoic evolution of continental humidity and paleoenvironment, deduced from the kaolinite content of oceanic sediments," pp. 171–87 (1987) with kind permission of Elsevier Science-NL, Sara Burgerhartstraat 25, 1055 KV Amsterdam, The Netherlands.

but there is a sharp transition to a predominance of illite in the overlying Rochester Shale. Noting that Upper Silurian rocks are evaporites, Meyer, Textoris, and Dennison (1992) speculated that the change in clay mineralogy signaled the transition to arid conditions. Although Meyer, Textoris, and Dennison (1992) were careful to observe that the shoreline shifted eastward coincident with the change in clay mineralogy, it seems likely that, barring unroofing of an entirely different suite of rocks, climate was the primary control.

Similarly, Hallam, Grose, and Ruffell (1991) documented variations in kaolinite through Late Jurassic-Early Cretaceous sequences in western Europe, emphasizing the decrease at many sites near the Jurassic-Cretaceous boundary. They noted diachroneity in the establishment of more arid conditions and hypothesized the development of highlands to the southeast that created a rain shadow.

Shallow-Water Clastic Sediments

The processes that control clastic sedimentation on continental shelves are so varied that distinguishing climate input is commonly very difficult. However, climate plays a role in the supply of sediment to the shelves and also directly affects some sediment-transport processes. The most important paleoclimatic indicators are storm deposits but, in some settings, the influence of climate on clastic input to the shelves has provided supplemental information about paleoclimate.

Influence of clastic input from the continents

Clastic sediments that accumulate on continental shelves reflect not only shelf conditions but also the climate of the adjacent continents. Gravel-sized clastics are most abundant where mean annual temperatures are well below 0°C, sand where temperatures are higher and rainfall relatively low, and mud where mean annual temperature is greater than 20°C and mean annual precipitation greater than 125 cm per year (Hayes 1967). These relationships are roughly related to the relative importance of chemical versus mechanical weathering. Because so many other processes affect clastic sedimentation on continental shelves, and the gravels at high latitudes might be relict from glaciation, the temperature-grain size relationship is weak. For example, Ziegler et al. (1979) distinguished between coarse and fine clastics in paleogeographic maps, but no apparent latitudinal relationship appeared in their maps.

Kvale et al. (1994) matched lamination patterns in Carboniferous rocks of the Illinois Basin with theoretical tidal and solar cycles and noted that the patterns deviated in a systematic way from the predicted pattern. Kvale et al. (1994) showed that these deviations likely were semiannual and represent seasonal sediment influxes, which they proposed resulted from seasonal rainfall.

Storm deposits

Storm deposits are recognized as beds of winnowed coarse grains or as graded beds. Storm-graded beds can be distinguished from turbidites by wave- versus current-dominated sedimentary structures, by evidence of multidirectional currents, and by grain-size distribution, that is, grains reworked in place (figure 3.26; Seilacher 1982). The sedimentary structure most commonly taken as indicative of storm influence is hummocky cross-stratification (Harms et al. 1975; Marsaglia and Klein 1983; Duke 1985; Nøttvedt and Kreisa 1987), but a number of other criteria also can help in the interpretation of storm deposits (Kreisa 1981).

Storm deposits resemble the idealized deposit illustrated in figure 3.26b (Kreisa 1981). Bases of storm beds are planar to undulatory, with wave

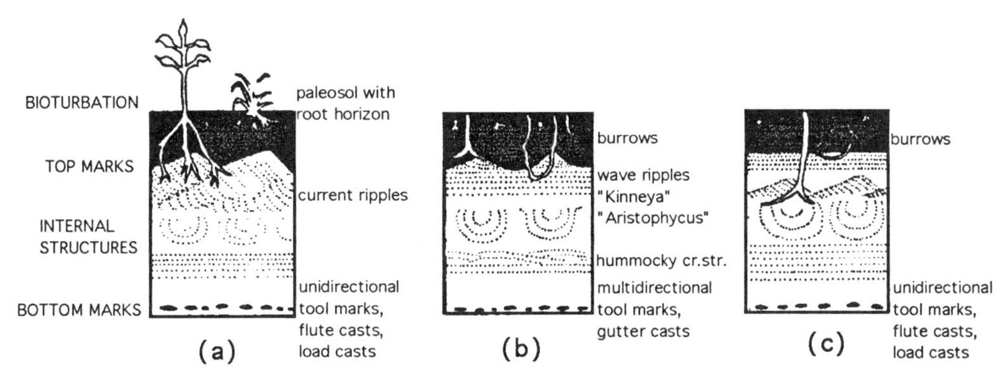

Figure 3.26. Comparison of idealized characteristics of flood *(a)*, storm *(b)*, and density current *(c)* deposits. *Black areas* are inter-event mudstones. From Seilacher (1982). Copyright © 1982. Reprinted by permission of Springer-Verlag New York from *Cyclic and Event Stratification,* G. Einsele and A. Seilacher, eds.

lengths ranging from 10 cm to several meters and erosional relief 1 mm to 18 cm expressed as shallow, elongate scars (called pot casts or gutter casts by other workers). Fossils are typically well preserved, uncorroded, and reworked (autochthonous), with few of the hallmarks of transport damage; indeed, the fossils may have material in or on them that is finer grained than the surrounding matrix. The coarse skeletal fraction is not graded and has no internal stratification or imbrication, and whole-fossil valves are oriented parallel to bedding, convex up in a 2:1 ratio. The laminated facies contains planar lamination, wave-ripple cross-lamination, hummocky cross-stratification, climbing wave-ripple cross-lamination, or a combination of these sedimentary structures, generally in the following order: plane and/or hummocky cross-stratification at base, overlain by climbing wave-ripple cross-lamination, overlain in turn by wave-ripple or matrix-rich planar lamination. The laminated beds commonly contain escape burrows (Kreisa 1981). Storm deposits vary by bathymetry and are thicker and commonly amalgamated closer to shore (Kreisa 1981; Aigner 1982). Other features include interbedded storm and fairweather beds, burrowed tops, lag-suspension couplets, and lenticular beds (Kreisa 1981).

Marsaglia and Klein (1983) and Duke (1985) compiled from the literature global data sets of storm deposits through time in order to determine the types of storms that result in the best geologic record and to document changes in storminess related to paleogeography and climatic change. Both studies recognized that most storms are one of two types, mid-latitude winter storms and hurricanes, and these will account for most of the geologic record of storms. Both studies assumed that the latitudinal distributions (table 3.3) and storm paths would have been the same in the past as they are today. Both studies relied on published information. Criteria used by Marsaglia and Klein (1983) were hummocky cross-stratification, intraformational conglomerates, erosional surfaces, shell lags, and mixed faunas, but they relied on the original authors' assessments of storm deposition. Duke's (1985) data set comprised reports of hummocky cross-stratification, which he regarded as the best storm indicator. Marsaglia and Klein (1983) distinguished deposits of winter storms from those of hurricanes by their paleolatitudinal positions and did not try to distinguish between them in regions of overlap (i.e., around 30° to 40° paleolatitude; table 3.3). On the other hand, Duke (1985) also used coastline orientation to resolve whether deposits were formed by winter storms or hurricanes in regions of overlap.

Notwithstanding strong debate on the sedimentology (Duke 1987; Klein and Marsaglia 1987; Swift and Nummedal 1987), the results of the two studies were not very different. Neither study revealed any patterns suggesting stronger periods of storminess in Earth history, despite the wide range of global climates sampled. Rather, paleogeography seemed to be the principal control on the location and abundance of storm deposits.

Iron-Bearing Minerals

Iron-bearing minerals most commonly discussed in paleoclimate studies are pyrite, chamosite,

Table 3.3.
Latitudinal distributions of present-day winter storms and hurricanes

Latitude	Storm systems
Above 45°	Frequent winter storms, very rare hurricanes
25°–45°	Common winter storms, occasional hurricanes
5°–25°	Occasional winter storms, occasional hurricanes
0°–5°	Very little storm activity of any kind

From Marsaglia and Klein (1983) in *Journal of Geology*, © 1983 by The University of Chicago.
"Occasional" means one to two times a year for winter storms, once every 3000 years for hurricanes.

Table 3.4.
Environments of formation of iron-bearing sedimentary minerals

Environment	Characteristic phases
Oxic ($C_{O_2} \geq 10^{-6}$)	Hematite, goethite, manganese oxides
Anoxic ($C_{O_2} \leq 10^{-6}$)	
Sulfidic ($C_{H_2S} \geq 10^{-6}$)	Pyrite, marcasite, rhodochrosite
Nonsulfidic ($C_{H_2S} \leq 10^{-6}$)	
Post-oxic	Glauconite, siderite, vivianite, rhodochrosite, no sulfide minerals
Methanic	Siderite, vivianite, rhodochrosite, earlier-formed sulfide minerals

From information in Berner (1981).
H_2S represents total sulfide; C is concentration in moles/l.

glauconite, siderite, and hematite. Primary hematite is evidence for oxidizing conditions at the time of formation; marine hematites are less common but no less unequivocal. Pyrite, chamosite, glauconite, and siderite are considered here. None of these minerals, including hematite, is a primary paleoclimatic indicator. Rather, they indicate conditions that can be related to climate.

Environmental conditions necessary for the formation of sedimentary iron minerals have commonly been distinguished on Eh-pH diagrams, but Berner (1981) recognized that in marine waters, pH is relatively constant and Eh is unmeasurable. Thus he proposed a classification of environments structured around oxygen and sulfide levels that provides a more practical guide to the evaluation of marine environments from iron-bearing minerals (table 3.4, figure 3.27). Environments are divided into oxic and anoxic, anoxic environments into sulfidic and nonsulfidic, and nonsulfidic environments into post-oxic and methanic. These categories relate to environments that develop during decomposition of organic matter, and they depend on the amount of organic matter and on the ionic species available for reduction

Oxic environments result in decomposition of organic matter by aerobic processes and oxidation of iron. Sulfate reduction to sulfides during decomposition of organic matter results in a sulfidic environment (also called euxinic). These environments are characterized by accumulation of H_2S and are considered strongly reducing. Post-oxic nonsulfidic environments occur where decomposition of organic matter is via oxygen, nitrate, manganese, and iron reduction, but where not enough organic matter is present to permit sulfate reduction; these environments are sometimes referred to as weakly reducing. Iron and manganese accumulate because of the lack of both oxygen and H_2S; the result is minerals such as glauconite, which has iron in both ferric and ferrous states (Berner 1981). Methanic environments are strongly reducing and occur when all H_2S is consumed as iron sulfides, allowing accumulation of iron and manganese and precipitation of siderite, vivianite, and other minerals as organic matter decomposition releases methane. Methanic environments are more common in nonmarine sediments, because in marine environments sulfate is so abundant it is rarely consumed before the organic matter is completely decomposed. Maynard (1982) extended Berner's (1981) classification to include descriptions of rocks that are commonly representative of each type of environment (see figure 3.27b). Siderite (iron carbonate) that forms in a methanic environment should have relatively high $\delta^{13}C$ (0‰ to +10‰), because in the reaction of organic matter to CH_4 and CO_2, light carbon is taken up preferentially by CH_4, and the CO_2 is incorporated in the carbonate, along with marine HCO^{3-}, which has a carbon isotopic composition around 0‰. Bearing in mind that siderite cannot form in oxic or sulfidic environments, a relatively low $\delta^{13}C$ (0‰ to −10‰) in siderite must indicate a post-oxic environment, because on decomposition all

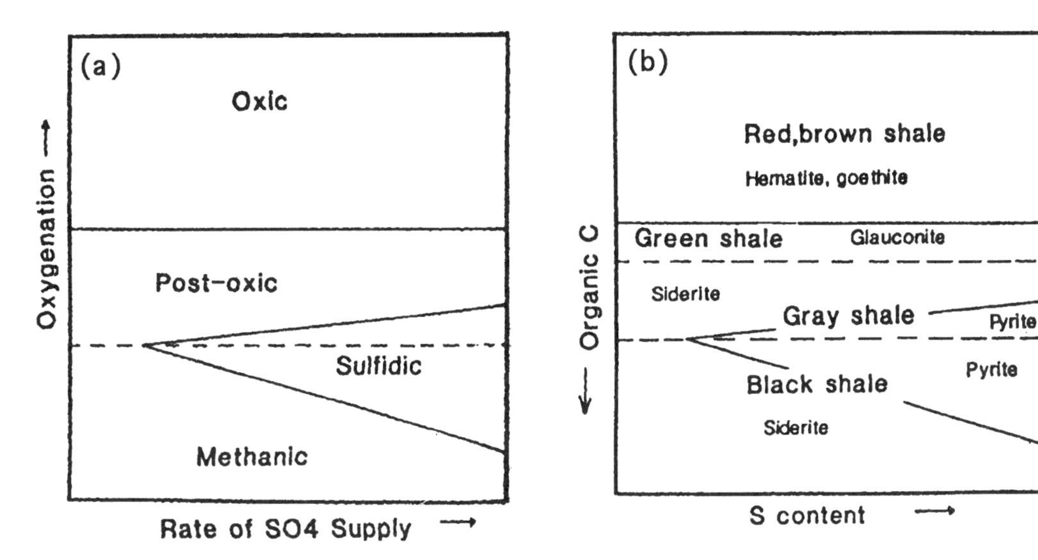

Figure 3.27. Diagenetic environments for the formation of iron-bearing minerals. *(a)* Graphical representation of Berner's (1981) classification of environments for the formation of iron-bearing sedimentary minerals (see table 3.4). *Dashed line* denotes presence/absence of oxygen. *(b)* Rock types expected in the fields defined by (a). *Upper dashed line* divides the post-oxic field into less and more organic-rich deposits. From Maynard (1982).

the organic carbon goes into CO_2 (Maynard 1982).

Pyrite

The importance of pyrite to paleoclimate studies has been mostly its role in the debate about the importance of anoxia to the preservation of organic matter. High S/C ratios have been taken as evidence for euxinic conditions, that is, the presence of H_2S in the sediments or water column, and have been used as a proxy for organic preservation. However, the relationship has not been upheld in subsequent studies (e.g., Calvert and Karlin 1991; Lyons and Berner 1992; Arthur and Sageman 1994). Therefore Raiswell et al. (1988) suggested the use of degree of pyritization (DOP) as an indicator of euxinic conditions. DOP is the ratio of Fe-sulfides to total reactive iron (sulfides + HCl-soluble iron). A DOP <0.45 indicates oxic conditions, a DOP of 0.45 to 0.75 indicates "restricted" conditions, and a DOP >0.75 indicates euxinic conditions. Calvert and Karlin (1991), however, showed that DOP can be iron limited. Thus pyrite does not hold up well as a paleoclimatic indicator, only as an indicator of reducing conditions in the sediments. Interestingly, however, DOP may be useful for understanding the preservation of organic matter. The preservation of hydrogen-rich organic matter may be enhanced in iron-poor environments (Tribovillard et al. 1994a), an observation that has implications for the preservation-production controversy.

Glauconite

Glauconite is an authigenic clay mineral that forms in weakly reducing conditions (see table 3.4, figure 3.27). It is kaolinitic when it forms but becomes increasingly smectitic (Odin 1988a). The general AIPEA-sanctioned definition is Fe-rich dioctahedral mica. The general formula is $K(R^{3+}_{1.33}R^{2+}_{0.67})(Si_{3.67}Al_{0.33})O_{10}(OH)_2$, with $Fe^{3+} >> Al$, and $Mg > Fe^{2+}$ (Bailey 1980). Much confusion surrounds the composition of glauconite, which in part lies in the observation that the final structure is not related to the initial composition (Odin 1973, 1988a). Rather, as authigenesis proceeds, the chemical evolution of the grains becomes increasingly dominated by exchange with sea water (Odin and Létolle 1980; Lisitsyna and Butuzova 1981; Clauer, Keppens, and Stille 1992) leading, for example, to potassium enrichment (Odin 1988a). A related mineral, chamosite, appears to be more easily characterized and less controversial (Odin 1973; but see Odin 1988b). Odin and Létolle (1980) recommended the term "glaucony" as the more specific term for the mineral glauconite, because of the widespread use of the latter word to describe any green grain, including some chamosites. This has not been widely adopted, but it should be because of the confusion that has arisen from indiscriminate use of the term "glauconite."

Glaucony is common in upwelling-zone deposits, forming part of the characteristic facies array discussed earlier in this chapter, but it is not an indicator of productivity. Rather, its formation as an authigenic mineral is favored by the conditions that are created by the high flux of organic matter to the sediments (Mullins et al. 1985), that is, a partial depletion of oxygen in the water column. Glaucony also can form in microreducing environments such as foraminiferan tests (Odin and Lamboy 1988). Glaucony occurs in both warm- and cool-water regimes (Odin 1973; Van Houten and Purucker 1984). Deposits result from winnowing and concentration of the green grains (Hein, Allwardt, and Griggs 1974; Van Houten and Purucker 1984; Glenn et al. 1994b), which can lead to confusion about the environment of formation (Amorosi 1997). The progressive maturation of glaucony grains with time can help identify deposits that were reworked when water conditions changed (Amorosi 1997).

Siderite

Siderite is mostly indicative of sulfate-poor waters, as discussed in the introduction to this section. However, Hallam (1984) linked siderite with iron oxyhydroxides and chamosite as a humid-climate indicator. This might be because it tends to form in waters that are low in sulfate, either freshwater or seawater that has been diluted by freshwater influx (see figure 3.27b). At best it can be used as one of many indicators suggesting such a climate.

Manganese

Manganese deposits take several forms, most of which are not linked to climate. These include deep-sea manganese nodule fields (e.g., Dymond and Veeh 1975) and late diagenetic, chert-hosted manganese ores (e.g., Hein, Koski, and Yeh 1987). However, sedimentary rock-hosted manganese ores and elevated manganese contents in sedimentary rocks have been linked to variations in oxygenation levels in the water column. The crux of these arguments is the observation that Mn^{2+} is strongly soluble in oxygen-depleted water (<2.0 ml/l O_2) and the difference in concentration of Mn in oxygen-depleted and oxygenated waters is several orders of magnitude (e.g., Dickens and Owen 1994). With large-scale expansion of the oxygen-minimum zone (OMZ), as proposed for oceanic anoxic events, large amounts of Mn^{2+} would accumulate in the water column. With reoxygenation, Mn^{2+} would precipitate as manganese oxide. An example of the application of this model to sedimentary rock-hosted manganese ores is illustrated in figure 3.28 (from Force et al. 1983; see also Cannon and Force 1983). Similar models, with slight variations, have been applied not only to manganese deposits on shelves and in epeiric seaways (Pomerol 1983; Frakes and Bolton 1984; Pratt, Force, and Pomerol 1991), but also to manganese flux rates in the deep sea (Dickens and Owen 1994).

Recently, the OMZ model for sedimentary rock-hosted manganese ore formation has been questioned for carbonate-hosted Mn on several grounds (Calvert and Pedersen 1996). First, de-

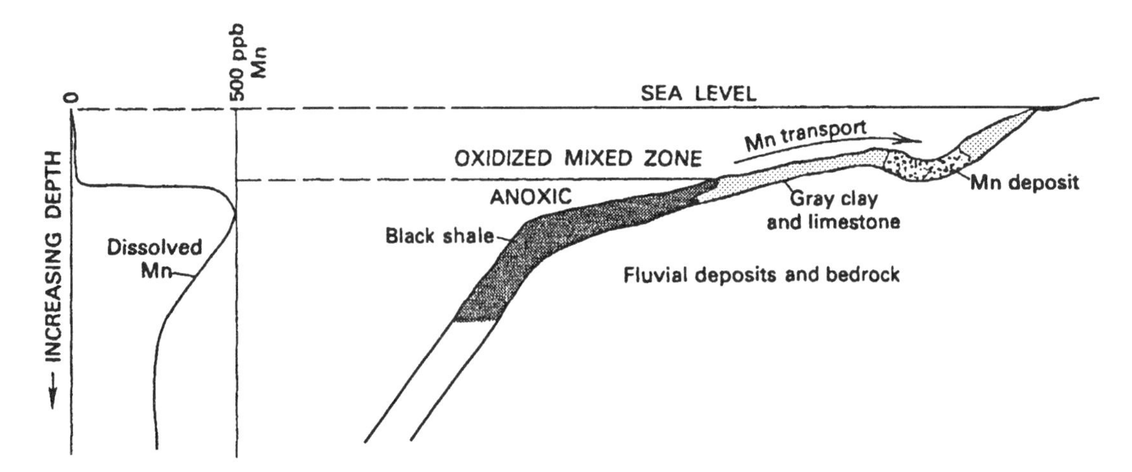

Figure 3.28. Application of the oxygen-minimum-zone model of manganese ore accumulation to sedimentary rock-hosted manganese ores. From Force et al. (1983).

spite the high concentrations of Mn^{2+} in oxygen-depleted waters, the concentrations are too low to account for the large accumulation of Mn in ore-grade deposits. Second, the principal phase, Mn-calcite, forms only in anoxic sediments that lie below oxygenated bottom waters, and accumulation is effected by recycling of Mn^{2+} between the anoxic and the oxic zones in the sediments; this was called the manganese pump by Calvert and Pedersen (1996). Manganese precipitates first at the sediment surface as Mn-oxyhydroxides, which are more soluble in anoxic pore waters than Mn-carbonates. The most efficient operation of the pump is where rapid accumulation of organic matter leads to consumption of oxygen at shallow depths in the sediment and accumulation of Mn-oxyhydroxides at the oxic-anoxic interface; eventually, as a result of burial, the interstitial concentration of Mn becomes high enough to cause precipitation of Mn-carbonates. The association of organic-rich shales and manganese deposits is, in Calvert and Pedersen's (1996) view, an indication that the organic matter accumulated rapidly under oxygenated bottom waters.

Barium and Barite

Elevated barium contents in sediments have been used as indicators of high biologic productivity since the observation by Goldberg and Arrhenius (1958) of a striking peak in Ba/TiO_2 at the equator in the Pacific Ocean (Paytan and Kastner 1996). In addition to being incorporated in foraminifera, barium precipitates as barite. Variations in Ba content in deep-sea sediments are used to supplement other information on productivity variations. For example, Zachos, Arthur, and Dean (1989) used barium mass accumulation rates and Ba/Al ratios to support their conclusion that biologic productivity collapsed in the North Pacific at the Cretaceous-Tertiary boundary; similarly, Thompson and Schmitz (1997) correlated Ba/Al ratios and $\delta^{13}C$ to interpret productivity changes in the Atlantic in the late Paleocene. From data such as those summarized in figure 3.29, Dymond, Suess, and Lyle (1992) produced an empirical equation for the calculation of new organic production from barium fluxes that was refined by Gingele and Dahmke (1994), who broke down total barium content into its various sediment constituents—30 ppm in carbonate, 120 ppm in biogenic silica,

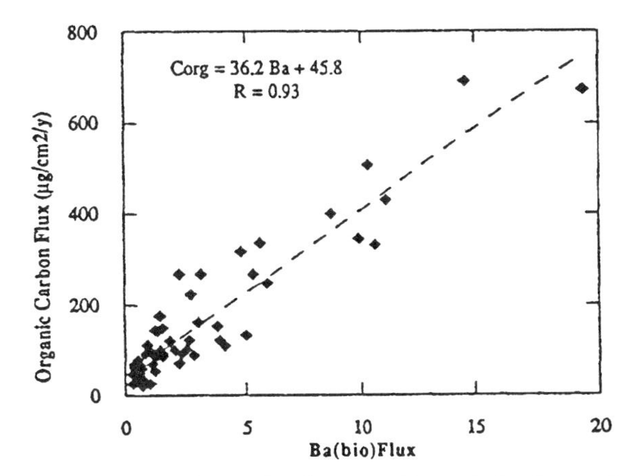

Figure 3.29. Relationship between biogenic Ba flux and organic-carbon flux to shallow-water sediment traps. From Dymond, Suess, and Lyle (1992). *Paleoceanography* 7:163–81, copyright by the American Geophysical Union.

60 ppm in particulate organic matter, and 400 ppm (range, 200 to 1000 ppm) in terrigenous detrital components—and established correction factors for the nonproductivity-related Ba components. Not surprisingly, clay-rich cores show a greater effect of correction for terrigenous Ba than cores that are mostly pelagic carbonate (see also Schroeder et al. 1997). More recently, François et al. (1995) pointed out that some of Dymond, Suess, and Lyle's (1992) samples may have contained refractory organic matter derived from the continents, and they modified the method to include only samples with low organic carbon-to-biogenic Ba ratios.

The use of Ba as a productivity indicator is ideal in pelagic sediments, where TOC, Ba concentration, and weight-percent barite ($BaSO_4$) are strongly correlated. In shelf cores, the signal is relatively weak and, where sediments are rapidly accumulating, can be completely obscured (Gingele and Dahmke 1994). Depth is also important in Ba accumulation, making site-to-site comparisons difficult. Gingele and Dahmke (1994) also discussed the mobility of Ba in pore waters (Arthur and Sageman 1994) and the fact that diagenetic barite layers can form at the oxic-anoxic interface in the sediments, particularly in highly productive, nearshore deposits, where organic input creates suboxic conditions at shallow depth in the sediment (Dymond, Suess, and Lyle 1992). The source for Ba in these layers is from below, so Ba cannot be used to estimate productivity below such layers.

Massive barite deposits are hydrothermal or

sedimentary; where they are sedimentary, they are regarded as indicative of past high biologic productivity (Murchey, Madrid, and Poole 1987; Coles and Varga 1988; Jewell and Stallard 1991; Jewell 1994). Some of the largest such deposits are in Nevada, in Paleozoic rocks, and are associated with organic-rich shales and phosphorite. Jewell and Stallard (1991) and Jewell (1994) proposed a model for accumulation of the Devonian barites that is the geographic equivalent of the formation of diagenetic barite layers from pore waters. In their model, barite particles settle into H_2S-rich waters and dissolve because sulfate is undersaturated as a result of sulfate reduction beneath the high-productivity zone (see also Arthur and Sageman 1994). As the sulfidic waters are advected alongshore out of the zone of highest biologic productivity and mixed with oxygenated water, barite precipitates at high rates.

Bedded barite does not occur in modern upwelling zones. However, some workers (e.g., Berry and Wilde 1978) have suggested that the Devonian oceans overall were oxygen depleted, meaning that sulfate reduction would have been more likely under conditions of high biologic productivity than they are today (Jewell and Stallard 1991). Calvert and Price (1983) documented a pattern of barium accumulation at the edges of the most organic-rich zone off Namibia, but the barium is mostly offshore, not downstream along the undercurrent, although a broad area of elevated Ba does lie south of the main organic-rich zone off Walvis Bay. The principal current at depth along much of the shelf is offshore-onshore (Dingle and Giraudeau 1993), a pattern opposite to that expected from a mechanism of Ba accumulation such as that proposed by Jewell and Stallard (1991) and Jewell (1994). Calvert and Price (1983) concluded that Ba off Namibia indicates open-ocean productivity, possibly because the open-ocean taxa involved are more efficient at transporting Ba.

Other Major and Trace Elements

Other major and trace elements have been studied intensively for their relationship to organic matter content and accumulation. Part of the impetus for these studies was the observation that many black shale units are enriched in metals and other trace elements. For example, the organic-rich Kupferschiefer in Europe derives its name from the very high concentration of copper. Certain trace elements, Cu among them, appear to have an affinity for organic matter. Others cited by at least some authors are Ag, As, Cd, Co, Cr, Mo, Ni, Pb, Sb, U, V, and Zn (Calvert and Price 1983; Brumsack and Thurow 1986; Calvert 1990; Arthur and Sageman 1994), but their means of incorporation into the sediment differ. For example, As and Mo appear to be greatly influenced by the formation of sulfides, whereas U and V bind to organic matter or phosphate directly (Calvert and Price 1983; Arthur and Sageman 1994). In the Black Sea, Zn, which is strongly related to organic matter in other deposits, shows no relationship to organic matter (Calvert 1990). An advantage of the use of trace elements, if they do indicate environment, is that they are apparently not affected by the degree of maturation of organic matter (Mongenot et al. 1996).

In a series of papers, Brumsack (e.g., Brumsack 1980, 1981, 1986; Brumsack and Thurow 1986) attempted to show that the types of metals accumulated are dependent on the mechanism of organic-matter accumulation, that is, whether the organic matter accumulated as the result of high biologic productivity or because accumulation was under an anoxic water column independent of high biologic productivity. Although some results seemed to point to a distinction between the two types of environments, and thus had relevance to the production-versus-preservation controversy over organic-matter accumulation, the results were inconsistent. Arthur and Sageman (1994) eventually concluded that no clear distinction could be drawn between the two types of environments based on trace element enrichment patterns.

Piper (1994) argued that metalliferous organic-rich rocks reflect more closely the original organic-matter composition and that they are not, in fact, enriched by unusual seawater chemistry or addition of metals from some other source. The rocks are metal rich because they are organic rich. This is true not because the metals were somehow attracted to the organic matter, as suggested by previous workers, but because they were part of the organic matter. He showed that the trace metal composition of metalliferous organic-rich rocks is virtually identical to the trace metal composition of modern marine plankton, and he suggested that nonmetalliferous organic-rich rocks have simply lost the original metals during diagenesis and weathering. Piper's (1994) study indicates that biologic productivity is as important for understanding the distribution of trace metals as it is for understanding the distribution of organic matter. Even moderately elevated productivity can lead to high metal contents (Piper 1994). Some metals are enriched by precipitation and adsorption from seawater, reflecting the redox conditions of the bottom waters. Piper cited rocks high in Mn, Cr plus V, and Mo as indicative of oxidizing, denitrifying, and sulfate-reducing conditions, respectively. He emphasized that the organic matter drives the redox reactions that lead to these enrichments.

Geophysical Methods for Interpreting Paleoproductivity and Total Organic Carbon

One of the limitations on studying the three-dimensional distribution of any rock characteristic is the paucity of subsurface data. Geophysical methods have contributed to the interpretation of subsurface structures and gross lithologic changes. Gradually, these methods have come to be applied to subtler variations, and particularly pertinent to this book is the use of geophysical techniques to study the distribution of organic carbon and indicators of biologic productivity in the subsurface.

Most of the productivity methods have been developed from geophysical logs compiled during deep-sea drilling cruises. Magnetic susceptibility has been proposed as a paleoproductivity indicator (Hartl, Tauxe, and Herbert 1995), the argument being that higher productivity increases the carbon rain, promoting iron and sulfate reduction, possibly in the microenvironment of sinking plankton tests, and dissolution of magnetite. However, other work, while finding a correlation between $\delta^{18}O$ and magnetic susceptibility measurements, failed to find a correlation between magnetic susceptibility and paleoproductivity indicators, or with eolian dust (Meynadier, Valet, and Grousset 1995). Sediment-density data from gamma-ray attenuation porosity evaluator (GRAPE) has been used to evaluate variations in terrigenous versus biogenic opaline deposits in the North Pacific for young sediments (95 Ka to the present) and to document the history of terrigenous sediment flux related to ice-rafting (Kotilainen and Shackleton 1995), but this technique has not been widely applied to paleoclimate studies in older rocks.

Knowing the three-dimensional distribution of organic carbon in the subsurface would be useful for interpreting productivity patterns, but the relatively scarce data from cores has limited our ability to approach productivity in this way. Recently, French geologists have developed a method of interpreting organic carbon contents using well logs, specifically sonic and resistivity logs (Carpentier, Huc, and Bessereau 1991). The method, called CARBOLOG®, takes into account the major components of the rock, that is, pore-water content, matrix composition, amount of organic matter, and clay content, including clay-bound water. On a sonic log, transit time is slower in water, organic matter, and clay than in other minerals, and thus a change in transit time indicates a change in the proportions of these constituents. Which constituents have varied cannot be determined from the sonic log alone. However, organic matter can be distinguished from water and clay using the resistivity log (which measures electrical resistivity). Water and clay have much lower resistivities than do organic matter and minerals besides clay. Thus an increase in transit time coupled with a decrease in resistivity indicates an increase in the water and/or clay content, whereas an increase in transit time coupled with an increase in resistivity indicates an increase in organic matter (Carpentier, Huc, and Bessereau 1991). So long as the logs can be calibrated with a few cores from the same basin, the method is accurate to within 20% or less (Tanck 1997).

Ikaite

Ikaite is an unstable mineral replaced by calcite in forms variously known as thinolite, glendonite, and other names (Brandley and Krause 1994). It forms as single, angular crystals or elongated rosettes, typically several centimeters long. Ikaite forms at temperatures as high as 8°C under high hydrostatic pressures (Shearman and Smith 1985; Stein and Smith 1986; Jansen et al. 1987; see also Brandley and Krause 1994), but, in shallow water, ikaite appears to be stable only at freezing temperatures (Suess et al. 1982; Kemper 1987; Kennedy, Hopkins, and Pickthorn 1987; Sheard 1990). Thus ikaite is regarded as an indicator of freezing conditions (Kemper 1987).

4

Terrestrial and Freshwater Biotic Indicators of Paleoclimate

Terrestrial biotic indicators of climate are divided into five groups: microfossils, including plant microfossils (pollen and spores), ostracodes, diatoms, and miscellaneous (charophytes and conchostracans); plant megafossils[1]; invertebrates; vertebrates; and trace fossils. Pollen and spores are included with microfossils rather than plants because they often co-occur with other microfossils, and paleoclimate interpretation of pollen is more commonly linked with other microfossils than with plant megafossils. Cuticle (the waxy outer layer of leaves), although collected in palynological preparations and analyzed microscopically, is best used in conjunction with leaf megafossils and therefore is included in the section on plant megafossils. Among the invertebrates, only mollusks have achieved any importance in pre-Quaternary paleoclimatology. Insects are important in Quaternary paleoclimatology, but sporadic efforts to interpret them with respect to climate in the older record have been unsatisfactory for a variety of reasons (e.g., see comments in Ponel 1995), despite their long geologic record (Labandeira and Sepkoski 1993). Plant megafossils and vertebrates have to date been the most informative for pre-Quaternary continental climates.

1. The term "megafossils" has become more prevalent in paleobotany than "macrofossils," which is the term applied to large invertebrates. I retain the conventional usages in this book.

CONTINENTAL MICROFOSSILS

The most paleoclimatically significant terrestrial microfossils are pollen and spores, ostracodes, and diatoms. Charophytes and other phytoliths and conchostracans (which are as large as 1 cm) have been used in some paleoclimatologic studies but have not been important. Paleoclimate studies using terrestrial microfossils, especially pollen, have been overwhelmingly concentrated in rocks of Quaternary or Holocene age, where they have provided most of the available information on terrestrial paleoclimates. The reason for this is that young fossils can be most confidently tied to extant taxa whose tolerances are known.

Pollen

Of the continental microfossils, pollen are by far the most important in pre-Quaternary paleoclimatologic studies. Use of pollen and spores for interpreting pre-Quaternary climates tends to follow either the nearest-living-relative (NLR) method or traditional biogeographic practices, depending on the age of the flora.

In a study of Miocene and younger palynofloras, Liu and Leopold (1994) and Leopold and Liu (1994) used the NLR method but also incorporated dominance and diversity into their interpretations. They (Leopold and Liu 1994) identified thermophiles, consisting of hardwoods from the modern northern hardwoods and mixed deciduous and mixed mesophytic vegetation types. The percent thermophiles then constituted a climate index, and they plotted this index for central Alaska (figure 4.1) and Siberia in the Miocene (Leopold and Liu 1994). Their temperature estimates showed a lower latitudinal temperature gradient between China and Alaska than is the case today (Liu and Leopold 1994). The trends in the climate index compared favorably with a deep-sea temperature curve for the North Pacific. In a subsequent study of Alaska and northwestern Canada in the Miocene, White et al. (1997) not only used the method employed by Liu and Leopold (1994) and Leopold and Liu (1994) but also employed percent ratios of pollen and spore taxa, mostly at the family and genus level, to estimate growing season temperature

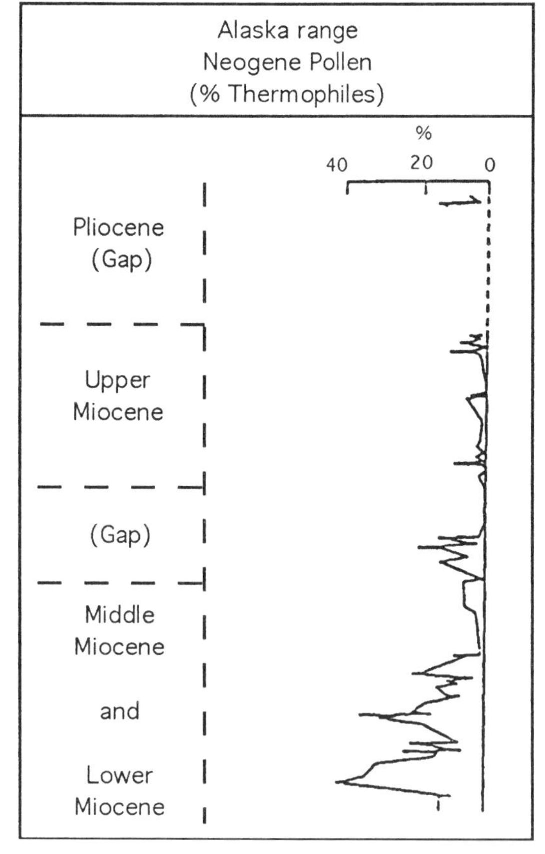

Figure 4.1. Climate index based on percent thermophilous pollen taxa in central Alaska and Siberia in the Miocene. A higher percentage of thermophilous taxa indicates warmer climate. Reprinted from *Quaternary International* 22/23, Leopold, E.B. and G. Liu, "A long pollen sequence of Neogene age, Alaska Range," pp. 103–40, Copyright 1994, with permission from Elsevier Science Ltd, The Boulevard, Langford Lane, Kidlington 0X5 1GB, UK.

(ratio of percent pollen from warm-climate trees to percent pollen from cool-climate trees), as well as ecological parameters such as canopy density (ratio of percent shrubs-plus-herbs to trees). The events recorded were correlated with global events derived from other records.

Liu and Leopold (1994) and Leopold and Liu (1994) could be confident of their taxonomic identifications, but the NLR method has also been used even when identifications are less certain. For example, in a study of Tertiary rocks of Arctic Canada (Kalkreuth, McIntyre, and Richardson 1993), pollen types were generally referred to modern taxa for convenience, with the use of modifiers such as "considered to be" or "probably" in reference to the taxonomic names. Despite this uncertainty, Kalkreuth, McIntyre, and Richardson (1993) could interpret from the palynoflora a mixed coniferous forest with a diverse angiosperm understory that "grow[s] in temperate climates with moderate rainfall," by analogy with modern forests. Taxodiaceous pollen, resembling those of *Metasequoia* (dawn redwood), some *Taxodium* (swamp cypress), and rare *Sequoia* (coast redwood) and *Glyptostrobus* (Chinese water pine), constituted 63% to 74% of most samples, with pine, spruce, and diverse angiosperm pollen constituting most of the rest of the palynoflora.

Studies on Mesozoic and older Cenozoic palynofloras tend to use traditional biogeographic methods. For example, Hubbard and Boulter (1983) used cluster analysis to identify four groups of pollen form genera in Paleogene rocks in Europe. They interpreted these as the Paleogene equivalents of groups now found in deciduous forest, fern and conifer forest, and paratropical rainforest; the fourth group was a mixed assemblage that they interpreted as reflecting the influences of transport. Hubbard and Boulter (1983) then determined paleoclimate by analogy with climatic controls on vegetation type, an approach similar to that used by Kalkreuth, McIntyre, and Richardson (1993).

Visscher and van der Zwan (1981) used a similar approach to identify two biogeographic units in Triassic rocks of the circum-Mediterranean region. One group is characterized by *Camerosporites secatus* and other conifer form genera and the other group by fern spores and cycad pollen. They interpreted these groups as xerophytic and hygrophytic, respectively, based on analysis of the floral assemblages from evaporites and coal beds, rather than on the ecology of modern representatives of those plant groups. Subsequently, van der Zwan, Boulter, and Hubbard (1985) used multivariate statistics on Early Carboniferous pollen and spores and their occurrence in coal beds and evaporites to define four types of climate: dry, moderately humid, humid, and very humid. Floral elements were considered humid when the principal components (PC, from principal components analysis) were high in coal beds, moderately humid when the PC were low in coal beds or high in coal beds and low in evaporites, neutral when the floral elements were absent from coal beds or PC were low in coal beds and low in evaporites, and dry when PC were high in evaporites or low in evaporites and absent from coal beds. The neutral floral elements were those that occurred everywhere. An interesting element of their analysis was the division of humid-climate floras into moderately humid, humid, and very humid, whereas remaining floral elements were either neutral (that is, without climatic significance) or dry. This illustrates an important general aspect of paleofloral studies—that dry climates tend to have fewer plants and fewer deposits in which plant remains are likely to be preserved and thus provide a smaller database on which to make floral distinctions. Another crucial point is that the sediment from which pollen are sampled can be central to paleoclimate interpretation, as it was in van der Zwan, Boulter, and Hubbard's (1985) study and others (e.g., Falcon 1989).

It is rare for single taxa of any group to have particular paleoclimatic significance. However, one pollen type, the Mesozoic genus *Classopollis*, has, indeed, proved to be particularly interesting and useful paleoclimatologically (Vakhrameyev 1982). *Classopollis* is unusual, especially among pollen taxa from the older geological record, in having been linked to other plant organs, specifically, the cones and leaves from plants in the family Cheirolepidiaceae. Based on associations with climatically significant rocks and other fossils (Vakhrameyev 1982), as well as on the morphology of the plants (Alvin, Spicer, and Watson 1978; Alvin, Fraser, and Spicer 1981; Francis 1983), *Classopollis* is regarded to be indicative of warm and, possibly, semi-arid and/or seasonal climates. In the southern portion of the former Soviet Union, the abundance of *Classopollis* varies dramatically through the Jurassic and Cretaceous, and these variations coincide well with changes in the distribution of other paleoclimatic indicators (Vakhrameyev 1982). The abundance also decreases from south to north along a paleolatitudinal gradient. Parrish et al. (1996) used *Classopollis* to interpret climate when little other paleoclimate information was available from boreholes in Australia.

Pollen and spores are easily transported by water, and their varying shapes, densities, and sizes mean that the grains also are hydrodynamically sorted (Muller 1959; Traverse and Ginsburg 1967; Mudie and McCarthy 1994). Saccate pollen, such as those derived from conifers, float readily and can be transported long distances. Spores are typically very small (<100 μm), and angiosperm pollen are typically smaller than gymnosperm pollen and much denser (Muller 1959; Traverse and Ginsburg 1967). Pollen are also produced in varying quantities by plants, depending on whether pollination is primarily by dispersal through the air or by insect or other animal vectors. Wind-pollinated plants typically produce far more pollen than do those that rely on animals. Studies that interpret short-term climatic change by changes in pollen counts must take these factors into account (Spicer and Parrish 1990a; Mudie and McCarthy 1994).

An additional problem related to transport is the fact that pollen are so durable they are easily eroded and redeposited, resulting in stratigraphically mixed assemblages (e.g., Norris and Miall 1984). Such mixing can be difficult to detect unless the ages of the mixed assemblages are significantly different (on the order of 10^6 years or more) or the older assemblage has undergone significant alteration in color. Two types of evidence and strategies that can be used to argue against mixing or to avoid mixed assemblages are limitation of samples to those assemblages that occur in nonmarine rocks and organic-rich horizons, and an orderly ecological transition of pollen and spore types, assuming the floras overall are well enough known to detect anomalies (Leopold and Liu 1994).

Nonmarine Ostracodes

The distribution of nonmarine ostracodes is dependent on temperature, light (which affects lacustrine vegetation), sediment texture, substrate type, water chemistry, predation, wave energy, and food supply (Palacios-Fest, Cohen, and Anadón 1994). The most important of these are temperature and aquatic chemistry, to which ostracodes are very sensitive (Delorme 1969; Cohen, Dussinger, and Richardson 1983; Forester and Brouwers 1985; De Deckker and Forester 1988; Gasse et al. 1990). Species abundance and diversity of ostracodes, as well as taxonomic composition, are controlled by water chemistry. Climate is only one control on lake chemistry (Eugster and Hardie 1978; De Deckker and Forester 1988; Forester 1991; Palacios-Fest, Cohen, and Anadón 1994).

Because ostracodes are so sensitive to water chemistry, very precise conclusions about paleolake chemistry can be reached if extant taxa are present in the samples and if the chemical tolerances of those taxa are known. Examples of chemical ranges of several modern species are shown in figure 4.2. Taxa whose tolerances are known have figured prominently in studies of younger deposits, which contain taxa that are still extant (Cohen 1981; and many others).

As useful as ostracodes are for tracing the chemical evolution of lakes, the most precise methods depend on knowledge of the environmental tolerances of individual species and thus require that the species represented in the fossil record be extant (Carbonel et al. 1988; De Deckker and Forester 1988). Some methods of paleoenvironmental interpretation, such as isotope and trace-element analyses, may be useful in older rocks. Isotope analysis in lacustrine ostracodes is subject to numerous limitations, and interpretation of trace-element results can also be dependent on knowledge of individual taxa (Chivas et al. 1993). Some methods, such as taphonomy (De Deckker and Forester 1988), are applicable to faunas of any age. In practice, paleoclimate studies using nonmarine ostracodes have rarely been done in rocks older than Pliocene (Carbonel et al. 1988).

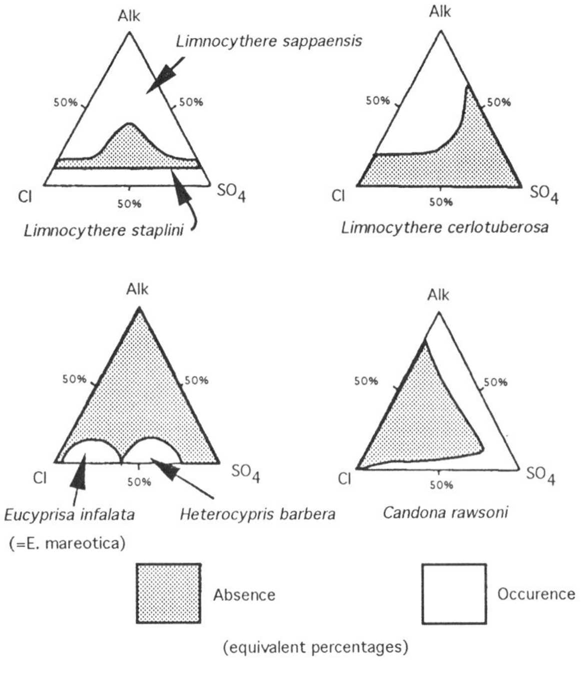

Figure 4.2. Chemical ranges of some modern nonmarine ostracode species. From Palacios-Fest, Cohen, and Anadón (1994), partially adapted from Forester (1986).

Some attempts have been made to link ostracode morphology to environmental control (table 4.1), but they are controversial (Carbonel et al. 1988). In addition to the characteristics listed in table 4.1, large tests are more common in cold water rich in dissolved nutrients, and ornamentation patterns are commonly more "intense" where ionic concentration, especially the Mg/Ca ratio but not salinity, is elevated (Carbonel et al. 1988). Nodosity, that is, outpocketings of the shell that involve the inner lining, is better developed in low-salinity environments and/or where abundant organic matter and silica are available. Forester (1991) argued that exotic carapace morphologies (see his figures 7 and 9) are typical of ostracodes living in large, stable lakes, which he inferred as implying a climate that was stable over geologic time. If these morphological variations and their environmental

Table 4.1.
Relationship between ostracode shell morphology and environment

Shape	Meaning
Almost rectangular	Harsh and unstable environments (Arctic ponds, ephemeral ditches)
Elongate, slightly triangular	As above
Elongate, pointed posterior	Persistent environments with low fluctuations (e.g., groundwater, sublittoral of old lakes)
Highly triangular	As above
Trapezoidal	As above
Widely arched dorsally	None stated

From information in Carbonel et al. (1988).

significance can be calibrated and different groups of ostracodes have the same morphological responses to climatic stimuli, ostracode morphology would have much potential in pre-Quaternary climate studies.

Other Nonmarine Microfossils

Like nonmarine ostracodes, nonmarine diatoms have been used extensively in studies of Quaternary climates but rarely in studies of older rocks (Bradbury 1988). Nonmarine diatomites are more common in humid, warm areas because abundance of nonmarine diatoms is likely to be controlled by silica supply, which is greater where weathering rates are faster (Tappan 1980). Silica accumulation rates are used as a proxy for productivity in freshwater diatomites, as they are in marine diatomites (Qiu et al. 1993).

Many diatom studies have been done in conjunction with pollen studies and, in those cases, much of the paleoclimate information comes from the pollen, rather than from the diatoms, although the diatoms provide complementary information. Where extant taxa are present and their tolerances known, nonmarine diatoms are as useful as nonmarine ostracodes in providing information about pH, chemistry, salinity, trophic status, depth, turnover characteristics, erosion and clastic influx, turbidity, turbulence, and stratification (Bradbury 1988). However, for lakes with extinct floras, paleoclimate is an input to the study of diatom ecology, not a a result (Bradbury 1988). Regardless, some workers have attributed certain nonmarine diatom floras to certain paleoenvironmental conditions (e.g., Gasse 1990).

Charophytes are the microscopic calcified reproductive structures of freshwater to brackish-water plants, and the species are strongly controlled by salinity (Wray 1978; Tappan 1980; Bachhuber 1992). As with other continental microfossils, they are most useful for environmental interpretation in Quaternary rocks that have species that are still extant (Kröpelin and Soulié-Märsche 1991; Soulié-Märsche 1993). However, Soulié-Märsche (1993) found that variation in size and aspect-ratio of one species in a Holocene deposit was relative to wide salinity fluctuations. If this can be shown to be a general response of charophyte taxa, it would be useful in paleoclimate studies in the older geologic record.

Conchostracans, also known as clam-shrimp, are valved arthropods that occur in rocks as old as mid-Cambrian (Frank 1988). Today they are confined to temporary bodies of freshwater, sometimes high in pH and salinity, in "subarid warm regions" (Frank 1988). They are commonly used as indicators of climates that were seasonal with respect to rainfall in the past.

TERRESTRIAL MACROFOSSILS: PLANTS

Terrestrial plants are potentially the most sensitive paleoclimatic indicators of the world's biota (Spicer 1989a). Unlike aquatic organisms, they are exposed directly to the atmosphere and, unlike terrestrial animals, they cannot modify their environment by their behavior. Leaves, in particular, are excellent indicators of the environments in which plants lived because, unlike other fossils, leaves cannot be transported far without being destroyed (e.g., Spicer and Greer 1986; Spicer and Wolfe 1987). Terrestrial plants, more than any other indicator except perhaps paleosols and coal beds, are sensitive to evapotranspiration rather than just temperature or just moisture. Plants are strongly influenced by soil conditions, but as these are also at least partly controlled by evapotranspiration, plant fossils and their associated paleosols may give similar information (but see Demko 1995).

Plant megafossils are most commonly leaves, seeds, or wood; flowers and other reproductive structures are not as commonly preserved. Paleoclimate information can be derived from plant megafossils in four ways: from their biogeography, from various aspects of their morphology, from taphonomy of plant organs, and from reconstructions of the vegetation.

Plant Megafossil Biogeography

The most problematic aspect of working with plant megafossils, which is particularly acute in phytogeographic studies, is the state of the systematics (e.g., discussions in Collinson 1986; Wolfe 1993). Plant fossils have been subject to the usual problems of monographic effects. In addition, fossil leaves in particular have commonly been given names of modern taxa, which implies that the taxa have been much longer lived and were much more widespread than is probably the case; this is especially true for older floras. Some who work on angiosperm leaves have abandoned traditional taxonomic names entirely, preferring instead to use descriptive form genera (Spicer 1986a,b). The problem is not limited to the angiosperms.

Notwithstanding these limitations, paleophytogeographic analyses can be useful for understanding paleoclimatic patterns if the data are used circumspectly. Literature data, especially figures, must be examined closely for probable synonymies and recorded appropriately. Ziegler (1990) used this approach in adapting descriptions of modern large-scale plant formations (biomes) to Permian plant megafossils (table 4.2). He used a combination of general concepts of morphologic adaptations in certain plants, their latitudinal distributions (see table 4.2), and their associated sediments to assign Permian floras to the biomes. For example, the predominately Chinese Cathaysian flora occurred in low paleolatitudes associated with coal beds. The gigantopterids, a major component of the flora, have been interpreted as lianas, which in modern vegetation are most abundant in tropical and paratropical forests. These floras were assigned to biome 1, evergreen tropical rain forest (see

Table 4.2.
Modern vegetational biomes and their latitudinal and climatic parameters

Biome	Present vegetation	Climate descriptor	Subdivision	Precipitation systems	Poleward climate boundary	Poleward latitude
0	None	Glacial	None	Orographic	None	90°
9	Tundra vegetation (treeless)	Arctic	None	Some summer frontal	0 month $\overline{T}$>0°C	c. 70°–60°
8	Boreal coniferous forests (taiga)	Cold Temperate	None	Summer frontal	<1 month $\overline{T}$ ≥ 10°C	c/ 66°–58°
7	Steppe to desert with cold winters	Midlatitude desert	None	None due to distance or rain shadows	<4 month $\overline{T}$ ≥ 10°C	51° ± 2°
6	Nemoral broadleaf-deciduous forests	Cool temperate	Western	Winter frontal, summer convective	<4 month $\overline{T}$ ≥ 10°C	c. 58°
			Eastern	As above	As above	44° ± 2°
5	Temperate evergreen forests	Warm temperate	Western	Winter frontal summer convective	Winter $\overline{T}$ <0°	46° ± 14°
			Eastern	As above	As above	36° ± 2°
4	Sclerophyllous woody plants	Winterwet	None	Winter frontal	≥10 month precip. >20 mm	38° ± 5°
3	Subtropical desert vegetation	Subtropical desert	Coastal	None, due to cold upwelling offshore	Limit of winter rains	32° ± 4°
			Inland	None, descending limb of Hadley Cell	Winter $\overline{T}$ <0°C	c. 33°
2	Tropical deciduous forest or savannas	Summerwet	Subtropical	Summer monsoon	<3 month precip. >20 mm	up to 25°
			Tropical	Summer extension of ITCZ	As above	15° ± 5°
1	Evergreen tropical rain forest	Everwet	Tropical	Coastal diurnal	<11 month precip. >20 mm	up to 25°
			Equatorial	ITCZ	As above	10° ± 7°

From Ziegler (1990) in *Palaeozoic Palaeogeography and Biogeography,* W. S. McKerrow and C. R. Scotese, eds., © 1990 The Geological Society.

ITCZ, Intertropical Convergence Zone; $\overline{T}$, mean temperature.

table 4.2). Biome 3, in contrast, is represented solely by evaporites and no plants; desert plants would not be expected to be preserved. The results of mapping the Early Permian biomes are shown in figure 4.3. If biogeographic studies encompass relatively small areas and short time intervals, the monographic problems of taxonomy are less severe, provided the workers are consistent in their identifications (Wing, Alroy, and Hickey 1995).

Plant Morphology

One need think only of the contrast between the rudimentary or absent leaves of desert plants and the gigantic leaves of rainforest-floor plants to know that much of a plant's response to its environment can be expressed in its leaves. Plants that have woody stems also record environmental conditions and variations in the morphology of the wood. Finally, plants have a few specialized adaptations for extreme conditions that, if preserved in the fossil record, can be particularly helpful in paleoclimate interpretation. Terrestrial plants have a limited number of strategies by which they can survive inimical conditions, limits set by basic plant morphology, physiology, and reproduction. This means that the same types of adaptations appear in different, distantly related taxa. If the same morphological characteristics are seen in two or more taxa from the same flora, those characteristics are probably environmentally, not genetically, controlled and are therefore powerful tools in paleoclimate study.

Leaf morphology

Certain characteristics of leaves have climatic significance (table 4.3). Important individual features are (1) the presence of very small leaves (leptophylly or aphylly); (2) sunken stomata and certain other cuticular features; and (3) very thin or very thick leaves, particularly of angiosperms (flowering plants), which are evidence for a synchronously deciduous or evergreen habit, respectively.

The leaves of angiosperms have proved especially useful for paleoclimate study. Numerous leaf characters (see table 4.3) have been quantitatively linked to climate, and the resulting empirical relationships have been applied to climate interpretation in Late Cretaceous and younger rocks. Application of these methods to older floras is not possible because of the rapid morphological changes that occurred during the earlier stages of the evolution of the angiosperms, but,

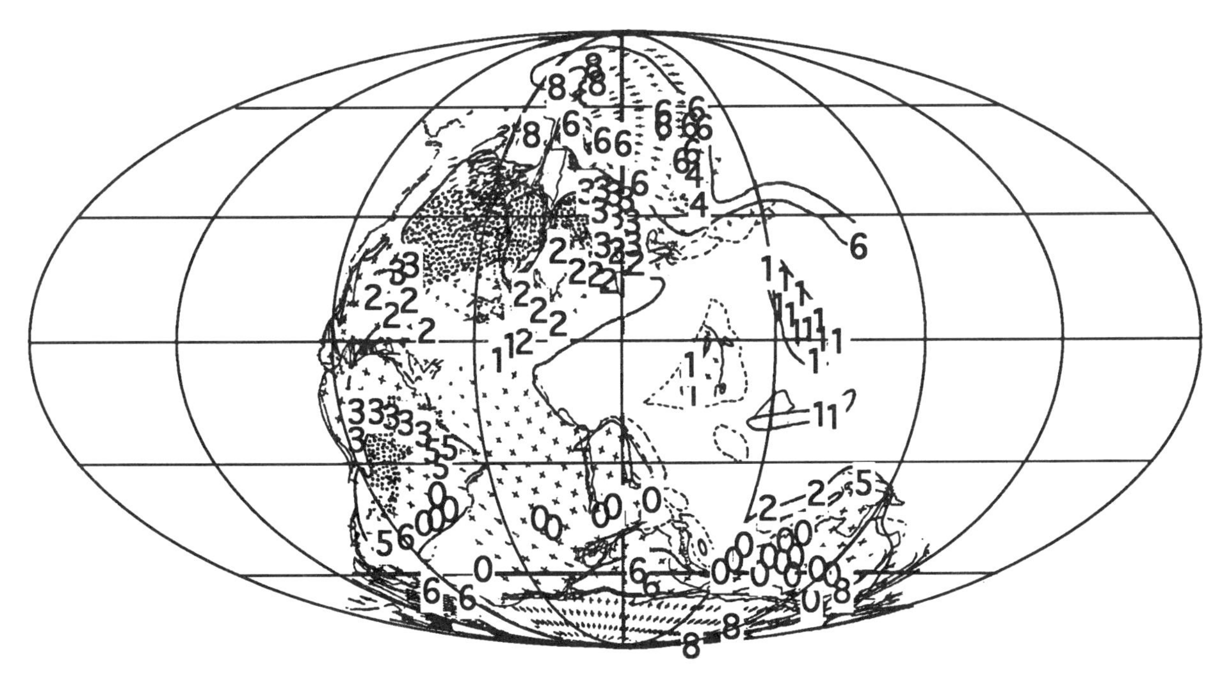

Figure 4.3. Early Permian vegetational biomes. Numbers correspond to the biomes listed in table 4.2 and their placement indicates sample localities. *Stippling* is the interpreted distribution of desert sandstones and evaporites. Biome 0 indicates the distribution of glacial deposits, biome 3 of evaporites, and biome 7 of desert eolian sandstones. From Ziegler (1990), in *Palaeozoic Palaeogeography and Biogeography,* W. S. McKerrow and C. R. Scotese, eds., © 1990 The Geological Society.

Table 4.3.
Characteristics of all leaves and of angiosperm leaves in particular used for paleoclimate interpretation

Climatically significant characteristics of all leaves
- Leptophylly or aphylly
- Sunken stomata
- Expanded petiole
- Thick cuticle with hairs
- Thin or thick (coriaceous texture)

Climatically significant characteristics of angiosperm leaves
- Aspect ratio
  - L:W <1:1[abcde]
  - L:W 1–2:1
  - L:W 2–3:1[d]
  - L:W 3–4:1
  - L:W >4:1

Leaf size
- Leptophyll 1 (0–5 mm^2)
- Leptophyll 2 (5–25 mm^2)[e]
- Microphyll 1 (80–400 mm^2)
- Microphyll 2 (400–1400 mm^2)[ab]
- Microphyll 3 (1400–3600 mm^2)[e]
- Mesophyll 1 (3600–9000 mm^2)[e]
- Mesophyll 2 (>9000 mm^2)[bde]

Shape
- Elliptical[b]
- Obovate[c e]
- Ovate

Margin type
- No teeth[ade]
- Compound teeth[cde]
- Acute teeth[be]
- Round teeth
- Teeth regularly spaced[e]
- Teeth closely spaced[e]

Base type
- Acute base[e]
- Round base[bc]
- Cordate (heart-shaped) base

Apex type
- Emarginate apex[ade]
- Attenuate apex[bde]
- Round apex

Other characters
- Spinose teeth[bc]
- Leptophyll 3 (25–80 mm^2)[a]
- % entire margin
- Lobed[e]

Angiosperm leaf characters except "Other Characters" were used in the CLAMP (Climate-Leaf Analysis Multivariate Program) analysis by Wolfe (1993).

Table 4.3. *(continued)*

[a]Characters found to be useful for interpreting mean annual temperature in a late Eocene flora from Florissant, Colorado (Gregory 1992).
[b]Characters found to be useful for interpreting growing season precipitation in late Eocene flora from Florissant, Colorado (Gregory 1992).
[c]Characters found to be useful for interpreting annual distribution of rainfall in a late Eocene flora from Florissant, Colorado (Gregory 1992).
[d]Characters found to be useful for interpreting paleoclimate in Eocene floras from interior North America (Wing and Greenwood 1993).
[e]Characters for which percent total variance was > 40% on either axis 1 (temperature) or axis 2 (precipitation) of CCA (Kovach and Spicer 1996).

by the Cenomanian (earliest Late Cretaceous), most or all of the characters listed in table 4.3 had evolved. Leaves from herbaceous plants are rarely preserved, so the reader should be aware that the following discussion refers to leaves of woody plants, that is, shrubs or trees. Leaf morphology has been used to estimate climate directly, and climate estimates based on leaf morphology from coastal and inland sites have also been used to estimate paleoaltitude. Because the altitude of an area is important in determining its climate, these methods are also discussed in this section. Geological methods of determining paleoaltitude were discussed in Chase et al. (in press); an alternative source of information, fish, is briefly discussed in the section on terrestrial animals.

Climate estimates from leaf morphology. A number of characteristics of angiosperm leaves are related to climate (see table 4.3). The leaves of angiosperms living in very wet climates commonly have elongated (attentuated) apices called drip tips, which may funnel water off the leaf to prevent epiphytic growth (Spicer 1989a). Leaves of angiosperms that live in dry and/or cold climates tend to be smaller than those from other climates and to have nonentire margins, that is, margins with teeth or spines. In contrast, those that live in warm and wet climates tend to have leaves that are large and have smooth margins. In addition to the morphology of the margin, the pattern of venation in the leaf helps distinguish these two types. The major veins in leaves with nonentire margins extend from the midvein to the edge of the leaf. The major veins in leaves with entire margins loop within the margins.

Bailey and Sinnott (1915, 1916) were the first

to note that the proportion of entire- to nonentire-margined leaves in modern floras seemed to be related to mean annual temperature (MAT) and to try to apply this relationship to the fossil record. In 1979, J. A. Wolfe published an extensive data set quantifying this relationship (figure 4.4) and establishing climatic limits for different types of forest (figure 4.5). Later, he increased the number of leaf characters used to show correlation between leaf morphology and climate (see table 4.3; Wolfe 1993, 1995). The method, using correspondence analysis (CA), is referred to as CLAMP (Climate Leaf Analysis Multivariate Program).

In the R-mode analysis of the data, the first two axes (figure 4.6a) explained just under 68% of the variability in the data set with CA (leaf characters only; Wolfe 1993; see also Kovach and Spicer 1996). Characters indicating the presence of various kinds of teeth cluster toward one end of axis 1 and the no-teeth character was toward the other end (figure 4.6a). This is consistent with the teeth/no-teeth character dichotomy on which the leaf-margin method (Wolfe 1979; see figure 4.4) was based, and axis 1 is likely related to temperature. The strongest division among characters along axis 2 was between the smaller and larger leaf sizes, with the leptophylls scoring the highest positive values and the mesophylls the highest negative values (see table 4.3). Indeed, leaf size was organized into a gradient from smallest (leptophyll 1) to largest (mesophyll 2) along axis 2. This axis most likely is related to precipitation, a conclusion supported by the strong score of attenuate apex with the larger leaf sizes. Interestingly, the smallest leaf size scored strongly positive on axis 1 (figure 4.6b) rather than negative, as might be expected (small leaves in cold or dry climates). This may be because the driest climates physiologically, the ones in which the smallest-leaved plants live, are warm as well as dry. Very wide leaves (L:W < 1:1) scored high on the negative side of axis 1, suggesting that the small leaves of dry, cold climates also tend to be round. Other characters scored relatively lower.

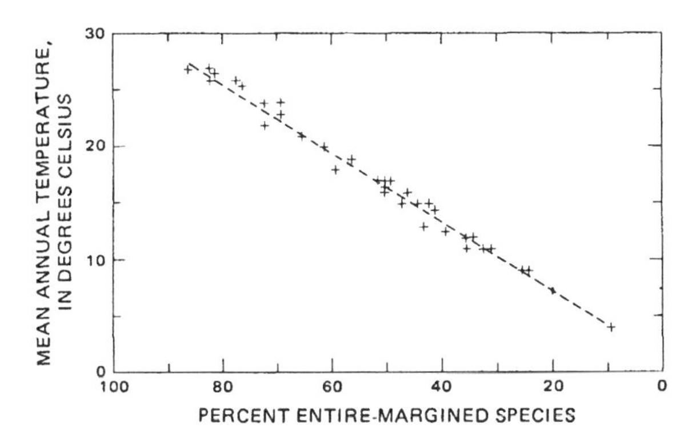

Figure 4.4 Correlation of percent entire-margined species with mean annual temperature, based on data from eastern Asia. From Wolfe (1979).

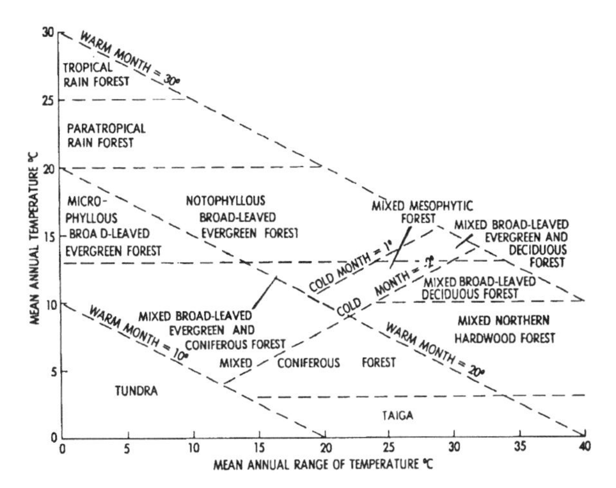

Figure 4.5. Nomogram showing the limits of mean annual temperature, mean annual range of temperature, and cold- and warm-month mean temperatures for various forest types, based on data from eastern Asia. Adapted from Wolfe (1979).

Wolfe (1995) concluded that of the temperature-related parameters indicated by leaf morphology, the most important was MAT, followed by cold-month mean temperature (CMM) and growing-season length (see also Wilf 1997). The precipitation results were much less clear (Wolfe 1995). One of the striking results of Wolfe's (1993, 1995) later analysis is that the original leaf-margin analysis proposed by Wolfe (1979) was so robust. Indeed, Wolfe (1993) concluded that percent no teeth is best for estimates of mean annual temperature, especially if the samples are divided into wet- and dry-climate samples. This can be accomplished from the relationships among the data points on a canonical correspondence analysis plot or from other information in the fossil record.

Estimates of mean annual precipitation (MAP) are relatively poor. The greatest scatter is in the colder climates (Wolfe 1993), where growing seasons are short. Precipitation that falls outside the growing season is irrelevant to the plants, so the correlation would be expected to be poor in those climates (Wolfe 1993). Therefore Wolfe

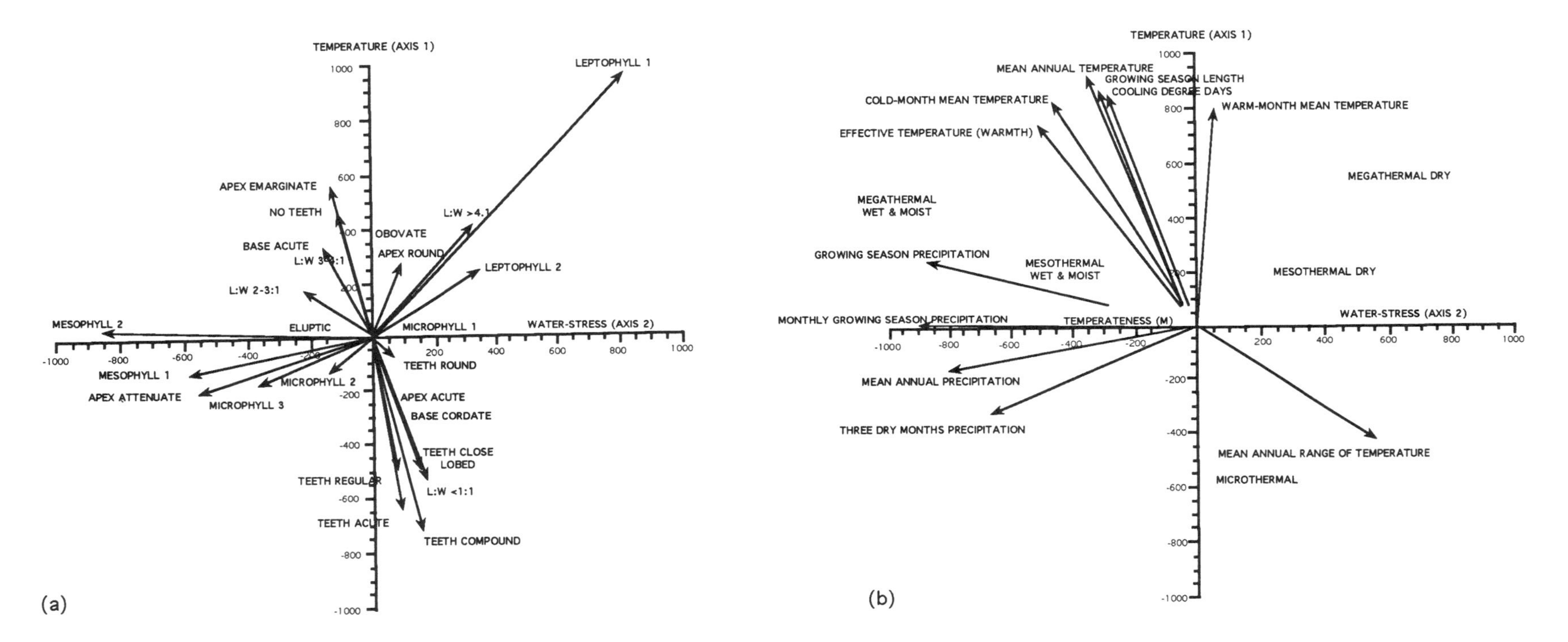

Figure 4.6. Template for determination of paleoclimate parameters from fossil leaf samples. *(a)* R-mode factor scores for modern floras. *(b)* Interpretation of factor scores in (a). The greater the distance of a sample along a vector, the more important is that climatic variable. See text for explanation. From Wolfe (1995). Adapted with permission, from the *Annual Review of Earth and Planetary Sciences,* Volume 23, ©1995, by Annual Reviews Inc.

(1993) examined the possibility of estimating mean growing season precipitation (MGSP). Growing season was defined as the number of months with mean temperatures above 10°C, which was chosen because the boundary between forested and treeless vegetation in cold regions is where the warm-month mean temperature is at least 10°C.

Although MGSP is more easily estimated than MAP, there are still some complexities. First, in hot and warm climates, the estimate of MGSP is good for climates with rainfall up to 145 cm MGSP, above which the correlation between the estimated and actual values breaks down. That is the amount of precipitation required to overcome water stress, and any additional precipitation is irrelevant to the plants. Second, in cold climates, plants do not exhibit a marked response to varying amounts of MGSP. Wolfe (1993) suggested that the cool growing-season temperatures and corresponding low evaporation in such climates compensate for the low precipitation in some localities. Another possible source of error is the omission of very large leaf sizes from the analyses (Wilf et al. 1997).

Wolfe (1993) identified the proportion of riparian species in the sample as the largest source of error, particularly in regions where seasonal drought occurs during the summer. Sample size, if greater than 20 species, was relatively less important. No significant difference existed in the accuracy of predicting climate parameters from samples of 20 species versus those with more than 100. Wolfe (1993) recommended 20 as the minimum number of species for temperature parameters (Wolfe 1971) and 30 as the minimum for precipitation parameters. Below 20 species, the estimates degrade significantly. A third factor influencing the estimates, particularly those strongly dependent on leaf size, is light regime. High-latitude floras tend to have a greater proportion of species with variable leaf sizes, tending toward larger sizes; this was also true in the early Tertiary (Wolfe 1980). In applying the method to fossil assemblages, other types of errors are important, including the vagaries of the systematics, phylogenetic controls on morphology, and the possibility that intraspecific variability will not be recognized.

Analyses similar to those of Wolfe (1993), but concentrating on leaf characters that appear to contain the most information, have been done by Gregory (1992, 1994; Gregory and McIntosh 1996) and Wing and Greenwood (1993). Rather than using correspondence analysis, Gregory (1992, 1994) calculated a correlation matrix from Wolfe's (1993) full data set to determine which leaf characters correlate with each other and with climate parameters. Gregory's (1992, 1994) multiple regression analysis for MAT was based on 29 characters; she omitted "teeth regularly spaced" and "teeth closely spaced" because they showed >95% correlation (negative) with "no teeth," which creates analytical problems. However, she also added "teeth spinose" and "leptophyll 3," which were in Wolfe's (1990, 1993) original data set but did not appear in those publications. The results were consistent with Wolfe's (1993) analysis. Using multiple regression analysis, Gregory (1992, 1994) then determined the characters that showed the strongest trends with climate parameters and determined paleoclimate from those characters only.

Wing and Greenwood (1993) used an approach similar to that of Gregory (1992, 1994) to analyze North American Eocene floras, except that they used the published data set from Wolfe (1993; that is, lacking the characters "teeth spinose" and "leptophyll 3"), and they limited the predictor variables to one per aspect of leaf morphology (that is, one character for margin type, one character for base type, etc.). They also limited the data set to sites with CMM >–2°C. They justified this limitation on the grounds that the nearest-living relative method provides evidence that the Eocene floras were never subjected to hard frosts and that correlation between climate and leaf physiognomy is different in very cold climates than in warmer climates (Wolfe 1993). Thus they felt that performing multiple regression analysis on frost-free sites only would give results more meaningful for Eocene floras.

The disadvantage to multiple regression analysis is that it assumes linear relationships between climate parameters and leaf characters, which is clearly not the case (Wolfe 1993), and it assumes that the characters are independent, which is also not the case (Kovach and Spicer 1996; Stranks and England 1997). Gregory (1992, 1994) argued that multiple regression analysis is superior to correspondence analysis because the latter weights all characters equally, whether they are related to climate or not, but her work preceded the reanalysis of Wolfe's (1993) data set with canonical correspondence analysis, which is a more flexible method than correspondence analysis (Kovach and Spicer 1996).

All workers stressed that samples of a fossil

flora should come from as many different subenvironments as possible (e.g., Spicer and Parrish 1986) because riparian vegetation, for example, tends to have a higher proportion of toothed-margined leaves than the regional vegetation as a whole. This is why the proportion of riparian species had such a strong effect on the statistical analyses. When this condition and that of sufficient sample size are met, paleoclimate estimates from leaf-character analyses are consistent with other information (Spicer and Parrish 1990a; Wing and Greenwood 1993) and give geographically coherent results (Wolfe and Upchurch 1986, 1987a; Wolfe 1994a).

It must be emphasized that these methods are best developed for Northern Hemisphere floras; the data set was primarily from North American floras. The fact that Southern Hemisphere floras might have to be calibrated differently is indicated by the differing slopes on leaf-margin/MAT graphs for Northern and Southern Hemisphere floras (Upchurch and Wolfe 1987a; Parrish et al. 1998). In addition, Wolfe's (1993) original data set included few tropical floras. This has limited application of the methods to floras from these regions (Jacobs and Deino 1996; Stranks and England 1997). Although the multiple-regression method yielded valid MATs for Kenya, seasonality of temperature, mean annual precipitation, and mean seasonal precipitation were overestimated (Jacobs and Deino 1996). Jacobs and Deino (1996) suggested the problem was with the multiple regression method, the lack of African floras in the original data set, or the high degree of intraspecific variability in the leaf characters in these floras.

In contrast, Stranks and England (1997) suggested that the relationship between climate and leaf physiognomy is nonlinear on a global scale; some of these nonlinearities were discussed by Wolfe (1993). They proposed a method, resemblance analysis, that relies solely on local calibration. The method starts with the correspondence analysis (Wolfe 1993) and calculates the Euclidean distance among the sites in the transformed data space. Then the climate parameters of nearest neighbors are used to estimate those of the unknown sample. Application to the fossil record is to compare the fossil flora with the modern ones to which it is most similar physiognomically, and to derive the paleoclimate parameters accordingly. The resemblance analysis method gave better results for the MATs of modern Australian and New Zealand floras than did CLAMP with the modern Australian and New Zealand floras added. This result suggests that simply adding more modern floras to the CLAMP data set may not improve the applicability of the method to floras that were not originally included.

Paleoaltitude estimates. Many of the samples that were outliers in Wolfe's (1993) and Gregory's (1994) analyses were from high altitudes. Although on the face of it paleoaltitude seems like a tectonic problem, it is in fact a paleoclimate problem, because how the climate signal from plants is interpreted depends on the altitude at which the plants grew. The problem was posed by Molnar and England (1990): how does one distinguish climatic cooling from cooling generated by uplift at the sample site?

Three methods for calculating paleoaltitude have been developed, two based on estimates of atmospheric lapse rates (figure 4.7) and one on atmospheric enthalpy (Gregory 1994; Forest, Molnar, and Emanuel 1995; Wolfe et al. 1997). The plant-based methods depend on assumptions about terrestrial lapse rates, that is, the decrease in temperature per 1000 m of altitude. One method projects coastal MAT inland and then uses present average local terrestrial lapse rates to calculate altitude from the difference in the projected sea-level MAT and the local MAT (Meyer 1992). The other method uses present average global terrestrial lapse rates to calculate altitude from a direct comparison of coastal and

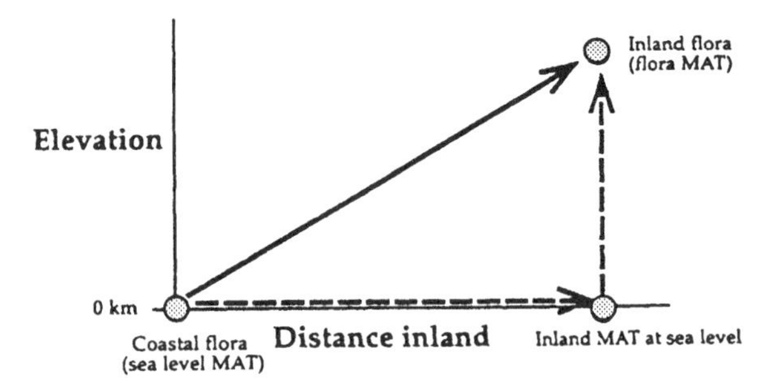

Figure 4.7. Illustration of two methods for calculating paleoaltitude. *Dashed line:* Sea-level mean annual temperature (MAT) is determined from a coastal flora and MAT inland is estimated using present terrestrial, sea-level temperature gradients. The interior sea-level MAT is then compared with the inland flora MAT and altitude determined from present average local terrestrial lapse rates. *Solid line:* Coastal-flora MAT is compared directly with inland-flora MAT using present average terrestrial lapse rates. From Gregory (1994) from *Palaeoclimates* 1:23–57 with permission of Gordon and Breach Science Publishers.

inland floras (Wolfe 1992). There are numerous pitfalls to both approaches (Gregory 1994; Gregory and McIntosh 1996). First, the coastal flora must have grown at sea level and must be from about the same latitude, although latitudinal temperature gradients can be very irregular. Second, global and local average lapse rates can be very different, and local lapse rates depend on latitude, altitude, position relative to the coast, and whether the site is on the leeward or windward sides of mountains (Meyer 1992; Wolfe 1992). The two methods give quite different results when applied to the Oligocene Florissant flora (Colorado, U.S.), 1.9 km and 3.1 km, respectively (Gregory and McIntosh 1996). The present altitude at Florissant is 2.5 km; these values are not inconsistent with values from oxygen isotopes in early and middle Eocene rocks of the Green River Formation (Utah, U.S.). Similar attempts to estimate paleoaltitude using fossil plants have given equally uncertain results (e.g., Povey, Spicer, and England 1994).

The third method of estimating paleoaltitude used Wolfe's (1993) data set to relate leaf morphology to mean annual enthalpy, a parameter that is more easily related to altitude (Forest, Molnar, and Emanuel 1995; Wolfe et al. 1997). The paleoaltitude estimate for Florissant using this method is 2.9 km (Gregory and McIntosh 1996), which falls between the values estimated from the other two methods, but closer to the method of Wolfe (1992). The method was updated more recently by Wolfe et al. (1997; see also discussion in Kovach and Spicer 1996). Stranks and England (1997) experimented with this parameter as well in formulating the resemblance analysis method discussed in the previous section.

Other uses of leaf morphological data in paleoclimate analysis. It is rare to find frost-damaged leaves in the fossil record, for several reasons. First, leaves that are most susceptible to frost damage are water rich and thick textured. Such leaves tend to be found on evergreen plants, which will not grow in regions subject to hard frosts. Second, leaves of seasonally deciduous plants do not suffer frost damage because they are thinner, usually shed before hard frosts ensue, and contain little water once they are shed. Thus an unusual occurrence of frost-damaged leaves in an otherwise warm-climate flora at the Cretaceous-Tertiary boundary in the western United States suggested to Wolfe (1991) an impact winter resulting from the purported bolide impact at that time.

Gregory (1997) investigated the possibility that leaf morphology is affected by varying CO_2 levels. In observations of white oaks raised in growth chambers at different CO_2 levels, she noted an increased length-to-width (L:W) ratio with higher CO_2. Based on this admittedly small data set, she suggested that MAT might have been biased by 1°C and MGSP by 20 cm when CO_2 levels were 2 to 3 times present levels. However, these errors are within analytical error of the established methods (e.g., Kovach and Spicer 1996) and are therefore insignificant.

Cuticle

Cuticle is the waxy, decay-resistant outer layer of leaves. Cuticle varies in thickness and morphology in response to both genetic and environmental factors. Cuticle includes impressions of the outlines of the outer cell layer and the morphologies of the stomata, which are openings in the surfaces of leaves that allow gas exchange. Shape of the guard cells, which control the size of the stomatal aperture, is preserved, and any structures that regulate gas exchange and water loss are usually part of the cuticle. In some cases, cuticle can be collected from leaf compression fossils and, once described, can be used to infer the presence of those taxa in the absence of megafossils. Even in the absence of a direct tie between leaf forms and cuticle, however, cuticle can be useful for interpreting paleoclimate. Cuticle provides advantages offered by neither pollen nor plant megafossils in that it is less subject to transport than pollen and more commonly preserved, and preserved in more facies, than plant megafossils (Upchurch 1995).

Characteristics of cuticle that can be related to climate are thickness, size and morphology of the stomata and associated structures, presence of stomata plugged with wax (nonfunctional stomata, characteristic of xeromorphic plants), and presence, abundance, and shape of hairs (Krasilov 1975; Spicer 1989a; Upchurch 1995). Evergreen plants that live in dry or cold climates or that have long periods of dormancy during the year tend to have thick cuticles, abundant hairs, and/or elaborate structures around the stomata that are designed to prevent water loss, although cuticle thickness is also characteristic of coriaceous (leathery, evergreen) leaves of the tropics, of those living in salt marshes, and of certain taxonomic groups (McElwain and Chaloner 1996). In some cases, whether the plant was evergreen or deciduous can be determined from cuticle

(Upchurch 1995). Where cuticle can be tied to particular leaf taxa, the presence of cuticle in palynological preparations can be used to expand the database on the geographic distribution of leaf types (for example, in cores; Wolfe and Upchurch 1987b).

Some evidence exists in the literature that cuticle might be useful for estimating relative paleo-CO_2 values (Beerling and Chaloner 1993, 1994; Van Der Burgh et al. 1993; Van de Water, Leavitt, and Betancourt 1994). The density of stomata in the cuticle of Pleistocene pine needles was generally higher between 30 and 15 Ka, decreasing about 17% during the 15 to 12 Ka deglaciation (Van de Water, Leavitt, and Betancourt 1994). This decrease was concomitant with a 1.5‰ decrease in $\delta^{13}C$ in the leaves over the same period and with the increase in CO_2 seen in ice-core data. A simplified explanation for these observations is that as CO_2 increases, leaves need fewer stomata to maintain the same levels of CO_2 input for photosynthesis. Plants may also discriminate against ^{13}C as CO_2 levels increase (Popp et al. 1989). Van de Water, Leavitt, and Betancourt (1994) cited many caveats, including the fact that gymnosperms are apparently less sensitive at the higher CO_2 levels of the postindustrial atmosphere and, presumably, would also have been relatively insensitive to the higher CO_2 levels of past greenhouse times.

In all these studies, the cuticle was associated with leaf fossils, which increases the confidence that cuticle from the same species is being compared. The principal limitation, in Van Der Burgh et al.'s (1993) view, is the discontinuousness of the fossil record of cuticle and leaves together. In a cautionary paper, Murray (1995) emphasized large differences in the responses of plants of a single species to CO_2 levels. For example, he cited one experiment that resulted in lower stomatal densities with higher CO_2 but larger leaves, and therefore a greater number of stomata per leaf, and other work showing changes on one side of the leaf but not the other. The high degree of variability in stomatal density has been known at least since the 1920s. Salisbury (1928) showed that stomatal densities relate most closely to water supply, but also to position on the leaf, size of the leaf, position of the leaf on the tree, and species. Notwithstanding these limitations, large variations in stomatal density in Devonian through Jurassic cuticles are consistent with variations in CO_2 predicted from geochemical models (Berner 1994; McElwain and Chaloner 1996).

Wood

Fossil wood is a repository of paleoclimate information that can be used to supplement conclusions based on other data (Fritts 1976, 1982; Creber 1977) and eventually might supply the same kinds of taxon-independent climate information provided by leaves and vegetational physiognomy (Wheeler and Baas 1991).

Growth and false rings. In warm climates where the growing season is year-round, trees either have no rings or have rings that are related to internal reproductive cycles and not to climate (Creber 1977). In seasonal climates, most trees have distinct growth rings; in eastern North America, all trees north of 42°N have growth rings (Woodcock and Ignas 1994). Growth rings are usually annual, and each ring consists of early wood and late wood. Late-wood cells have small lumina (the spaces within the cells) and thick walls, and late wood appears dark in hand sample. In contrast, early wood cells have large lumina and thin walls, and early wood is lighter than late wood. Growth rings are delineated by an abrupt transition from late wood, which forms late in the growing season, to early wood, which forms early in the next growing season (see also discussion in Wheeler and Baas 1991, 1993). The line separating the two kinds of wood may represent several months of no growth or, in very high latitude trees, sometimes several years of no growth. Growth-ring characteristics that are useful in pre-Quaternary woods are mean ring width, the relative widths of late wood and early wood, and presence or absence of false rings.

All other things being equal, wide rings indicate long growing seasons except when the tree is young. Young trees grow quickly and, because they are of small stature, the same amount of wood growth will result in wider rings than when the tree is larger (Fritts 1976). Rings with a higher proportion of late wood are indicative of cooler growing seasons or more precipitation late in the growing season than those with less late wood (Creber 1977; Creber and Chaloner 1984). Very narrow late wood is particularly useful, as it indicates the specific condition of a favorable but rapidly terminated growing season. This type of wood has been found in two types of paleoclimates, as interpreted from independent data. In one regime, the summers were warm, permitting rapid growth, but the light regime, which controls the onset and cessation of growth, changed very rapidly. This situation occurred in very high paleolatitudes (e.g., Jefferson 1982; Creber and Chaloner 1984; Parrish and Spicer 1988a; Kumagai et al. 1995). At high latitudes, day length changes rapidly, from 24-hour days to 24-hour nights within 6 weeks at 85° latitude. Day length is a signal for many trees to start and stop growth, so the signal at high latitudes is particularly strong.

The second type of climate that results in this type of growth ring is a low-latitude monsoonal climate such as occurs in southern Asia today or on Pangea in the Late Carboniferous to Jurassic (Parrish 1993). Rings in fossil woods from low-latitude Pangea vary greatly in width but all have very narrow late wood (Keller and Hendrix 1997; J. T. Parrish, unpublished data). A characteristic of strong monsoonal climates is a transition from abundant to no monthly rainfall within a month, and plants in monsoonal climates commonly are deciduous. Thus the rapid cessation of water supply in low latitudes had the same effect on trees as the rapid onset of winter light conditions at very high latitudes. Abrupt transitions from early wood to late wood indicate rapid onset of water limitation (see review of early wood-late wood controls in Creber and Chaloner 1984). A complication in the study of woods from trees that grew in dry climates is that such trees can lay down several rings per year, representing new leaf flushes after rainfall (Glock and Agerter 1963).

The use of growth rings for paleoclimate interpretation has two important limitations. In all cases, growth-ring characteristics are useful in a comparative sense, not in any absolute sense. In other words, the information on ring width is much more useful if two floras are compared, from two levels within a section, for example (e.g., Spicer and Parrish 1990b), than if they are reported independently. No limit of ring width is characteristic of warm versus cold climates, and ring width is not quantitatively related to the length of the growing season, because growth-ring characteristics are partly controlled by genetics and partly by local growing conditions, as well as by climate (Fritts 1976; Creber 1977). Another limitation of the use of growth-ring characteristics in paleoclimate interpretation is the desirability of having at least two wood taxa in each flora that share the observed characteristics. This increases the likelihood that the observed characteristics are climatically and not genetically controlled.

False rings indicate temporary interruptions of growth during the growing season (Fritts 1976) and are best expressed in branches (Chapman 1994). False rings are composed of small, thick-walled cells but can be distinguished in most woods by the gradual increase of cell sizes on the outside of the false ring, as opposed to the abrupt increase that occurs at the boundary of a true ring. False rings are caused by temporary water stress, including drought and freezing, waterlogging of the roots, and/or insect attack (Fritts 1976). It is impossible to tell from the false rings themselves what their cause is, but the probable cause can be inferred from other information, for example, the presence of frost damage or insect damage of associated leaves (Creber and Chaloner 1984; Spicer and Parrish 1990b). Whether drought or root waterlogging is the probable cause has to be inferred from the overall climate and sedimentary regime.

Annual variability of ring width is a guide to the interannual variability of climate, especially if similar profiles appear in all samples (Creber 1977; but see Parrish and Spicer 1988a). An index that summarizes the history of interannual variability, mean sensitivity, also appears in paleoclimate studies (e.g., Parrish and Spicer 1988a) but is of limited use because it is highly dependent on the position of the tree in the forest (Chapman 1994); mean sensitivity is lower in trees from the center of the forest than in those from the edges (Fritts 1976). Even where the standing forest is partially preserved (e.g., Jefferson 1982; Francis 1983, 1984, 1991; Ammons et al. 1987), a lack of knowledge of the full extent of the forest prevents assessment of the significance of mean sensitivity with regard to position within the forest versus regional climatic conditions.

In practice, availability of wood in the geologic record that is suitable for the study of growth rings is a limiting factor (Parrish and Spicer 1988a). Although growth rings occur in all parts of the tree, the ideal sample for the study of growth rings is from an adult tree, in the trunk well above the stump flare and below any trunk division (Chapman 1994). Root wood and wood from low on the trunk has little latewood and the rings may be indistinct; paleoclimate interpretations should not be made on such woods (Chapman 1994). Branches can be recognized by their small size and growth rings that are wider on one side than the other; care must be taken in making paleoclimate interpretations from branches because this wood is not necessarily representative of growth in the tree as a whole (Chapman 1994).

Characters of angiosperm woods. In an extensive survey of fossil angiosperm wood anatomical characters, Wheeler and Baas (1991) examined trends through time. Those characters that, from discussions in Wheeler and Baas (1991, 1993), appear to have the most promise for paleoclimate interpretation are listed in table 4.4. Many wood anatomical characters are at least generally correlated with climate in modern floras but are not listed in table 4.4 because Wheeler and Baas (1991) regarded the climate correlation to be weak or fortuitous, that is, the variations in the wood anatomical characters are better explained by wood anatomical evolution than by climate. A number of trends had major breaks between the Cretaceous and Tertiary and, based on modern climate-wood relationships, various features gave inconsistent climate indications for the Cretaceous. Wheeler and Baas (1991) therefore offered the general conclusion that many wood anatomical characters are likely to be most useful for paleoclimatic interpretation in Tertiary fossil woods only. Trends in a few characters in table 4.4 did not develop until the early Tertiary; these are indicated as reliable for the mid- and late Tertiary only, and the absence of the characters cannot be taken as absence of the indicated climate (Wheeler and Baas 1991, 1993).

Limitations to Wheeler and Baas's (1991, 1993) analysis apply to the use of wood anatomy in general for paleoclimate interpretation. First, the sample size for any individual time period is relatively small. Second, the data for each time interval are unevenly distributed in space, and the existing samples may not be representative of global climate. Third, most of the species of fossil wood are based on single specimens. Finally, wood is generally conservative, meaning that individual species or, in some groups, genera cannot be discerned on the basis of wood anatomy. Despite these problems, angiosperm wood anatomy has great potential for contributing to paleoclimate interpretation.

Angiosperm wood contains vessels that are large in cross section compared with the tracheids; vessels are the primary fluid-conducting

Table 4.4.
Wood anatomical characters most likely to provide information and their potential paleoclimatic significance

Anatomical feature	Paleoclimatic significance
Growth rings	Absence indicates warm, wet, nonseasonal climate
Ring porosity	Presence indicates seasonal (temperature and/or moisture) or warm and dry climates; sometimes indicates deciduousness; late Tertiary only
Vessel groupings	Increase in clusters indicates increase in winter-wet and dry climates*; mid- or late Tertiary only; *tangential arrangements, warm temperature and winter rainfall*
Vessel diameter	In large trees, large sizes (>200 μm) indicate wet, warm climates, especially in Tertiary, doubtful for Cretaceous; *small diameter, cool temperatures (in diffuse-porous woods)*
Vessel frequency	In trees (as opposed to shrubs), high frequency (≥40/mm^2) indicates cool seasonal to polar, alpine, or dry climates; commonly inversely proportional to vessel diameter; *< 20/mm^2, minimum limit for precipitation of 900 mm/yr; >40/mm^2, cool temperatures*
Vessel element length	Problematic because of measurement difficulties in fossil woods
Tracheids	Presence indicates seasonally arid and arid climates*
Helical vessel wall thickening	Common in markedly seasonal (temperature and/or moisture) climates; mid- or late Tertiary only

From information in Wheeler and Baas (1991, 1993).
*Indicates potential significance but more data are needed.
Observations in italics are from a detailed study on climate-character relationships in the eastern U.S. (Woodcock and Ignas 1994).

structures. Many of the characters listed in table 4.4 are characters of the vessels, and variations in the size and density of vessels can be related to climate, at least for woods that are mid-Tertiary or younger in age. The climatic significance of some vessel characters has been studied in detail for the eastern United States (Woodcock and Ignas 1994), and those observations are incorporated in table 4.4.

Plants that live in cold and/or dry climates are subject to embolisms within the vessels (Poole 1994). Thus trees living in such climates tend to have vessels that are grouped (so that if one develops an embolism, the rest can continue to conduct fluid for that part of the tree), small vessel diameters, high vessel density, and conductive tracheids. Trees living in warm, wet climates have large, isolated vessels and nonconductive tracheids. Two indices that quantify the likelihood of a tree developing embolisms, and that, in turn, provide a guide to the type of climate in which the tree lived, are vulnerability and conductivity (Carlquist 1977, 1988). Vulnerability can be expressed as $V = d/D$, where d is the mean vessel diameter and D is the mean number of vessels per square millimeter. Conductivity can be expressed as $C = (r^4/10^6) \times D$, where r is the mean radius of the vessels; note that the ability of a tube to conduct is not equal to the cross-sectional area but to r^4. Vulnerability and conductivity calculated for Late Cretaceous North American woods are consistent with other information (Wolfe and Upchurch 1987a; Wheeler, McClammer, and LaPasha 1995).

Adaptations for extreme conditions

Few areas of the continents are so cold, dry, or wet that plants cannot grow there (Spicer 1989a). To cope with the most extreme conditions, plants have evolved life cycles that permit survival in strongly seasonal climates or specialized structures that permit survival despite constant inimical conditions. Life-cycle responses are deciduousness, if the plants are woody, and herbaceousness. Leaves are expensive organs to maintain because they respire at night, losing water through their stomata. Woody plants that live in environments with seasonal or sporadic adverse conditions have two strategies for coping with those conditions. They either discard their leaves when conditions are unfavorable, a process called deciduousness, or they have adaptations that allow the retention of leaves with minimal water loss and risk of frost damage. Aphylly, that is, the absence of leaves, and leptophylly, the presence of small, scale-like leaves, are common adaptions in many groups for water stress. Water stress can include freezing, so leptophylly is found in plants that live in very cold as well as very dry climates. The leaves of evergreen plants that survive periodic inimical conditions also tend to have thick cuticles. The leaves of herbaceous plants die at the end of the growing season and the plants survive as spores, seeds, or underground storage organs.

Woody plants can be divided into three intergradational categories, depending on how leaves are lost (Spicer 1989a). These are obligately synchronously deciduous plants, which shed all leaves at the same time every year regardless of the environmental conditions in which they live (or are placed by humans); facultatively synchronously deciduous plants, which may shed all leaves at the same time when conditions become unfavorable, but otherwise are evergreen and lose leaves sporadically; and obligate evergreens, plants that cannot shed leaves synchronously and that die when unfavorable conditions occur. These life cycles and the evolutionary history of their appearance in different groups clearly could have implications for interpreting paleoclimate but, unless close relatives are still extant, inferring life cycle from morphological characteristics can be difficult.

A leaf is shed at a specialized structure at the base of the petiole. This structure has a characteristic appearance on shed leaves and the stems from which they are shed that is easily recognizable in the fossil record. Whether there are differences in these zones, depending on whether the plants are obligate evergreens, has apparently not been quantified. However, branches or stems of some deciduous woody plants (e.g., *Ginkgo*, maples, and the extinct cycad *Nilssonia*; personal observation and Kimura and Sekido 1975) are characterized by what is termed long shoot/short shoot morphology. The short shoot refers to a structure that is formed from tissue remaining on the stem after abscission of successive years' worth of the leaves. The short shoot is typically a few millimeters or centimeters long and has a characteristic ridged texture from the abscission scars. The long shoot is the stem itself. Leaves of deciduous plants, angiosperms in particular, tend to be thin and rigid, and many have long petioles (Krasilov 1975). Deciduous conifers commonly show a reduction of leaf size toward the base of the abscised shoot (Spicer and Parrish 1986).

Aerenchymatous roots are specialized structures that allow trees to grow in standing water (Krasilov 1975). The roots grow from the trunk above the water line and take the form of stilt-like structures, as in mangroves, or buttresses, as in bald cypress; these roots serve the respiratory function that cannot be served by the submerged roots (Krasilov 1975). Some hygromorphous plants have greatly thickened lower stems (Krasilov 1975). Some plants have spongy stems, which are called manoxylic, that cannot survive prolonged freezing. Modern cycads have manoxylic stems and for this reason are limited to warm climates. This fact has been a source of some confusion in interpreting paleoclimates, created by the assumption that ancient cycads were also frost sensitive. However, some ancient cycads were vine-like and deciduous and thus could survive much harsher conditions than their modern counterparts (Kimura and Sekido 1975). In contrast to hygromorphic plants, the stems of xeromorphic plants are commonly short, thick, and sheathed in heavy tissue (Krasilov 1975). All of these features have been described from fossil plants (Krasilov 1975) and could be used to help interpret paleoclimate.

Taphonomy

Understanding the taphonomy of plant parts is critical to understanding their use as paleoclimatic indicators (Ferguson 1985; Gastaldo 1986; Spicer and Greer 1986; Spicer 1989b, 1991). The role of taphonomy is particularly important for paleoclimate in helping determine the degree to which a particular fossil flora has been sampled, whether the plants were deciduous, and in reconstructing the vegetation.

Leaves have been shown to sample local vegetation and to be subject to destruction within a very short distance from where they were shed (Ferguson 1985; Burnham 1989; Spicer 1989b); this increases their value in determining the local paleoclimate but also emphasizes the need to sample the complete spectrum of depositional and growth environments. Unlike leaves, wood can be carried long distances, and it is important to determine how far from its growth site (and growth climate) it is likely to have been transported. If wood occurs in marine rocks, it could have been transported thousands of kilometers, and paleoclimate information from such samples would be almost impossible to put into context. Presence of branch bases or root balls on the trunk and absence of abrasion are evidence that a fossil log was deposited not far from its growth site.

An especially useful application of taphonomy comes from the observation that leaves that are shed at the end of the growing season form leaf mats. In the fossil record, leaf mats are preserved as bedding planes covered with overlapping leaves that are all in about the same state of preservation. This mode of preservation is the result of the leaves being shed synchronously into the depositional system *and* rapidly buried. This type of occurrence is crucial because the leaf structures indicating deciduousness are not conclusive evidence that the plant was synchronously deciduous. Only synchronous deciduousness is indicative of seasonal climate, and information about this habit can be gained only from taphonomy. Taphonomic information on leaves is also important for inferring the cause of false rings.

Other types of information used in paleoclimatic interpretation are overall diversity of various vegetational components and the type of preserved wood, which aids reconstruction of the arboreal, as opposed to the shrubby, elements (Parrish and Spicer 1988b). It is important to emphasize, as will be shown in the following section, that reconstruction of the vegetation for paleoclimate studies usually requires building a picture of the vegetation as a whole rather than the individual elements of the vegetational mosaic (streamside, interfluve, and so on), although understanding each element, their relationships to each other, and their place in the overall depositional environment is also important. For example, Demko (1995) showed in the Triassic Chinle Formation (western United States) that plant fossils are preserved only in sediments deposited in incised valleys, where the water table was confined. Based on other information, the Chinle Formation (Dubiel et al. 1991) was interpreted to have been deposited in a strongly seasonal climate with respect to rainfall. The high water tables in the incised valleys permitted development of the wet conditions required for the fossilization of plants. Once the valleys were filled with sediment, the water tables were no longer confined, and the intense oxidizing conditions of the dry season destroyed whatever plant material might have been deposited. Knowing this has important implications for use of the vegetation in interpreting the regional climate.

Vegetational Physiognomy

Vegetation is strongly controlled by climate, and the physiognomy of the vegetation—the forest, as opposed to the trees—is responsive to climate independent of the taxa that compose the vegetation. Wolfe (1993) emphasized the difference between a flora, which is the aggregate of the taxa in an area, and vegetation, which is independent of the particular taxa it comprises. For example, temperate rain forests in northwestern North America and in southeastern Australia are very similar in appearance and structure even though the taxa in those forests are very different. In both regions, the dominant trees are large and the ground cover dense, but the dominant trees in North America are conifers whereas those in Australia are eucalypts. When Wolfe (1979) quantified the relationship between mean annual temperature and the percent entire-margined taxa of angiosperms, he also classified vegetation types and quantified the constraints imposed on their distributions by climate (see figure 4.5). In well-preserved fossil floras, it is possible to reconstruct the vegetation and, from the physiognomy, place constraints on the paleoclimate. Data used to reconstruct the vegetation include diversity of the various components of the flora, the type of wood preserved (if any), and the diversity of pollen relative to the diversity of the megafossils. Taphonomy is critical to vegetational reconstruction (Spicer and Parrish 1986; Wnuk and Pfefferkorn 1987; Gastaldo, Demko, and Liu 1990; DiMichele and Phillips 1994; Demko 1995; Allen 1996).

Wolfe (1985) mapped changes in the distribution of vegetation types during the Tertiary Period, an exercise that revealed one potential weakness in the method. High-latitude forests in the Tertiary were not quite like any other high-latitude forest in that the broad-leaved component was characterized by unusually large leaves (Wolfe 1980). Wolfe (1985) found this observation to be sufficiently striking that he created a new classification that does not appear on figure 4.5, polar broad-leaved deciduous. Changes in the distribution of forest types in the Northern Hemisphere between the latest Paleocene-early Eocene and late middle Eocene are shown in figure 4.8. These changes suggest a cooling of the Northern Hemisphere during this time. The tropical-paratropical semideciduous floras on figure 4.8b suggest a monsoonal climate and may have modern analogues along the west coast of Mexico and in the Caribbean (Wolfe 1985).

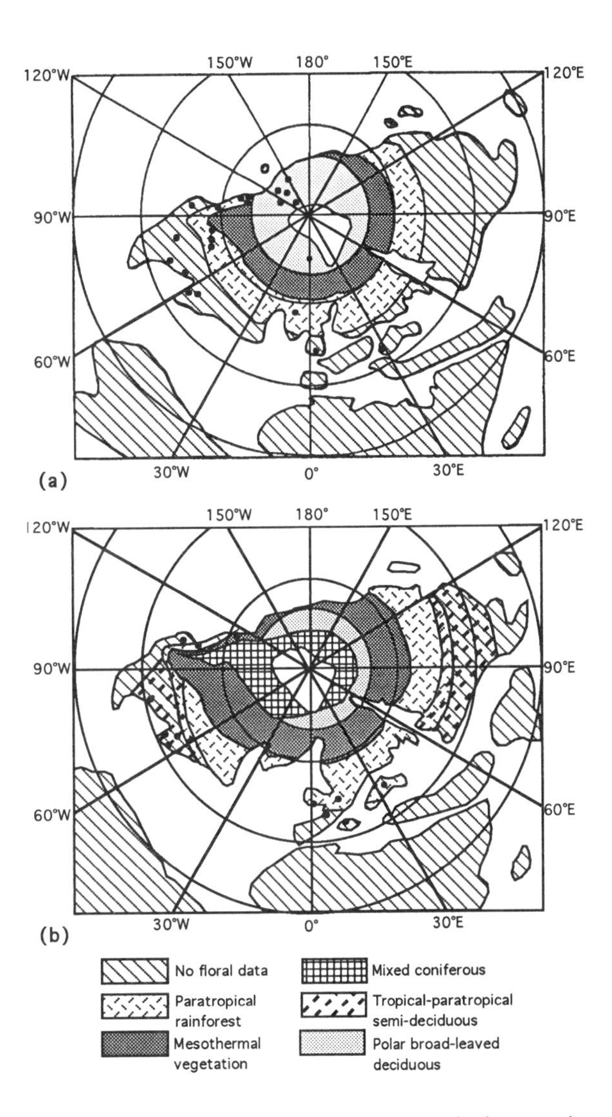

Figure 4.8. Distribution of forest types in the latest Paleocene-early Eocene *(a)* and late middle Eocene *(b)*. Data points indicated by *dots*; distribution inferred where data are lacking. Mesothermal vegetation includes mixed mesophytic, mixed broad-leaved deciduous evergreen and coniferous, notophyllous broad-leaved evergreen, and microphyllous broad-leaved evergreen forests (see figure 4.5). From Wolfe (1985) in *The Carbon Cycle and Atmospheric CO_2: Natural Variations Archean to Present*, pp. 357–75. Geophysical Monograph 32, copyright by the American Geophysical Union.

Nearest-Living-Relative Method in Paleobotany

Despite the demonstrable problems equating modern leaf taxa with ancient leaf taxa (e.g., Spicer 1986b), the NLR method (also called the floristic method in studies of fossil floras; Wolfe 1993) has been used commonly in the interpretation of paleoclimatic tolerances of fossil plants. Although the method may be useful in young rocks, particularly if many taxa are used and fossil assemblages resemble modern ones (e.g., Leopold and Liu 1994), in older rocks it can be very misleading (see discussion in Wolfe 1971). Already cited was the example of using the presence of cycad leaves to infer warm climate, even in sediments deposited at very high paleolatitudes (e.g., Smiley 1969).

Regardless, the NLR method can be useful if applied judiciously. For example, in a study of North American Eocene floras that also used angiosperm leaf-margin analysis, Wing and Greenwood (1993) bolstered their conclusion that climate was warm and humid with the tolerances of modern counterparts to the palms, gingers, tree ferns, and cycads found in the Eocene floras. They established criteria by which one might gain confidence in using the NLR method:

1. close taxonomic and morphologic relationships between the fossil plants and their modern counterparts;
2. several groups of fossil taxa that have modern relatives and that all give the same interpretation;
3. living relatives diverse and widespread, not relictual; and
4. anatomical or physiological features that constrain climatic tolerances (for example, manoxylic wood; large, unprotected buds; and soft, water-rich tissues).

Wing and Greenwood (1993) used palms, gingers, and cycads to place a lower limit on the temperature fluctuations during the coldest months in the Eocene. An important element of this approach was the assumption that not all the groups represented are likely to have evolved in the same direction to change their tolerances (Dorf 1970).

Another example of the use of the NLR method is the application of modern mangrove ecology and tolerances to such floras in the past (Adams, Lee, and Rosen 1990). Mangroves, which comprise several taxa, are specialized for living just at the tide line (Steers 1977). Modern mangroves have approximately the same latitudinal distribution as modern coral reefs, occurring between 32°N and 38°S, and they also show a diversity gradient within those limits (Walter 1977). Several workers have assumed similar patterns for ancient mangroves (Adams, Lee, and Rosen 1990; Westgate and Gee 1990).

TERRESTRIAL MACROFOSSILS: ANIMALS AND TRACE FOSSILS

Freshwater and terrestrial bivalves and gastropods, freshwater fishes, and various terrestrial tetrapods have all been used in paleoclimatology, and in all groups, both biogeographic information and stable isotope data have been important. Trace fossils include both indicator taxa and analysis of trace-fossil tiering.

Freshwater and Terrestrial Mollusks

As with marine mollusks, the application of freshwater and terrestrial mollusks to paleoclimatology has been in the biogeography of the organisms and in analysis of the stable isotopic compositions of their shells.

Biogeography

Freshwater and terrestrial mollusks are much less diverse and some taxa are much longer-lived than their marine relatives, so biogeographic patterns for these animals have been of limited use compared with marine mollusks. The largest freshwater bivalves are the unionids. The use of unionids as paleoclimatic indicators depends on analogy with the tolerances of modern forms. For example, modern unionids do not estivate and prefer clear, oxygenated water, and thin-shelled forms most often occur in standing, rather than running, water. Good (Good, Parrish, and Dubiel 1987; Parrish and Good 1987) used these characteristics to support the interpretation that the Monitor Butte Member of the Triassic Chinle Formation (western United States) was deposited by perennial, clean, well-aerated streams and lakes, which has some implications for paleoclimate and paleohydrology. The NLR method has also been applied to land snails, for example, in early Tertiary faunas of the southwestern United States-northwestern Mexico (Roth and Megaw 1989).

Stable isotopes in terrestrial and freshwater mollusks

Interpretation of the isotopes of carbon and oxygen in terrestrial and freshwater mollusks is less straightforward than in their marine counterparts, and this is particularly true for terrestrial mollusks. Although workers routinely collect data on carbon isotopes, they commonly provide no interpretation. The following discussion thus concentrates on oxygen isotopes and other chemical signatures of freshwater and terrestrial mollusks.

Most of the stable isotope work on terrestrial snails has been on modern and Quaternary fossils (e.g., Yapp 1979; Magaritz and Heller 1980; Magaritz, Heller, and Volokita 1981). These studies indicated that evaporation is an overriding control on the $\delta^{18}O$ composition of terrestrial snail shells and that such information might be useful for determining evaporation gradients; none attempted to determine paleotemperatures. In contrast to land snails, freshwater clams as old as Carboniferous have been subjected to stable isotope studies (e.g., Brand 1994). D. L. Dettman (personal communication) has found that aragonite fractionation in modern unionids is the same as in marine bivalves and that shell growth ceases below 12°C. In addition, they found no differences in isotopic fractionation among modern species. Unaltered aragonite in fossil shells from freshwater mollusks is detected by the presence of nacreous layers, x-ray diffraction confirmation of composition, and yellow-green luminescence (Dettman and Lohmann 1993).

In late Paleocene and early Eocene unionids from the Powder River Basin (Montana and Wyoming), microsampling revealed seasonal amplitudes of $\delta^{18}O$ variation as high as 9.8‰. In the late Paleocene, the total range of surface-water composition interpreted from several shells was $\delta^{18}O$ −8‰ to −23‰ SMOW, and in the early Eocene, the range was −8.5‰ to −16‰. A stratigraphic presentation of these results is shown in figure 4.9. Using the reasoning discussed in chapter 1 (figure 1.9), Dettman and Lohmann (1993) concluded that the very low values in the Paleogene must reflect the contribution of snow melt to the waters in which the bivalves grew. In the temperature relationship for aragonite, shell-$\delta^{18}O_{PDB}$ is equal to water-$\delta^{18}O_{SMOW}$ at 20.9°C, with a 1‰ deviation from the value for every 4.7°C deviation. Over the range of temperatures between 0° and 30°, the water must therefore range from 4.4‰ more negative than the shell to 1.9‰ more positive. Dettman and Lohmann (1993) emphasized that the snow could not have fallen within the basin because that would have required conditions too cold for the clams to grow (and therefore lay down the shell material with very low values).

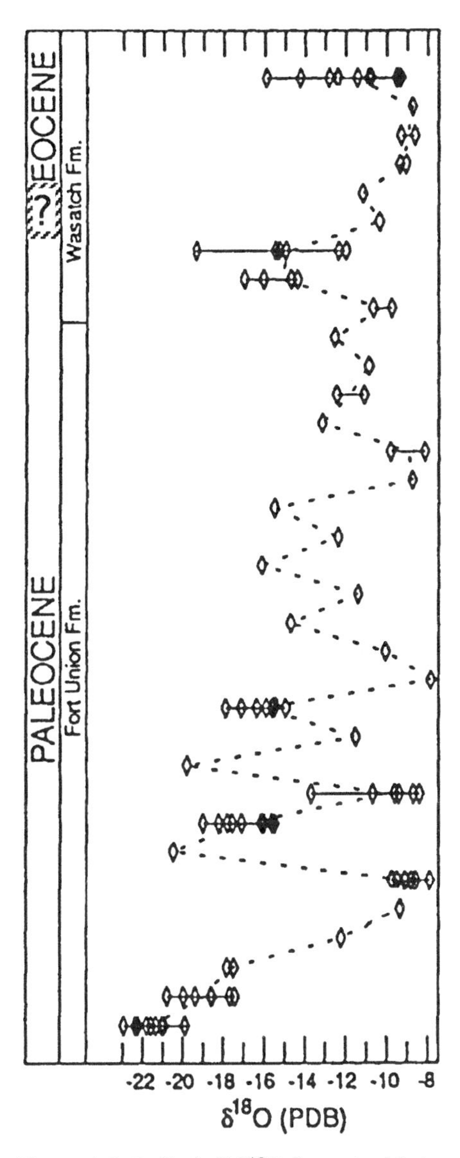

Figure 4.9. Bulk shell $\delta^{18}O$ for unionids in the Powder River Basin, western United States. *Dotted line* connects average values. Vertical position does not correspond to the thicknesses of units; each sample represents at least 3 years growth. From Dettman and Lohmann (1993) in *Climate Change in Continental Isotopic Records*, pp. 153–63. Geophysical Monograph 78, copyright by the American Geophysical Union.

Vertebrates

Terrestrial and freshwater vertebrates are most commonly linked to climate through their biogeography. However, uniformitarian arguments about physiology, anatomical features and related ecology, taphonomy, and stable isotopes also have been important in the use of these organisms for paleoclimatologic interpretations.

Biogeography

The use of terrestrial and freshwater vertebrate biogeographic information in paleoclimatology must take into consideration the patchiness of preservation of vertebrate assemblages, the mobility of the organisms, nonclimatic controls on distribution and diversity, and taphonomic controls, including differential geographic preservation of vertebrates. In general, terrestrial and freshwater vertebrate biogeography contributes little to paleoclimate interpretation because the distribution of the organisms is strongly dependent on factors unrelated to climate. Many studies have attributed aspects of the distribution of freshwater and terrestrial vertebrates to paleoclimate, but nearly always in light of other paleoclimatic indicators and nearly always as one of several alternative explanations.

Taken together, several groups of dinosaurs, therapsids, and Triassic tetrapods show a consistent, bimodal pattern of mid-latitude maximum diversity from the Triassic through the Cretaceous (J. M. Parrish, unpublished compilation); labyrinthodont amphibians show the same distribution in the Triassic (Cifelli 1980). This pattern might be interpreted to be controlled by climate, indicating that these groups were restricted to mid-latitude, temperate climates. However, four observations call this interpretation into question: (1) most of the fossiliferous localities are in mid-latitudes; (2) continental area in low paleolatitudes was less than in mid-latitudes (Parrish 1985), leaving less room for potential localities (Raup 1976); (3) continental regions that were in low latitudes in the late Paleozoic and Mesozoic are now in areas of South America and Africa that are harder to access and/or more poorly studied than vertebrate localities from mid-paleolatitudes (Cifelli 1980); and (4) bone, like everything else, weathers faster in tropical soils (Behrensmeyer 1978), so if the vertebrates did live in low latitudes, it is possible there would be no record of them. Thus the tropical portions of the original biogeographic range of vertebrates might be missing for geographic or taphonomic reasons having nothing to do with climatic controls on the distributions of the animals. However, if low-latitude, even equatorial, climates were dry, as has been suggested for the Jurassic (Hallam 1993), this latter point would not apply.

During the Triassic, most of the continental area was assembled into the supercontinent Pangea. For this reason, the Triassic might be regarded as an ideal time to study for the effects of climate on the biogeography of terrestrial organisms because geographic barriers were minimal and differentiation of the faunas would therefore be more likely related to climate. In the Early Triassic, the global vertebrate fauna was cosmopolitan and widely distributed latitudinally (Parrish, Parrish, and Ziegler 1986). By the Middle Triassic, the global fauna was more differentiated (Tucker and Benton 1982). The southern Pangean faunas were dominated by therapsids (mammal-like reptiles), whereas the northern faunas were dominated by amphibians and aquatic reptiles, with rare therapsids and archosaurs. This difference has been interpreted as climate controlled, that is, that climate was wetter and warmer in the north (Tucker and Benton 1982), but there are two problems with this interpretation. First, no plausible mechanism exists for very different climate in northern and southern Pangea at this time (Parrish 1993). Second, the archosaur and therapsid taxa were cosmopolitan. These two factors suggest that the faunal differences were perhaps more related to chance differential preservation of biofacies, that is, greater preservation of aquatic environments in the north, than to climate. By the Late Triassic the global fauna was cosmopolitan again. Therapsids tended to occur in lower latitudes in the Late Triassic than they did earlier, however, and given the relatively widespread distribution of localities, this part of the pattern might be related to climate. Parrish, Parrish, and Ziegler (1986) suggested that the shift of therapsids from higher to lower latitudes might indicate cooling of the equatorial region, but no strong evidence exists that climate had any control on vertebrate distribution in the Triassic.

The fact that strong patterns are not seen in the Triassic can be interpreted as an argument for climate *not* being a major influence on reptile biogeography at that time. One possibility is that reptiles were diversifying and dispersing at the very time when geographic barriers to genetic exchange and dispersal were minimal, and therefore differentiation related to environmental factors, including climate, was also minimal. Another possibility is that temperature and/or precipitation gradients were low. Little evidence exists in the geologic record for a strong latitudinal temperature gradient in the Triassic. Finally, the cosmopolitanism of the faunas could simply be evidence that the animals were capable of moving long distances. Unless they are tied to some specific, climate-controlled resource, the mobility of terrestrial vertebrates may be an impediment to their use as paleoclimatic indicators.

In the absence of clear patterns that can be linked to climate, biogeographic arguments have been closely tied to assumptions about the physiology of the organisms, using the nearest-living-relative (NLR) method. For example, extinct crocodilians have been commonly used to indicate warm climates, a practice that took on particular significance with the discovery of alligatorines in Paleogene rocks on Ellesmere Island (West, Dawson, and Hutchison 1977). Throughout the Tertiary in North America, crocodilians were associated with other warm-climate indicators, such as salamanders, giant tortoises (Estes and Hutchison 1980), and palms (Wing and Greenwood 1993). When the data on terrestrial vertebrates are good, the ranges of crocodilians can be confidently constrained. Markwick (1994) showed that their geographic range in North America was far greater in the Eocene than later (figure 4.10). He concluded that present-day tolerances can be used to indicate paleoclimate if they are consistent with information from stable isotopes, paleobotany, and other sources.

Perhaps the most enduring application of the NLR method has been the interpretation of the environmental tolerances of lungfish. Modern lungfish live in warm climates that are highly seasonal with respect to rainfall, and they are capable of enduring long periods of dormancy (estivation) during dry seasons. Because the morphology of the group has changed so little, it has been commonly assumed that their presence indicates the same kinds of environmental tolerances (e.g., Shaeffer 1970). Fossil lungfish are indeed found in units that are considered indicative of warm, seasonal climates on other grounds, for example, in the Triassic of North America (Dubiel et al. 1991) and of Australia (Parrish et al. 1996).

Modern small mammals are sensitive to climate and, through use of transfer function-like statis-

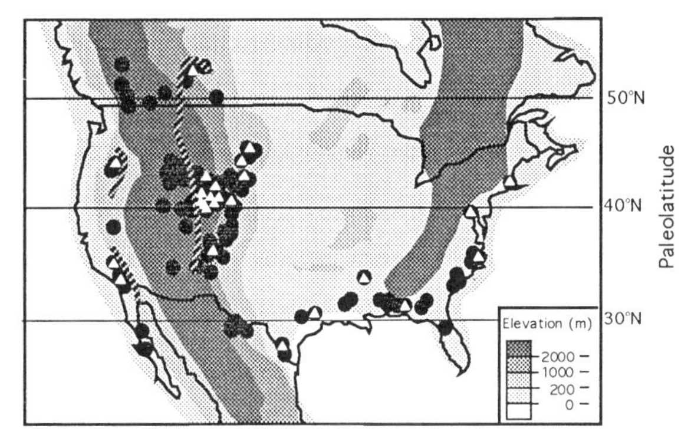

Figure 4.10. Distribution of all vertebrate localities (*circles*) and crocodilians (*triangles*) in the Eocene of North America. From Markwick (1994) in *Geology*. Reproduced with permission of the publisher, the Geological Society of America, Boulder, Colorado, USA. Copyright © 1994 Geological Society of America, Inc.

tics, can provide paleoclimatic information for the recent past (e.g., Montuire et al. 1997). The presence of numerous small mammals associated with the crocodilians in the Paleogene rocks of the Ellesmere Island, whose modern relatives are warm-climate dwellers, has commonly been taken as additional evidence for warm climates there. However, as pointed out by McKenna (1980), "mammals are mobile and readily adaptable creatures, so that it is dangerous to conclude very much from mammalian data regarding previous climates, especially if the extrapolations reach back to the Eocene." McKenna (1980) cited as more reliable the paleobotanical data (Schweitzer 1980; Wolfe 1980). In the Tertiary of Pakistan, mammalian faunal changes are correlated to, but not regarded as indicative of, climatic and environmental changes (Barry et al. 1995).

Although the use of diversity gradients has foundered on the considerations discussed at the beginning of this section, related information from smaller areas might be useful in paleoclimate interpretation. Legendre (1987) found an intriguing correlation between the species-to-family (S:F) ratio in nonpredatory mammalian faunas plus bats in France and the $\delta^{18}O$ curve from benthic foraminifera for the Atlantic Ocean (see figure 2.14). He suggested that S:F ratios might therefore be used as paleotemperature indicators, likening the variations in species richness to those seen with latitude. However, he also stated a number of caveats, that the faunas must come from a small area and a continuous stratigraphic sequence in order to minimize the types of taphonomic effects that have probably affected larger-scale studies.

Freshwater fishes present the inverse problem to the mobility of terrestrial vertebrates. Freshwater fishes are limited geographically where land connections are lacking, regardless of climate. Thus connections among their waterways controlled their biogeography far more than climate (Turner and Tarling 1982) and continental drift was an overriding control on biogeographic change for these organisms (Young 1990).

In summary, terrestrial and freshwater vertebrate biogeography is strongly affected by many factors other than climate (Ostrom 1970; Hallam 1994). The most reliable biogeographic studies are those founded on data from numerous localities that were not separated by geographic barriers. In addition, the evolutionary history of the group must be considered and specific evidence sought for limitation by climate.

Physiology

Physiological arguments for climatic limitations on vertebrates have probably created the greatest controversy in the relationship between terrestrial vertebrate biogeography and paleoclimate. The two most important issues center around the physiologies of ancient vertebrates and their tolerances for cold.

The first issue is the well-publicized debate on the warm-bloodedness of dinosaurs and therapsids (review in Farlow 1990). The terminology for the debate includes two pairs of words that are commonly used interchangeably, but which do not have quite the same meanings. "Warm-blooded" animals are referred to as endotherms, "cold-blooded" animals as ectotherms. Ectotherms depend on external heat to drive metabolic processes. Some ectotherms are poikilothermic, meaning that body temperature is completely dependent on and matches environmental temperatures. In other ectotherms, body temperature does not track environmental temperature. Instead, these animals maintain body temperature within a narrow range through various behaviors, such as basking. Many, but not all, fishes are poikilothermic; reptiles, on the other hand, are generally not poikilothermic.

Endotherms are capable of generating body heat metabolically. Even some fishes are endothermic. Leatherback turtles use heat generated by activity and maintained by their large size to survive in the North Atlantic (Paladino, O'Connor, and Spotila 1990). "Warm-blooded" animals in the usual sense are homeotherms; that is, body temperature remains within a narrow range and is internally regulated. Some mammals are not true homeotherms, in the sense that their body temperatures may not be fully regulated by metabolic processes but, rather, are at least partly dependent on environmental temperatures. Endothermy and ectothermy are end-members of a wide range of physiological states that are expressed even in modern animals (e.g., papers in Thomas and Olson 1980).

The debate on warm-bloodedness in dinosaurs and in therapsids was brought to the fore by Bakker (1975), publicizing and building on work by Ostrom (1964, 1966, 1970). Bakker (1975) assumed that high-latitude continental areas in the Permian were cold and that therapsids therefore had to be endotherms. He sought evidence for endothermy in dinosaurs as well, and presented the following evidence:

1. bone histology—dinosaurian bone is similar to mammalian bone;
2. predator-to-prey ratios—fossils of prey species (herbivores) outnumbered those of predators (carnivores) by a ratio of 10:1, consistent with the supposed high metabolic needs of endotherms versus ectotherms and far above predator-to-prey ratios in modern faunas, in which ectotherms are the top predators;
3. biogeography—dinosaurs lived at very high latitudes; and
4. morphology and gait characteristics—shoulder, pelvic, and limb anatomy shows that the animals were capable of rapid movement.

Questions have been raised about all these points (Farlow 1990). The porous bone histology of mammalian and dinosaurian bone may reflect ontogeny rather than physiology. Determination of predator-to-prey ratios is so subject to assumptions and possible sources of error that Farlow (1990) doubted their utility for determining physiology. The high-latitude vertebrate faunas cited by Bakker (1975) also contain labyrinthodont amphibians, which were less arguably ectotherms (e.g., Ostrom 1970). In addition, ectotherms are not necessarily intolerant of cold climates (Paladino, O'Connor, and Spotila 1990). The active habit of the animals suggests they were capable of migrating to avoid the harshest conditions (Parrish et al. 1987), and some dinosaurs show possible adaptations for living in dim light (large eyes in Cretaceous hypsilodontids of Australia; Rich and Rich 1989). Although gait does not constitute a priori evidence for endothermy (Farlow 1990) and, indeed, Bakker (1975) regarded the sprawling therapsids as endothermic by virtue of their occurrence at high latitudes, active gait and endothermy tend to go together because the metabolic demands of an active habit are more easily met with an endothermic physiology (Ostrom 1970). Farlow (1990) argued that perhaps dinosaurs had variable metabolic rates, dependent on ontogeny or season. This argument has received some support from $\delta^{18}O$ studies of dinosaur bone, which suggest an intermediate endothermy for some groups (Barrick, Showers, and Fischer 1996).

Despite the limitations, the NLR method has been applied to physiological tolerances of some groups as an indicator of their environmental tolerances. Giant tortoises, which all belong to a single genus, *Geochelone*, are an example for which the NLR method is relatively easy to accept. The genus is diverse (11 species) and no differences in temperature tolerances exist among the species. As pointed out by Smith and Patterson (1994), "if all the known members of a natural group are restricted to certain temperature conditions, then a fossil member can be inferred, by parsimony, to possess the group's temperature tolerances." This is in contrast, for example, to the wide climatic range of the lizard genus *Varanus* (Ostrom 1970). A similar argument can be made for crocodilians (Cassiliano 1997), with the exception of the subfamily Alligatorinae, which ranges into areas that have occasional below-freezing temperatures. Alligators' tolerance to cold, which is aided by their ability to become dormant, is dependent on access to water. Crocodilians, which are poikilothermic as well as ectothermic, have both anatomical and behavioral adaptations that permit the animals to maintain body temperature within a narrow range, varying by as little as 2° to 6°C. The argument here is that physiology is a conservative trait, that if modern forms are ectothermic poikilotherms, then extinct forms (to the family level, at least, which would encompass all crocodilians) probably were, too. Because crocodilians are dependent on environment for metabolic energy, and because of the limited temperature range within which their metabolic processes can function, they are restricted to areas that can provide those temperatures year-round. Cassiliano (1997) noted that although crocodilians have complex behavior that helps them maintain a constant body temperature, behavior can modify the environment only so much without long-distance migration; crocodilians are unlikely ever to have migrated because of the relative inefficiency of their gaits.

Anatomy and ecology

Direct anatomical adaptations for climate (e.g., hair and fat) are rarely preserved. Anatomical features that have been preserved are sometimes attributed to climate, but often ambiguously (Ostrom 1970). For example, the dorsal fin of the therapsid *Dimetrodon* has been proposed as a heat-transfer structure (Tracy, Turner, and Huey 1986) or as a display organ (Bakker 1971). Confidence interpreting anatomical characteristics increases when corroborating evidence exists. For example, mammoths are associated with glacial landforms and deposits, so their hairiness is more confidently attributed to climate.

Nevertheless, the relationship between anatomy and overall ecology and behavior may provide insights to paleoclimate. For example, Janis (1984) compared the Tertiary ungulate faunas in North America with modern African faunas. Her study relied on the relationship between the type of vegetation (which is related to climate) and the distribution of various types of ungulates, which can be identified from hard parts. In Africa, the diversity of horned ruminants (e.g., gnus) and equids (horses and zebras) is closely correlated with vegetation type. Where leafy vegetation is abundant, horned ruminants are more diverse; where grass is the dominant vegetation, equids are more diverse. In the Tertiary of North America, equids were more diverse, leading Janis (1984) to conclude that the Tertiary of North America had more grassland and therefore climate was drier and/or more seasonal than that of present-day Africa, which has a higher diversity overall of ruminants. The interpretation of grassland vegetation is supported by palynological studies (e.g., see discussion in Wang, Cerling, and MacFadden 1994).

Additional information comes from the types of ruminants in the faunas (Janis 1984). Ruminants tend to have horns only if they defend feeding territories, usually in woodlands where they do not have to range long distances to get enough food. African ruminant taxa are 70% horned, whereas <30% of the ruminants in the Tertiary of North America had horns. Because woodland vegetation also signifies more rain than grassland, the low percentage of horned ruminants in the Tertiary of North America supports the interpretation that the North American Tertiary climates were drier and/or more seasonal. Note that the paleoclimate interpretation is really from the plants and vegetational physiognomy, but that the vegetation can be determined from the vertebrate fossils, which might be better preserved than plants in some environments.

A relationship between body size and temperature has been formalized as Bergmann's rule, which states that individuals within a warm-blooded species increase in size with increasing latitude (and therefore decreasing temperature; see, for example, discussion in Mayr 1963). Use of this relationship has not been particularly useful for paleoclimate studies because it requires samples of the same species over a broad latitudinal range to detect a pattern. However, variations on the relationship between body size and climate have been used. For example, Legendre (1986) noted that ranked species plots of body weight, called cenograms, have characteristic forms in modern faunas, depending on the environment in which they live. Faunas with an even distribution of body weights among the species, except for a small number of species at very high body weights, occur in tropical rainforest environments. Cenograms from more open environments have gaps corresponding to a lack of species at mid-range body weights. In addition, the slopes of the curves become steeper toward more arid and high-latitude faunas, indicating successively fewer species overall and fewer large species.

Climatic change can be related to body size change only if three conditions are met (Morgan et al. 1995). The most important is climatic change demonstrated from other data, implying that body size alone cannot be a reliable paleoclimatic indicator. The other two conditions are that the observed changes in body size should be consistent with observed principles of physiology, energetics, and trophic preference (for example, if the fauna consists entirely of herbivores, a problem with sampling might be indicated) and that changes in trophic structure should accompany changes in size.

In an exercise similar in some ways to that of Janis (1984), Gunnell et al. (1995) used trophic structure among vertebrates to interpret Tertiary faunas and environments of the United States. Based on analogy with the trophic structure among mammals from various environments today, they concluded that vegetation in the Paleogene fluctuated between closed, humid-climate forests and more open, drier woodlands. Trophic structure of the vertebrate fauna in the Neogene indicated that savanna woodlands were the typical habitat at that time. Thus the climate can be inferred to have fluctuated between relatively wet and/or cool and warm and/or dry, ending in the Neogene in a climate that was relatively warm and seasonal with respect to rainfall.

Body size and shape of freshwater fishes are related to temperature and stream gradient (G. R. Smith, personal communication; see also Smith and Lundberg 1972). Although the relationship has not yet been quantified, preliminary indications are that freshwater fish morphology might provide a taxon-independent approach toward interpreting paleoclimate and paleoaltitude (G. R. Smith, personal communication).

Taphonomy

As with plants, how animals are preserved can provide clues to the climatic regime in which they lived and died. The Triassic Chinle Formation in the western United States (Dubiel et al. 1991) and Cretaceous floodplain deposits in Alberta (Canada) and Mongolia (Jerzykiewicz and Sweet 1987) were interpreted as having been deposited in semi-arid climates on similar grounds: presence of well-developed caliche and deposits of ephemeral ponds and streams. Both units also contain articulated skeletons of dinosaurs that have recurved necks resulting from postmortem contraction of the large nuchal (head-supporting) ligament (Parrish 1989). If the animal died on land and was not washed into a water body, this type of preservation is most likely to be the result of prolonged drying at the surface (Weigelt 1989).

Elder and Smith (1988) used taphonomy of fish to challenge the notion that the Green River Formation (Tertiary; Rocky Mountains, U.S.) was deposited in a shallow playa. They observed that at >16°C, fish float after death because of the rapid production of gasses by decay, and the skeletons become disaggregated. Articulated skeletons, such as those found in the Green River Formation, are thus more likely to have been derived from fish that lived in waters colder than 16°C. Depth of the water also aids preservation in that the bubbles of decay gasses are small where the water is deep. Elder and Smith (1988) argued that the large accumulations of intact skeletons in the Green River Formation indicate (1) mass death, caused by summer algal blooms or by overturn in the fall, bringing H_2S to the surface, and (2) settling of the fish into cold, deep waters, where bacterial gas generation was suppressed and scavengers were discouraged by the anoxic waters.

Stable isotopes in terrestrial and freshwater vertebrates

The stable isotopic composition of the phosphate in bones and teeth is related to the physiology of the animal and the isotopic composition of the water ($\delta^{18}O$) or food ($\delta^{13}C$, $\delta^{15}N$) that the animal ingested, all of which can be related to climate. In mammals, body temperature is relatively constant and the biogenic phosphate precipitates in equilibrium with body water, so the $\delta^{18}O$ composition of the phosphate in bones and teeth ($\delta^{18}O_{PO4}$) is directly related to the original isotopic composition of the water (Luz, Kolodny, and Horowitz 1984; Luz and Kolodny 1985; Bryant, Luz, and Froelich 1994). Relating the isotopic composition of the environmental water ($\delta^{18}O_w$) to temperature requires making assumptions about climate-related fractionation processes, such as evaporation and orographic fractionation. Two additional assumptions are common, that rainwater ($\delta^{18}O_{precip}$) is equal to $\delta^{18}O_w$ of the local meteoric water, and that $\delta^{18}O_{PO4}/\delta^{18}O_w$ fractionation for fossil animals is the same as that of their modern relatives.

The relationship between $\delta^{18}O_{PO4}$ and $\delta^{18}O_w$ for several groups of mammals is shown in figure 4.11. Fractionation between drinking water ($\delta^{18}O_w$) and $\delta^{18}O_{PO4}$ is generally linear, although the slopes and intercepts are different for different taxa, even within the same group, for example, horses (Bryant, Luz, and Froelich 1994). $\delta^{18}O_{precip}$ and $\delta^{18}O_{PO4}$ are also related ($r = 0.66$), especially if relative humidity is included ($r = 0.9$; see also Ayliffe and Chivas 1990); humidity influences the depletion of leaf water ^{16}O through its influence on evapotranspiration.

Data from bones and teeth show a lot of scatter (Bryant, Luz, and Froelich 1994; Sánchez Chillón et al. 1994). Bone is constantly resorbed and reprecipitated during the life of the animal, whereas tooth enamel is not, so use of both

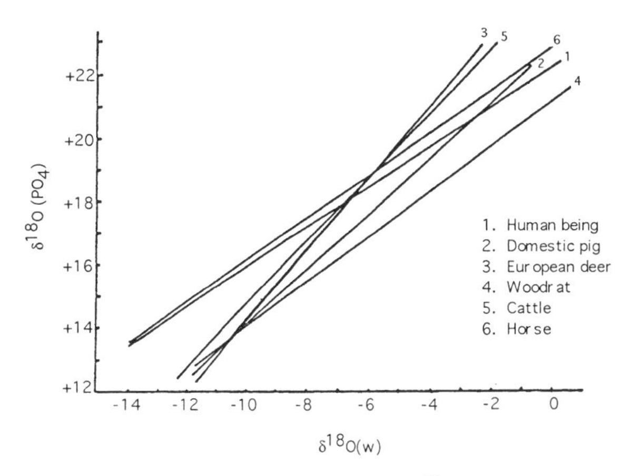

Figure 4.11. Relationship between $\delta^{18}O_{PO4}$ (bone and tooth phosphate) and $\delta^{18}O_{water}$ for several groups of mammals. Compiled by Mou Yun from data in Bryant, Luz, and Froelich (1994), D'Angela and Longinelli (1990), and Longinelli (1984).

sources of data can be misleading, bone because the composition may reflect only short-term paleoclimatic conditions and enamel because it may incorporate wide paleoclimatic fluctuations (Bryant, Luz, and Froelich 1994). Even tooth enamel, however, is subject to variability not related to climate (Bryant et al. 1996a). For example, the position of the tooth in the jaw and the timing of mineralization of the tooth, that is, when it formed during the lifetime of the individual, exert strong controls on its isotopic composition (Bryant et al. 1996b); Fricke and O'Neil (1996) recommended limiting analysis to small, late-forming teeth. Bryant et al. (1996a) concluded that variation in horses, at least, is so great as to obscure all but long-scale climatic changes.

Fish bone phosphate also records the $\delta^{18}O$ of environmental water (Kolodny, Luz, and Navon 1983). Kolodny, Luz, and Navon (1983) proposed that temperature could be determined by comparing the isotopic compositions of fish and mammal bones from the same deposit. The mammal bone would preserve the isotopic composition of environmental water independent of temperature fluctuations, and the fish bone would preserve both the isotopic composition of the water and any temperature fluctuations. The difference between the mammalian and fish bone $\delta^{18}O$ would be temperature and/or salinity. This method has recently been applied to coexisting fishes and sirenians in the Miocene of France (Lécuyer et al. 1996). Variability among taxa in fractionation of $\delta^{18}O$, both genetically and environmentally, make application of this approach difficult (Bryant, Luz, and Froelich 1994; see also Luz, Kolodny, and Horowitz 1984).

A similar approach was taken by Koch, Zachos, and Dettman (1995), who used information available from paleosol carbonates, carbonate from the apatite in mammal teeth, and bivalve shells to calculate temperature. Because of the complexity of the relationship between environmental water and body water in mammals, Koch, Zachos, and Dettman (1995) applied a range of values for the drinking water-apatite carbonate relationship, using $\delta^{18}O_{apatite} = \delta^{18}O_{drinking\ water} - 27.7‰$ for modern pigs from Maryland and $\delta^{18}O_{apatite} = \delta^{18}O_{drinking\ water} - 32.7‰$ for a zoo elephant in Chicago. The $\delta^{18}O$ value determined from the bivalves is assumed to be the drinking water $\delta^{18}O$ in these equations, and the resultant value, plus the constant 37°C body temperature for mammals, was used to estimate meteoric water $\delta^{18}O$. In soil carbonates, assuming they formed throughout the year at shallow depths where the average temperature is approximately equal to the mean annual temperature, the fractionation relationship is

$$1000\ \ln\alpha^{calcite}_{water} = \frac{2.87 \times 10^6}{T^2} - 2.89, \quad (4.1)$$

where

$$\alpha^{calcite}_{water} = \frac{(1000 + \delta^{18}O_{calcite})}{(1000 + \delta^{18}O_{water})} \quad (4.2)$$

The meteoric water $\delta^{18}O$ value derived from the mammals is used in these equations to estimate mean annual temperature. Temperature estimates through time in the Powder River and Bighorn Basins using this method are shown in figure 4.12. The estimates approximately match paleobotanical and climate-model estimates of mean annual temperature. Koch, Zachos, and Dettman (1995) pointed out at least two potential problems with these estimates. First, the water is reasonably assumed to represent the water the mammals drank, but the soil water is not the same water and might have been heavier because of evaporation. Second, the methods assume that tooth enamel and soil carbonates recorded the same events, but teeth are actually laid down over fewer years than soil carbonates.

Using a different approach, Kolodny and Luz (1991) were able to put some constraints on the temperature and isotopic composition of a Devonian lake (Orcadian Basin, United Kingdom), which lay at approximately 20° to 30°S. The lake contained both surface- and bottom-dwelling fishes, and Kolodny and Luz (1991) hypothesized that the isotopic difference between the two groups was dependent on temperature. Assuming surface-water temperatures of 37°C, which would be possible in a subtropical lake, the $\delta^{18}O_w$ would have been between −7‰ and −3‰. If the bottom-water temperatures are assumed to have been 4°C, a reasonable value considering the continental interior setting, the isotopic composition of the water would have been closer to −6‰, making the surface waters 15° to 20°C. This general approach has been used by several workers (e.g., Brand 1994). It has the advantage of narrowing the possible temperature ranges, but it does rely on some broad assumptions about the climate.

Fishes precipitate aragonitic structures called otoliths, which are commonly preserved in lacustrine sediments. Patterson, Smith, and Lohmann (1993) and Smith and Patterson (1994) determined the fractionation factor for these structures, the slope of which was similar to the marine aragonite fractionation but offset from that curve. In ectothermic, poikilothermic fishes, including marine forms, the relationship is independent of taxon for the taxa studied (Patterson, Smith, and Lohmann 1993). Otoliths are banded, and Patterson, Smith, and Lohmann (1993) showed that a detailed $\delta^{18}O$ profile across a transverse section records seasonal temperature fluctuations.

The major use of carbon isotopes in terrestrial vertebrates has been relating changes in $\delta^{13}C$ in the animals to those in soil carbonates (e.g., Cerling, Wang, and Quade 1993) and to evolutionary events in the vertebrates. Vegetation consists mostly of plants having one of two types of metabolism, called C_3 and C_4. The carbon isotopic composition of C_4 plants, which are adapted for drier climates, is higher than that of C_3 plants, and this is reflected in the carbon isotopic composition of the tooth enamel of herbivores. The evolution of hypsodonty (rootless, continuously growing teeth) in herbivorous mammals had been tied to the development of widespread savannah (that is, C_4) grasslands. However, the evolution of hypsodonty around 15 Ma in North America and earlier in South America precedes a major shift in soil $\delta^{13}C$ at about 8 Ma (Cerling, Wang, and Quade 1993; MacFadden et al. 1994;

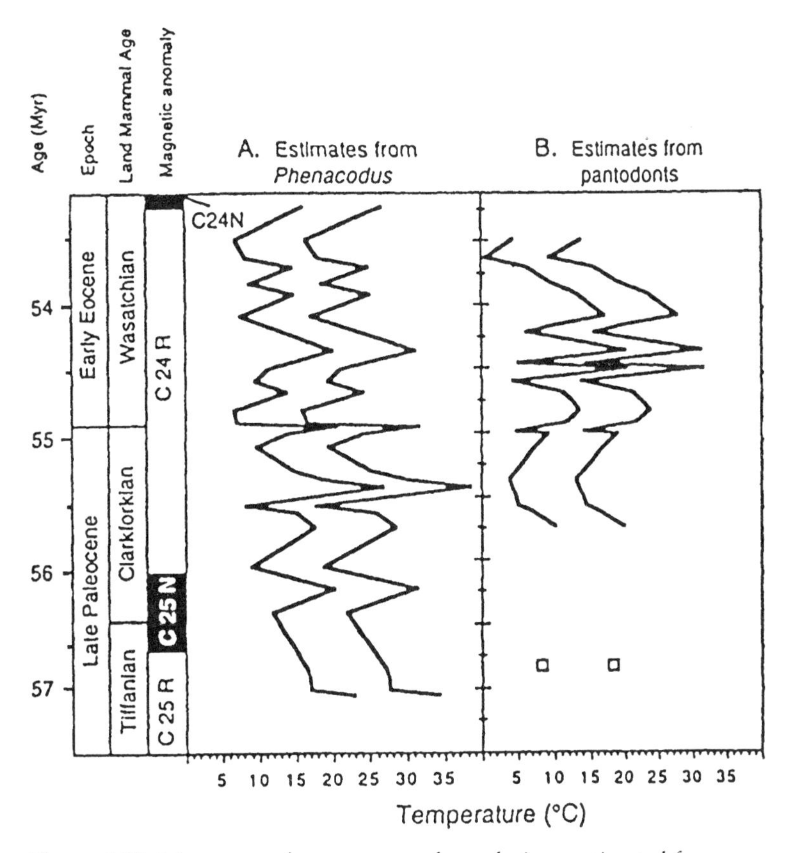

Figure 4.12. Mean annual temperature through time, estimated from paleosol-carbonate $\delta^{18}O$, and meteoric water $\delta^{18}O$ from the mammals *Phenacodus* and pantodonts, Powder River and Bighorn Basins, western United States. See text for explanation. Reprinted from *Palaeogeography, Palaeoclimatology, Palaeoecology* 115, Koch, P.L., J.C. Zachos and D.L. Dettman, "Stable isotope stratigraphy and paleoclimatology of the Paleogene Bighorn Basin (Wyoming, USA)," pp. 61–89 (1995) with kind permission of Elsevier Science-NL, Sara Burgerhartstraat 25, 1055 KV Amsterdam, The Netherlands.

Wang, Cerling, and MacFadden 1994; MacFadden, Cerling, and Prado 1996). In South America, therefore, hypsodonty evidently evolved with a spread of C_3 grasslands (MacFadden, Cerling, and Prado 1996).

As with any other material analyzed for stable isotopes, bone is subject to diagenesis. Characteristics of bone that has undergone significant diagenesis are elevated strontium, uranium, and rare earth elements (e.g., Denys et al. 1996; Kolodny et al. 1996). Indeed, two of the leaders in the applications of bone-phosphate isotopic analysis, Y. Kolodny and B. Luz, recently sounded a cautionary note about such work, and particularly about the application of the analyses to interpretation of dinosaur physiology (Kolodny et al. 1996). They provided a number of criteria by which unaltered bone might be recognized, none of which was certain. Principal among their recommendations, in line with their and earlier work, were avoidance of bone in preference to tooth enamel because tooth enamel is, in life, much more highly crystalline, and avoidance of any material that has elevated levels of strontium, uranium, and rare earth elements. They noted that preservation of microstructure is not a sufficient criterion.

TRACE FOSSILS

Trace Fossils Made by Animals

Terrestrial trace fossils provide information about behavior that can be related to paleoclimate and that is not necessarily provided by body fossils (Hasiotis and Dubiel 1993a). Some trace fossil assemblages are useful as indicators of different levels of saturation of the substrate by water. The most commonly cited trace fossil in this regard is *Scoyenia*, which has been cited as indicative of environments undergoing periodic drying (Breed and Breed 1972; Bromley and Asgaard 1979). This interpretation has been challenged by Hasiotis and Dubiel (1993a), who argued that the trace fossil is indicative of moist to saturated substrates in immature paleosols and of marginal fluvial and lacustrine deposits. In Triassic red beds of Greenland, a *Rusophycus* assemblage occurs in thin, ripple-laminated sandstones with mud drapes, conchostracans, and plant parts and thus was identified as characteristic of periodically desiccated river deposits (Bromley and Asgaard 1979).

Where the trace fossil can be linked by inference or observation to a particular organism, interpretation of the significance of the trace is strengthened. For example, *Cylindricum*, which is similar to the better-known marine trace fossil *Skolithos*, occurs in crevasse-splay and levee deposits in the Triassic Chinle Formation (western United States; Hasiotis and Dubiel 1993a). It occurs in groups of 10 to 100 individuals and resembles burrows made by modern burrowing and nesting insects. Hasiotis and Dubiel (1993a) interpreted the traces as brood nests of unknown flying insects, which would take advantage of exposed, newly deposited sediments. Thus *Cylindricum* was interpreted as indicative of lowered water tables. Crayfish burrows (*Camborygma*; Hasiotis and Mitchell 1989; Hasiotis and Dubiel 1993a,b, 1995; Hasiotis and Mitchell 1993) are especially useful for paleoenvironmental interpretation. Modern freshwater crayfish move with the water table, so characteristics of the burrows are dependent on its depth and stability (Hasiotis and Dubiel 1993a). These characteristics can provide information about water-table fluctuations that may be related to paleoclimate.

The abundance and tiering of trace fossils and their positions relative to variations in paleosols can be useful for interpreting paleoclimate (Bown and Kraus 1983; Hasiotis and Dubiel 1994). Bown and Kraus (1983) noted that most trace fossils in the Willwood Formation (Eocene; Wyoming, U.S.) were restricted to specific sediment and paleosol types and to identifiable horizons, and they suggested that this was related to local paleohydrology. Hasiotis and Dubiel (1994) developed this idea and provided detailed descriptions of the tiering (vertical partitioning) of trace fossils within paleosols of the Chinle Formation (figure 4.13). The types, dimensions, and positions of the burrows allowed them to estimate the depth of the water table. Interpreting paleoclimate from tiering in terrestrial trace fossils is relatively new compared with marine trace fossils, and the full range of tiering styles has not been documented. However, the work that has been done shows that tiering can be expected to vary with different types of soil conditions, which, in turn, are dependent on climate.

Root Traces and Rhizoliths

One element of the trace-fossil tiering concept developed for terrestrial trace fossils by Hasiotis and Dubiel (1994) is the presence of rhizoliths and root traces. The morphology of these features is related to local paleohydrology and therefore at least partly to climate.

Klappa (1980; Esteban and Klappa 1983) classified rhizoliths into five types: root molds, root casts (sediment or cement-filled root molds), root tubules (cemented cylinders around root molds), rhizocretions (diagenetic accumulation around living or dead plant roots), and root petrifactions. Rhizoliths are usually calcite but occasionally dolomite, gypsum, gibbsite, or silica. Excellent photographs of various types are found in Klappa (1980) and Blodgett (1984) and complete descriptions in Retallack (1990).

Root traces and rhizoliths occur in a variety of arrangements within the sediments, but two general types are particularly useful for interpreting local paleohydrology for potential application to

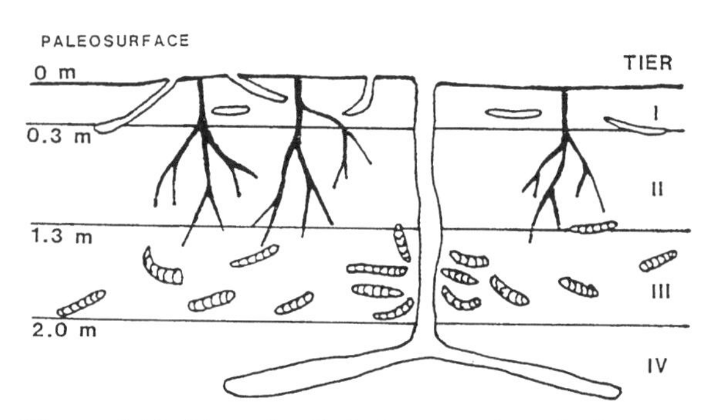

Figure 4.13. Trace-fossil tiering in paleosols from the Shinarump Member of the Chinle Formation (Triassic; western United States). From Hasiotis and Dubiel (1994).

paleoclimate problems. These are roots occurring as horizontal mats and those that are long and predominately vertical (Cohen 1982; Mount and Cohen 1984; Retallack 1990). In the Koobi Fora Formation (Plio-Pleistocene, Kenya), horizontal root mats are observed in laminated siltstones and vertical root casts in sandstones and conglomerates. Cohen (1982; Mount and Cohen 1984) related these observations to drainage conditions and therefore water availability for the plants. Both types of root systems occur in the Koobi Fora, which shows that local paleohydrology can vary irrespective of climate, which is a limitation of this method. However, a predominance of one type or another through several different facies could provide some evidence for overall climatic conditions.

SCOTT W. STARRATT

5

Terrestrial and Freshwater Lithologic Indicators of Paleoclimate

This chapter covers lithologic paleoclimatic indicators that are formed on land or in freshwater. Most of the methods are qualitative, although the paleoclimatic ranges for many of these indicators are narrow, and stable isotope methods are beginning to provide some quantitative information. Although the indicators discussed herein are in some ways more indicative of atmospheric climatic patterns than are marine indicators, the interpretation of many terrestrial indicators is complicated by the fact that they can be influenced as much by tectonics as by climate. In addition, although the stable isotopic composition of the oceans can be assumed to have been constant within a fairly narrow range, this assumption cannot be made for the stable isotopic composition of meteoric waters.

Evaporites are included in this chapter because, although the largest deposits formed from evaporated seawater, the formation of the deposits themselves was a function of terrestrial climate, that is, a direct response to atmospheric influences. This also allows lacustrine and marine evaporites to be discussed together.

EOLIAN DEPOSITS

Eolian deposits comprise wind-blown sediment or fossils and their associated strata. Eolian deposits occur in both terrestrial and marine rocks, but because eolian material in the deep sea reflects conditions on the continents, all types will be discussed in this chapter. Eolian deposits include eolian sandstones and associated rocks, loess, eolian dust in the deep sea, and volcanic ash (table 5.1). Eolian sandstones are a general indicator of low moisture supply and provide the most direct information about paleowind directions. In addition, the presence of eolian sandstones indicates a minimum wind strength, that which is required to transport sand. Eolian sandstone depositional systems are sensitive to changes in moisture supply. Loess implies conditions associated with continental glaciation. Eolian dust in the deep sea has been used to infer changes in moisture supply on the continents and to determine wind direction and speed. Volcanic ash deposits provide information about wind direction and, possibly, strength.

Table 5.1.
Summary of eolian deposits and paleoclimate information derived from them

Deposit	Paleoclimate information
Eolian sandstone	Wind direction, aridity
Eolian depositional systems	Wind direction, changes in moisture supply
Loess (dust)	Changes in moisture supply, temperature (indirectly)
Eolian dust in deep sea	Wind direction and stength, changes in continental climates
Volcanic ash	Wind direction

Eolian Sandstones and Eolian Sandstone Depositional Systems

Presently, eolian sand deposits cover approximately 6% of the area of the continents, and 97% of these deposits are in sand seas that occupy low mid-latitude arid regions. Although sand dunes, most notably, coastal dunes, can occur in humid climates, sand seas, also called ergs, are well accepted as indicative of arid or semi-arid climates. Data compiled by Gyllenhaal (1991) show that modern ergs between 40° north and south latitudes are all in climates with mean annual precipitation of 90 cm or less, and that most occur in climates with 30 cm or less, with the mode at 0 cm (figure 5.1a). These climates have no, or short, rainy seasons (figure 5.1b).

Ergs are accumulations of eolian sand that cover at least 125 km^2. Smaller accumulations are called dune fields or, if no dune bedforms are present, sand sheets (Pye and Tsoar 1990). Ergs at the lower size limit are uncommon, as 85% of the wind-blown sand occurs in ergs that are larger than 32,000 km^2 (Wilson 1973). Figure 5.2 illustrates the distribution of active ergs compared with the distribution of desert climates, defined as regions with <15 cm mean annual precipitation (Wilson 1973; Kocurek 1996).

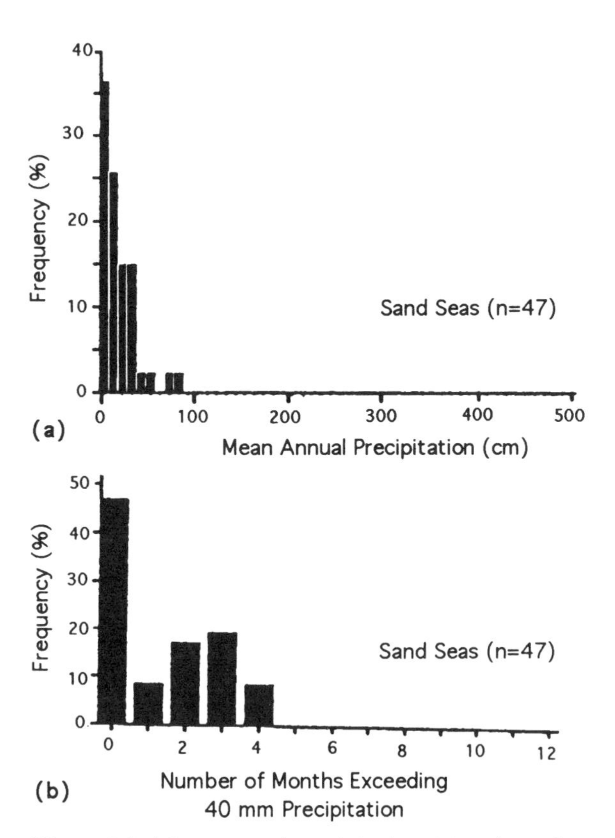

Figure 5.1. Mean annual precipitation *(a)* and number of months exceeding 40 mm precipitation *(b)* in areas of modern ergs (N = 47). From Gyllenhaal (1991).

Controls on the formation of ergs and dunefields

The basic requirements for formation of ergs and dunefields (Pye and Tsoar 1990; figure 5.3) are a large supply of sand (Mainguet 1978), sufficient wind energy for transport or in situ reworking of sand, and suitable topographic and climatic conditions maintained over a long period. The threshold velocity for transporting sand in air is approximately 4 m/sec (Bagnold 1941). The distribution of global wind energy is not uniform and is highest in coastal regions and in the poleward extremes of continents (Pye and Tsoar 1990). The potential for sand transport is great in circumpolar regions not only because wind energies are higher, but also because the air is denser at higher latitudes and thus can move grains easier at the same air speed (Selby, Rains, and Palmer 1974). Most desert areas have relatively low windiness because fronts do not easily penetrate the mid-latitude zonal high pressure and because convective storms, which derive much of their strength from latent heat release during condensation of water vapor, are relatively rare because of the low relative humidities.

Windiness in coastal regions is largely responsible for the presence of dunes in humid climates, including areas that have more than 2 m of rainfall per year (Pye and Tsoar 1990). Wind energy is high enough in these regions that the amount of rainfall is irrelevant. In general, coastal dunes have more slumps, root traces, and saturated-sediment, compressional deformation features (Pye and Tsoar 1990) than do arid-climate dunes. However, none of these features is diagnostic, and the interpretation of humid-climate, coastal dunes in the geologic record relies on other information as well, including, but not limited to, paleogeography, reconstruction of the size of the dunefield, paleosol studies, and floral studies. Because coastal dunes are not climatically significant, they will not be treated further in this book.

Most, though not all, modern ergs occur in structural or geomorphic basins (Pye and Tsoar 1990). Sand transported by wind will be deposited at any place along the transport path where wind flow decelerates enough to allow set-

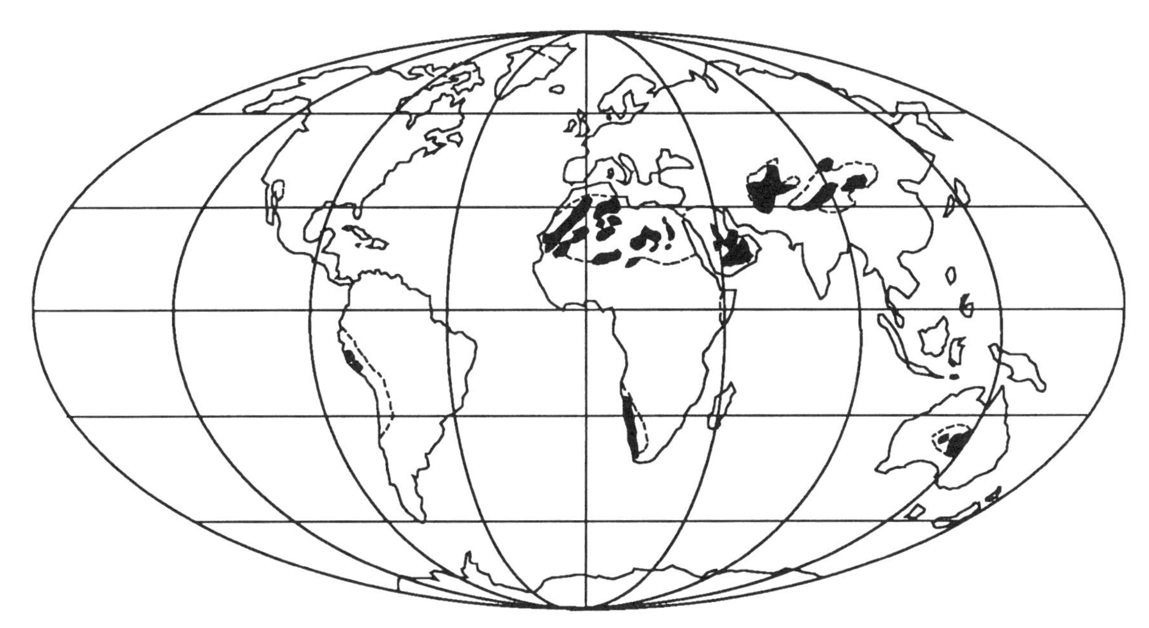

Figure 5.2. Distribution of active ergs (*black*) and dry climates (*dashed lines*), defined as regions with <15 cm mean annual precipitation. Reprinted from *Sedimentary Geology* 10, Wilson, I.G., "Ergs," pp. 77–106 (1973) with kind permission of Elsevier Science-NL, Sara Burgerhartstraat 25, 1055 KV Amsterdam, The Netherlands.

tling of grains (Wilson 1973). Accommodation space must exist for significant depositional accumulation, which limits eolian sandstone deposits to basins that have active addition of accommodation space, through changes in base level (Kocurek 1996).

Modern sand dunes

Pye and Tsoar (1990) divided eolian sands and their associated deposits into two groups that are formed by two different sand-transporting processes. The first group includes eolian lag deposits, sand sheets, and interdune deposits. In these, the fine-grained sediments commonly have been removed by winnowing, although they can be trapped by moisture, salt, and vegetation. Transport is as bedload by surface traction, and the grains are concentrated as residual sheets. In general, the grain sizes are large, although sand sheets may have a significant silt component; a characteristic of sand sheets is poorly sorted sediment. In contrast, dune sands are moderately to well sorted, with grains in the size range of 50 to 70 mm. Transport is by saltation and avalanching.

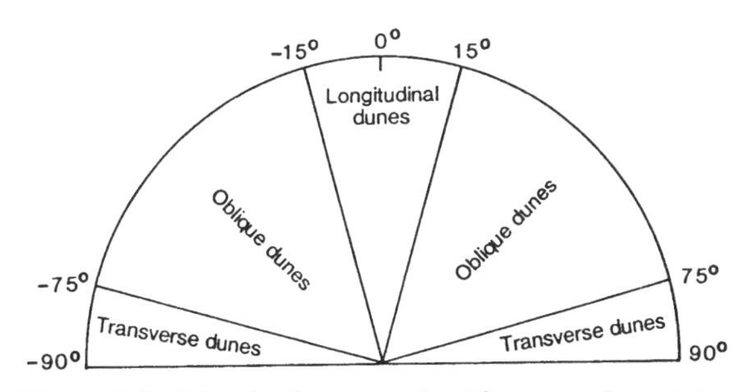

Figure 5.3. Morphodynamic classification of straight- or slightly sinuous-crested dunes, related to the angle between the dune trend and the resultant of the sand-transporting winds. From Hunter, Richmond, and Alpha (1983) in *Geological Society of America Bulletin.* Reproduced with permission of the publisher, the Geological Society of America, Boulder, Colorado, USA. Copyright © 1983 Geological Society of America, Inc.

Sand dunes are classified by morphology or morphodynamics and, as emphasized by Kocurek (1991), the terms are distinct. Morphological terms are descriptive only. Dunes are classified morphologically into linear, crescentic, and star dunes, with variations in the descriptions depending on the specific morphology and complexity. For example, crescentic dunes include barchan and parabolic dunes. Morphodynamic terms include transverse, oblique, and longitudinal and refer to the relationship between the direction of the dominant sand-transporting wind and the trend of the crestline of the dune (see figure 5.3). A general relationship exists between the type of dune and the nature of the sand-transporting winds. This relationship links the morphology of dunes and their morphodynamics, and an understanding of it is critical to interpretation of paleowind directions. Rubin and Ikeda (1990) demonstrated that under unimodal winds, dunes will always be transverse. However, perfectly transverse dunes are uncommon because unimodal sand-transporting winds are uncommon. Instead, a dune will have a morphology that maximizes the transverse nature of the dune with respect to all the sand-transporting winds that affect it. Or, in the words of Kocurek (1991), "[for example,] an oblique bedform is oriented for the most transverse position to each wind (weighted by wind energy) but is also oblique with respect to the average vector." Very detailed relationships between sand-transporting winds and dune morphology have been demonstrated in, for example, the Namib Sand Sea by Lancaster (1983) and the Algodones Dunes (California, U.S.) by Sweet et al. (1988) and Sweet and Kocurek (1990).

Eolian sandstones

Dunes in their original three-dimensional geometry are rarely, if ever, preserved in the older geologic record. Features of eolian deposits that are observed are foreset beds, wind-ripple deposits, scalloped cross-strata, and the surfaces that bound them (Kocurek 1991, 1996). Bounding surfaces are erosional and occur within or between sets of cross-strata (Kocurek 1996). A great deal of attention has been paid in the past 10 to 15 years to defining the nature of the bounding surfaces. This is important for understanding not only the relationship between paleowind and the resulting sedimentary features in the geologic record, but also for understanding changes in eolian depositional systems and the extent to which those changes were driven by paleoclimatic change. In the following section, "foreset beds" refers to the lamination formed by individual laminae deposited during single, steady wind events; other workers sometimes use the term "compound crossbeds" to refer collectively to foreset beds and the various types of bounding surfaces (e.g., Rubin and Hunter 1983).

Bounding surfaces were first described systematically by Brookfield (1977), and the terminology has since been modified by numerous workers, most recently, Kocurek (1996). Bounding surfaces are divided into four types (Kocurek 1996): interdune surfaces (first-order surfaces of Brookfield), superposition surfaces (second-order surfaces of Brookfield), reactivation surfaces (third-order surfaces of Brookfield), and super surfaces. Super surfaces are formed during changes in conditions in eolian depositional systems and are discussed in that section; the others relate to changes in the dunes themselves.

The different types of surfaces are illustrated in figure 5.4, and it is important to note especially the relationship between the different kinds of surfaces and the bedforms. Interdune surfaces are flat lying or convex up and truncate foreset beds and other dune structures (Pye and Tsoar 1990; Kocurek 1996). Superposition surfaces are low- to moderately dipping, flat or convex-up, and bound sets of foreset beds (Pye and Tsoar 1990; Kocurek 1996; figure 5.4c). They are formed during migration of dunes that are superimposed on larger bedforms; in practice, several ranks of superposition surfaces may exist in complex systems (Kocurek 1991). Reactivation surfaces are small scale and separate and are subparallel to foreset beds (Kocurek 1991, 1996; figure 5.4a). They are formed by changes in local wind direction and/or velocity, or by modification of airflow by the dune itself, and thus represent reactivation on a dune lee face (Brookfield 1977; Kocurek 1991, 1996).

For interpreting paleowind directions, interdune or superposition surfaces are the most important of the bounding-surface types. In simple transverse dunes, sand transport and dune migration are in the same direction and the crossbeds will dip in that direction (figure 5.4a); in that case, wind direction can be interpreted from dip of the foreset beds. A more complicated case is presented in figure 5.4c. In this hypothetical case, the principal sand transport is toward the east, as reflected in the orientation of the larger bedforms. However, a secondary sand-transporting wind has created bedforms that are superimposed on the larger one. These examples emphasize the importance of reconstructing bedforms for interpreting paleowind directions.

The work of Rubin and his colleagues (Rubin and Hunter 1982, 1983; Rubin 1987; Rubin and Ikeda 1990) has provided a theory-based but practical set of models that show the relationship between transport and bedforms and the predicted signatures in the geologic record. Rubin (1987) produced a pictorial atlas of patterns in the geologic record that would be expected given different bedform types and transport characteristics; the atlas includes photographs of outcrops that illustrate these patterns.

Two points made by Rubin (1987) are particularly important in the context of this book. First, neither a bedform nor a bedding classification

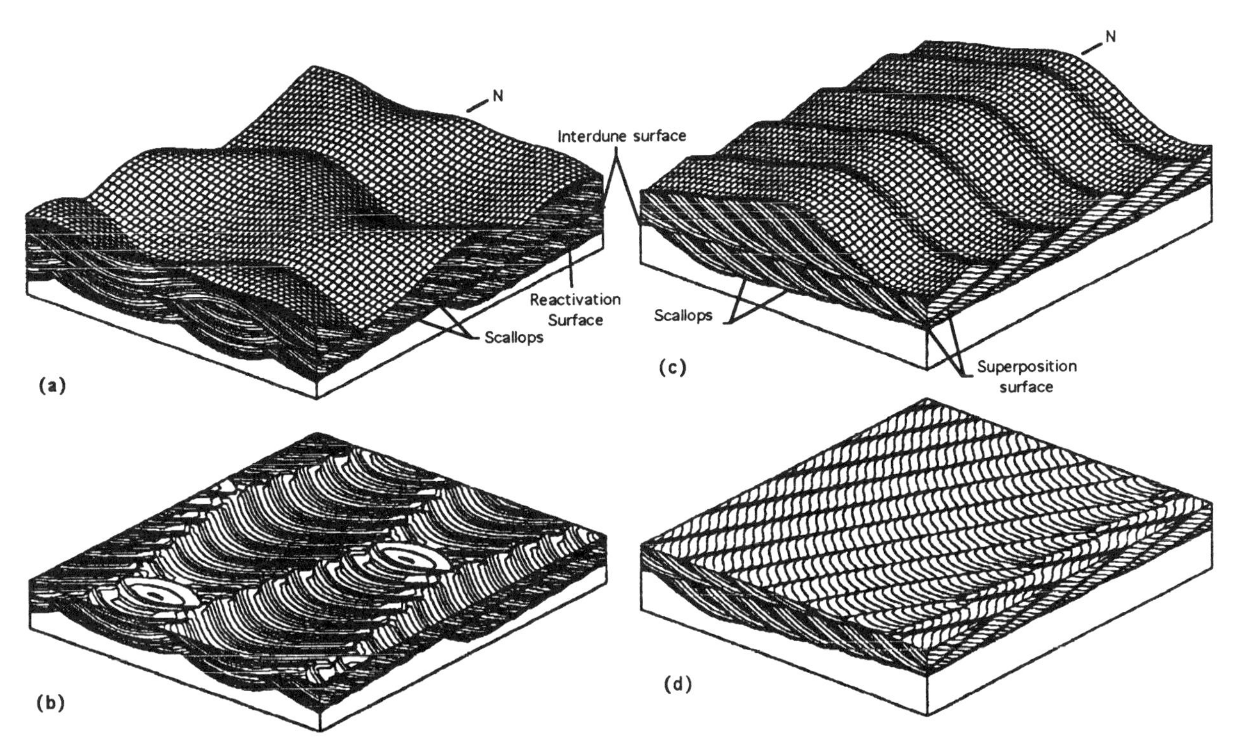

Figure 5.4. Bounding surfaces in relation to dune bedform morphology. *(a)* Sinuous-crested dunes, in which the sinuosity between dunes is out of phase. Note scalloped crossbeds. *(b)* Block diagram showing the three-dimensional arrangement of crossbed foresets and bounding surfaces for (a). In this case, the crossbeds and the bounding surfaces dip in opposite directions. These are transverse dunes (see figure 5.3), so wind was toward north. In this case, the crossbed sets are predominately oriented within 45° of downwind, whereas the bounding surfaces show no predominate orientation. *(c)* Large, straight-crested dune with superimposed smaller, straight-crested dunes. The migration vectors of the large and small dunes are perpendicular. *(d)* Block diagram showing the three-dimensional arrangement of bounding surfaces for (c). In this case, orientations of the crossbeds vary by about 90°. The resultant of the migration vectors is parallel to the trend of the superposition surfaces on the horizontal part of the block diagram and parallel to the wind. Adapted from Kocurek (1996) "Desert aeolian systems," in H. Reading, ed. *Sedimentary Environments: Processes, Facies and Stratigraphy,* Blackwell Science Ltd, adapted in turn from Rubin (1987).

system alone is adequate for interpreting sediment-transport processes from the geologic record. This is because bedform classification systems exclude bedding and cannot incorporate changes through time, and bedding classification systems incorporate changes through time but do not include the three-dimensional characteristics of the bedform. Bedform information is needed to interpret transport (see figure 5.3), but bedding is what is preserved in the geologic record. This is why Rubin's (1987) approach is so helpful—it incorporates bedforms, which are not directly recorded in the geologic record, but it also shows the resulting bedding, which is preserved. Many workers now use his models to duplicate observed bedding in order to interpret the bedform and transport directions. The second point is that sand-transport direction, as determined from bedding, and migration directions of bedforms are not necessarily parallel to the wind direction (Rubin and Ikeda 1990). This is discussed further in the following section.

Determining paleowind directions from eolian sandstones. The simulations by Rubin (1987) illustrate that wind directions must be interpreted after determining the bedform(s) in which the sand was transported. The maximum dip of foreset beds in a package of eolian sandstones indicates the dominant paleowind direction only in transverse dunes (two-dimensional dunes, variable or invariable in the terminology of Rubin and Ikeda 1990; Kocurek 1991). To use this method, one must be sure that exposure is such that the maximum foreset dip can be measured and that the bedforms were transverse (e.g., the very simple case in figure 5.4a). The relationship between dune morphology and wind (see figure 5.3) means that crestline orientations could be good indicators of composite regional winds (Rubin and Ikeda 1990) but, because eolian sandstone deposition involves erosion as well as accumulation, crestlines are rarely, if ever, preserved in the geologic record.

Some information on bedform type may be gained from the style of foreset stratification (figure 5.5). There are three types of strata in foreset beds: grainfall, grainflow, and wind-ripple laminae. Grainfall occurs when sand grains are transported across the windward side of a dune and deposited on the lee face. Grainflow occurs when the angle of repose is exceeded and the grains flow down the lee face. In grainfall and grainflow, gravity is the driving process. Wind-ripple laminae form when ripples migrate across the lee face of the dune, and traction is the driving process; wind-ripple laminae are diagnostic for eolian sandstones. Figure 5.5 illustrates some stratification types and their significance. Although certain types of stratification are associated with certain types of dunes, it is important to emphasize that the stratification types reflect the relationship between the dune crest and the wind. Complex dunes may have a variety of stratification types (Kocurek 1996).

Although some eolian sandstones are exceptionally well preserved and allow reconstruction of the bedform directly (e.g., Clemmensen 1987), Kocurek (1991) emphasized that any particular dune can behave differently at different times, which makes reconstructing dune morphology from the geologic record difficult. Even the highly complex stratification patterns predicted by Rubin (1987; see figure 5.4) resulted from bedform geometries that were regular compared with many actual dunes. Kocurek (1991) also pointed out that distinguishing superposition, interdune, and super surfaces can be difficult. Another problem is that deposition of eolian sandstone commonly involves substantial erosion of underlying bedforms as new bedforms are deposited (Kocurek 1991, 1996).

Marzolf (1983) employed the simplest method of determining wind direction, measuring maximum dips on foreset beds and obtaining vector

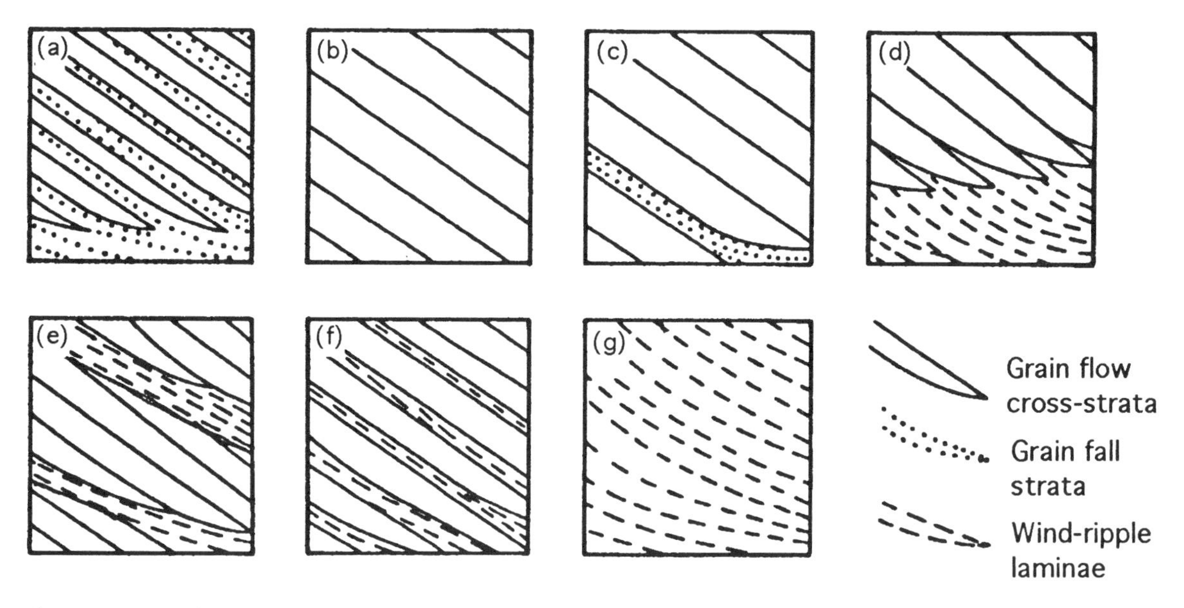

Figure 5.5. Stratification types in eolian sandstones. *(a–c)* Stratification expected in transverse dunes (see figure 5.3), consisting of *(a)* alternating grainfall and grainflow, *(b)* all grainflow, or *(c)* predominately grainflow with occasional grainfall strata. (a) indicates a smaller dune or higher winds than (b); the grainfall in (c) might indicate a storm. *(d)* Stratification expected in oblique dunes (see figure 5.3), indicating simultaneous grainflow and wind-ripple migration. *(e)* Alternating periods of grainflow and wind-ripple deposition, indicating alternating transverse and oblique winds and a change in the dune shape. *(f)* As in (e), but with no change in dune shape resulting from the alternation between transverse and oblique winds. *(g)* As in (d), in which subsequent dune migration (not shown) has eroded the grainflow layer; oblique or longitudinal (see figure 5.3) lee face characterized by wind-ripple transport only; or a low-relief, transverse lee face in which the wind ripples migrated downslope. Large simple sets of grainflow cross-strata (e.g., b and c) may reflect regional wind, and such cross-strata can serve as a first approximation of the regional-scale representation of paleowind. From Kocurek (1991). Reprinted with permission, from the *Annual Review of Earth and Planetary Sciences,* Volume 19, ©1991, by Annual Reviews Inc.

resultants. Stereographic projections of his data, including those from the Navajo Sandstone (Jurassic; Utah, U.S.), presented in figure 5.6, show foreset-bed azimuths that plot in arcs spanning 180° or less and are commonly concentrated in the center of the arc. The vector resultants across the depositional area of the Navajo are consistent. Marzolf (1983) also examined variation in the data and concluded that changes in vector-resultant foreset-bed orientations are independent of the style or scale of cross-bedding. This study illustrates that cross-strata alone can be useful for interpreting large-scale wind patterns (Kocurek 1991).

Chrintz and Clemmensen (1993) combined cross-bed measurements with models of the bounding surfaces to interpret paleowind directions in the Permian Yellow Sands, United Kingdom. They measured foreset-bed attitudes, and the data produced a squeezed bimodal compass rose, with one mode at 313° and one at 171°, both of nearly equal importance. Bounding surfaces produced a similar geometry, but with the modes closer to 180° apart. The resultant sand drift was 238°. Chrintz and Clemmensen (1993) modeled the cross-bed sets and bounding surfaces, using the methods described by Rubin (1987). The model that most closely reproduced their data consisted of sinuous, linear dunes with some reversal on linear draas (draas are very large bedforms). They suggested a bidirectional wind regime, north and southeast, for the draa overall, and seasonal winds, northwest-north and southeast-east, for the superimposed linear dunes.

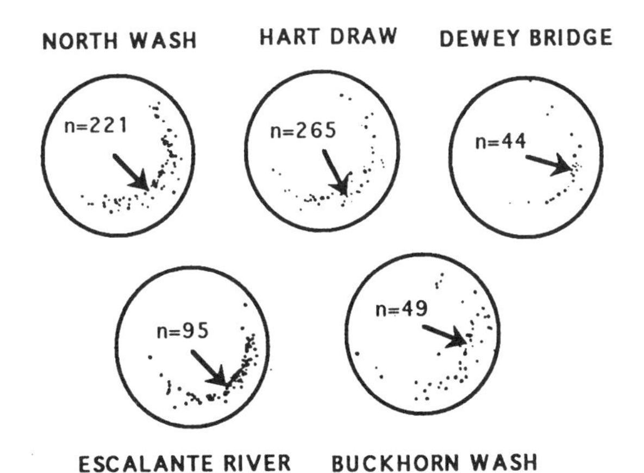

Figure 5.6. Stereographic projections of foreset-bed orientations and dips (*dots*) from the Navajo Sandstone (Jurassic; Utah, U.S.) and the resultant vectors; north at top. From Marzolf (1983).

One weakness of these and similar studies (e.g., Blakey and Middleton 1983) is that none has an independent check on the paleowind determinations other than internal consistency. Methods of independently checking such determinations are few. One method would be to determine whether other indicators of eolian transport, if available, show the same directions. If such deposits are not preserved, the only method available for independently checking paleowind determinations would be to compare them with winds predicted by climate models. Marzolf's (1983) study, which covered a relatively large area and showed consistent results over that area, was the most amenable to comparison with global climate models. Studies such as that by Chrintz and Clemmensen (1993) are not as amenable to comparison with global climate models because the climatic patterns they record are much smaller than a single grid cell for which winds are calculated. The only study to date to compare detailed climate models with a large data set on paleowind determinations from eolian sandstones was that by Parrish and Peterson (1988) for the Colorado Plateau, U.S. Like the data set of Marzolf (1983), Peterson's (1988) consisted almost entirely of cross-bed dip resultants. Also like Marzolf's (1983) data, Peterson's (1988) showed strong consistency on large spatial scales.

Paleowind strength and cyclicity. Small-scale cyclicity in the bedding of eolian sandstones has raised the question of whether it is a natural consequence of the sedimentary processes (autocyclic) or driven by some external forcing (allocyclic). Hunter and Rubin (1983) categorized cyclicity in eolian sandstones into two types, fluctuating-flow cyclicity, implying cyclic changes in wind direction and/or strength, and superimposed-bedform cyclicity, which forms when smaller bedforms move over larger ones (table 5.2). Concordant cycles, in which foreset beds have an element of cyclicity and no internal bounding surfaces (i.e., reactivation surfaces), are commonly expressed as alternations between wind-ripple and grainflow strata (see figure 5.5f); these are most likely fluctuating-flow cycles (Hunter and Rubin 1983). Compound cyclic crossbedding of fluctuating-flow origin might look like figure 5.5e.

Hunter and Rubin (1983) interpreted the small-scale cyclicity in beds of the Navajo Sandstone as annual because the wind speeds that would be required to advance the dunes by the

Table 5.2.
Features useful for distinguishing cyclic crossbedding of superimposed-bedform and fluctuating-flow origins.

	Cyclic crossbedding of superimposed-bedform origin	Cyclic crossbedding of fluctuating-flow origin
Concordant cyclic crossbedding	Complete concordance (i.e., in both transverse and longitudinal cross section) uncommon, possibly nonexistent where crossbeds dip at least at angle of repose	Complete concordance commonly produced
	Cycle width[a] commonly small	Cycle width[a] great unless plan-form curvature is great
	Probably not distinguishable in longitudinal cross section	
Compound cyclic crossbedding	Cycle whose crossbedding indicates formation by feature migrating along a larger lee slope; produced readily	Cycles whose crossbedding indicates formation by feature migrating along a larger lee slope; produced only where plan-form curvature of that slope is great
	Cycle length[a] commonly great	Cycle length[a] generally small
	Cycle width[a] commonly small	Cycle width[a] great unless plan-form curvature of larger lee slope is great
	Features produced by reverse flow compatible with formation by lee eddy simultaneously with forward flow	Features produced by reverse flow may indicate origin at times when forward flow did not occur

From Hunter and Rubin (1983).

[a]Cycle width and length are the lateral extent of the cycle in transverse and longitudinal cross sections, respectively. A longitudinal cross section is one parallel to the migration direction of the feature that produced the crossbedding.

observed amounts are reasonable and adequate and annual cycles are regular, whereas daily cycles would require unreasonably high wind speeds, and storm and multiannual processes would probably create cycles that are more irregular. Since Hunter and Rubin's (1983) paper, it has been shown (e.g., Parrish and Peterson 1988) that the Colorado Plateau, U.S., was under the influence of a strongly seasonal monsoonal climate, with alternating wind directions and strengths between the seasons.

Eolian sandstone depositional systems

Eolian sandstones record wind directions. Desert eolian depositional systems as a whole, on the other hand, record not only winds but other paleoclimatic conditions as well. These depositional systems are very sensitive to climatic change, particularly changes in moisture supply.

Central to interpreting the effect of climate on eolian sandstone depositional systems is understanding their structure and the processes operating within them. In the simplest conception, an erg consists of eolian sand bedforms and the deposits in between the bedforms, usually referred to as interdune deposits. Interdune deposits are extremely variable and strongly dependent on the proximity of the water table to the air-sediment interface. Interdune deposits may be deflationary or may represent eolian, fluvial, or lacustrine deposition (e.g., Lancaster and Teller 1988). Interdune deposits are the most sensitive parts of the eolian depositional system to groundwater levels, which has led eolian workers to categorize interdune deposits into dry and wet, unstablized (active) and stabilized (inactive; Kocurek and Havholm 1993; figure 5.7).

The study of change in dunefields or ergs is commonly conceptually organized around super surfaces (Talbot 1985; Kocurek and Havholm 1993; Kocurek 1996). Super surfaces result from changes in accommodation space (Kocurek 1988; Kocurek and Havholm 1993). Accumulation and preservation space are determined by topography (allowing deceleration of flow and deposition) and by position of the water table (figure 5.8), and changes in accumulation and preservation may result from changes in sea level, climate, or tectonics. Accumulation is also determined by sand supply. Stabilization by vegetation can result from climatic change or from a rise in base level associated with a sea-level rise.

Super surfaces may be difficult to distinguish from interdune surfaces. Criteria for distinguish-

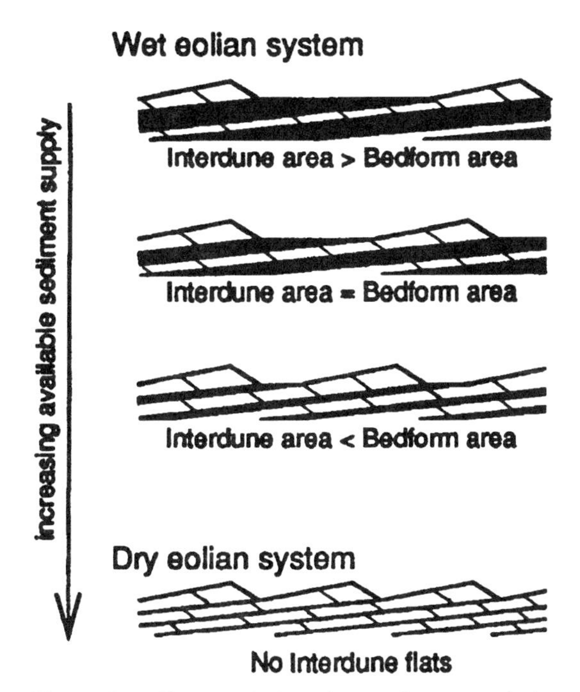

Figure 5.7. Characteristics of actively accumulating eolian systems in a gradation from wet to dry eolian systems. From G. Kocurek and K.G. Havholm ©1993, reprinted by permission of the American Association of Petroleum Geologists.

ing super surfaces include root structures, paleosols, evaporites, polygonal fractures, preferential cementation, lag surfaces, zibars (low-relief bedforms that lack slip faces; Nielson and Kocurek 1986), and granule ripples (Kocurek 1988; Lancaster 1993), but these will occur only if the super surface was exposed for any length of time. A super surface can also be simply an unconformity, for example, marking the transition from a wet to a dry eolian system (figure 5.9).

Ancient eolian sandstone depositional systems. What is the paleoclimatic significance of dry and wet eolian systems? Dry eolian systems do not necessarily indicate drier climate because climate is not the only control on water-table levels, especially in coastal plain settings. The concentration of data on modern sand seas in the dry end of the climatic spectrum (see figure 5.1) suggests that wet and dry eolian depositional systems are not strongly differentiated climatically. However, changes in an eolian depositional system, such as those illustrated in figure 5.9, may be indicative of paleoclimatic change. The challenge is to distinguish paleoclimatic change from changes in sea level or tectonics, a challenge that applies to many terrestrial paleoclimatic indicators.

Changes in eolian deposition and, specifically, fluvial-eolian interactions, were used to interpret paleoclimatic change by Herries (1993), who studied the Kayenta Formation and the Navajo Sandstone (Jurassic; Colorado Plateau, U.S.) where they intertongue. Herries (1993) described two scales of fluvial-eolian interactions. Small-scale interactions are those that took place between the Navajo erg and the adjacent Kayenta fluvial system. Herries (1993) interpreted these as the result of either autogenic or very short term atmospheric processes (e.g., thunderstorms). Large-scale interactions between the fluvial and eolian systems were interpreted by Herries (1993) to be climatically driven phases of erg construction and destruction. These phases are represented in the geologic record as three cycles transitioning from dominantly fluvial deposition to dominantly eolian deposition before permanent establishment of the Navajo Sandstone erg. Herries (1993) interpreted these as drying-upward cycles. This conclusion was partly based on evidence of erg stabilization and erosion by rivers at the top of each cycle, consistent with a change to a wetter climate. Cycles at this scale also might be indicative of more remote climatic control, such as glacio-eustasy (Atchley and Loope 1993) or orbitally forced climate cycles (Clemmensen, Øxnevad, and de Boer 1994).

Most workers have concentrated on super surfaces indicating cessation or interruption of sand accumulation (Havholm et al. 1993), but Clemmensen and Tirsgaard (1990) endeavored to define those surfaces representing establishment of erg depositon. They called these surfaces sand-drift surfaces and limited the term to surfaces formed during the transition from subaqueous to eolian deposition. Sand-drift surfaces are characterized by sharp, usually erosional lower contacts, bedding with dips that are horizontal to low-angle, and common scours and lag grains. The immediately underlying deposits may contain early cementation features, desiccation cracks, and evaporites. In cyclic sequences, the surfaces are typically centimeters to meters apart vertically. Sand-drift surfaces extend laterally on the scale of meters to kilometers, consistent with other work (e.g., Langford and Chan 1989). In defining sand-drift surfaces, Clemmensen and Tirsgaard (1990) stated that the eolian deposits that overlie sand-drift surfaces are usually developed as sand sheets, and thus the term might not be applicable to the onset of erg deposition as documented, for example, by Herries (1993).

Despite the limitation to Clemmensen and Tirsgaard's (1990) definition, the concept is im-

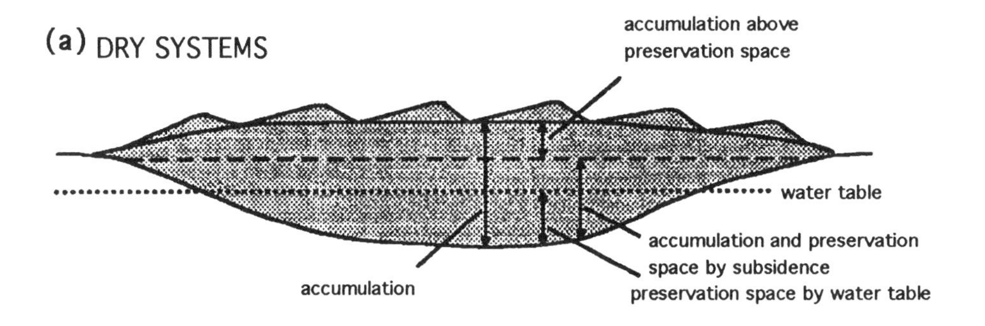

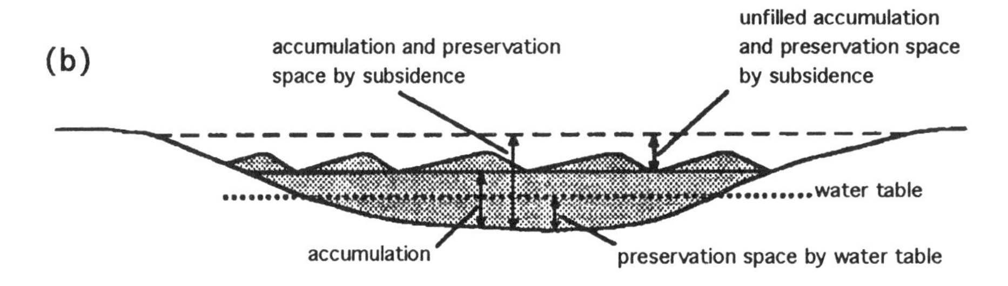

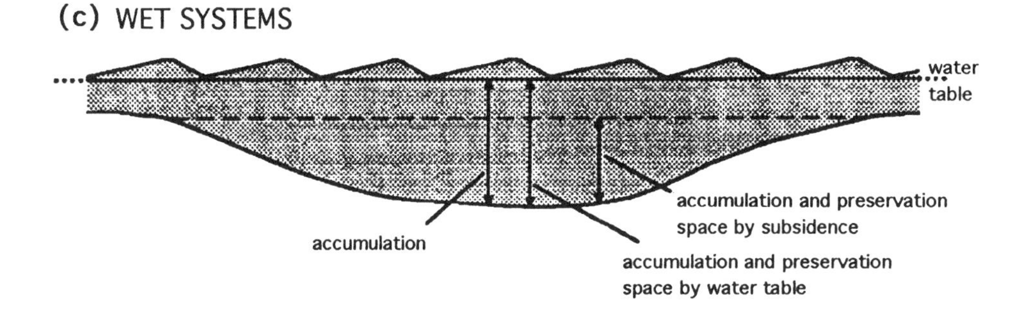

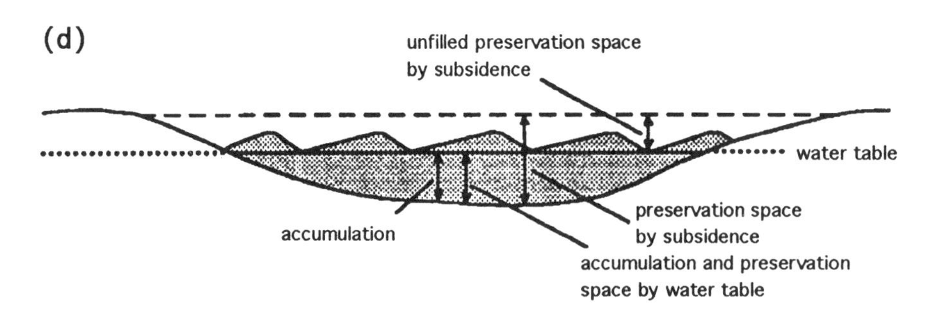

Figure 5.8. Schematic diagram showing the components of accumulation and preservation space for dry *(a,b)* and wet *(c,d)* eolian depositional systems for cases in which the preservation space is filled (c) or overfilled (a) or not yet filled (b,d). Note the basin topography, which allows for deceleration of wind flow and deposition, and position of the water table in each case. In addition, note that in dry systems, accumulation can occur above the preservation space, whereas accumulation in wet systems is more constrained. Accumulation above the preservation space is unlikely to be preserved in the geologic record. From G. Kocurek and K.G. Havholm ©1993, reprinted by permission of the American Association of Petroleum Geologists.

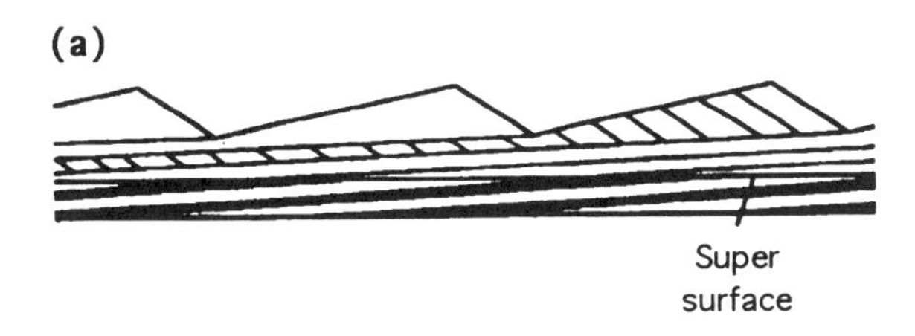

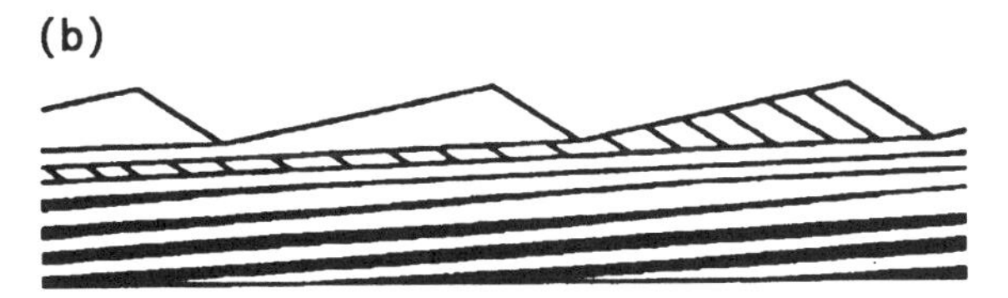

Figure 5.9. Examples of the transition from accumulation in a wet eolian system to accumulation in a dry one. *(a)* Super surface marks a period of deflation with a falling water table until dune growth covers the interdune flats and accumulation in a dry system begins. Note the distinct unconformity. *(b)* Gradual shrinking of the sizes of interdune flats with dune growth in a dry system, in effect burying the interdune flats. No super surface forms. From G. Kocurek and K.G. Havholm ©1993, reprinted by permission of the American Association of Petroleum Geologists.

portant for interpreting paleoclimates because the examples they gave are cyclic sequences that in some cases may have been controlled by cyclic climatic change. The examples they cited include the middle Buntsandstein (Triassic; Europe), where sand-drift surfaces overlie lake deposits and represent contraction of the lake. A second example was the Eriksfjord Formation (Middle Proterozoic; Greenland), in which fluvial beds containing sand-drift surfaces are interbedded with thin lake beds. Clemmensen and Tirsgaard (1990) attributed the interbedding to paleoclimatic change, with lake transgression corresponding to a humid phase and fluvial deposition representing more arid intervals. A third example was the Corrie Sandstone (Permian; Scotland), which consists of alluvial-fan deposits containing sand-drift surfaces. Clemmensen and Tirsgaard (1990) interpreted this unit as representing alluvial fan-erg margin interactions. An alternate approach to determining whether the changes are paleoclimatic or tectonic is to examine changes in adjacent basins (e.g., Meadows and Beach 1993).

Sand sheets. Sand sheets are blankets of sand that do not have significant development of dune bedforms, are uniform in thickness, and have relief generally <5 m and slope angles less than 6°; commonly, a sand sheet has a coarse lag deposit on the upper surface (Pye and Tsoar 1990) and incorporates coarse-grained layers (Fryberger, Ahlbrandt, and Andrews 1979). Sand sheets do not necessarily occur within the context of the cyclic deposits described by Clemmensen and Tirsgaard (1990). The general climatic significance of sand sheets, as distinguished from systems characterized by dunes, is that climate was wetter (Schwan 1988), especially if the sand sheet formed away from the coastline (see also Kocurek and Nielson 1986; Crabaugh and Kocurek 1993). A low supply of sand can also contribute to sand sheet formation (Gunatilaka and Mwango 1987). Diagnostic features for coastal sand sheets would be large evaporite lenses and bioturbated mounds with organic layers, evaporites, and bimodal cross-bedding, which indicates flow separation around the plants; the dominant wind is indicated by a vector between the dip sets (Gunatilaka and Mwango 1987).

Other climatically significant eolian sand deposits

Under certain circumstances, eolian and evaporitic systems may include dunes composed of sand-sized clay pellets. Although these are unusual deposits, rarely reported in the fossil record, the climatic and sedimentologic conditions required for their formation are so specific that they are worth including here. Clay pellet dunes were reviewed by Bowler (1973). Commonly, the dunes are lunette shaped (that is, similar to parabolic sand dunes in that the cusps extend upwind) and border salt flats or playa lakes. The dunes are composed of clay pellets, which are derived from mechanical disintegration of mud chips. The clay pellets accumulate during seasonal hot, dry conditions, under unimodal wind regimes.

Eolian sediments can accumulate on snow-covered surfaces and then be deposited as the snow melts. Such deposits are called niveo-eolian (Ballantyne and Whittington 1987). Like clay pellet dunes, niveo-eolian deposits are unusual, but the conditions required for their formation are such that identification of the deposits in the geologic record would provide valuable and otherwise unavailable information on paleosnow-cover.

Loess

Loess deposits are accumulations of silt-sized material that was transported and deposited by winds. Young loess, that is, deposits accumulated during the Pleistocene, has three diagnostic characteristics. The range of grain sizes is very narrow, 20 to 60 μm, with a prominent mode at 20 to 30 μm (Pye 1987); the deposits are unconsolidated and porous; and the beds are massive (Pye 1987). Pye and Tsoar (1990) divided eolian dust into two categories, loess and far-travelled eolian dust. Loess grains have a minimum size of 10 to 20 μm (Tsoar and Pye 1987; Pye and Tsoar 1990) and a maximum of 70 μm. Pye and Tsoar (1990) called these settling grains, that is, grains whose settling velocities are large compared to the shear stress imparted by the winds. Grains <10 to 20 μm in size are nonsettling grains and are deposited during rain storms. The nonsettling grains commonly are transported over the ocean and are incorporated into deep-sea sediments.

Pleistocene loess deposits accumulated during major glacial advances, and the largest deposits have a periglacial distribution. These characteristics have led to the conclusion that the grains in loess were produced principally by cold-weather processes, that is, frost weathering and glacial and fluvioglacial grinding (Smalley 1966), but some loess was probably derived from desert sediments far from glacial action. The requirements for formation of substantial loess deposits are (1) source areas consisting of unvegetated surfaces covered with sediment that has a high silt-to-clay ratio and (2) frequent, strong, turbulent winds. These conditions were likely to have been widespread during glacial episodes, and thus loess is commonly regarded as an indirect indicator of glaciation and of dry, cold climates in periglacial regions. The point that loess is an indirect indicator is an important one because the largest loess deposits in the world are probably composed of desert loess from the cold deserts of Central Asia (Pye 1987). For accumulation of loess, Tsoar and Pye (1987) emphasized the importance of suitable traps, which would include (1) topographic obstacles, (2) areas of moist ground, and (3) vegetated surfaces. They noted that in the Pleistocene, dust was blown from deserts and deposited in the adjacent semi-arid regions, and the thickest units are closest to the source.

Two obstacles exist to interpreting the paleoclimatic significance of pre-Pleistocene loess. First, Pleistocene loess is easily recognizable because the deposits are porous and unconsolidated. These features are so characteristic that they are considered diagnostic. Clearly, any loess deposit buried by subsequent sedimentation is not going to retain these features as it is compacted and lithified. The two remaining features, massive bedding and narrow range of grain sizes, might not be enough to distinguish an older loess deposit from massive siltstone deposited by some mechanism other than wind. The second obstacle is an understanding of loess-forming processes. Pleistocene glacial periods appear to have created very widespread conditions favorable for loess formation, and these conditions extended beyond the present deserts (Pye 1987). However, in the present interglacial, vast quantities of dust are still being produced in desert regions, and though it is volumetrically less than the Pleistocene dust, the question arises whether at times in the past, when deserts were more widespread, desert loess could have been just as voluminous in the absence of large-scale glaciation. The use of loess as a paleoclimatic indicator ultimately will depend on detailed investigations of the relationship between desert dust production and accumulation and glacial climate, especially as periglacial loess appears to indicate cold, dry climate in the region of deposition, but coeval peridesert loess may indicate wetter climate (Gerson and Amit 1987). It will be important to eliminate the possibility that deserts could produce enough dust for formation of loess deposits independent of whatever mechanisms increased dust production in deserts during the Pleistocene glacials.

Reports of pre-Pleistocene loess deposits are relatively scarce. The most likely explanations for this paucity of reports are that (1) loess deposits formed but were not preserved and (2) the deposits exist but are not recognized as loess. If substantial accumulation of dust requires dry continental interiors, as might be supposed from the distribution of Pleistocene loess, preservability of such deposits would be relatively low. Depositional settings in general, including those in continental interiors, are characterized by sediment modification and reworking by rivers, wind, or groundwater fluctuations, all of which would make preservation of loess deposits as such most unlikely.

The term "loessite" describes lithified loess (Pye 1987). Loessites reported by Soreghan (1992a,b) in Arizona and New Mexico, U.S. are probably reworked, but they closely fit the characteristics of loess. These Pennsylvanian loessites are sheet-like, thin (<5 to 15 m), have sharp but not erosive bases, and are massive, in addition to being composed of silt-sized grains typical of loess. In addition, the siltstones occupy only the most northern (present coordinates), upwind portions of the basins. Upwind of the basins were extensive ergs, sabkhas and mudflats, and arid- to semi-arid-climate fluvial systems, all of which would have been excellent sources for loess (Pye 1987). The siltstone is the only siliciclastic material in parts of the section in the Pedregosa Basin (Arizona, U.S.), and thus it is unlikely to have been derived from proximal fluvial sediments because the other grain-size fractions are missing. Finally, episodes of loessite deposition correlate with episodes of eolian activity north of the basins (Soreghan 1992a). In addition to being one of the few pre-Pleistocene loess deposits, the Pennsylvanian loessites described by Soreghan (1992a) are important for two reasons. First, they are non-Waltherian parts of what is normally a marine or marginal-marine sequence (Soreghan 1992b); that is, the loessites occur only during sea-level lowstands and are not part of the lateral facies array. Thus their occurrence in those sequences is clearly related to climate and not solely to changes in base level. Second, they are peridesert, not periglacial, deposits. Their climatic significance may be similar to that of loess deposits in the Negev Desert, which appear to have formed primarily during the glacial episodes of the Pleistocene (Gerson and Amit 1987; Pye and Tsoar 1987). Middle Pennsylvanian-Permian loessites were also reported by Johnson (1989) from the Maroon Formation in Colorado, U.S., upwind from the Upper Pennsylvanian deposits reported by Soreghan (1992a). He was able to discern that the Maroon Formation siltstones mantle irregular paleotopographic features; such mantling is characteristic of Pleistocene loess.

Soreghan's (1992a) and Johnson's (1989) sections were terrestrial or very shallow marine, but significant amounts of eolian material can be transported in deeper marine sediments close to the source. Such a situation occurs off the Sahara (Fischer and Sarnthein 1988). The result is that deeper marine rocks are dominated by silt, and the shallow-water sediments have a significant component of eolian sand. The eolian sand feeds sand wedges on the shelf and from there is distributed through channels as grainflow and in turbidity currents. The stratigraphy that results is unusual for marine sediments, consisting of channel sands and draped silts; clay is very low in abundance. Fischer and Sarnthein (1988) noted that this is very similar to the Brushy Canyon Formation in the Delaware Basin (Permian; Texas, U.S.). The principal difference between the Brushy Canyon Formation and the sediments off the Sahara is that the modern silts are massive, whereas the Delaware ones are laminated. Fischer and Sarnthein (1988) noted that if water in the Delaware Basin were density stratified, as it almost certainly was during at least part of its history, the primary lamination resulting from deposition would have been preserved.

Eolian Dust in the Deep Sea

Eolian dust in the deep sea has been useful for interpreting changes in wind strength and direction and climatic change on the continents (Rea 1994). Far-travelled dust has a grain size generally <10 to 20 μm (Pye 1987), although modal grain sizes as large as 25 μm have been reported more than 500 km offshore northwest Africa (Sarnthein et al. 1982). Distinguishing eolian dust from water-transported silt-sized grains is difficult, but detrital, opal-free quartz and nonauthigenic feldspars, kaolinite, illite, and other clays are generally considered to be eolian if deposited farther than 1000 to 2000 km from the shelf (Johnson 1979; Rea and Janecek 1981; Duce et al. 1991; Rea 1993). Mineral assemblages dominated by smectite and mixed-layer clays are regarded to be volcanic (Rea and Janecek 1981; Weber et al. 1996). Biogenic material from terrestrial organisms, such as freshwater diatoms or opal phytoliths, which are silt-sized grains of opaline silica in the tissues of plants such as grasses, also are found in open-ocean sediments. Finally, the geochemistry of the dust can provide very specific information on provenance (Kyte et al. 1993; Nakai, Halliday, and Rea 1993; Weber et al. 1996).

Three characteristics of wind-blown dust are important for paleoclimate interpretation: grain size, amount transported, and composition. In modern samples, grain size does not change significantly more than 1000 to 2000 km beyond the source (Rea, Leinen, and Janecek 1985; see also Rea and Hovan 1995). On this basis, changes in eolian grain size through time can be used to indicate changing atmospheric conditions, specifically wind strength (Rea, Leinen, and Janecek 1985; Rea 1994; see also Rea and Hovan 1995). The amount of dust transported depends on availability of material from the source. Larger amounts are commonly taken to mean greater aridity in the source region (Prospero 1981). Grain size, which indicates wind intensity, and mass accumulation rates, which indicate availability, vary independently (Rea, Leinen, and Janecek 1985).

Most of the work on eolian dust in the deep sea is in Quaternary and Holocene sediments. This work has addressed a variety of problems: wind strength versus increased aridity at the source as an explanation for increase in dust transport during the last glacial maximum, variations during glacial-interglacial cycles, and relationship of biogenic components in dust to on-land climates (see, for example, Sarnthein et al. 1981, 1982; Pokras and Mix 1985; Rea and Leinen 1988; Hovan et al. 1989; Pye and Zhou 1989; Clemens and Prell 1990; Anderson and Prell 1993; deMenocal, Ruddiman, and Pokras 1993). For the pre-Pleistocene record, when neither the oceanic nor the continental record is as extensive, it may be even more difficult to distinguish the contributing factors.

Conclusions of earlier work on eolian dust in the older geologic record of the deep sea (e.g., Leinen and Heath 1981; Rea and Janecek 1981; Janecek and Rea 1983; Schramm 1989) were tentative because they were based on relatively few samples taken over a broad area. More recently, the larger number of sites studied has begun to permit better-constrained interpretations. For example, figure 5.10 shows data sets from the North and South Pacific. The timing of a Miocene maximum in dust flux (Leinen and Heath 1981; but see Janecek and Rea 1983) is different in the two hemispheres, and the accumulation rates were lower in the Southern Hemisphere. One conclusion drawn by Rea (1989) is that Northern and Southern Hemisphere atmospheric circulation as recorded in these sites was independent (see also Rea 1986). Comparison of sites in the Indian and North and South Atlantic oceans led Rea (1993) to revise an earlier interpretation (Rea and Janecek 1981) of the importance of sea level, as none of these records shows any influence of a major lowstand that occurred in the Oligocene. The problem of spatial distribution of studies also was illustrated by the differences in results of Anderson and Prell (1993) and Clemens and Prell (1990), who studied near-coastal and open-ocean sites in the Indian Ocean.

An additional problem is the limitations on dating of events. The accuracy of mass accumulation rate data and the timing of events are dependent on the accuracy of the time scale used (Hovan and Rea 1992; Arnold, Leinen, and King 1995). In addition, it should be noted that each sample from cores with low sedimentation rates probably represents at least 2000 yr of sedimen-

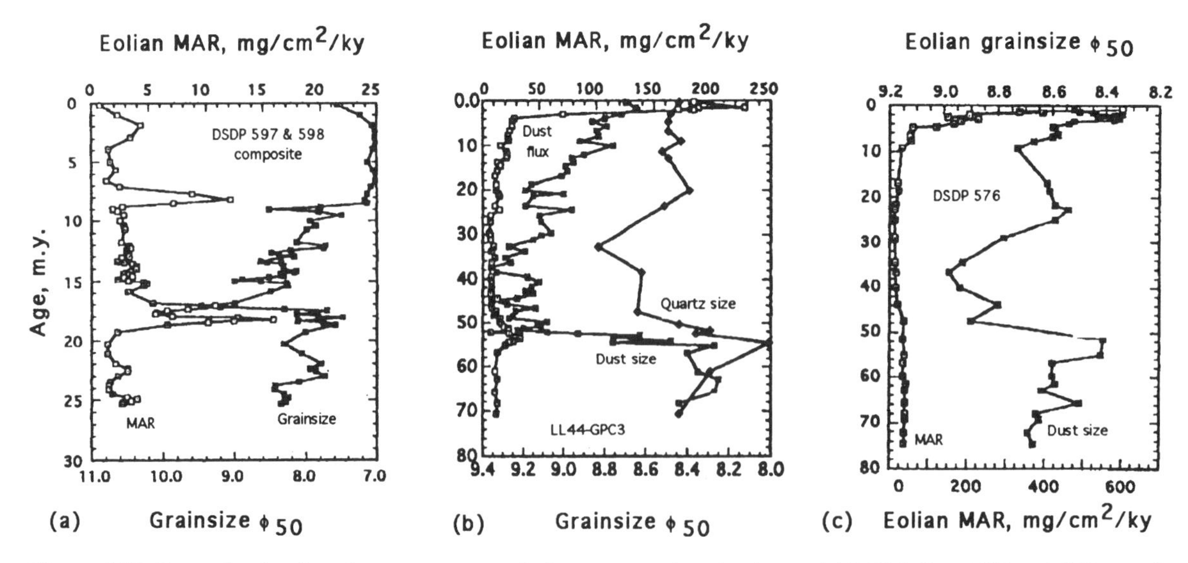

Figure 5.10. Records of eolian dust mass accumulation rates and grain size at *(a)* DSDP Sites 597 and 598, southeastern Pacific, *(b)* core LL44-GPC3, north-central Pacific, *(c)* DSDP Site 576, northwestern Pacific. Note that vertical scales are not the same. Adapted from Rea (1994) *Reviews of Geophysics* 32:159–95, copyright by the American Geophysical Union.

tation owing to mixing by infauna (Rea, Leinen, and Janecek 1985).

Volcanic Ash (Bentonite)

Deposits of volcanic ash, commonly called bentonites, also can be indicators of wind direction and possibly are underutilized as such. These differ from other eolian deposits in that each bed, particularly if the bed is thin, represents a single eruption and thus is virtually instantaneous on geologic time scales. The time scale of eruptions is commonly comparable to short-term climatic or even meteorological time scales, that is, hours or days.

Each bentonite represents a single event during which the ash was distributed by the wind that was blowing at the moment; thus a single bentonite may not reflect the regional prevailing winds. However, the predominate direction indicated by a sequence of bentonites probably will correspond to the prevailing wind. This is because there is a higher probability that the prevailing wind will be blowing during any given eruption. For example, if the prevailing wind blows 60% of the time, there is a 60% probability that an eruption will occur when the prevailing wind is blowing. Because the time scales of volcanic eruptions are short, two types of information that cannot be gathered from other eolian deposits are potentially available from bentonites. These are the different (e.g., seasonal) winds that blew in a particular region and the variability in the strength of the winds. Although this information is potentially available from eolian sandstones, it should have been apparent from the earlier discussion that extracting such information from eolian sandstones can be difficult. In contrast, a succession of thin bentonites is likely to record many of the winds and the proportion of deposits showing any particular direction might be a record of the importance of each wind.

The most extensive use of bentonites as paleowind indicators has been in rocks of the Cretaceous Western Interior seaway in the United States (Slaughter and Earley 1965; Elder 1988). Elder (1988) studied in detail four bentonite beds near the Cenomanian-Turonian boundary (figure 5.11). Three of the beds thicken toward northwestern Arizona and extend eastward (paleocoordinates) into Colorado. A fourth bed thickens toward northwestern Montana and extends south-southeast (paleocoordinates). Elder (1988)

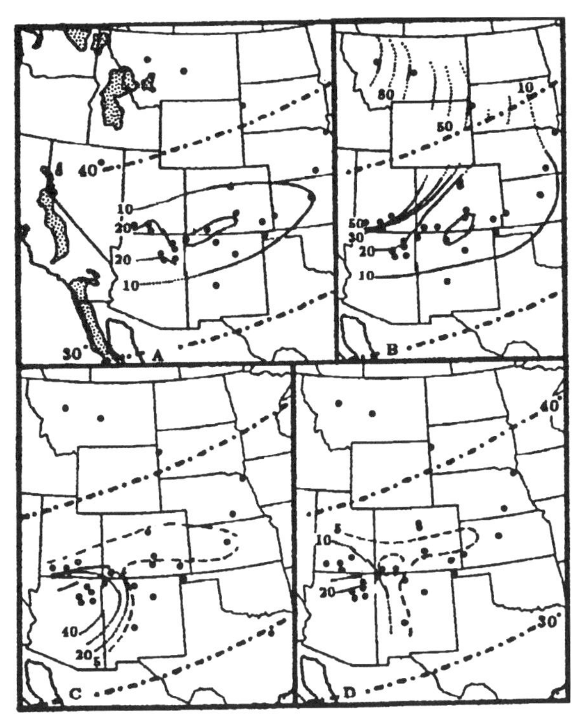

Figure 5.11. Isopach maps (in centimeters) of bentonite beds in Cretaceous rock of the Western Interior seaway (western North America). *Solid lines,* 10 cm contour intervals; *dashed line,* 5 cm contour intervals; *dotted lines,* possible projections of contours; *dot-dash lines,* 30° and 40° parallels at 90 Ma. *Stippled pattern,* major Cretaceous batholiths (shown only once); *dots,* localities. From Elder (1988) in *Geology.* Reproduced with permission of the publisher, the Geological Society of America, Boulder, Colorado, USA. Copyright © 1988 Geological Society of America, Inc.

concluded that two different volcanic sources are indicated for these beds. The orientation of the eastward-trending beds is consistent with westerlies predicted for that region by Parrish and Curtis (1982) and Barron (1990, figures 13 and 15). Ash in the bed that originated farther north was apparently erupted during a time of northwesterly winds. Based on this small sample of four beds and the model results, one can conclude that the prevailing winds in this region were westerlies. Studies of volcanic ash have also been used to determine the positions of continents, assuming certain general atmospheric wind patterns (Hildebrand 1988; Parrish and Samson 1988, see also Samson et al. 1989; Huff, Bergström, and Kolata 1992, see also Chesnut 1993 and Huff, Bergström, and Kolata 1993).

Volcanic ash is composed largely of shards of glass, and it weathers very quickly. During the weathering process, silica in the ash is lost to formation of authigenic clays, especially smectite. Marchant et al. (1996) found pristine volcanic ash at altitudes above 1000 m in the Dry Valleys region of Antarctica that gave $^{40}Ar/^{39}Ar$ dates as old as 15.15 Ma and as young as 4.33 Ma. They concluded from this unusually old, unaltered volcanic ash that the climate in the region remained well below 0°C at those altitudes for at least the last 15 Ma.

EVAPORITES

Evaporites are minerals formed when seawater or lake water evaporates at a rate faster than it is replenished. Marine evaporites usually form in semi-enclosed basins, coastal lagoons, or open coastal regions where evaporation rates are high. Salinas, or saline pans, are periodically or perennially flooded and commonly have halite. Sabkhas are deflated clastic deposits at least partially cemented by evaporite minerals. Sabkhas are flooded only sporadically, and halite might be deposited but rarely preserved (Handford 1991). Evaporites are indicative of climates in which evaporation exceeds marine influx or precipitation plus runoff.

Evaporite Mineralogy and Precipitation

The types of minerals formed in an evaporite basin depend principally on the ionic composition of the waters and on how concentrated the brines become; a relatively minor but nevertheless important effect is the relative humidity of the atmosphere. Seawater has been relatively constant in ionic composition through time, a conclusion drawn partly from the observation that marine evaporites have precipitated in a fairly standard sequence as the brine becomes more concentrated (Holland 1984). In contrast, lacustrine evaporite sequences are complex and dependent on ionic concentrations that vary widely among lakes (Eugster and Hardie 1978; Hardie, Smoot, and Eugster 1978).

The general evaporite sequence from seawater is carbonates → sulfates → chlorides. However, as discussed by Dean (1978) and many others, the sequence of evaporite minerals forming from seawater is complicated by factors such as mixing within the basin. The mineral within each of these groups that actually precipitates depends on such factors as variability in the concentrations of sulfate and phosphate ions. For example, magnesite will form as the initial carbonate mineral only under extreme evaporative conditions (Dean 1978). The study of ancient evaporites is further complicated in that some minerals, for example, dolomite and polyhalite ($K_2Ca_2Mg(SO_4)_4 \cdot 2H_2O$), can be diagenetic.

Some evaporite minerals or forms of minerals may be indicative of specific climatic conditions, based on observed modern occurrences. So-called chickenwire nodular anhydrite, when it forms in sabkha sediments, appears to require temperatures >35°C for nucleation and >20°C to be maintained (Kinsman 1976); however, this will be useful only in a demonstrated sabkha environment, as nodular anhydrite apparently also has formed in deep-water evaporite basins (Dean, Davies, and Anderson 1975). Glauberite ($Na_2SO_4 \cdot CaSO_4$) is most common in hot climates, for example, in Death Valley (California, U.S.), whereas mirabilite ($Na_2SO_4 \cdot 10H_2O$) forms in cooler climates (e.g., Caspian Sea or as a winter precipitate in the Great Salt Lake; Dean 1978; Smith 1979; Sonnenfeld and Perthuisot 1989; Warren 1989); the critical temperature in some systems for mirabilite is 15°C (Smith 1979). Restrictions on relative humidity for formation of halite and sylvite imply that an occurrence of thick marine deposits of these minerals indicates a very arid climate.

Evaporite deposition is aided by (1) a deficit in the water budget, (2) high temperatures, (3) active wind circulation, and (4) low humidities (Warren 1989; Harwood and Kendall 1993). The deficit in the water budget means that evaporation must exceed influx, either through precipitation, river input, or marine input. Influx of water to evaporite basins comes in the form of precipitation, influx of seawater or river water, or influx of groundwater. High temperatures and active wind circulation greatly increase evaporation. Formation of evaporite minerals from very concentrated brines may be inhibited simply by high atmospheric humidities. Mean coastal humidites are about 60% to 70% (Handford 1991). For example, the maximum relative humidity at which halite will form is 76% (Handford 1991), and sylvite (KCl) will not form if the relative humidity is 65% or higher (Sonnenfeld and Perthuisot 1989; Harwood and Kendall 1993). In addition, evaporation rates decrease under constant atmospheric conditions as the brine becomes more concentrated.

In addition to simply concentration, residence time of the brine is important. Residence time is the balance between rates of influx and rates of reflux (Kendall 1988). Where reflux rates are large compared with influx rates, the brine will have a short residence time and only less-soluble evaporite minerals will form. In contrast, where reflux rates are small compared with influx rates, more time is available for evaporation to concentrate the brine and the more-soluble evaporite minerals will form. The most favorable conditions for evaporite formation are moderate to long residence time, moderate to high salinity, and a high brine level.

Evaporites and Climate

Distribution of evaporites

Evaporite deposits tend to occur in low mid-latitudes, under the descending limb of the Hadley circulation. Evaporites can potentially form in very high latitudes as well, where precipitation is low; brines, for example, have been reported in Antarctica (Wilson 1979). However, evaporation rates are so low in the polar regions that evaporites formed there are unlikely to be geologically important (Sonnenfeld and Perthuisot 1989).

The geographic distribution of modern evaporites is similar to that of ergs (compare figure 5.12 with figure 5.2). Evaporites form in deserts and occupy many of the desert areas not covered by eolian sand. Evaporites are more common than ergs in monsoonal climates, that is, climates that are warm and seasonal with respect to rainfall. In these areas, the water influx during the wet season is not enough to dissolve the previous dry season's deposits, and it provides the source for the salts deposited during the next dry season. Most of the large modern evaporite deposits are forming in such seasonal environments (Sonnenfeld and Perthuisot 1989), but in none of these environments is the wet month very wet. Indeed, for gypsum, the climates under which evaporites are forming are very similar with respect to mean annual precipitation and seasonality to that in which ergs occur (figure 5.13; compare with figure 5.1).

The existence of thick evaporite sequences indicates at least seasonally arid and probably warm climate over an extended period. A minimum amount of time required to accumulate a large deposit can be estimated from laminated sequences such as the Castile and lower Salado Formations (Permian; Texas, U.S.). These contain more than 200,000 laminations (Dean, Davies, and Anderson 1975) and, if these laminations are annual, the deposit required a minimum time of as many years to accumulate. Because such laminations are not necessarily

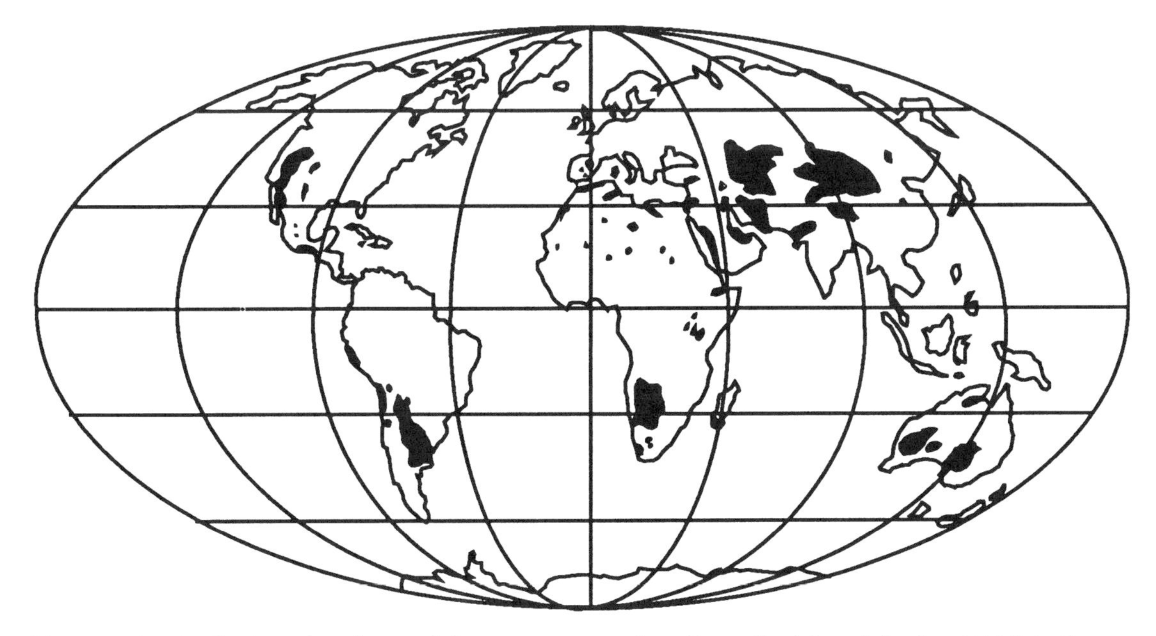

Figure 5.12. Distribution of modern and Quaternary evaporites. Generalized from Schreiber (1988) in *Evaporites and Hydrocarbons,* by B. Charlotte Schreiber, ed. Copyright 1988, Columbia University Press. Reprinted with permission of the publisher.

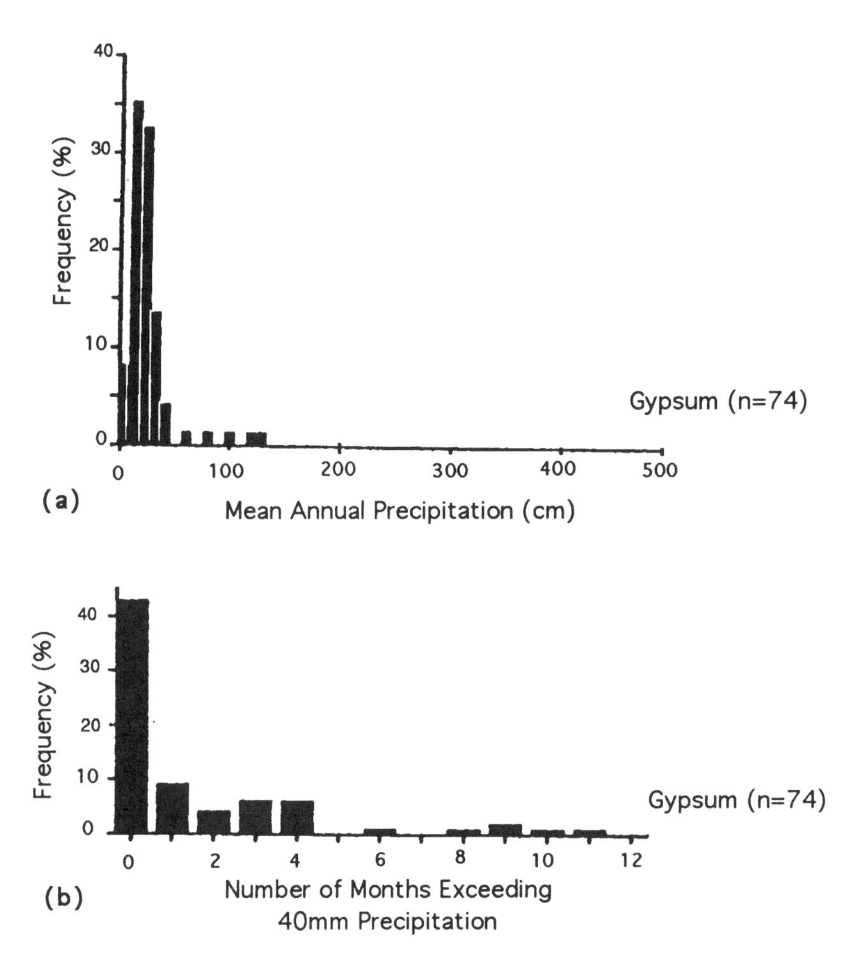

Figure 5.13. Mean annual precipitation *(a)* and number of months exceeding 40 mm precipitation *(b)* in areas of modern gypsum deposits (*N* = 74). From Gyllenhaal (1991).

annual (Braitsch 1971; Warren 1989), the actual time required may be much greater.

Topographic relief on a basin helps create a climate dry enough to form evaporites, and evaporites are common in both modern and ancient rift basins. Relative humidity decreases in a descending air mass because the air becomes warmer. If the air mass descends below the altitude of its origin, or if it has lost moisture flowing over mountains before descending, the air may become dry enough to evaporate large amounts of water. Air masses moving into rift zones must first traverse the elevated rift shoulders (see table 1.1) and are likely to lose some moisture in the process. If the rift basin floor is at, near, or below sea level, those air masses are capable of inducing high evaporation rates. This is seen in East Africa today, whose lakes have some of the better-studied lacustrine evaporite sequences. The same process occurred in the opening South Atlantic during the Early Cretaceous. The stratigraphic succession is generally: terrestrial rocks, overlain by marine evaporites, overlain by normal marine rocks (Evans 1978). The transition from terrestrial rocks to marine evaporites indicates the collapse of the rift floor (Doyle et al. 1977), which allowed influx of seawater and brought about the topographically related paleoclimatic conditions favorable to the formation of evaporites. Continued rifting finally led to flooding of the basin, so that the influx of seawater exceeded evaporation.

Distribution of evaporites in the geologic record

Evaporites are perhaps the least controversial of the paleoclimatic indicators, and consequently their distribution in the past has been part of nearly every study that has looked at global paleoclimates (e.g., Briden and Irving 1964; Robinson 1973; Drewry, Ramsay, and Smith 1974; Parrish, Ziegler, and Scotese 1982; Witzke 1990; Scotese and Barrett 1990; and many others). The paleolatitudinal distribution of evaporites through time has remained close to the latitudinal distribution today, although it should be noted that the distribution today is partially skewed in the Northern Hemisphere by the Asian monsoon and the continentality of Asia. However, some important variations in the latitudinal distribution of evaporites through time can be noted (Parrish, Ziegler, and Scotese 1982; Parrish and Barron 1986). Monsoonal circulation is thought to have been important during the late Miocene (Prell, Murray, and Clemens 1992) and during the Pangean interval (Parrish 1993), and monsoonal climates in drier regions are ideal for the formation of evaporites (Sonnenfeld and Perthuisot 1989). A characteristic of Asia- or Pangea-scale monsoonal climates is that the eastern equatorial regions of the continents in the vicinity of the monsoon circulation tend to be drier than would be expected in a more zonal climate (Parrish 1993). Compared with other times in Earth history, evaporites deposited in the Late Carboniferous-Jurassic and Miocene cluster closer to the equator.

The distribution of evaporites in the geologic record is not uniform (Gordon 1975; Railsback 1992); this is indicated both by tallies of evaporite deposits, including correction for erosion, and by fluctuations in sulfur isotopes in marine sulfate minerals. Although some workers have used the changes in evaporite deposition through time to suggest major changes in global climate (e.g., Frakes 1979), the distribution of evaporites through time may have more to do with the distribution of suitable sites for deposition. Railsback (1992) showed that sea level and the latitudinal distribution of land could have exerted the major control on evaporite deposition.

Cycles in evaporites

As might be expected from the relationship between brine concentration and evaporite-mineral precipitation, evaporites are sensitive indicators of changes in brine concentration, and thus sensitive to any climatic changes. Laminations are common in thick evaporite deposits. The presence of laminations with extensive lateral continuity has been taken as conclusive evidence that evaporites can form in deep-water basins (Dean, Davies, and Anderson 1975), a subject that was controversial for many years after it was first proposed by G. Richter-Bernberg (cited by Pannekoek 1965; see also Schmalz 1969; Warren 1989; Sonnenfeld and Perthuisot 1989) and has continued to be controversial (Handford 1991; Clauzon et al. 1996). Laminations are typically alternating carbonate and anhydrite (Dean, Davies, and Anderson 1975) and typically millimeter-scale or smaller. In some basins, they have been interpreted as seasonal, with the carbonate indicating the wetter season and the anhydrite indicating the drier season (Braitsch 1971; Anderson et al. 1972; Anderson and Dean 1995). Lamina compositions can change through a unit, and the changing compositions of the laminae may indicate multiannual or longer-term climate changes (Anderson and Kirkland 1960; Anderson and Dean 1995).

Cyclicity on a scale larger than laminae also may indicate climate change (e.g., Anderson et al. 1972). In the Paradox Basin (Utah, U.S.), cyclic Late Carboniferous deposits contain black shale, carbonate, anhydrite, halite, and rare potash salts (Hite 1970). The cycles are consistent in sequence but vary in composition; schematic cycles from different parts of the basin are shown in figure 5.14 (Raup and Hite 1992). The evaporites grade laterally into carbonate sequences, and the black shales are basin-wide. There are at least three interpretations for the composition of the cycles in the Paradox evaporites. Hite's interpretation (Hite 1970; Hite and Buckner 1981; Raup and Hite 1992) was that the black shales formed during transgressions. A cap of fresher seawater would flow into the basin over the brine, causing freshening of the basin waters and precipitation of lower-solubility evaporites. The stable stratification resulting from the less-dense seawater would finally lead to stagnation in the underlying water column and deposition of the black shales. Regression (see figure 5.14, evaporite cycle) led to concentration of the water and precipitation of evaporites and, finally, desiccation of the basin and an erosional unconformity atop the highest-solubility salts. This model calls on the observation that organic productivity is high in highly saline brines (saturated with respect to halite or sylvite) and assumes continued high productivity during the lowest-salinity stages in the development of the basin. The clastics in the black shale and the abundant terrestrial organic matter were introduced to the basin during black shale deposition (Hite, Anders, and Ging 1984).

An alternate model calls for freshening to occur because of climatic change and increased river influx leading to deposition of the black

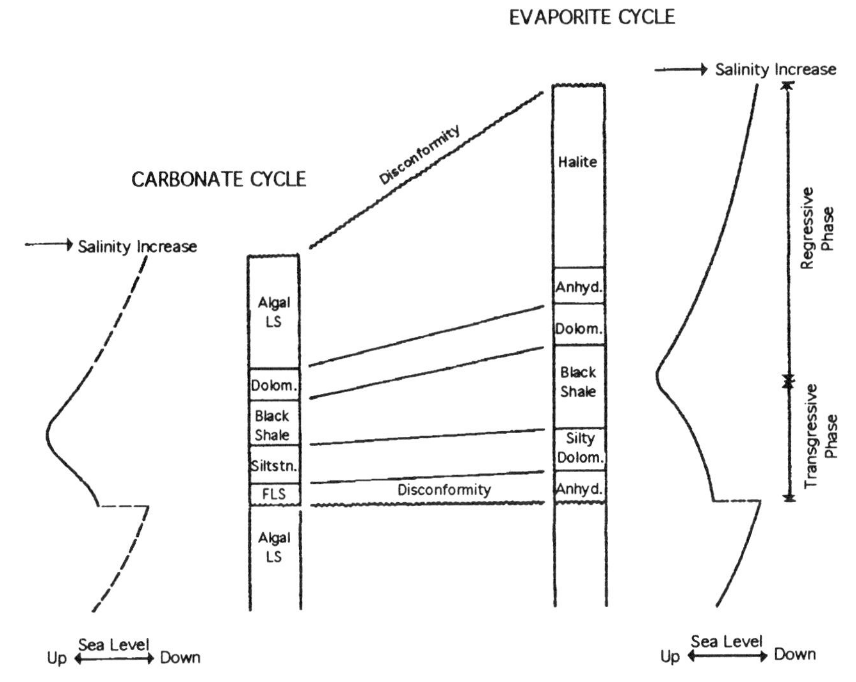

Figure 5.14. Schematic cycles in the Paradox Basin evaporites (Late Carboniferous; Utah, U.S.). Anhyd., anhydrite; LS, limestone; Dolom., dolomite; FLS, foraminiferan limestone. From Hite and Buckner (1981).

shales (Hite, Anders, and Ging 1984). Increased river influx would not only introduce clastics and abundant terrestrial organic matter but also could boost productivity by creating estuarine circulation, a type of upwelling that leads to high productivities in estuaries. This would require some connection with the ocean and would have to have operated in conjunction with a sea-level rise. Climate during sea-level highstands probably was more humid at this time in western North America (Johnson 1989; Soreghan 1992a), so this model is consistent with other geologic evidence. $\delta^{18}O$ profiles, showing lower values in the clastics, support the interpretation that the black shale units in the Paradox Basin were deposited during times of freshwater influx (Magaritz 1987). Magaritz (1987) also interpreted dolomite laminae in the anhydrites as indicative of seasonal influx of freshwater.

The third model emphasized brine residence times and required no fluctuations in sea level (Kendall 1988). In this model, a black shale served as an aquaclude at the bottom of the basin, preventing the brine from exiting the basin as groundwater flow and increasing brine residence times. As the basin filled with deposits of progressively more soluble evaporites, the water level rose above the edges of the black shale deposit, and brine began to reflux from the basin, shortening brine residence times and leading to deposition of progressively less soluble salts. The problem with this model is that it does not account for the next cycle of black shale. In addition, sea level and climate are known from abundant other evidence to have been fluctuating at this time in Earth history, so the relatively complex Kendall (1988) model is not necessary to explain the Paradox Basin evaporite cycles. However, it might apply to other evaporite basins that formed during times of less fluctuating external conditions. An example is illustrated in figure 5.15 for a sequence in western Canada. The dashed curve represents the conventional interpretation, which would require a sea-level rise to explain the transition from the Muskeg anhydrite to the Sulfur Point dolomites and overlying limestones, followed in turn by a sea-level drop for deposition of the Watt Mountain nonmarine clastics, followed by a sea-level rise to explain the evaporites and overlying open-marine carbonates of the Fort Vermillion and Slave Point Formations, respectively. Kendall (1988) suggested instead that the sequence represents a simple regressive-transgressive pattern. In this model, the Muskeg was deposited during relatively high sea level, when seawater influx at the surface was

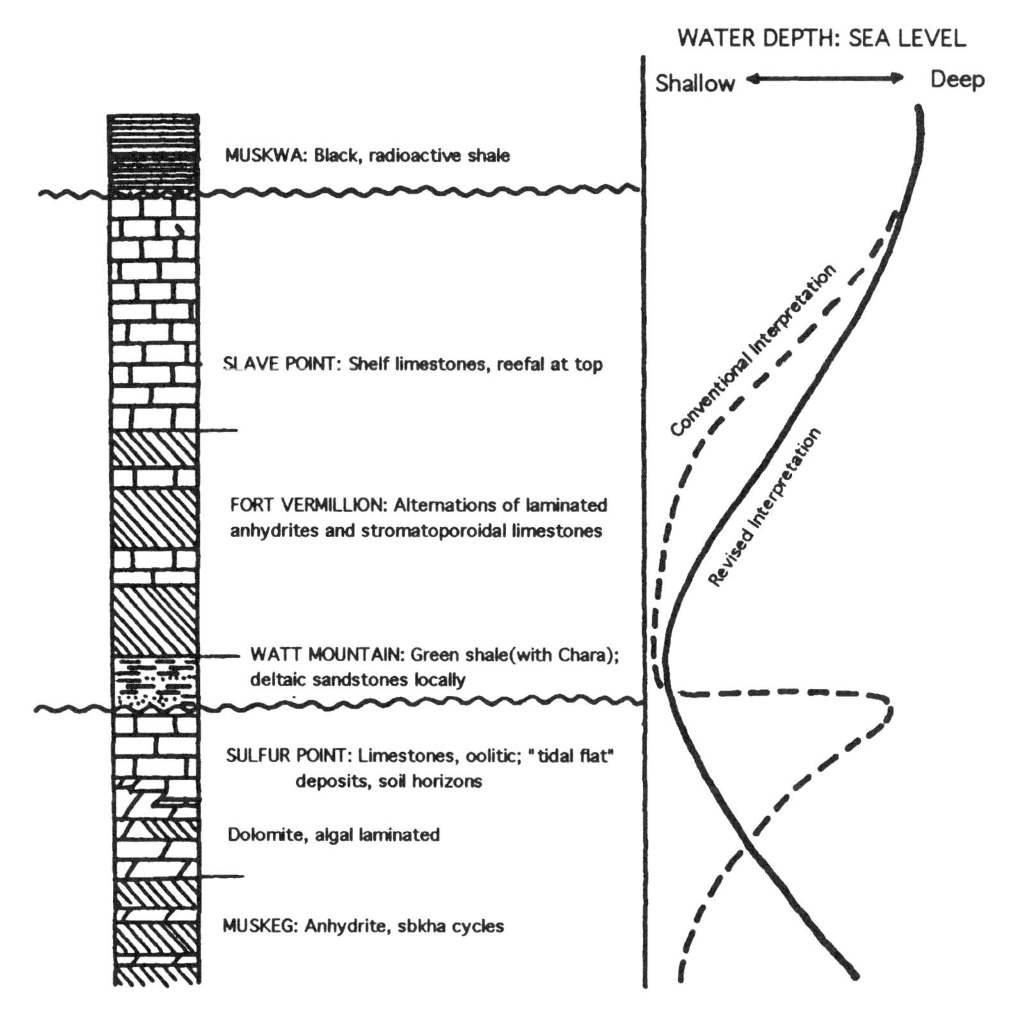

Figure 5.15. Hypothetical section through Middle Devonian evaporites and associated rocks in Alberta, Canada. The right-hand side of the diagram shows two alternative hypotheses for evaporite deposition: controlled by sea level (*dashed line*) or by brine residence times (*solid line*). See text for explanation. From Kendall (1988) in *Evaporites and Hydrocarbons*, by B. Charlotte Schreiber, ed. Copyright 1988, Columbia University Press. Reprinted with permission of the publisher.

greater than brine reflux, in this case, over a sill at the basin margin (similar to the reflux of Mediterranean water to the Atlantic over the sill at the Straight of Gibraltar). In this situation, the brine residence time would have been long. The Sulfur Point was deposited as sea level dropped, which decreased brine reflux relative to influx by closing off the sill, although influx could have persisted through seepage. Finally, the basin water level was low enough to permit deposition of the nonmarine sediments.

The significance of the various models of evaporite cycles is that some are related to climatic variability and others are not. Although evaporites are good indicators of dry paleoclimates, evaporite cycles cannot be assumed to indicate paleoclimatic change and the nature and alternate causes of the cycles must be analyzed.

ZEOLITES AND ASSOCIATED AUTHIGENIC MINERALS

Basins that undergo evaporation and brine formation may not deposit and preserve evaporite minerals of the types discussed in the previous section, but under certain sedimentological conditions, the evaporative nature of the basins may be revealed by the presence of authigenic minerals such as Na-silicates, borosilicates, K-feldspars, and zeolites. The presence of these minerals can be diagnostic, but because the minerals are cryptic in nature, they have not been exploited often in basin descriptions.

A lateral zonation is common in zeolite deposits and is related to evolution of the brine (Smith 1979). Diagenesis of preceding phases establishes the zonation of zeolites. Six zeolite minerals are common in saline, alkaline lake deposits (Surdam and Sheppard 1978): analcime, clinoptilolite, phillipsite, erionite, chabazite, and mordenite. Common associated deposits include dolomitic mudstone, crystal molds of gaylussite ($Na_2CO_3 \cdot CaCO_3 \cdot 5H_2O$) or pirssonite ($Na_2CO_3 \cdot CaCO_3 \cdot H_2O$), freshwater organisms near the inlet, tufa or spring deposits, and lake-margin deposits such as algal stromatolites, oolite and pisolite, and flat-pebble conglomerates (Surdam and Sheppard 1978). Gaylussite and pirssonite can be diagenetic minerals (Smith 1979). The zeolite deposits are laterally zoned, with volcanic glass in the outer zone and authigenic potassium feldspar in the inner zone (e.g., Pleistocene; Lake Tecopa, California, U.S.; Sheppard and Gude 1968).

All evaporite minerals and authigenic zeolites are subject to rapid dissolution and/or diagenesis, even through the duration of lake sedimentation. Thus nonmarine evaporites are not commonly recognized in the geologic record (Smoot and Lowenstein 1991). The recognition of a saline lake deposit in the geologic record may thus depend sometimes on subtle characteristics within the mudstones. One example of the application of zeolites to the older geologic record is in the Brushy Basin Member of the Morrison Formation (Jurassic; San Juan Basin, U.S.; Turner and Fishman 1991). This unit contains sediments of a very large (1.8×10^4 km^2) saline, alkaline lake identified on the basis of sedimentology, facies relationships, and zonation of clay minerals, zeolites, and authigenic feldspars (figure 5.16). The chemical facies are in tuffs and consist of smectite in the outer zone, surrounding zones of clinoptilolite, analcime plus potassium feldspar, and authigenic albite, which occupies the center. Mixed layer illite/smectite clays also are zoned, from highly smectitic in the outer zones to highly illitic in the center. In the Brushy Basin Member, tuffs altered to clinoptilolite are orange-pink (10R7/6), those altered to analcime and potassium feldspar are commonly dark green (5G4/2) with lighter-colored zones or patches, and both facies are more indurated than unaltered tuff beds, which helped Turner and Fishman (1991) with field identification of the facies. However, these kinds of distinctive features in the field are evidently unusual, as they have not been mentioned in the literature on other deposits.

Turner and Fishman (1991) concluded that not only was the lake saline and alkaline but also that it was shallow and subject to periodic desiccation. The lake deposits lack shoreline and delta facies. Fluvial sandstones and tuffs are interbedded, commonly with ripup clasts in the sandstones from the underlying tuffs. The sandstones are sheet-like or laterally discontinuous. These relationships suggested to Turner and Fishman (1991) that the streams and lake were ephemeral and that the sandstones were deposited directly on dried lake deposits. Desiccation cracks, transported mud chips, and clay-pellet aggregates also are evidence of the ephemeral nature of the lake. However, scattered horizons containing root traces, lacustrine limestones, and plant fossils indicate that the lake was periodically filled.

PALEOSOLS AND RELATED SEDIMENTARY ROCKS

A major thrust in geomorphology is the relationship between landforms and climate. Unfortunately, landforms become increasingly obscured further back in time, so the paleoclimate information that can be derived from them is unavailable for most of geologic time. However, weathering surfaces are commonly preserved in the geologic record and can provide information about paleoclimate. Karst surfaces indicate removal of bedrock through dissolution, sometimes related to climatic processes. Paleosols—ancient soils, sometimes called buried soils (Yaalon 1971)—are rocks that record in situ weathering of bedrock or sediments. Paleosols include such well-studied deposits as red beds and laterites. Paleosols occur in most terrestrial depositional sequences, and thus the potential database for interpretation of terrestrial paleoclimates is enormous (Roeschmann 1971). Recognition of this potential and the need to exploit it has been a driving force behind the study of pre-Quaternary paleosols in the last decade or so (Reinhardt and Sigleo 1988; Retallack 1990), and even relatively well-studied paleosols have been receiving new attention with regard to their paleoclimatic significance. Finally, some sedimentary rocks are formed from weathering products

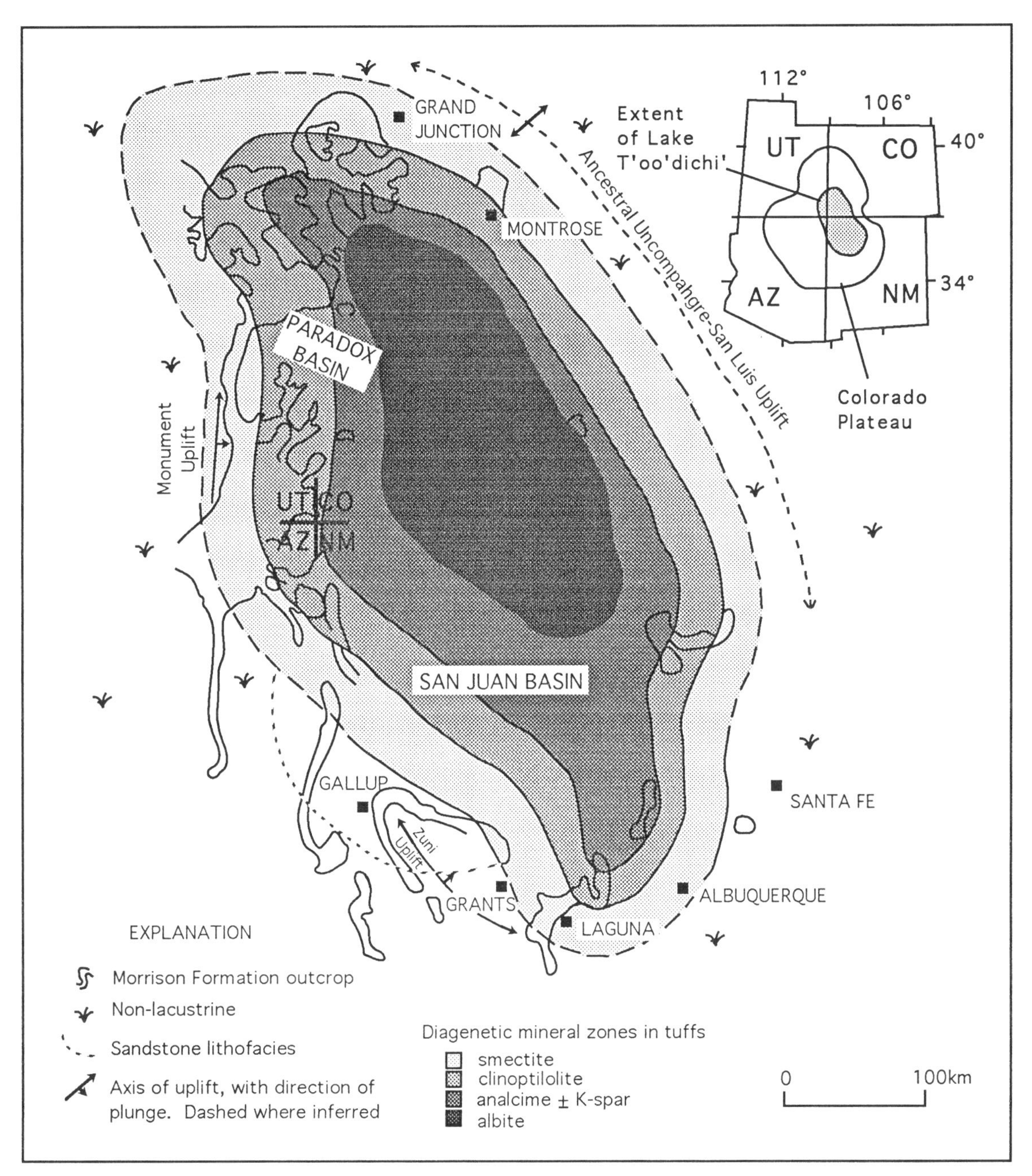

Figure 5.16. Diagenetic-mineral and fluvial-sandstone facies of Lake T'oo'dichi' (Morrison Formation, western United States), a Jurassic saline, alkaline lake. From Turner and Fishman (1991) in *Geological Society of America Bulletin*. Reproduced with permission of the publisher, the Geological Society of America, Boulder, Colorado, USA. Copyright © 1991 Geological Society of America, Inc.

or are related to weathering at a distance from the site of deposition or precipitation of chemical constituents. These rocks are also discussed here. Classification systems for paleosols attempt to follow those for modern soils, for which an elaborate classification scheme exists (Soil Survey Staff 1975, 1990). However, many of the characteristics that distinguish modern soil types are not readily preserved in the geologic record (Mack, James, and Monger 1993). For example, two important characteristics of modern soils are base saturation level and bulk density. Base saturation is the percent base cations (for example, Na^+) of total exchangeable cations (e.g., Birkeland, 1974). Clearly, such properties would be easily altered with even minimal diagenesis and burial. In addition, the classification of many modern soils includes specific climatic conditions. An example cited by Mack, James, and Monger (1993) is the definition of Aridisol[1], which requires an arid moisture regime, defined as soil that is not moist for periods of 90 days or more during the time when soil temperature is 8°C or higher. If the definition of paleosols were based on such climate information, that information would have to come from paleoclimate models, but no paleoclimate models are that accurate (Mack, James, and Monger 1993), and if they were we would not need paleosols as paleoclimatic indicators. Paleoclimate should not be used as part of the interpretation of a paleosol; rather, a paleosol should be identified on the basis of other criteria to provide information for interpreting paleoclimate (Retallack 1990).

Paleosols have characteristics that are the result of exposure and weathering and, importantly, characteristics acquired after burial (Retallack 1990; Mack, James, and Monger 1993). Paleosol horizons will always be different from the underlying rock in chemistry or texture and, characteristically, color. Paleosols are commonly richer in carbonate, silica, organic matter, and/or sesquioxides (Fe_2O_3, Al_2O_3). If the carbonate content is high, the paleosol is called a calcrete or caliche; if high in silica, silcrete. Textural differences include the presence of nodules or concretions; massive, hackly, or jointed texture; and/or root traces. Paleosols are commonly more highly colored than the surrounding rock. Coloration is typically brown, red-orange, green, or mottled. Brown or black colors can indicate a higher organic matter content; red, orange, or purple colors indicate the presence of iron sesquioxides; green indicates the presence of reduced iron; and mottling usually indicates that the water table was high or fluctuating. Paleosols are typically found in sedimentary rocks, but they also can be surfaces on metamorphic or igneous rock, for example on lava flows. In paleosols developed on sedimentary rocks, root traces are common and primary sedimentary structures are generally destroyed (Retallack 1988).

Paleosols that formed under the same climatic conditions might look very different depending on the stage of soil development at the time of burial, which stops pedogenic processes unless the cover is thin (Kraus 1997). By studying well-exposed rocks of the Eocene Willwood Formation (Big Horn Basin, Wyoming, U.S.) and working out relationships among paleosols as illustrated in figure 5.17, Kraus and Bown (1988) documented a developmental sequence for the paleosols (figure 5.18). A paleosol that was buried early in its stage of development might give a much different impression of the environmental conditions of formation, including climate, than a mature one. If exposures are limited, the paleoclimate interpretation might be different, depending on which part of the developmental sequence was exposed. A characteristic of pre-Quaternary paleosols that helps in interpretation, particularly in comparative studies, is that the pre-Quaternary geologic record tends to preserve the same general part of each land-surface complex, namely, the areas of deposition.

The systematic study of pre-Quaternary paleosols and climate was pioneered by Retallack (1981, 1983, and subsequent publications). Retallack (1988) formalized terminology for the recognition of paleosols in the field and adapted the modern soil classification scheme to paleosols (see also Retallack 1990). He listed characteristics of paleosols that might serve as specific indicators of climate and, unlike many workers, made the distinction between climates that are moderately or extremely seasonal (Retallack 1990). *Rainfall* indicators include the presence or absence of calcium carbonate, with absence indicating relatively wet climates, and depth to the calcic horizon (Retallack 1990; but see Mack, James, and Monger 1993). Clay mineralogy is re-

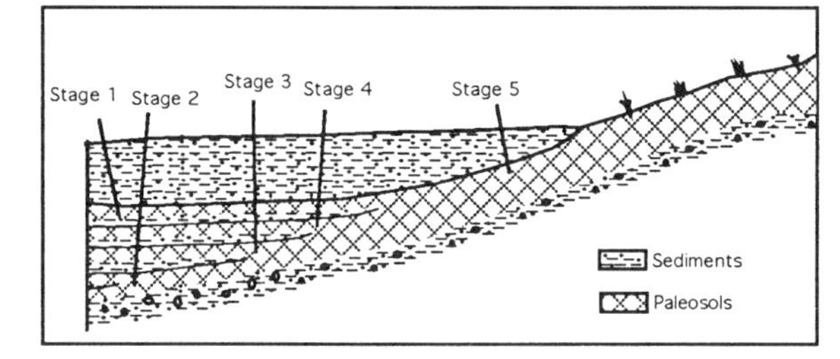

Figure 5.17. Paleosols buried at different stages of development. Note the relationship among the paleosols on the exposed surface, the correlative paleosol buried by the last inundation of sediment, and the partially developed and correlative paleosols that were buried by earlier sedimentation events. The stage numbers refer to the stages described in figure 5.18. The part of the paleosol that remains exposed in this diagram would be unlikely to be preserved in the geologic record, so this diagram also shows that nearly all paleosols preserved in the older record formed in depositional, not erosional, settings.

1. Soil orders are formal names.

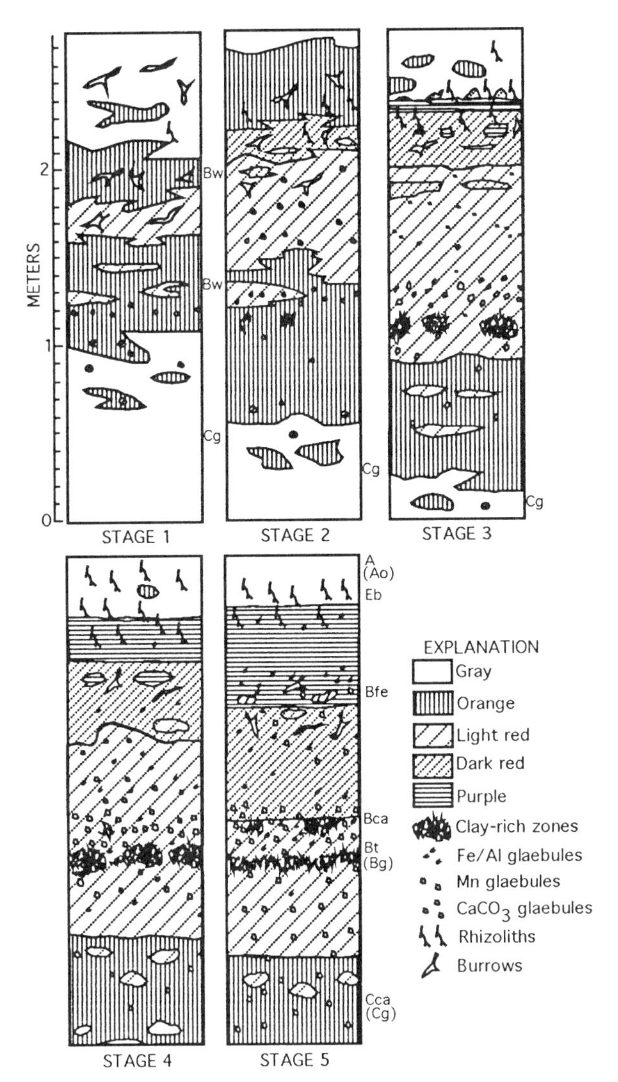

Figure 5.18. Paleosol developmental sequence in the Willwood Formation (Big Horn Basin, Wyoming, U.S.). The possible lateral relationships among the stages are illustrated in figure 5.17. From Kraus and Bown (1988) in *Paleosols and Weathering Through Geologic Time: Principles and Applications*, J. Reinhardt and W.R. Sigleo, eds. Reproduced with permission of the publisher, the Geological Society of America, Boulder, Colorado, USA. Copyright © 1988 Geological Society of America, Inc.

lated to rainfall, and evaporites associated with paleosols are indicative of dry climates (see also McCahon and Miller 1993). *Temperature* indicators listed by Retallack (1990) include tough, inert, spherical micropeds, which are distinctive of tropical soils, and permafrost and other periglacial features. Retallack (1990) also included karst as indicative of warm climates, but karst formation is probably not temperature dependent. Indicators of *seasonality* of rainfall include pseudoanticlines (gilgai), concretions, charcoal (indicating periodic drying of the vegetation), and a root pattern consisting of a near-surface network of fine roots and stout, deeply penetrating roots. Seasonality of temperature that includes freezing during the cold months is indicated by evidence of frost cracking or frost heave. Additional features of paleosols that are indicative of paleoclimate but that were not listed by Retallack (1990) are discussed in the following sections.

Mack, James, and Monger (1993) proposed a classification scheme that is in many ways more practical for application to paleosols than the modern soil classification scheme. The combination of detailed field criteria presented by Retallack (1988, 1990) and the classification by Mack, James, and Monger (1993) makes the use of paleosols as paleoclimatic indicators easier for many geologists, although it results in a less detailed classification (see James, Mack, and Monger 1993; Retallack 1993). The paleosol classification proposed by Mack, James, and Monger (1993; table 5.3) is based on six features commonly preserved in the geologic record: organic-matter content, horizonation, redox conditions, in situ mineral alteration, illuviation of insoluble minerals and compounds, and accumulation of soluble minerals. Mack, James, and Monger (1993) eliminated as diagnostic characteristics color, chemically unstable minerals, thicknesses of horizons, and other characteristics that might be substantially altered by later sedimentologic and diagenetic processes. Examples of their soil descriptions applied to paleosols of various ages are shown in figure 5.19; modifiers used in the descriptions are explained in table 5.4. An approach that accomplishes some of the same goals as Mack, James, and Monger's (1993), but within the framework of modern soil classification, is the use of pedotypes (Retallack 1994). Pedotypes are also purely descriptive, being "a kind of soil or paleosol as recognized in the field" (Retallack 1994) and based on type profiles, similar to type sections for rock units. Using this approach, combined with paleontological data, Retallack (1994) reconstructed the paleoenvironments and paleoclimates of the units at the Cretaceous-Tertiary boundary in Montana. If the pedotypes defined there and elsewhere prove robust in units throughout much of the geologic column worldwide, they are likely to eventually provide as detailed a paleosol classification as the one for modern soils.

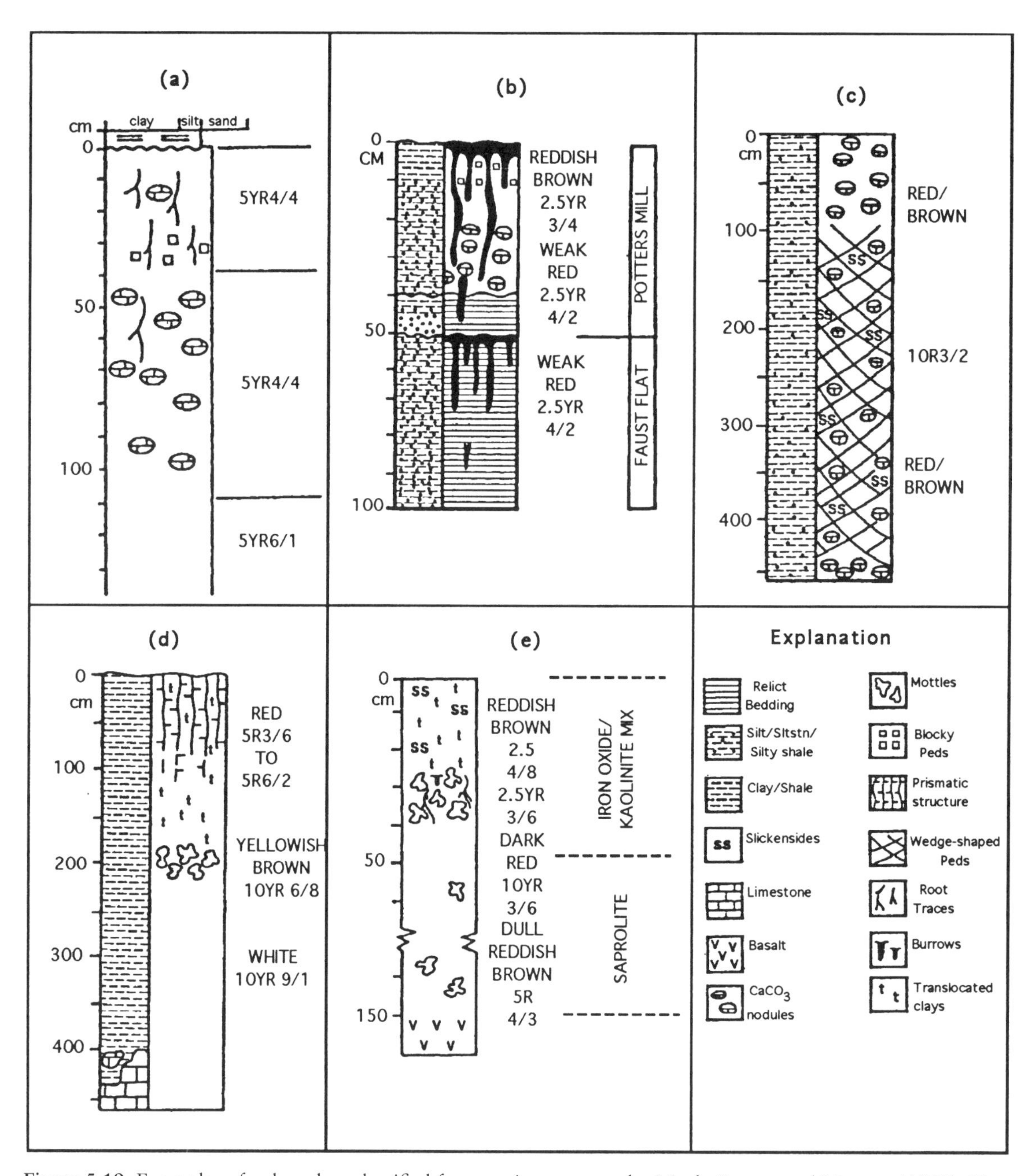

Figure 5.19. Examples of paleosols reclassified from various sources by Mack, James, and Monger (1993). Not all possible soil types are illustrated and other features, not illustrated, can also be useful in paleoclimate interpretation. *(a)* argillic Calcisol, Camp Rice Formation (Pleistocene; southern New Mexico, U.S.); *(b)* Protosol, Faust Flat Member, originally classified as Entisol, and Calcisol, Potters Mills Member, originally classified as Inceptisol, Juniata Formation (Ordovician; Pennsylvania, U.S.); *(c)* calcic Vertisol (this was also the original classification), Clifton Down Mudstone (Carboniferous, United Kingdom); *(d)* Argillisol (original classification; Cretaceous, southeastern United States); *(e)* argillic Oxisol (originally classified as laterite; Cretaceous, Negev Desert). Stage 5 paleosols in the Willwood Formation (see figure 5.18) were classified as Spodosol (Kraus and Bown 1988), and reclassified by Mack, James, and Monger (1993) as argillic, gleyed, calcic Spodosol. Descriptors defined in table 5.4. From *Geological Society of America Bulletin.* Reproduced with permission of the publisher, the Geological Society of America, Boulder, Colorado, USA. Copyright © 1993 Geological Society of America, Inc.

Table 5.3.
Classification of major modern soil types and of paleosols

Recent soil orders	Paleosol type	Descriptions
Alfisol*	See Argillisol, Calcisol	Gray to brown soils in variable climates; mid-latitudes, e.g., NW Europe, central Saskatchewan, Texas Gulf of Mexico region
Aridisol*	See Calcisol, Gypsisol	Low in organic matter, sometimes with a salt bloom at the surface occurring in climates that are dry > 6 mo of the year, e.g., Sahara, central Australia, western Argentina
Entisol*	See Protosol	Soil on recent sediment in variable climates, e.g., northern Labrador, NW Australia
Histosol	Retained	Modern: organic soils forming in areas with poor drainage in climates where precipitation > evaporation, e.g., Scotland, Hudson's Bay, western Ireland and England Paleosol: coal
Inceptisol*	See Protosol, Calcisol	Immature soil, usually in climates where precipitation > evaporation, e.g., parts of India, Alaska
Mollisol*	See Spodosol, Calcisol	Very organic-rich soils in seasonal climates, e.g., central Asia, Yucutan Peninsula
Oxisol	Retained	Modern: deeply weathered soils, rich in sesquioxides, forming in tropical to subtropical climates where precipitation > evaporation, e.g., central Africa, Madagascar, Amazon River basin Paleosol: chemically unstable minerals altered to clay and sesquioxides. Oxic horizon must (1) have been subsurface, (2) have authigenic, but not diagenetic, 1:1 clays, (3) have authigenic, but not diagenetic, sesquioxides formed in situ, (4) have < 10% unstable grains (feldspars, volcanic rock fragments, ferromagnesium minerals)
Spodosol	Retained	Modern: soils with subsurface accumulation of amorphous material overlain by a sandy bleached horizon forming in cool climates where precipitation > evaporation, e.g., eastern Scandinavia, northern Siberia, NE Canada Paleosol: distinct spodic horizon formed by illuviation of organic matter and iron oxides (i.e., down-profile increase in both organic matter and iron oxides)
Ultisol*	See Argillisol	Soils with subsurface clay accumulations forming in variable climates, e.g., NE Australia, SE US, SE China
Vertisol	Retained	Modern: soils with high content of expandable clays formed in climates that are seasonal with respect to rainfall, e.g., eastern Australia, west-central India Paleosol: profile homogenized by pedoturbation, mostly shrinkage and swelling of expandable clays; such clays need not still be present
	Calcisol	Subsurface pedogenic calcic horizon
	Gypsisol	Pedogenic gypsum as surface or subsurface horizon, precipitated in vadose zone
	Gleysol	Two types: (1) surface or subsurface horizon that was consistently in reducing conditions, as evidenced by gray and green mottling, organic matter, pyrite; (2) horizon that experienced alternating oxidation and reduction, as evidenced by red, yellow, brown mottling in gray and green matrix, nodules of iron and manganese oxide
	Argillisol	Argillic horizon, i.e., subsurface horizon of illuvial clay
	Protosol	Weak development of horizons unrelated to homogenization by pedoturbation

From information in Mack, James, and Monger (1993).

*The modern soil name has not been carried over to the classification of paleosols. However, brief descriptions for these soils are provided. Note that the definitions of modern soil types include information about climate.

Table 5.4.
Modifiers for description of paleosol orders

Albic	Presence of eluvial (E) horizon quartz arenite underlying originally organic horizon and overlying clay-rich horizon
Allophanic	Presence of allophane or other amorphous Si and Al compounds
Argillic	Presence of illuvial clay
Calcic	Presence of pedogenic carbonate
Carbonaceous	Presence of dark organic matter but not coal
Concretionary	Presence of glaebules with a concentric fabric
Dystric	Low base status as indicated by the paucity of chemically unstable grains such as feldspars and volcanic rock fragments
Eutric	High base status as indicated by the abundance of chemically unstable grains such as feldspars and volcanic rock fragments
Ferric	Presence of iron oxides
Fragic	Subsurface horizon that was hard at the time of soil formation (for example, root traces and burrows terminate or are diverted at this horizon)
Gleyed	Evidence of periodic water logging, such as drab hues; mottles of drab colors and yellow, red, or brown; or presence of pedogenic pyrite or siderite
Gypsic	Presence of vadose gypsum or anhydrite
Nodular	Presence of glaebules with an undifferentiated internal fabric
Ochric	Presence of light-colored A horizon
Salic	Presence of pedogenic silica
Vertic	Presence of decimeter-scale desiccation cracks, wedge-shaped peds, hummock-and-swale structures, slickensides, or clastic dikes
Vitric	Presence of relict or actual glass shards or pumice

Adapted from Mack, James, and Monger (1993) in *Geological Society of America Bulletin*. Reproduced with permission of the publisher, the Geological Society of America, Boulder, Colorado, USA. Copyright © 1993 Geological Society of America, Inc.

Red Beds

Red beds are not a type of paleosol. Red paleosols can comprise a variety of types with only the red color in common, including, in Mack, James, and Monger's (1993) classification, Oxisols, Vertisols, Calcisols, Gypsisols, Spodosols, and Argillisols. However, the term "red bed" is so widespread in the literature that a section devoted to red beds is necessary.

Clastic red beds

Red beds are iron- (usually hematite-) stained, clastic sedimentary rocks. The intensity of the red color depends on two factors, grain size and time since deposition (Walker 1967). Fine-grained rocks are commonly redder than coarse-grained rocks because the relative concentration of hematite is higher. In addition, newly deposited sediments are not red, but under the right conditions, including climate, they can become red in a short time (10^3 to 10^4 years; Walker 1976). Although Van Houten (1982a), for example, suggested that reddening occurs over long periods, Bronger and Bruhn (1989) found no correlation between age and reddening in a study of Indian lateritic deposits. Rather, with increasing aridity, the hematite-to-goethite ratio generally increases, and with drying and an increase in temperature, goethite will convert to hematite (see also Trolard and Tardy 1987).

A classification of red beds was offered by Van Houten (1982a):

1. desert red beds—mostly sandstones;
2. evaporitic tidal flat and lagoonal red beds—claystone, clay possibly derived from distant, wet regions;
3. savannah red beds—red paleosols, with red and drab beds or lenses, and fluvial rocks;
4. marine red beds—red marine shales and claystones, sometimes including oolitic hematite or weathered glauconite (Parron and Nahon 1980).

With the possible exception of the fourth type, red beds would appear to be indicative of climates that are warm and dry or seasonal with respect to rainfall. The hematite in terrestrial red beds forms by postdepositional dehydration or by precipitation in alkaline groundwaters (Walker 1967, 1976). Both processes are more likely to occur in desert environments, hence the original idea that red beds form under arid conditions. In addition, iron may be supplied to desert sediments in transported dust that infiltrates the pore spaces (Walker, Waugh, and Crone 1978; McFadden 1988). In rare cases, red beds may be composed of detrital hematite, as suggested for evaporite-flat red beds, but this mechanism is apparently not so common as previously thought (Blodgett, Crabaugh, and McBride 1993).

Currently, red bed formation is thought to be favored in warm climates with alternating wet-and-dry conditions (Walker 1967; Robinson 1973; Bárdossy and Aleva 1990; Dubiel and Smoot 1994). This indication extends from desert red beds, which form under prolonged aridity with brief wetter intervals (Walker 1967), through monsoonal "savannah" red beds, which form in climates with roughly equal-length wet and dry seasons (Dubiel et al. 1991), to red Oxisols such as bauxite, which form in wet climates with short dry seasons (Bárdossy and Aleva 1990). The alternation of wet and dry conditions in all cases is thought to be necessary for the formation of hematite and other sesquioxides. Robinson (1973), Parrish (1993), and many others have pointed out that red beds are most common in the geologic record in the Permian and Triassic, when the supercontinent Pangea created a strongly monsoonal climate. The formation of hematite is favored by higher temperatures as well as by drying (Tardy and Nahon 1985; Trolard and Tardy 1987).

Blodgett, Crabaugh, and McBride (1993) and Dubiel and Smoot (1994) discussed common features in red beds that can be used to interpret the climates under which they were formed (table 5.5). These authors strongly emphasized that no single feature is diagnostic of climate, with the exception of syndepositional evaporites. Rather, all the features of the soils and associated deposits must be considered in the analysis.

It is not uncommon, particularly in the record from the Carboniferous Period, to find red beds associated with coal beds, the so-called delta-plain red beds of Turner (1980). The red beds are typically thin and laterally discontinuous, they occur in the fine-grained parts of upward-fining units, and they usually show evidence of exposure such as mudcracks, rootlet horizons, and pedogenic features, especially carbonate nodules (Turner 1980). Evidence for rapid development of the red color is the presence of both red and green intraclasts in intraformational conglomerates (Turner 1980). These types of red beds are particularly characteristic of delta-plain environments that, based on other evidence, were deposited in moist, tropical climates (Turner 1980), although the presence of carbonate nodules calls this interpretation into question.

Table 5.5.
Features of sedimentary rocks that can be useful in interpreting climate from red beds

Diagnostic soil types: calcretes, silcretes, laterite, vertisols
Soil characteristics: gilgai (pseudoanticlines), slickensides, morphology and distribution of calcareous and ferruginous nodules; mottling: gley indicates waterlogged soil, mottling associated with root traces or burrows indicates fluctuating water table; depth to calcic horizon; horizon thickness and color
Frost wedges
Type of stream sediments, that is, perennial or ephemeral
Type of alluvial fan deposits
Floodplain features that provide information about the frequency and size of flood events, for example, mudcracks, rhizoliths (root casts), invertebrate ichnofossils
Root and invertebrate traces that provide information about depth to water table: degree of verticality, bimodality of depth, that is, deep and shallow root traces; ecological tiering in invertebrate traces
Evaporites
Eolian deposits
Charcoal

Summarized from information in Dubiel and Smoot (1994).

In studying a section of Carboniferous delta-plain red beds, Besly and Fielding (1989) refuted conclusions that reddening in the section indicates climatic drying because no climate change could be inferred from the paleosols. Red sediments first appear as mottles in drab sediments, and they gradually become more prominent upsection, ending as ferruginous, "sesquioxide-rich" paleosols that probably qualify as laterites. Field relationships show that coal-bed-bearing strata and red beds are intertonguing, supporting the conclusion of Besly and Fielding (1989) that reddening occurred in parts of the depositional system where the sediments were better drained. They concluded that the red beds were not related to climate, but an argument can be made that the better-drained soils record climate more faithfully than the coal beds; indeed, poorly drained areas in general do not yield easily interpretable paleosols (Kraus 1997). If the red beds or calcareous horizons alternate with the coal beds, as they apparently do in the Sydney Basin of Nova Scotia (Carboniferous; Canada), they might be indicative of cyclicity in hydrology that is related to climate cycles (Tandon and Gibling 1994).

Oolitic hematite and chamosite

In marine rocks, oolitic ironstone (reviewed by Petránek and Van Houten 1997) is composed of chamosite or the similar mineral berthierine, well crystallized by diagenesis and not represented in modern sediments (with one possible exception; Odin et al. 1988). The minerals are commonly weathered to hematite or goethite, and terrestrial oolitic ironstones might be originally ferrous. Oolitic ironstone is common in coal-bed-bearing sequences (e.g., Boardman 1989), but it also occurs in Ordovician rocks, before land plants and peat swamps evolved. Oolitic ironstone also occurs in marine rocks, which has engendered some controversy about whether the ooids formed in situ or were transported (Odin et al. 1988). Although both autochthonous and allochthonous deposits exist (Young 1989), a model of formation near river mouths seems to be favored for the marine deposits (Odin 1988).

General agreement seems to exist that the iron is derived from weathering, hence the relationship of the marine deposits to terrestrial climate. Two hypotheses have been formulated to explain lacustrine and terrestial deposits of oolitic ironstone. The first is similar to the model for formation of the deposits in marine rocks, that is, transport of iron by rivers and precipitation of the iron-bearing minerals near the river mouths (Lemoalle and Dupont 1973). The paleoclimatic implications of this model are high humidity and temperatures upstream and aridity, leading to the formation of a saline lake, downstream. The second hypothesis is that the ooids are pedogenic and, where they occur in lacustrine rocks, were transported (Myers 1989; Siehl and Thein 1989; Young 1989). The paleoclimatic significance of this model is that the ooids are formed in regions with marked wet and dry seasons and the ooids are then reworked and concentrated.

Oxisols and Karst

Oxisols: laterite and bauxite

Laterites are iron- and aluminum-rich paleosols that indicate intense weathering. They consist of highly oxidized minerals of iron and aluminum, hence the name Oxisols (sometimes spelled Oxysols). "Laterite" is a general term for the most highly oxidized types of Oxisols and includes iron-, aluminum-, and kaolinite-rich end members (figure 5.20). Iron- and aluminum-rich end members both require periodic desiccation; poorly drained soils in constantly rainy environments will be kaolinite (Bárdossy 1982; Bárdossy and Aleva 1990). Laterites can be eroded and redeposited as detrital laterites (Bestland et al. 1996).

Numerous workers have studied the climatic conditions of laterization and bauxitization (Bárdossy and Aleva 1990). In the opinion of Bárdossy and Aleva (1990), the processes require the same climatic conditions, with bauxitization representing a more extreme progression of weathering (see also Bárdossy 1982). They summarized the conditions under which bauxitization is occurring today as

1. annual rainfall greater than 1200 mm, distributed over 9 to 11 wet months alternating with 1 to 3 dry months, and
2. mean annual temperature greater than 25°C.

Nicolas and Bildgen (1979) added a third criterion:

3. maintenance of favorable climatic conditions for an extended period, probably a minimum of 10^4 years.

Bárdossy and Aleva (1990) analyzed the distribution of modern bauxites and climate in order to address controversy about the necessity for warmth and seasonality of rainfall, superimposing maps of the 22°C isotherm, the 1200 mm isohyet, seasonality, and the distribution of bauxites (figure 5.21). Their compilation shows that seasonality of rainfall and warmth favor bauxite formation. The relatively narrow and quantitatively defined climatic requirements for the formation of bauxite caused Price, Valdes, and Sellwood (1997) to recommend its use as a test of paleoclimate models.

Some disagreement exists about the temperature requirements for laterite. High-latitude laterites are common in the Permo-Triassic (Van Houten 1982b), when climate was seasonal over large areas of the globe (e.g., Parrish, 1993), and in the Late Cretaceous-early Tertiary, when

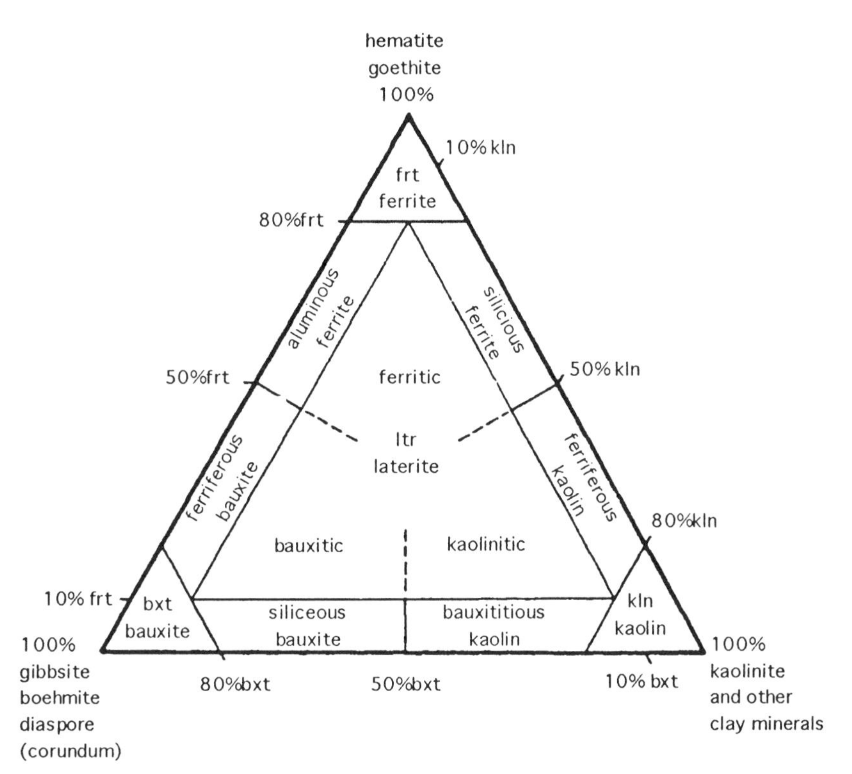

Figure 5.20. Classification of laterites. bxt, bauxite; frt, ferrite; kln, kaolinite. Reprinted from *Lateritic Bauxites*. Developments in Economic Geology 27, by Bárdossy, G. and G.J.J. Aleva (1990) with kind permission of Elsevier Science-NL, Sara Burgerhartstraat 25, 1055 KV Amsterdam, The Netherlands.

global climate was warmer. However, laterites also occur at quite high latitudes in the Miocene and Pliocene (Van Houten 1982b). These high-latitude laterites, and the Permian ones, are all on basalt (J. T. Parrish, unpublished compilation). Climates in these regions at these times are known by numerous criteria to have been cool (e.g., Taylor et al. 1992). This suggests that the climatic constraints for laterite formation are less severe on basalts, which are already richer in iron and aluminum than many other potential host rocks (this conclusion is questioned by G. Bárdossy, personal communication). However, among Bárdossy's (1982) criteria for laterization was the requirement of an annual sum of mean daily temperatures of 4000° or more. This would correspond to a mean daily mean temperature of 11°C, which is cooler than the tropical climates indicated by other workers. This minimum temperature condition for laterization might be possible in cool climates that are not strongly seasonal with respect to temperature. Laterites on basalts should be investigated more closely before such rocks are used as paleoclimatic indicators.

Karst

"Karst" is the term for a land surface whose morphology is largely controlled by dissolution, ranging from scattered sinkholes (and commonly underlain by caves) to remnant pinnacles. Karstic surfaces are commonly regarded to be indicative of warm climate because the karstification process proceeds more rapidly in regions of high meteoric carbon dioxide (but see Bischoff et al. 1994). The abundance of this gas in groundwaters is dependent on plant cover, productivity, and respiration rates, all of which are higher in warm climates, but this does not preclude formation of karst in cool climates (Choquette and James 1988), as CO_2 is more soluble in cold water, and vegetation, which adds CO_2 to groundwater, can be abundant in cool climates as well. Calcrete is not present in cold-climate karst, but calcite does precipitate onto clasts in the soil horizon (Choquette and James 1988).

Although the water table is important in the formation of karst, karst is not necessarily indicative of high rainfall or high groundwater lev-

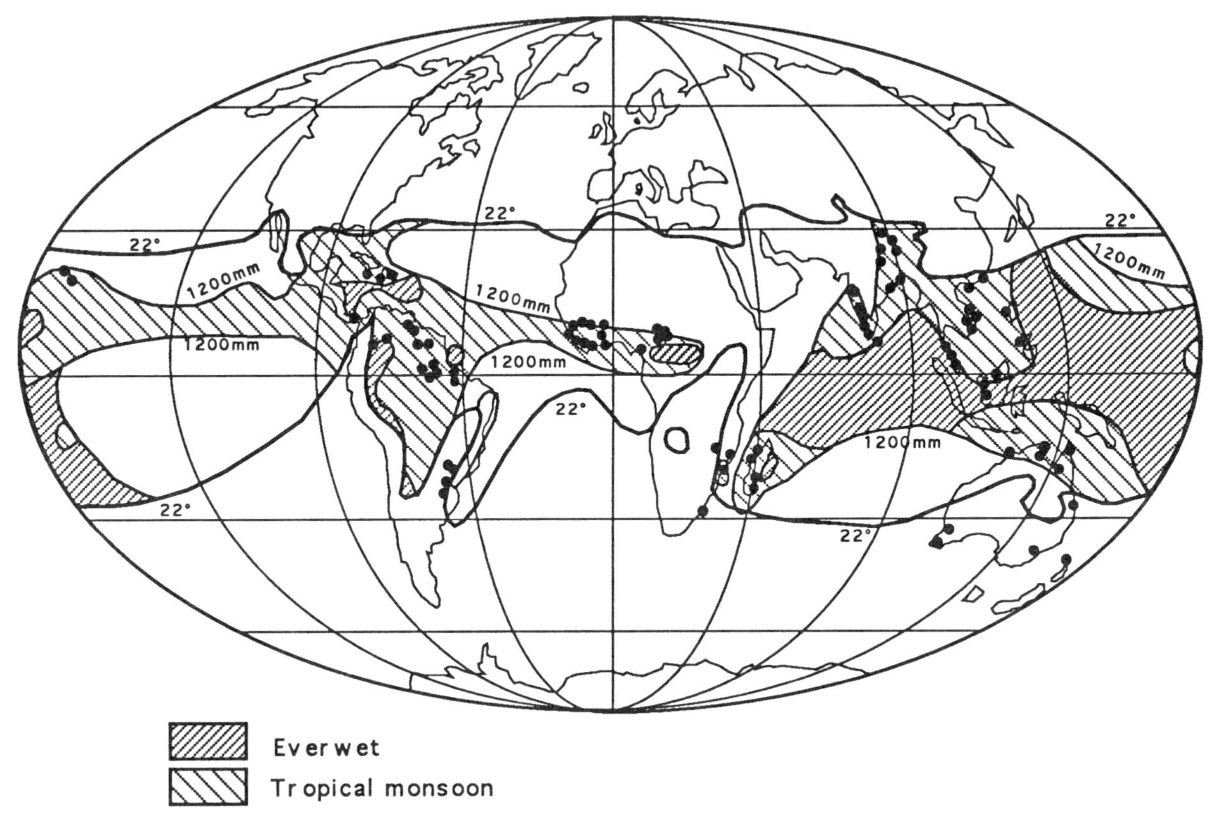

Figure 5.21. Distribution of modern bauxites (*dots*) in relation to tropical monsoonal climate and tropical everwet climate. These climates are defined by the 22°C isotherm and ≥1200 mm annual rainfall; monsoonal climates have a significant dry season (≥1 month no rain), everwet climates have no significant dry season. Modified from *Lateritic Bauxites*. Developments in Economic Geology 27, by Bárdossy, G. and G.J.J. Aleva (1990) with kind permission of Elsevier Science-NL, Sara Burgerhartstraat 25, 1055 KV Amsterdam, The Netherlands.

els (Choquette and James 1988; Wright 1988). Minor karstification may be accompanied by formation of calcrete and rhizocretions. Cementation of the carbonates in these cases is minor, with little vadose cement except needle-fiber crystals and sparse, blocky, low-iron phreatic cement. These features are indicative of karst formed in a semi-arid environment (figure 5.22). In contrast, extensive karst, accompanied by coarse, sparry, vadose and phreatic cementation, is indicative of humid climate.

Vertisols

Vertisols are excellent indicators of warm climate that is seasonal with respect to rainfall (Blodgett 1985a,b), but they will not form unless the host sediment contains a substantial amount of expandable clays (smectite and mixed-layer smectite/illite; Robinson and Wright 1987); modern Vertisols have more than 35% clay (Blodgett 1985a). Vertisols are diagnosed by pseudoanticlines and pseudoslickensides, which form during repeated expansion and contraction of the clays. Strongly seasonal, semi-arid climates will generate calcic Vertisols (Blodgett 1985b; see figure 5.19c). For example, vertisols and calcic vertisols are common in the Chinle Formation (Triassic; western United States; Dubiel 1987, 1992; Dubiel, Skipp, and Hasiotis 1992). Climate models predict a strong monsoonal influence in the region of Chinle deposition (Parrish and Curtis 1982; Kutzbach and Gallimore 1989) and evidence from eolian sandstones and from the evolution of sedimentary environments in the region provides independent evidence of this influence (Parrish and Peterson 1988).

Calcisols, Gypsisols, Natric Paleosols, Silcrete

The carbonate in Calcisols is commonly called caliche. The term "calcrete" is sometimes used interchangeably with "caliche" (Goudie 1983), but in many papers it appears to be limited to highly indurated layers; in paleosols, caliche is

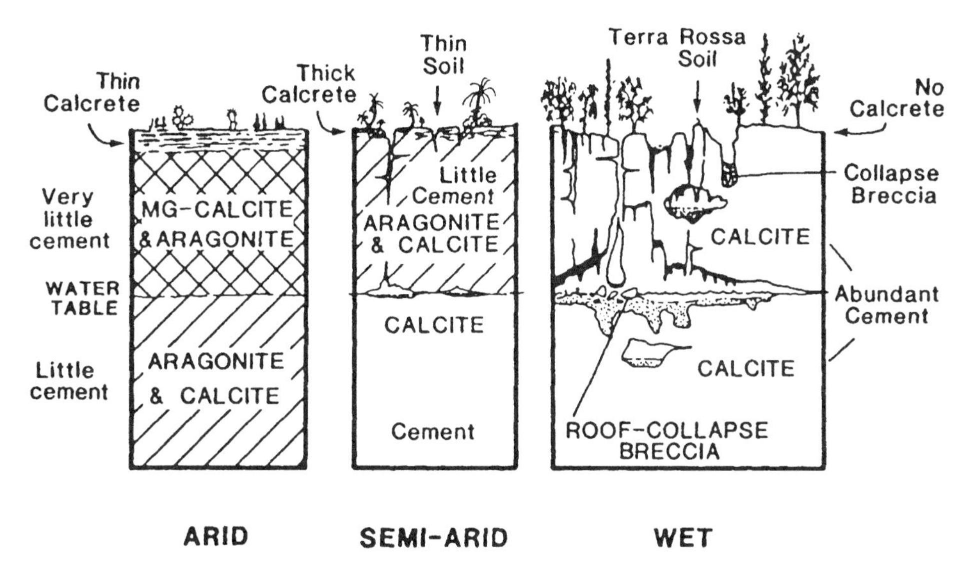

Figure 5.22. Climatic significance of different karst features. From James and Choquette (1984).

sometimes referred to as a petrocalcic horizon. Esteban and Klappa (1983) described caliche as hardpan (continuous and highly indurated) or as platy, nodular, or chalky caliche; they also provided excellent color photographs illustrating various features. Variations in caliche morphology in floodplain deposits are illustrated in figure 5.23 (Lehman 1989). Calcretes (sensu lato) are characteristic of warm climates with limited precipitation (Goudie 1983). Annual rainfall is generally 400 to 600 mm, although on a limestone substrate, calcrete can form in climates with 1000 to 1500 mm annual rainfall. The clays in calcrete horizons from warm climates are commonly more than 80% attapulgite, palygorskite, and/or sepiolite; these clays are commonly taken as paleoclimatic indicators when they occur in marine sediments. Goudie (1983) also noted that calcrete can form in cold climates, as freezing lowers the solubility of calcium carbonate (see also Cerling 1984; Dijkmans et al. 1986). Machette (1985) recommended abandoning the

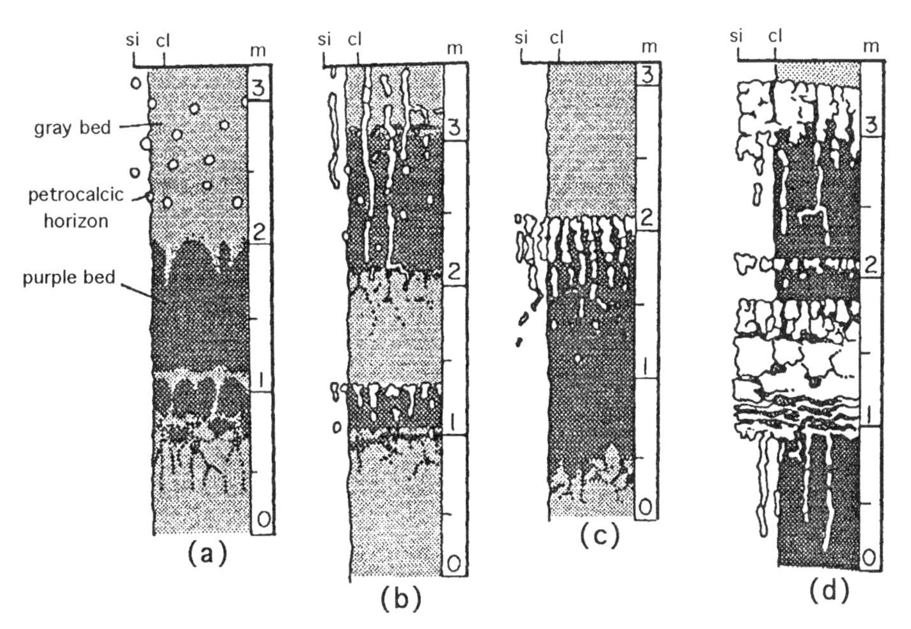

Figure 5.23. Variations in the morphology of petrocalcic horizons in floodplain deposits. From Lehman (1989) in *Geological Society of America Bulletin.* Reproduced with permission of the publisher, the Geological Society of America, Boulder, Colorado, USA. Copyright © 1989 Geological Society of America, Inc.

term "caliche" and using either "calcic soil" [Mack, James, and Monger (1993) proposed the abbreviated "Calcisol"] or, for the more indurated forms, "pedogenic calcrete." This recommendation was partly to emphasize that the soil carbonate should not simply be the result of redistribution of carbonate already in the system.

A significant problem with the interpretation of pedogenic carbonate and paleoclimate is the necessity of distinguishing pedogenic carbonate from that formed in lakes, along the water table, as a product of later diagenesis, or by other nonpedogenic processes. Paleoclimatologists thus must take care to look for features that might suggest other than a surficial origin. Nonsurficial calcretes are commonly penetrative, anastomosing, and discontinuous and may be calcified root mats (Wright, Platt, and Wimbledon 1988; Rossinsky, Wanless, and Swart 1992). In comparson with surficial calcretes, $\delta^{18}O$ may be slightly but significantly lower in nonsurficial calcretes, probably because evaporation rates are greater at the soil surface (Wright and Vanstone 1991; Rossinsky, Wanless, and Swart 1992). Wright, Platt, and Wimbledon (1988) noted that surficial carbonates can be formed by lichens. These deposits have fenestral and biogenic fabrics that might make them almost indistinguishable from stromatolites or tufa. Even soil carbonate nodules could be suspect; the presence of soil carbonate nodules in coeval fluvial channels is one indication that they were pedogenic and not later diagenetic (Dubiel 1992; Dubiel, Skipp, and Hasiotis 1992). Smith (1994) described calcic paleosols from the St. David Formation (Pliocene-Pleistocene; Arizona, U.S.) that were intermediate between true calcic paleosols and nonpedogenic carbonates. He termed these hydromorphic paleosols. The calcic horizon has a sharply defined base, similar to groundwater carbonates, and upward-decreasing carbonate content, but they are also rooted and associated with mudstone-bearing pedogenic fabrics. He interpreted these as indicating a paleo-water table close to the surface.

Paleosols indicating evaporative climates are gypsisols, which contain gypsum, and natric paleosols, which show evidence of having been influenced by sodic salts (probably halite). Gypsisols are relatively rare in the geologic record (Mack, James, and Monger 1993), possibly because the salts are so subject to dissolution and diagenesis (Retallack 1990). Gypsisols commonly contain palygorskite, sepiolite, and mirabilite (Retallack 1990). The source of gypsum or solutes that can form gypsum is salt-rich dust, for example, from desiccated lakes or playas (Wells, McFadden, and Dohrenwend 1987). Natric paleosols have strongly developed columnar, polygonal peds with domed tops that are characteristic of modern natric soils (McCahon and Miller 1997). In Permian rocks of Kansas (Council Grove Group), paleosols with these features are associated with desiccation polygons, teepee and boxwork structures, and beds and crystal molds of gypsum; calcic paleosols are also common in the section (McCahon and Miller 1997).

Silcrete is much less common than calcrete, and no typical silcrete profile exists (Summerfield 1983a). Assigning an age to silcretes exposed today is difficult, but Summerfield (1983a) has concluded that most are relict, that is, not forming under present climatic conditions. Summerfield (1983a,b) identified two types of silcrete in Africa, in the Cape coastal zone and in the Kalahari desert. The Cape coastal zone silcretes are characterized by deep weathering profiles and are high in TiO_2, perhaps brought about by high organic activity (Summerfield 1983a); high TiO_2 concentrations are indicative of intense weathering (Force 1991). Calcrete is associated with silcrete in the Cape coastal zone, either overlying the silcrete or within it, in which case the silcrete is brecciated. Some silcretes have associated authigenic glaebules and colloform features. Summerfield (1983a,b) interpreted these to have formed in a humid climate. In contrast, the Kalahari-type silcretes are not associated with weathering profiles and have low TiO_2. Some are probably silicified calcretes, and these contain abundant chalcedonic vug fillings. They are also commonly associated with eolian sands, calcrete, and playa sediments. Summerfield (1983a,b) concluded that eolian input was the most important source of silica for these silcretes, and that they formed in an arid or semi-arid climate.

Stable Isotopes in Paleosols

Stable isotopes of oxygen and carbon in soil carbonates are beginning to prove useful for interpreting paleoclimate, paleoatmospheric CO_2, and fossil plant ecology. However, the controls on $\delta^{13}C$ and $\delta^{18}O$ contents in soil carbonates are complicated. The principal controls on $\delta^{13}C$ in soils that are free of detrital carbonate are the proportion of atmospheric CO_2 in the soil CO_2, which is usually negligible, the photosynthetic metabolism of the plants, and soil respiration rates (Cerling 1984). The principal controls on $\delta^{18}O$ are temperature and $\delta^{18}O$ of the meteoric and groundwaters, which are in turn affected by evaporation and the $\delta^{18}O$ of the precipitation and runoff. Most work on stable isotopes has been on soil carbonates and most of the attention has been concentrated on carbon isotopes, with oxygen isotopes interpreted in only a very general way. Some work, however, has been done on oxygen and hydrogen isotopes in other minerals.

Stable isotopes in paleosol carbonates

The metabolism of the plants that dominate the vegetation has become a major focus in paleoclimate studies using paleosol carbonates because it is so closely tied to climate and paleoatmospheric pCO_2. Modern plants have three types of photosynthetic metabolism, called C_3, C_4, and CAM; the first two are by far the most important. Most plants have C_3 photosynthesis. C_4 and CAM plants are adapted for hot and/or dry climates. Modern C_4 plants are dominant in tropical savannahs and temperate grasslands and important in dry steppes, semidesert scrublands, and tropical and temperate deserts (Cerling and Quade 1993). C_4 photosynthesis is also favored under conditions of low atmospheric pCO_2. C_4 plants have characteristic morphologic features, called Kranz anatomy, that are sometimes preserved in the fossil record (e.g., Cerling, Wright, and Vanstone 1992). The earliest evidence of these features in fossil plants is Miocene in age (Thomasson, Nelson, and Zakrzewski 1986; Ehleringer et al. 1991). CAM plants form a relatively minor constituent of modern vegetation and are not considered important.

The carbon isotopic composition of the respired CO_2 from C_3 plants is about −27‰ to −26‰ $\delta^{13}C_{PDB}$, although in regions of water stress, the $\delta^{13}C$ of C_3 plants can be greater (Cerling, Wang, and Quade 1993). Respired CO_2 from C_4 plants averages −13‰ $\delta^{13}C_{PDB}$. In contrast, atmospheric CO_2 is +6‰. In almost all soils, the isotopic difference between the plant organic matter and soil carbonate is 14‰ to 15‰, so that a soil carbonate forming in a region with a pure C_3 vegetation would have a carbon isotopic composition of about −13‰ and in a region with pure C_4 vegetation, about +2‰ (Cerling 1992). Because the difference in these values is large, it is possible to use mixing lines to estimate the proportions of C_3 and C_4 plants in the vegetation. These proportions can then be used to interpret paleoclimate.

Apart from the complexity of soil-carbonate isotopic composition, the principal problems for interpreting paleosol carbonates are overprinting and diagenesis (Cerling 1984). Carbonates from areas with relatively high sedimentation rates are more likely to preserve a pristine signal because they are more rapidly removed from the zone of pedogenesis, where they might be affected by later climatic change. Cerling (1984) suggested that the presence of micrite might be an indicator that diagenesis did not occur. Later studies showed that $\delta^{13}C$ is relatively unaffected by diagenesis but that $\delta^{18}O$ is strongly affected (Cerling 1991).

Several studies of the isotopic composition of paleosol carbonates have been conducted by Cerling and his colleagues with the purpose of testing the idea that the carbonates can be used for paleoclimate studies (Cerling and Hay 1986; Cerling, Bowman, and O'Neil 1988; Cerling et al. 1989; Quade, Cerling, and Bowman 1989; Cerling, Wang, and Quade 1993; Smith et al. 1993; Quade, Solounias, and Cerling 1994; Quade and Cerling 1995). A dramatic shift in $\delta^{13}C$ occurs in the Siwalik Sequence in Pakistan (Miocene) and in Nepal, at about 7 Ma (Quade, Cerling, and Bowman 1989; Cerling, Wang, and Quade 1993; Quade et al. 1995; Quade and Cerling 1995; figure 5.24). The trend in $\delta^{13}C$ suggests a shift from predominately C_3 to C_4 vegeta-

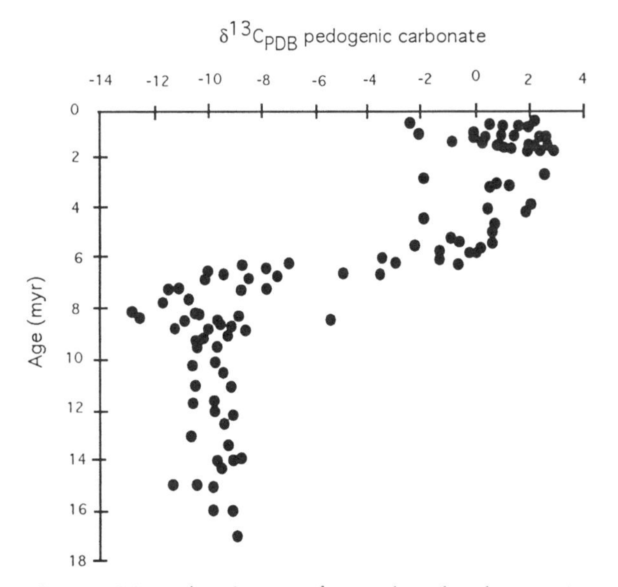

Figure 5.24. Carbon isotopes from paleosol carbonates in the Siwalik Sequence (Miocene, Pakistan). The shift at about 7.4 Ma indicates a change from predominately C_3 to C_4 grasslands. From Quade, Cerling, and Bowman (1989). Reprinted with permission from *Nature* 342: 163–6. Copyright 1989 Macmillan Magazines Limited.

tion, which would be consistent with onset of the Asian monsoon. Cerling, Wang, and Quade (1993) showed that the shift in Pakistan has a counterpart in North America, suggesting that the indicated change was global. They favored a decrease in CO_2, which would favor C_4 plants, as the cause for this global change, possibly driven by the Himalayan uplift (Quade, Cerling, and Bowman 1989; Raymo and Ruddiman 1992; Quade et al. 1995). The shift on land correlates with shifts in marine sediments and could be explained by either an increase in the proportion of C_4 plants worldwide or a decrease in total terrestrial biomass. A striking $\delta^{13}C$ shift is not evident in soil carbonates or tooth enamel from South America (MacFadden et al. 1994); rather, the shift was more gradual. Whatever the cause of the rapid change to C_4 grasses, it was evidently limited to the Northern Hemisphere.

If one assumes that C_4 plants did not evolve before the Miocene, the isotopic composition of soil carbonates can also be used to estimate paleo-pCO_2 of the atmosphere, because a higher concentration of atmospheric CO_2 would lead to a higher proportion of atmospheric CO_2 in soil gasses and thus higher isotope values (Cerling 1991, 1992; Cerling, Wright, and Vanstone 1992). The assumption of no C_4 plants is important because C_4 plants respire CO_2 that is closer to the isotopic composition of atmospheric CO_2, blurring the distinction between soil and atmospheric CO_2, and because C_4 plants tend to live in warmer environments, meaning that respiration rates are higher and atmospheric input to soil CO_2 correspondingly lower. The assumption is particularly important because desert soils, if occupied by C_3 vegetation, would be more accurate for estimating pCO_2 when atmospheric pCO_2 was relatively low (Cerling 1991; figure 5.25). In contrast, nondesert soil carbonates would be more accurate for higher values of pCO_2. Gleyed soil carbonates are less useful because the range of $\delta^{13}C$ is small over a large range of atmospheric pCO_2.

Two other assumptions must also be made in order to estimate pCO_2 from soil carbonates, that $\delta^{13}C$ of the atmosphere has remained constant at about +6‰ and that $\delta^{13}C$ of C_3 plants has remained constant (Cerling 1992). The latter is known to not always be true, because under water stress and warm temperatures, C_3 plants will have $\delta^{13}C$ of −24‰, as opposed to the normal $\delta^{13}C$ of −26‰ (Cerling 1992). In addition, application of the method to warm-climate soils

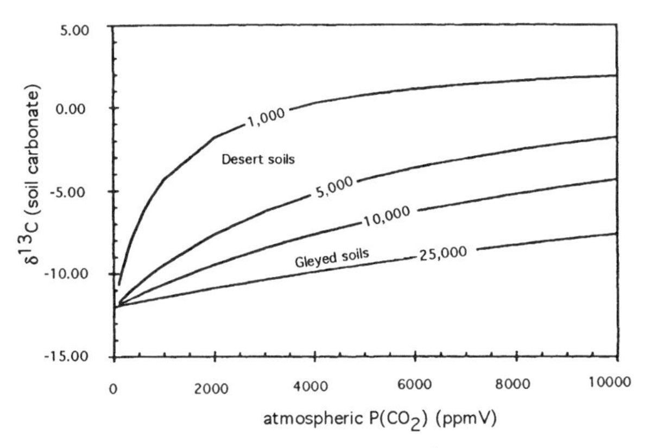

Figure 5.25. Relationship between $\delta^{13}C$ of soil carbonate and pCO_2 for different types of soils. Note that the curve is steeper for desert soils at low atmospheric pCO_2 and steeper for nondesert soils at higher atmospheric pCO_2. The isotopic composition is a function of atmospheric pCO_2 and soil-derived CO_2 (isopleths in ppmV). From Cerling (1991) in *American Journal of Science* 291:377–400.

is hard to verify because today all warm soils are occupied by at least some C_4 plants.

Using the assumption of a lack of C_4 plants prior to the Miocene, Cerling (1991) estimated paleo-pCO_2 for several intervals. The estimates are consistent with values predicted independently by Berner (1991) using a geochemical model that is based on geological and biological controls on CO_2 levels. Using the same techniques, Mora et al. (1991) estimated Late Paleozoic pCO_2 at values that also were consistent with Berner's (1991) prediction. If the C_3 plants were under water stress and their isotopic composition greater, as discussed by Cerling (1992), the pCO_2 estimates would drop from a range of 2000 to 3000 ppmV to a range of 1500 to 2500 ppmV (Cerling 1992).

The assumption of no C_4 vegetation prior to the Miocene has been questioned by Wright and Vanstone (1991). They studied calcified root mats, which formed well below the land surface and should record primarily root-respired CO_2. In their results and those of other workers, the $\delta^{13}C$ values are high, about −1‰ to −4‰, much higher than would be expected if the vegetation were dominated by C_3 plants. Such high values would be interpreted to indicate very high pCO_2, as reported, for example, by Mora et al. (1991). However, if Wright and Vanstone's (1991) assumption that the root mats were mostly isolated from the effects of atmospheric CO_2 is correct, the high values would suggest the presence of C_4 (or CAM) plants in the vegetation. To support this conclusion, Wright and Vanstone (1991) noted that some modern representatives of the taxa that were likely to have occupied the Carboniferous landscape—lycophytes, ferns, and gymnosperms—do use CAM photosynthesis.

The necessity of making an assumption about the proportion of C_4 plants is alleviated if organic matter is present in the paleosols, because the organic matter is a direct record of the $\delta^{13}C$ of the flora, and carbon isotope variations from that record are more likely to be related to atmospheric pCO_2 (Cerling 1992).

In general, there is a good relationship between $\delta^{18}O$ of soil water and meteoric water and between that of meteoric water and soil carbonate (Cerling and Quade 1993). However, the $\delta^{18}O$ value is subject to modification by evaporation, infiltration of meteoric waters of a different isotopic composition, seasonal variation, and diagenesis. A general lack of knowledge of the relationship between oxygen isotopic composition of soil waters and the formation of soil carbonates at present hinders their use for paleotemperature determinations (Cerling and Quade 1993). Nevertheless, Quade et al. (1995) argued that preliminary paleotemperature determinations are possible, if one assumes that the lowest $\delta^{18}O$ values represent soil carbonates formed in waters least affected by evaporation, and that the isotopic composition of soil waters today is representative of that of the past.

Stable isotopes in other paleosol minerals

Not all paleosols contain carbonates, so stable isotopes of other minerals might be useful if they can be shown to be related to paleoclimate. Goethite (FeOOH), boehmite (AlOOH), gibbsite ($Al(OH)_3$), and paleosol phosphate have potential for such analysis. The iron and aluminum minerals have the added advantage of containing both oxygen and hydrogen, which allows further constraint of the composition of meteoric waters. In addition, the minerals tend to occur together, allowing development of more accurate estimates of temperature through mutual solution of the fractionation equations (Yapp 1993). δD (fractionation of deuterium and hydrogen) and $\delta^{18}O$ have also been used together on authigenic silica interpreted to have formed as a weathering product on the Proterozoic land surface (e.g., Kenny and Knauth 1992; see also Knauth and Epstein 1976).

Yapp and Poths (1992) estimated atmospheric pCO_2 from the goethite in the Neda Formation, which contains a small carbonate component, as $Fe(CO_3)OH$, using a method similar to that used in soil carbonates. They calculated pCO_2 by generating profiles of the concentration of the carbonate component and $\delta^{13}C$ and applying a mixing equation of soil CO_2 and atmospheric CO_2. The estimated values were as high as 16 times the modern value, consistent with Berner's (1991) models.

COAL BEDS

Coal is rock containing >70% by volume of carbonaceous material (Bates and Jackson 1984). The precursor to coal is peat, defined as "an unconsolidated deposit of semicarbonized plant remains . . . carbon content is about 60%" (Bates and Jackson 1984). Peat becomes coal through diagenesis under burial temperatures and pressures. To economic geologists, the term "coal" is limited to those deposits that are economically valuable, that is, that have a very low content of inorganic material (ash, <10%). In practice, however, many geologists report coals without determining the diagenetic maturity or the ash content, so the term "coal" has a broader meaning than the economic one. Here, I use "coal bed" to indicate this less formal usage, but the reader should be aware of the distinction even for the following discussion.

Peat deposits are essentially organic-rich soils, included with Histosols, and coal beds fall into Mack, James, and Monger's (1993) paleosol order of the same name (see table 5.3). However, coal beds loom large in the paleoclimate literature, probably because they are so important economically. Because of this, they represent a rather special case of paleosols treated separately here.

The impetus behind the vast knowledge of coal and other terrestrial carbonaceous deposits has come from economics, and many of the approaches to interpreting paleoclimate from coal have been developed by economic geologists. The terminology they have developed is complex and peculiar to coal geology. A review of the terminology is necessary for the discussion of the climatic significance of coal beds.

Coal geologic and petrographic terms are summarized in tables 5.6 through 5.8 and come from a variety of sources; several classification systems exist. Terms describing the environment of peat formation (table 5.6) commonly are not just descriptive but also carry implications for the genesis

Table 5.6.
Terminology for describing peat-forming environments

Peat that is built up above the general ground surface and for which rain is the principal water supply: ombrotrophic mire, raised bog, blanket bog, bog forest[ab] (Moore 1987); raised mire; domed peat (Neuzil et al. 1993); high moor (Teichmüller 1989); terrestric (Moore 1989)
Peat that is approximately level with the general ground surface and for which ground water is the principal water supply: rheotrophic mire, fen (general term for rheotrophic mires *or* for a mire that is seasonally dry at surface), carr (temperate swamp), swamp, swamp forest[a] (Moore 1987); low-lying mire; planar peat (Neuzil et al. 1993); topogenic mire, low moor (Teichmüller 1989); telmatic (Moore 1989); swamp (dominated by trees), marsh (dominated by herbaceous plants), minerotrophic peatland (generic term) (Cameron, Esterle, and Palmer 1989)
Peat that is underlain by water: floating mire; floating bog (if built up above water surface), floating swamp (if level with water surface; Moore 1989); not recognizable in geologic record (see text)
Accumulation of terrestrial organic matter with a high content of inorganic matter, enough that coal derived from the peat would be uneconomic: marsh, saltmarsh, wetlands, with high inorganic content (Moore 1987)

The terms listed in each category are not necessarily exactly equivalent (e.g., see Moore 1989).
[a]The term "forest" is used only when trees are a significant portion of the vegetation.
[b]Cameron, Esterle, and Palmer (1989) did not limit "bog" to peat built up above the water table; rather, they considered "bog" to be any wet, soft land underlain by peat; they equated the term with "mire."

of the peats or processes in the ecology or hydrology of the peat-forming environments. Most of the terms are formulated around modern peat occurrences, so that extension of the terms to coal beds adds an extra layer of interpretation that is always genetic. Petrographic terms can be either descriptive, genetic, or both. The megascopic appearance of coal is described with terms collectively referred to as lithotypes (table 5.7). The four main divisions of lithotypes are vitrain, clarain, durain, and fusain (Stach et al. 1982). These are composed of single microlithotypes or mixtures of microlithotypes, which in turn are composed of single types of macerals or mixtures of macerals. Vitrain is bright coal, clarain is semibright or composed of very thin (millimeter-

Table 5.7.
Coal lithotypes, their constituent microlithotypes, and their maceral compositions

Lithotype	Microlithotype	Maceral Groups
Vitrain		
Bright coal	Vitrite	Vitrinite
Semi-bright or dull coal (dull if much liptinite)	Clarite	Vitrinite, liptinite
Durain		
Dull coal	Durite	Inertinite, liptinite
Mostly dull coal	Trimacerite	Vitrinite, liptinite, inertinite
Dull coal	Inertite	Inertinite
Dull coal	Liptite	Liptinite
Clarain		
Bright, dull coal (dull if much inertinite)	Vitrinertite	Vitrinite, inertinite
Fusain		
Dull coal (see text)	Fusite, semifusite (these are subgroups of inertite)	Fusinite, semifusinite (inertinite)

From information in Stach et al. (1982) and Teichmüller (1989).

scale) bands of different microlithotypes, durain is dull coal, and fusain is soft, fibrous coal with a silky luster similar to charcoalified wood (Stach et al. 1982). The three main divisions of macerals are huminite/vitrinite, liptinite (also called exinite), and intertinite, and each division is divided into many subdivisions (see tables 5.7, 5.8).

Two indices that quantify relationships among maceral types have been used to define coal-bed facies and segregate depositional settings: the Gelification Index (GI) and the Tissue Preservation Index (TPI; Diessel 1992). These are defined as follows:

$$GI = \frac{tv + ti}{dv + gv + gi + di} \quad (5.1)$$

$$TPI = \frac{v + gi}{ti + di} \quad (5.2)$$

where v is total vitrinite, tv is telovitrinite (telinite and telocollinite), dv is detrovitrinite (desmo-

Table 5.8.
Important subdivisions of the major coal maceral classes and their significance

Maceral class	Subdivision	Interpretation
Vitrinite	Desmocollinite	Clear groundmass showing cell walls on etching; indicates high degree of carbonization, more woody vegetation
	Telinite	Cell walls
	Telocollinite	Structureless tissue showing cell walls on etching; indicates lower degree of carbonization, more woody vegetation
Liptinite	Liptodetrinite	Particulate liptinite; indicative of the type of vegetation
	Cutinite	Derived from waxy leaf cuticle; indicative of the type of vegetation
Inertinite	Fusinite	Charcoal and charcoal-like particles; indicating fire and/or higher degree of carbonization
	Semifusinite	Intermediate degree of carbonizaiton of cell walls; indicating higher degree of carbonization than vitrinite, lower than fusinite
	Macrinite	Inertinite groundmass of oxidized desmo- or telocollinite; indicating higher degree of carbonization
	Inertodetrinite	Fragmented inertinite
	Micrinite	Granular groundmass

Composite of information from McCabe (1984), Stanton et al. (1989), and Teichmüller (1989).

collinite), *gv* is gelovitrinite (derived from colloids), *ti* is teloinertinite (cell tissues that have not been gelified), *di* is detroinertinite (inertodetrinite, micrinite), and *gi* is geloinertinite (macrinite). GI is, in effect, a ratio between vitrinite and inertinite and, according to Diessel (1992), a measure of the continuity of moisture availability, with high GI indicating more continuous wet conditions and low GI indicating intermittent drying (through evaporation or freezing). TPI is a measure of the degree of humification, that is, tissue destruction; TPI is also affected by the proportion of wood in the plant material that made up the peat because herbaceous plants are degraded much more readily. Low values of TPI indicate greater tissue destruction.

Related indices are the Vegetation Index (VI) and the Groundwater Influence Index (GWI; Calder 1993). VI incorporates more explicitly than TPI information about the proportion of woody versus herbaceous vegetation in the peat. The ratio comprises, in the numerator, macerals thought to be derived from lignin-rich forest vegetation (telinite, telocollinite, fusinite, semifusinite, suberinite, resinite) and, in the denominator, macerals thought to be derived from herbaceous, cellulose-rich (desmocollinite, inertodetrinite) and marginal aquatic (alginite, sporinite, cutinite, liptodetrinite) vegetation. GWI is similar in purpose to GI but incorporates information on the amount of mineral matter as an indication of mineral influx. The ratio is gelocollinite plus corpocollinite plus mineral matter to telinite plus telocollinite.

DiMichele and Phillips (1994) were critical of conclusions based on GI and other techniques incorporating ratios of vitrinite to inertinite. The basis for their criticism was that GI is strongly dependent on the original tissue type and that fire, which contributes to the inertinite component, can indicate short-term climatic change and not just mire evolution. They further noted that charcoal (fusinite) cannot be gelified and is derived from a process completely independent of the peat diagenetic processes that produce other types of inertinite. Finally, different types of plants differ in combustibility, and all of these factors make interpretation of the ratios very difficult. Nevertheless, GI/TPI plots are commonly used in coal-bed interpretation. Other types of maceral plots and indices were discussed by Diessel (1992).

Peat Formation

Although some coal beds are formed from allochthonous material, for example, logjams in streams, most coal beds are autochthonous, that is, derived from peat that was formed in mires from vegetation that was adapted to waterlogged soil; the former are not paleoclimatically significant and will not be discussed further.

The formation of peat usually takes place

under anaerobic, acidic conditions (Teichmüller 1989), which are created by groundwater levels that are at or above the ground surface and by acids derived from plant material. How much peat forms depends on (1) exclusion of clastic sediments, (2) the interplay between changes in the groundwater level and the accumulation of organic matter, which directly affects (3) the balance between plant productivity or organic input and decay (Moore 1987), and (4) the supply of nutrients (Cecil et al. 1985). Climate plays an important role in both the hydrologic budget and in the productivity and decay of organic matter. In relatively well-drained areas or dry climates, peat can form only where mean annual temperatures are low and decay is therefore inhibited. In contrast, peat can form in wet climates at any mean annual temperature (Schopf 1973). However, the balance between plant productivity and decay means that the highest accumulation rates are in warm, not hot, climates. Peat formation is also favored where mean annual precipitation is relatively high in both temperate and tropical climates. In tropical peats, even though production drops off above about 3.2 m mean annual rainfall, loss also drops off, so that in both tropical and temperate environments, peat formation is increasingly favored as mean annual precipitation increases.

It is commonly stated that peat forms where precipitation exceeds evaporation for the year (e.g., Parrish and Barron 1986). To address this assertion, Gyllenhaal (1991) calculated the Thornthwaite index for low-latitude regions in which peat is forming today. Figure 5.26 shows Gyllenhaal's (1991) results. Most peats, particularly those close to the equator, occur above the Thornthwaite envelope, reflecting formation in regions of high precipitation. However, many peats are forming in regions where the potential evapotranspiration is roughly equal to the precipitation (within the envelope on figure 5.26). In general, peats do form where precipitation is equal to, or exceeds, evaporation, but high rainfall is not a good predictor of peat distribution today. Instead, peat is most closely related to what Ziegler et al. (1987) termed precipitation continuity, that is, the number of months with rainfall >20 mm (Ziegler et al. 1987) or 40 mm (Gyllenhaal 1991; figure 5.27).

Coal beds originate in two types of mires, planar mires and raised mires (McCabe 1984, 1991a; table 5.9). Planar mires occupy topographic low spots in a landscape, and the surface of the mire is approximately level with the surface of the groundwater table. Planar mires are thus fed by groundwater (topogenous; see table 5.6). Vegetation is diverse, although it may be

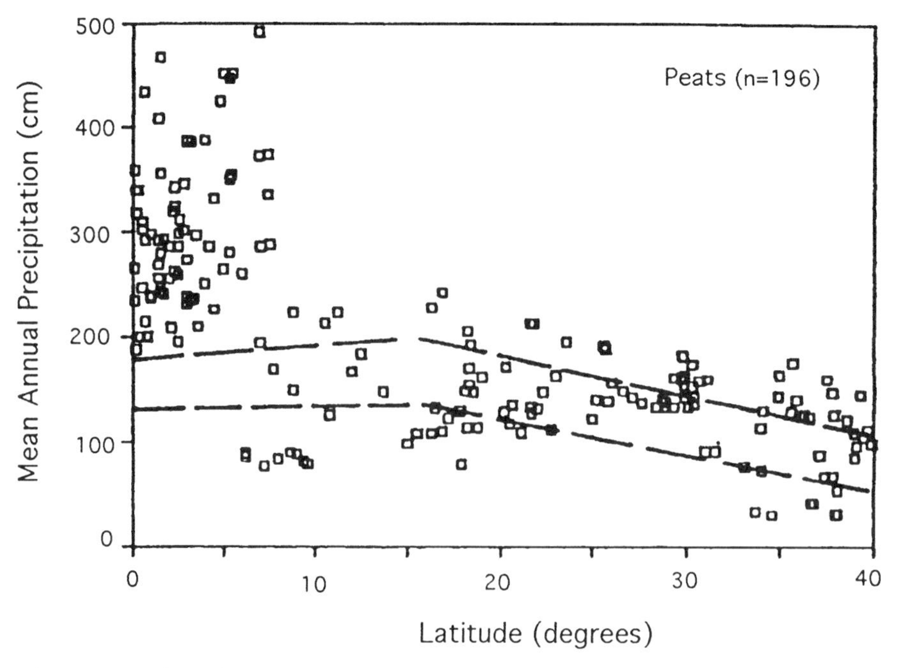

Figure 5.26. Thornthwaite trend and mean annual precipitation in localities (N = 196) in which peat is forming today. From Gyllenhaal (1991).

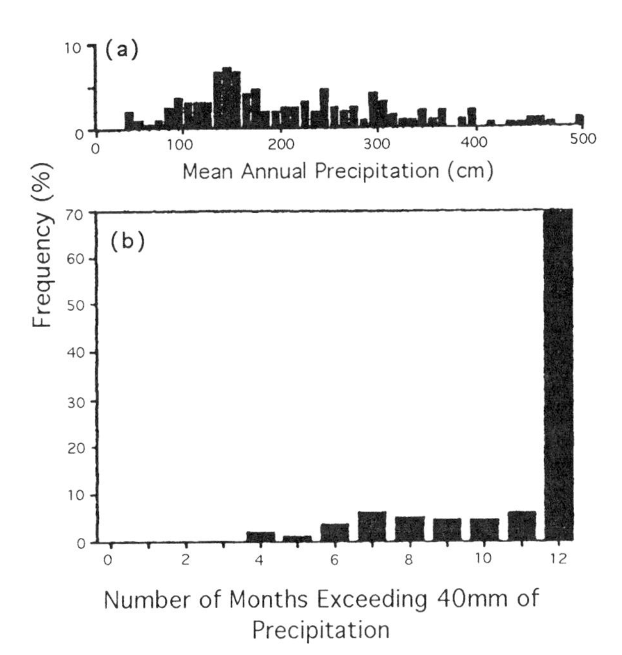

Figure 5.27. Histogram of modern peat occurrences (N = 196) plotted by mean annual precipitation *(a)* and the number of months in which precipitation is >40 mm *(b)*. From Gyllenhaal (1991).

less diverse than the vegetation in surrounding clastic communities. Coal beds formed from planar mires may have some clastics, usually clay-sized material, although the more clastics the mire contains, the less likely it is to become coal rather than just carbonaceous shale. Raised mires are composed of peat that has built as much as 14 m above the local, general water table (McCabe 1984). The water table level within a raised mire is above that of the surrounding area because rain is retained within the peat (e.g., figure 5.28). Vegetation at the top of the mire is depauperate and may be stunted because the nutrients are supplied only in rainwater (Cameron, Esterle, and Palmer 1989).

Eustatic sea-level changes and tectonics are thought to be important controls on peat formation on continental margins. A number of models have been put forth linking peat formation and sea-level change, particularly sea-level rise (Ryer 1984; Lawrence 1992; McCabe and Shanley 1992; Roberts and McCabe 1992; Heckel 1995; and references therein). However, a number of

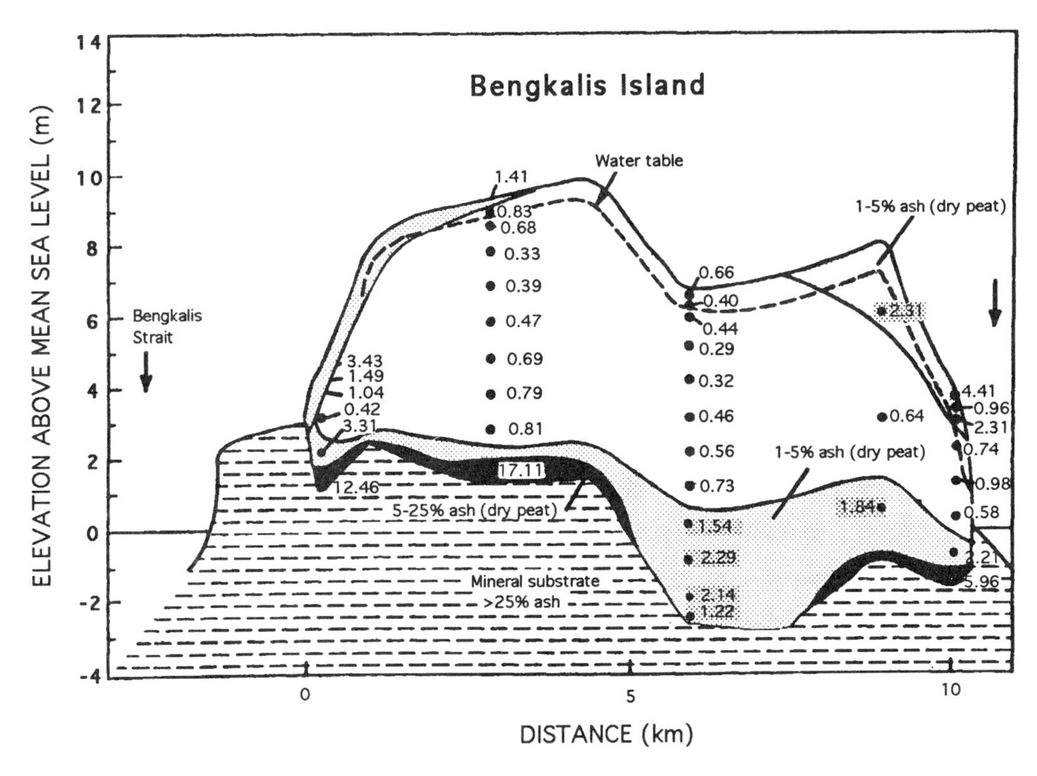

Figure 5.28. Distribution of mineral matter (ash, *dots* and associated numbers) in a typical raised mire (Bengkalis Island, Indonesia), showing level of water table. From Neuzil et al. (1993) in *Modern and Ancient Coal-Forming Environments,* J.C. Cobb and C.B. Cecil, eds. Reproduced with permission of the publisher, the Geological Society of America, Boulder, Colorado, USA. Copyright © 1993 Geological Society of America, Inc.

Table 5.9.
Characteristics of raised mires and low-lying mires as end members of commercial-quality coals

Characteristic	Raised mires	Low-lying mires
Climate	Everwet tropical	Seasonal tropical
Water source	Ombrogenous	Topogenous
Nutrient content	Oligotrophic	Mesotropic*
Surface morphology	Domed	Planar
pH	<4	4 to ~7
Eh	?	?
Floral communities	Low diversity; zoned; xeromorphic	High diversity; random; luxuriant
Microbial activity	Low (cellulose preserved)	High (cellulose degraded)
Mechanism of degradation	Primarily chemical	Primarily microbial
Ash content	Low, uniform	High, variable
Sulfur content	Low, uniform	High, variable
Nitrogen content	Low, uniform	High, variable
Cation exhange capacity	High	Low
Specific conductivity	Low	High
Base saturation	Low	High
$[Ca^{2+}]$	Low	High
Fiber content	Fibric	Hemic to sapric
Biogenic sulfide	Low	High
Biogenic methane	Low	High

Reprinted from *International Journal of Coal Geology* 5, Cecil, C. B., et al., "Paleoclimate controls on Late Paleozoic sedimentation and peat formation in the central Appalachian basin (U.S.A.)," pp. 195–230 (1985) with kind permission of Elsevier Science-NL, Sara Burgerhartstraat 25, 1055 KV Amsterdam, The Netherlands.

*Revision according to C. B. Cecil (personal communication).

authors have questioned the importance of sea level in the formation of peat (e.g., Cecil and Englund 1989), especially the importance of sea-level rises. For example, Nemec (1992) pointed out that rising sea level causes fluvial systems to become aggradational, leading to channel instability and avulsion, which is detrimental to the establishment of mires. Winston and Stanton (1989) noted that major transgressions will drown the mire environments.

Many workers (e.g., Weaver and Flores 1989; Smyth 1989; Krassilov 1992) regard tectonics as the single most important controlling factor in the formation of coal beds. Like thick evaporites, thick peats require an actively subsiding basin in order to form. However, evaporites can keep forming if the rate of subsidence exceeds the rate of sedimentation, even if the basin is filled with water, whereas a mire will be drowned if the rate of subsidence exceeds the rate of peat formation. Therefore a relatively delicate balance exists between peat accumulation rates, which vary according to climate, and basin subsidence rates (McCabe 1991b). Although coal beds are found in many types of basins, foreland basins are the best settings for peat preservation (figure 5.29). Strike-slip basins tend to subside too fast and are unlikely to accumulate coal beds in higher latitudes unless temperatures and therefore plant productivity are high. Strike-slip basins in Washington State contain coal beds of Eocene age (Johnson 1985), but the Eocene is regarded to be a time of unusual warmth, and productivity would have been high. In contrast, peats could accumulate in any climate on passive margins, but they are less likely to be thick because of the low subsidence rates (McCabe 1991b; McCabe and Parrish 1992). If peat accumulation is limited by subsidence, peat forming on such margins is also more likely to be destroyed before burial because of the long exposure time at the surface.

Coal Beds as Paleoclimatic Indicators

Much of the vast literature devoted to the interpretation of coal beds is aimed at the local or

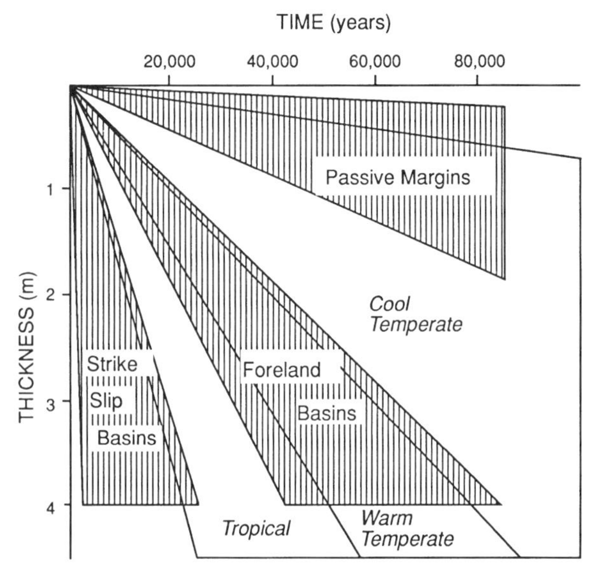

Figure 5.29. Relationship between basin subsidence rates (*shaded areas*) and peat accumulation rates (*unshaded areas*). From McCabe (1991b).

basinal environment of formation of peat and not explicitly at climate. Much of the literature refers to "wet" and "dry" environments of formation, but these terms almost always apply to the position of the water table rather than to climate, and although the two may be related, caution must be exercised in extracting paleoclimate information.

Two types of information from coal beds have been explicitly linked to climate in a way that permits use of individual coal beds or sequences of coal beds as paleoclimatic indicators, independent of the other information that might be available from the rocks of the basins in which they occur. These are certain aspects of the chemistry of the coal beds and the identification of raised mires—neither method is without controversy. At a global scale, certain observations about the distribution of peat and coal beds in space and time permit use of coal beds as a guide to continent- or global-scale paleoclimatic patterns. At a basinal or regional scale, however, in the absence of information about coal-bed chemistry or mire type, the controls on the distribution of coal beds in the geologic record are so complex that, in general, coal beds have been used as paleoclimatic indicators only in combination with information about sedimentation style, presence or absence of other chemical sediments, paleosols, and biota.

The following sections cover the paleoclimatic significance of coal bed characteristics and distribution, in order of increasing spatial and temporal scale. Cited studies incorporate climate in one of two ways. Some assume the climate a priori from the vegetation or paleolatitude of the deposits and then discuss other controls on variations in coal beds. The others derive paleoclimate information from the system as a whole, including the coal beds.

Coal-bed chemistry and paleoclimate

The chemical arguments for the interpretation of paleoclimate from coal beds come principally from the work of Cecil and his colleagues (Cecil et al. 1981, 1982, et seq.). Their work was partly in response to studies purporting to show that high-sulfur coals were influenced by marine sedimentation. They showed, instead, that variability in sulfur content is greater among freshwater coal beds than between those ascribed to freshwater versus marine influence (Cecil et al. 1982), and that it is highly correlated with calcium carbonate content. Cecil et al. (1981, 1982) argued that peat that forms economic coal cannot form in seawater because ash and sulfur enrichment is too great there. Thus they concluded that all economic coals were originally freshwater peats. They further concluded that, if all economic coal beds were derived from freshwater peats, ash content must be indicative of climate, and they proposed the following model, which predicts three types of peat:

1. anaerobic (permanently waterlogged) peat with pH <4.5, which would give rise to low-ash, low-sulfur, vitrinite-rich coals;
2. anaerobic with pH >4.5, which would give rise to high-ash, high-sulfur, liptinite-rich coals; and
3. intermittently aerobic peat, which would give rise to low-sulfur, moderately high-ash, inertinite-rich coals.

The division at a pH of 4.5 represents the difference between environments in which bacterial degradation is significant and those in which it is negligible, because at pH <4.5, bacteria cannot survive. Sulfur and ash are low because under acid conditions, iron reduction to pyrite occurs only if redox potentials are higher than those found in peat, sulfate-reducing bacteria are suppressed, and acid-soluble elements, such as Ca^{2+}, are leached (Cecil et al. 1982). The first type of peat would be characteristic of raised mires; the second type, of planar mires; and the third type, of planar mires in climates with lower annual rainfall or a dry season.

According to the Cecil et al. (1982) model, the minerals reflect the geochemical conditions of weathering, soil formation, sedimentation, and early diagenesis (Cecil et al. 1985). Siderite in the absence of limestone indicates acid conditions; quartz arenites are the product of intense weathering (see also Donaldson, Renton, and Presley 1985). Upper Middle Pennsylvanian to uppermost Pennsylvanian rocks commonly include lacustrine limestone with indications of subaerial exposure, that is, microkarst, brecciation, and subaerial crusts, as do the kaolinite deposits. Cecil et al. (1985) suggested that these latter features represent well-drained soils, possibly resulting from decreased annual rainfall. They summarized the climate as in figure 5.30. Note that Cecil et al. (1985) did not attribute all the lithologic changes to climate.

Coal-bed petrography and paleoclimate

Vitrinite-rich coal beds are generally regarded to have been deposited in wet conditions, usually

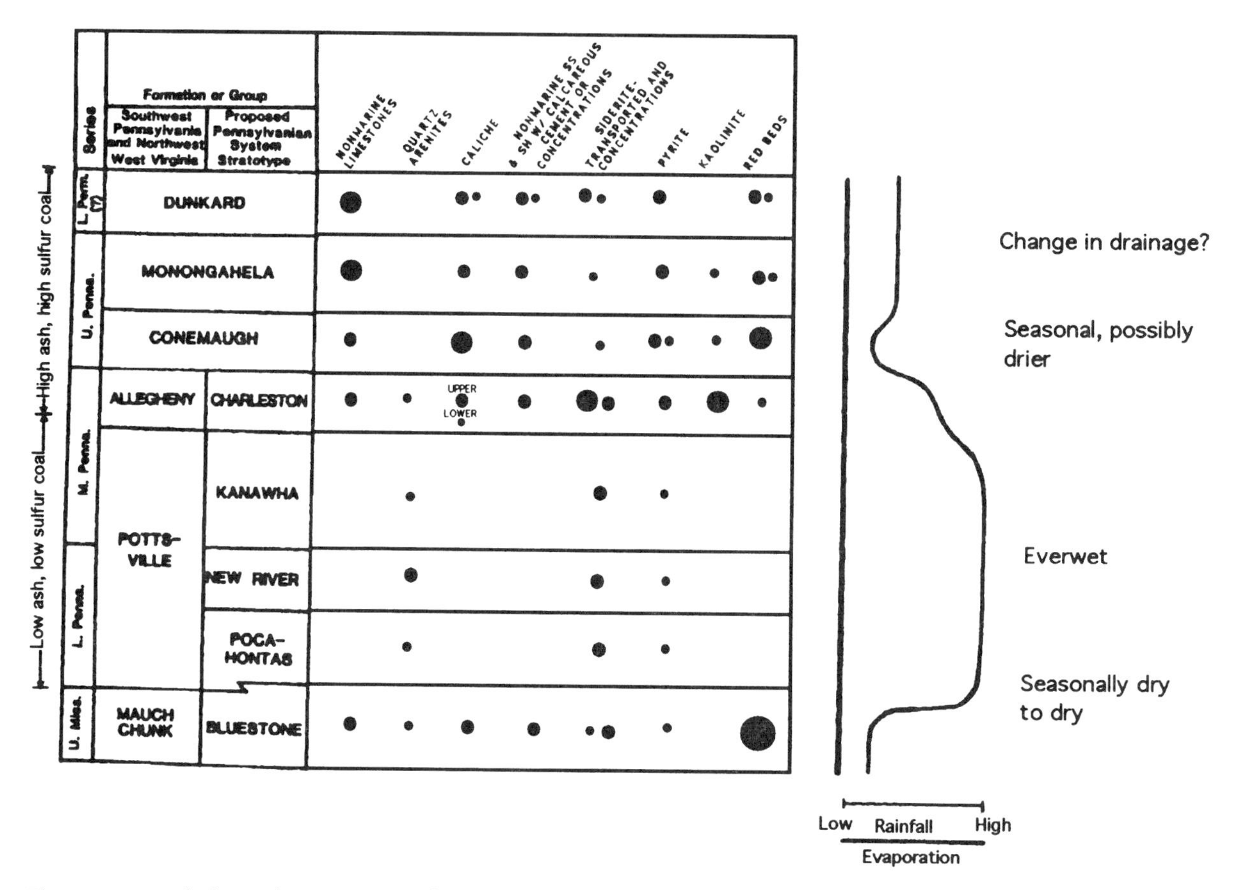

Figure 5.30. Lithologic characteristics of strata and interpreted climate curve for Carboniferous and Lower Permian rocks of the eastern United States. Reprinted from *International Journal of Coal Geology* 5, Cecil, C.B., et al., "Paleoclimate controls on Late Paleozoic sedimentation and peat formation in the central Appalachian basin (U.S.A.)," pp. 195–230 (1985) with kind permission of Elsevier Science-NL, Sara Burgerhartstraat 25, 1055 KV Amsterdam, The Netherlands.

meaning high water tables, especially if the coal beds have clay partings and inclusions of syngenetic pyrite (Teichmüller 1989). Inertinite-rich coal beds are generally regarded to have been deposited in dry conditions, usually meaning relatively low or fluctuating water tables (Teichmüller 1989). A special case that may be reasonably confidently tied to climate is abundance and cyclicity of inertinite and charcoal (Lamberson et al. 1996). In the Gates Formation (Early Cretaceous, Canada), inertinite occurs in much higher concentrations than is seen even in modern peats that are periodically burned. Lamberson et al. (1996) interpreted these concentrations as cycles of drying, rather than variations in subsidence rate, largely because of the cyclicity.

Several workers have proposed that petrographic characteristics can be tied to temperature. For example, Diessel (1992) noted that the proportion of undecomposed cell tissues (telocollinite and related macerals) is higher in high-latitude peats today than in tropical peats. Cutinite, a specific type of liptinite derived from leaf cuticles, should be high in temperate peats formed in mires dominated by deciduous trees because seasonal leaf fall adds large quantities of such tissue to the peats each year. Cutinite may be associated with semifusinite and inertodetrinite, which is interpreted as the result of freeze-drying of the soft tissue of the leaves (Taylor, Liu, and Diessel 1989). Independent evidence for cold winters and strong seasonality with respect to temperature might include strong growth rings in wood in the coal or associated sediments. These characteristics have not been applied in reverse to interpret such a climate for coal beds in regions lacking supporting paleoclimate evidence.

The petrographic composition of the coal bed also is dependent on the vegetation of the mire, which in turn is dependent on climate, but also on geologic age (Cameron, Esterle, and Palmer 1989; Smyth 1989). Thus with the possible exception of fire frequency, Smyth's (1989) remarks on the utility of coal-bed petrography hold true: "In any given coal-bed sequence, variations in organic matter type can be correlated with paleodepositional environments [in the broad sense]. However, relationships established for one area do not necessarily apply to another."

Raised mires and paleoclimate

Raised mires are climatically significant because they form only in climates that have constant rainfall. Climates with a significant dry season cannot support raised mires because the mire tops would dry out and degrade, resulting in the deaths of the plants that make up the organic matter in the peat. The rainfall does not have to be abundant so long as it is steady and evaporation rates are relatively low. The degree of doming of a raised mire is related to humidity, with wet tropical peats the most domed, cool temperate peats the next, and warm temperate peats the least domed; in cool temperate peats, the tendency toward doming is greatest close to the coast (Cameron, Esterle, and Palmer 1989).

In modern raised mires, vegetation at the top is low diversity and small statured because the water table is more subject to fluctuations with changes in rainfall and because nutrients are less available for the plants. Thus evidence not only for vegetational change but also for morphological change independent of taxonomic change might be evidence for doming. Accompanying those changes would be textural and chemical changes that could be expected to be preserved with compaction and coalification. Raised mires show an upward decrease in mineral content (see figure 5.28) and in the content of syngenetic sulfides (Neuzil et al. 1993). In addition, raised mires show a trend in phosphorus from relatively high close to the underlying substrate, to very low in the middle of the peat, to higher values again at the top of the peat (figure 5.31; Neuzil et al. 1993). This pattern results from the ability of the plants that are adapted to raised mires to very efficiently recycle nutrients from the underlying peat as they grow and the peat accumulates. Such an adaptation would be critical for the plants because nutrient influx from outside the mire is negligible.

Despite the striking differences between modern raised and planar mires, distinguishing between them in the geologic record is not straightforward. Compaction in peat ranges from a minimum of 1.4:1 to 30:1, depending on the composition of the plants that make it up (Ryer and Langer 1980; Winston 1986). In comparison, muds compact in a ratio of about 3:1 and sands 2:1 or less. Thus by the time a mire has

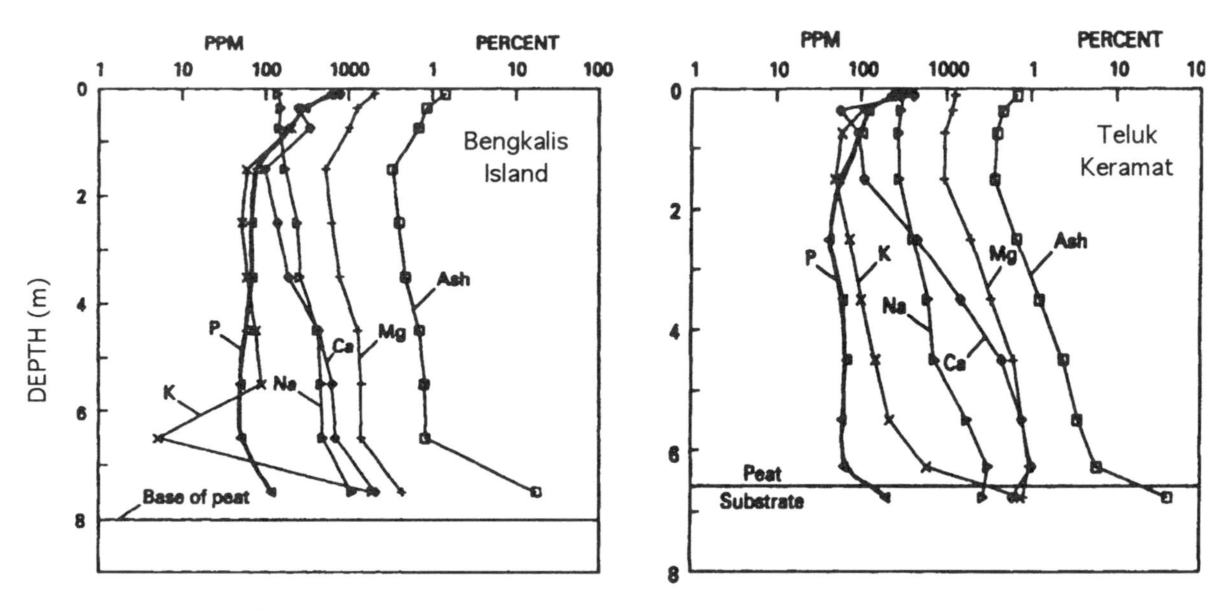

Figure 5.31. Profiles of phosphate and other elements and total mineral matter (ash) in tropical, raised mires in Indonesia. From Neuzil et al. (1993) in *Modern and Ancient Coal-Forming Environments,* J.C. Cobb and C.B. Cecil, eds. Reproduced with permission of the publisher, the Geological Society of America, Boulder, Colorado, USA. Copyright © 1993 Geological Society of America, Inc.

been buried, its original geometry may be lost, although in some exceptional cases, the geometry of the bed might be a reflection of doming. A raised mire might be inferred indirectly from stratigraphic relationships with surrounding rocks. For example, Flores (1993) noted that sandstone and adjacent coal bed thicknesses in some Tertiary coal beds of the western United States are the same, up to 60 m. Given the much greater compaction ratios of peat compared with sandstone, this relationship implies that the coal beds were derived from raised mires between entrenched channels (Flores 1993).

In summary, the principal criteria for identifying a raised mire are vertical changes through the coal bed in the vegetation (e.g., Bartram 1987; Fulton 1987; Winston 1988; Winston and Stanton 1989; Eble and Grady 1993; but see DiMichele and Phillips 1994), chemistry (Bartram 1987; Eble and Grady 1993; Neuzil et al. 1993), and petrography (Bartram 1987; Winston 1988; Diessel 1992; Eble and Grady 1993; Flores 1993; but see Esterle and Ferm 1990).

Interpretation of local environments, basin analysis, and paleoclimate interpretation

Most basin- and region-scale studies concentrate either on interpreting whether the peat formed in a particular depositional environment, or on the relationship of the mires to the river channels. Thus many features of coal beds that are cited by some workers as indicating a change in climate might be cited by others as evidence for a change in the vegetation and by still others as evidence for a change in the locus of peat formation within the depositional system or in the depositional system as a whole. One implication of this variability in interpretation is that at this scale, coal beds have to be interpreted on a case-by-case basis for climate information. Another implication is that coal beds must be interpreted not only using a variety of methods, but also within their overall sedimentary, tectonic, and paleogeographic settings.

One basin-scale study used coal bed characteristics almost exclusively to interpret changes in climate, but even this study derived most of the information from the mire vegetation rather than from coal-bed petrography or chemistry. The study was conducted over a span of more than a decade by T. L. Phillips and his colleagues and was summarized in Phillips and DiMichele (1990). Phillips and his colleagues concentrated on the information on mire plants obtained from coal balls, which are carbonate concretions that formed in peats before compaction and coalification and thus preserve the original structure of the peat constituents. Coal balls are common in Upper Carboniferous coal beds, especially those of the Illinois Basin. Unfortunately, coal balls are virtually nonexistent in coal beds of other ages. This means that one important control on coal-bed petrography according to Teichmüller (1989)—the taxonomic composition of the vegetation, which would have changed dramatically through time—cannot be studied as directly as in the Upper Carboniferous coal beds. Thus direct comparisons among coal beds of different ages with respect to the influence of taxonomic composition are not easily accomplished.

The climate curve proposed by Phillips and Peppers (1984) for the Late Carboniferous is shown in figure 5.32. Large trees and high plant diversity characterized both the Westphalian D and Stephanian coal beds, indicating stable swamps, but the peats formed in the Stephanian show more evidence of degradation. The interpreted vegetational patterns are related principally to taxonomic composition but also are at least partly supported by other data. For example, Harvey and Dillon (1985) noted that changes in inertinite content, which they interpreted as evidence for low water tables, reflected the interpretations of drying by Phillips and Peppers (1984). In addition, some of the features of the climate curve also were mirrored in the curve constructed by Cecil et al. (1985) using lithologic and coal-quality data (see figure 5.30); the major discrepancy between the two curves is the absence of the first drier interval of Phillips and Peppers (1984) in the curve of Cecil et al. (1985). Phillips and Peppers (1984) and DiMichele and Phillips (1994) also discussed how the ecology of the plants bears on environmental and climate interpretations.

Following are typical examples of detailed basin- and region-scale sedimentologic and stratigraphic studies that incorporate climate into interpretations of coal-bearing sequences. The reasons for including these studies are (1) to characterize the principal focus of most studies of coal beds at this scale, (2) to point out that climate is generally included as an a priori assumption, (3) to show how the same types of information on coal beds can be interpreted differently depending on the stratigraphic, sedimentologic,

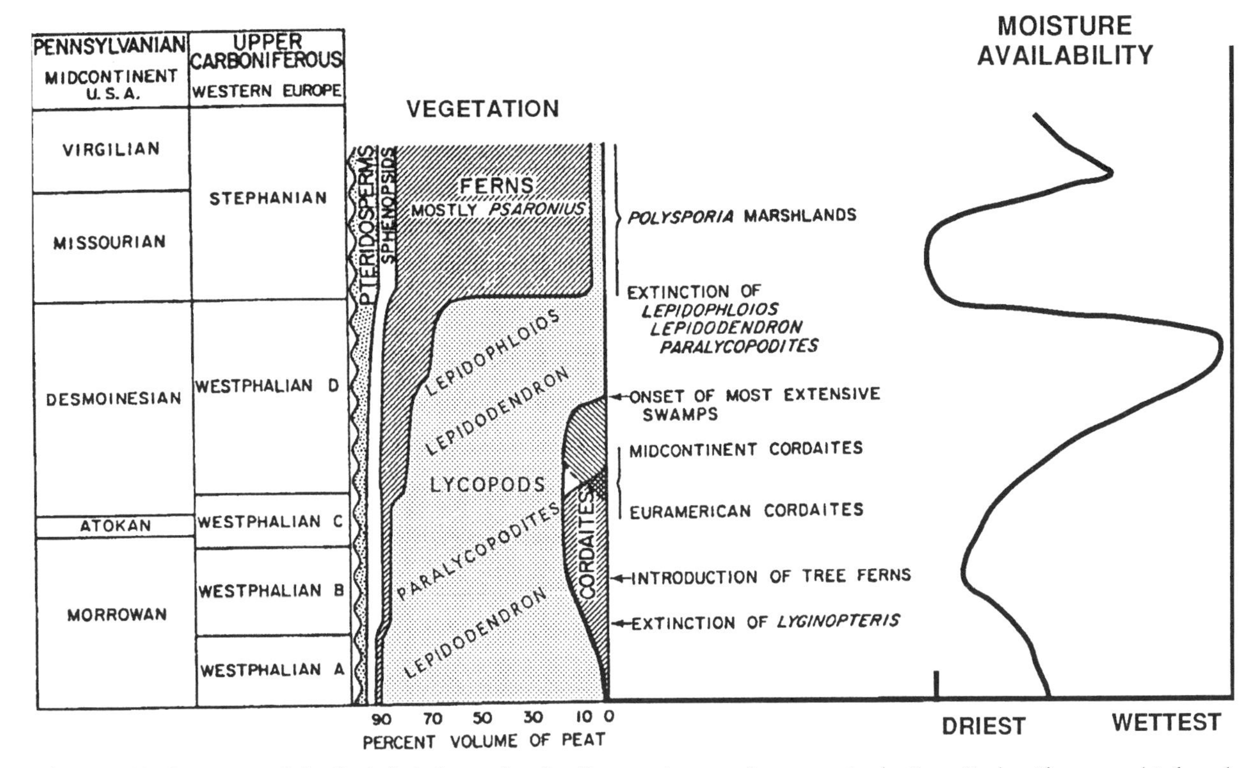

Figure 5.32. Summary of the floristic information for Euramerican coal swamps in the Late Carboniferous and inferred moisture availability. "Wettest" and "driest" are relative terms only; relative heights of wet peaks were determined from the volume of coal in the eastern United States. Reprinted from *International Journal of Coal Geology* 3, Phillips, T.L. and R.A. Peppers, "Changing patterns of Pennsylvanian coal-swamp vegetation and implications of climatic control on coal occurrence," pp. 205-55 (1984) with kind permission of Elsevier Science-NL, Sara Burgerhartstraat 25, 1055 KV Amsterdam, The Netherlands.

and tectonic context, and (4) to demonstrate why, at this scale, coal beds are not good paleoclimatic indicators.

Witbank coalfield (Karoo Sequence, Carboniferous-Permian; South Africa). The first example is a study of the Witbank coalfield, which has four seams. Holland et al. (1989) took as a starting point conclusions by previous workers (Falcon 1989) that the pollen and plant megafossils indicate a periglacial environment for the lower seam and a postglacial, cool temperate climate for the upper three seams. They showed that seams 1, 2, and 4 all have high inertinite contents, whereas seam 3 is higher in vitrinite. They concluded that seams 1 and 2 were formed in the upper delta plain, in a glaciofluvial environment (see also Taylor, Liu, and Diessel 1989). These seams are contemporaneous with braided fluvial deposits and underlain by glaciofluvial deposits and diamictites. Seam 4 was also formed in an upper delta plain, though glacial influence had waned. In contrast, they concluded that seam 3 was formed on the lower delta plain; it caps an upward-coarsening deltaic sequence and is laterally impersistent, a feature that was apparently controlled by differential compaction. In this case, the climatic information was primarily from the plants, secondarily from associated glacially influenced sedimentary rocks (see section on glacial deposits), and not at all on variations in the coal beds.

A broader geographic comparison of the Permian coal beds in the Karoo foreland basin was made by Cadle et al. (1993). The Permian coal beds are in the Vryheid Formation of the Ecca Group, which is mostly marine. In general, the peats formed when subsidence rates were low, allowing partial aerobic degradation of the peat. The coal beds overall are high in inertinite (specifically semifusinite, macrinite, inertodetrinite, and micrinite) and ash. Cadle et al. (1993) attributed these characteristics to temperate climate with seasonal precipitation (which they took as a given from the same sources cited by Holland et al. 1989), elevated environments (e.g., upper delta plain), or swamps that were frequently inundated (which elevates pH and Eh). They interpreted later Permian and Triassic red beds, culminating in the Stormberg eolian sandstones, to indicate hotter, drier conditions; however, the Upper Triassic Molteno Formation indicates briefly wetter conditions, allowing accumulation of thin coal beds. The Molteno Formation coal beds overlie sandy, ephemeral stream deposits of the Banboesberg Member. Previous paleobotanical work (Falcon 1989) indicated that climate was warm temperate with seasonal rainfall during deposition of the Molteno Formation, in contrast to the warm, arid climate of the underlying deposits. Peat formed in alluvial plain swamps that were removed from the locus of fluvial activity, described by Cadle et al. (1993) as "ponded embayments which were inundated periodically by flood water from adjoining ephemeral braided streams." Coal beds in the Molteno Formation are high in mineral matter, 25% to 35%, and they are also rich in vitrinite. In this case, the coal-bed petrography contributed to the climate interpretation, although climate was not the only possible explanation for the petrographic features that Cadle et al. (1993) observed.

Springhill coalfield (Cumberland Basin, Carboniferous; Nova Scotia, Canada). The second example is a study of the Springhill coalfield in the Cumberland Basin, Nova Scotia, which was a low-latitude Carboniferous intermontane basin (Calder 1994). The Springhill sequence was deposited in an actively subsiding basin and consists of >4000 m of rocks deposited in 4 Ma during the Westphalian A-B. The coal beds were interpreted as derived from planar mires based on (1) the stratigraphic affinity of coal beds and fan deposits, (2) the strong gelification of tissues, (3) the prevalence of mineral partings, and (4) the dominance of arborescent lycopods, co-dominance of calamites (sphenopsids) and cordaites (gymnosperms), and the presence of the lycopod, *Paralycopodites*, an assemblage that is common in coal beds during the first Late Carboniferous dry interval, described by Phillips and Peppers (1984).

The coal-bearing part of the section consists of a series of cyclothems. These cyclothems are of the same order as those described elsewhere in North America that have been attributed to sea-level changes associated with the waxing and waning of the southern Pangean ice sheet. Apparently, no independent evidence for changes brought about directly by climatic versus eustatic sea-level change (Soreghan 1992b) exists in the Springhill coalfield, so either explanation, or both, for the cyclothems is plausible (Heckel 1995). The cyclothems are embedded in an upward-fining and upward-reddening basin-fill sequence. Calder (1994) concluded that this sequence is controlled by declining subsidence rates and the consequent changes in drainage, although he also suggested that an overall drying climatic change could be responsible.

In these studies, climate was discussed as a controlling factor, but interpretations of the climate came from other types of information, namely vegetation, other sedimentary rocks, and the latitudinal positions of the basins. Characteristics of the coal beds were explained by climate but not used to interpret the climate. If the characteristics of the coal beds are used to interpret depositional environment, it becomes almost impossible to interpret climate from the same data. One would have to interpret depositional environment independent of the coal bed characteristics, hold depositional environment constant, and then observe which coal-bed characteristics must be the result of climate and climatic change. Such a study apparently has never been done and in practice might be impossible because the characteristics of the coal beds may be the only information available about subtle variations in the hydrology of the depositional environment. This is why coal beds are not good paleoclimatic indicators at this scale.

Calder (1994) emphasized, as have other workers (e.g., Nemec 1992), the complexity of conditions under which coal beds form, the requirements of tectonics, local climate, hydrology, and interactions among these that leave only narrow "windows." Because the system is so complex, few attempts have been made to model, even conceptually, these windows.

Global and large-scale temporal distribution of peat and coal beds and relationship to patterns of climate and paleoclimate

Three approaches have been taken to the study of continent- or global-scale climate and the distribution of coal beds. The oldest approach has been to regard coal beds as indicative of wet climate and to reconstruct climate from their distribution. This approach has been used implicitly or explicitly by numerous workers, notably Briden and Irving (1964), Robinson (1973), Habicht (1979), and, more recently, Scotese and Barrett (1990), despite objections from some workers, notably Schopf (1973). The other two approaches have been developed in tandem but have yet to cross over. One is to model paleoclimate and compare climatic patterns with the distribution of coal beds (Parrish, Ziegler, and Scotese 1982; Rowley et al. 1985; Patzkowsky et al. 1991; Otto-Bliesner 1993). The other approach has been to document as precisely as possible the conditions under which peat is forming today in order to refine coal beds as paleoclimatic indicators (Gyllenhaal 1991; Lottes and Ziegler 1994).

Ziegler et al.'s (1987) general survey of peat distribution led to the conclusion that precipitation continuity is important, that is, that climates that are seasonal with respect to rainfall are not generally favorable for peat accumulation. This conclusion was supported by the results of Gyllenhaal's (1991) survey of modern peat-forming environments (see figure 5.27). Ziegler et al. (1987) noted two exceptions to this general rule. In high-latitude regions, the cold and dry seasons correspond, and all biological activity is suspended during the winter. Peat accumulation is slow, but so is degradation. In lower latitudes, if climate is seasonal, the season in which the most rainfall occurs is important, and a summer maximum is the most favorable because the water supply is highest when evaporation is highest and, because temperatures are warm, biological productivity is highest. Ziegler et al. (1987) emphasized that orographic effects can be important to maintaining precipitation continuity.

Lottes and Ziegler (1994) started with the premise that "climate plays an essential role in peat bog formation." They defined precipitation continuity as the number of months with >40 mm of rainfall, and they combined data on the present distribution of rainfall with data on the growing season, defined as the number of months with mean temperature >10°C. Precipitation is the limiting factor except at high latitudes, where temperature is the limiting factor. Lottes and Ziegler (1994) found that 82% of modern peats occur in regions in which 80% to 100% of the months fit the criteria of precipitation >40 mm and temperature >10°C. They did not analyze the conditions under which the remaining 18% of peats were forming, but their data do provide a measure of the degree to which climate might play a dominant role in peat formation.

Gyllenhaal (1991) did address the dry tail of peat distribution, that is, peats now forming in regions where rainfall is <80 cm/yr (see figure 5.27). Most of these occurrences are of two types: (1) coastal peats in salt marshes and mangrove swamps where a high water table is maintained by sea level (e.g., in California, U.S., and Mozambique), and (2) lakes and riverine swamps where a high water table is maintained by runoff from headwaters in distant mountains or equatorial regions (e.g., in Sudan). All of these peats are thin with high ash contents and are likely to form carbonaceous shale instead of coal in the geologic record; indeed, they may not fit the correct definition of peat. Almost 70% of modern peat localities receive >40 mm of rain every month, but because typical evapotranspiration rates in the tropics exceed 100 mm per month, many of these localities have moisture deficits during some months.

If a lack of seasonality of precipitation is important for peat accumulation, the intimate association of peats and laterites in many parts of the world (e.g., Besly and Fielding 1989) would appear to be paradoxical. Gyllenhaal (1991) specifically addressed this point, noting that a locally high water table can be maintained in a seasonal climate because the surface area of a mire can be small relative to the drainage basin, and evapotranspiration from the mire may not be great enough to create a deficit against influx from the much larger area. Thus although the adjacent regions may experience fluctuations in water supply that are great enough to create a seasonal moisture deficit and formation of laterites, the mire itself may not experience enough of a moisture deficit to inhibit the formation of peat. In these systems, however, the mires would be expected to be planar.

Gyllenhaal (1991) defined several assemblages of paleoclimatic indicators along a gradient of mean annual precipitation, three of which include peats (figure 5.33). Thin high-ash coal beds associated with caliche and/or minor gypsum can be expected where mean annual precipitation (MAP) is 25 to 130 cm/yr; coal beds with moderate to high or variable ash and sulfur content but not associated with caliche can be expected where MAP is 130 to 250 cm/yr; and low-ash, low-sulfur coal beds (likely derived from raised mires) can be expected where MAP is >250 cm/yr, with precipitation at least subequal to evapotranspiration every month of the year. Thus thick, low-ash, low-sulfur coal beds would appear to be good indicators of an everwet climate; however, this should not be taken to mean that thickness variations can be used to indicate climatic variations, although they may indicate

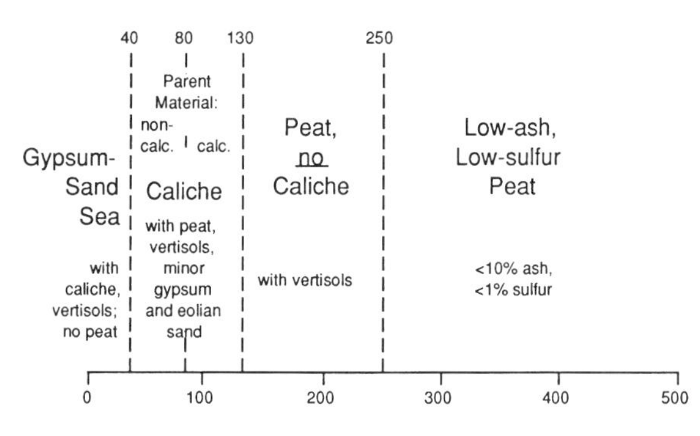

Figure 5.33. Assemblages of paleoclimatic indicators along a precipitation gradient. From Gyllenhaal (1991).

persistent reestablishment of conditions favorable to peat accumulation (Shearer, Staub, and Moore 1994).

Influence of Peat Formation on Climate

Some calculations of atmospheric CO_2 levels through time have depended on estimates of burial and weathering of organic carbon (Ronov 1964; Budyko and Ronov 1979; Lasaga, Berner, and Garrels 1985; Berner 1990a, 1991, 1992), and the distribution of coal beds through time and space has been an important part of these calculations. Sloan et al. (1992) suggested that extensive and rapid peat formation and burial could explain global warming in the late Paleocene-early Eocene, which is regarded to be an important greenhouse interval in Earth history. They noted opposite trends of bulk $\delta^{13}C$ and benthic foraminiferal $\delta^{18}O$ and the relationship of these trends to changes in long-term sea level. They doubled previous estimates of the area of wetlands for that time and calculated a tripling of methane production. They then suggested that the increased methane production would favor formation of methane-rich polar clouds, which can trap heat in the polar regions. A complication to this model is the observation that methane production is minimal in acid swamps (Cecil et al. 1982), so the types of wetlands would be critical to Sloan et al.'s (1992) model.

LACUSTRINE DEPOSITS

Lakes are sensitive recorders of short-term climatic change. The reasons for this sensitivity are the continental setting of lakes, the fact that they are shallow and therefore subject to the influences of the atmosphere, and the fact that they are usually depositional environments, meaning that every change in the lake has the potential to be recorded in the sediments. Closed lakes, that is, those without surface outlets, are particularly sensitive to climatic change because they are strictly dependent on runoff and precipitation versus evaporation in the drainage basin and lake (Kutzbach and Street-Perrott 1985; Street-Perrott and Harrison 1985).

Lakes record climate in several ways. First, changes in the levels of closed lakes are regarded as indicative of changes in climate. The only non-climatic controls on such lake deposits are the rate of outflow resulting from seepage, which is dependent on the geology underlying the lake and is not generally subject to change, particularly cyclic change, and changes in runoff resulting from stream piracy. Especially compelling are studies showing synchronous change in the levels of lakes spread over a large area. Second, the general character of sediments deposited in the lake can be a clue to climate. This is particularly true of anoxic lakes and saline lakes. Third, chemical changes in the organic matter and carbonates can track climatic change. The use of organic geochemistry for interpreting lake levels and paleoclimate has been developed for Quaternary and Holocene lakes (e.g., Krishnamurthy, Bhattacharya, and Kusumgar 1986; Talbot and Livingstone 1989; Meyers 1990, 1994; Giresse, Maley, and Brenac 1994), but has not been commonly extended into the pre-Quaternary record. Finally, lake sediments are commonly laminated, and the composition and variation of the laminations can be controlled by climate.

Most paleoclimate studies of lakes have been on Quaternary deposits. The principal disadvantage of lake deposits for pre-Quaternary paleoclimate studies is that lakes are transitory. Large lakes might last 2 to 3 million years and lake systems as long as 40 million years (Olsen 1986), but most lakes last for only 10^2 to 10^5 years. Thus pre-Quaternary lake deposits will, in general, provide only a window on short-term climatic change at any particular time that is difficult to interpret in a larger temporal or spatial context. Distinguishing between local and regional influences might be impossible. In addition, lacustrine deposits are extremely variable, so much so that few rules exist that apply to all lakes, and each lake must be interpreted independently. Despite these limitations, studies on older lake deposits have yielded important information about paleoclimate.

A recurring theme in this chapter applies to lacustrine deposits as well, and that is the problem of distinguishing between tectonic and climatic influences. Lake deposits as a whole are heterogeneous, so that a sedimentary sequence interpreted as tectonically controlled in one lake might be attributed to climatic change in another (Cohen et al. 1997). Indeed, sedimentary styles can vary widely even within a single lake. An example of this variation is that predicted for parts of Lake Tanganyika, a rift lake in East Africa (Cohen 1990; figure 5.34). The example shows two cycles of lake-level change, possibly induced by climate. Sedimentary transitions are predicted to be rapid because lake levels tend to change rapidly, but they have very different characters depending on the lake margin type. On the shoaling margin, the units would tend to be limited in thickness but have broad lateral distribution because of the low bathymetric gradient, whereas the opposite is true on the escarpment margin. The character of the sequences also is controlled by sediment supply. Where lacustrine sediments include a chemical component such as carbonate, covariation of chemical and terrigenous components may indicate climate control, whereas variation in the terrigenous component alone may be related solely to tectonics (Drummond, Wilkinson, and Lohmann 1996).

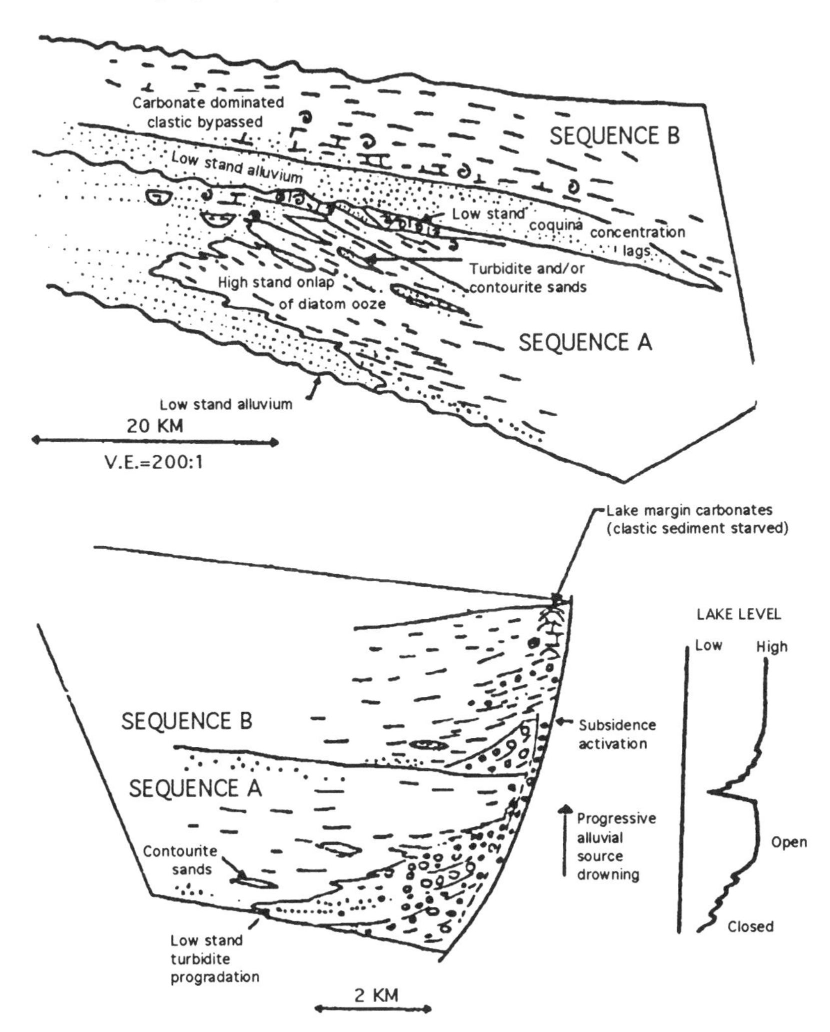

Figure 5.34. Hypothetical stratigraphic sequences on the shoaling *(top)* and escarpment *(bottom)* margins of Lake Tanganyika, showing sequences related to changes in lake level as indicated in the lake-level curve. The escarpment margin is the location of the fault in the half-graben filled by this part of the lake. Modified from A.S. Cohen ©1990, reprinted by permission of the American Association of Petroleum Geologists.

Lake-Level Studies

Interpretations of cyclic change in lacustrine deposits have tended to focus on changes in lake level, primarily because in well-exposed and well-documented systems, the cyclic changes include evidence of repeated flooding and desiccation in parts of the system. Although such studies are referred to as lake-level studies, implying depth, the climatic sensitivity of lakes has less to do with depth than with lake surface area (Benson et al. 1990). Some of the most comprehensive studies of the relationship between lake levels and climate have been those conducted with various coworkers by Street-Perrott (e.g., Street-Perrott and Harrison 1984, 1985; Kutzbach and Street-Perrott 1985).

Because lacustrine sedimentation is so much more sensitive to climate in closed lakes, distinguishing closed from open lakes is an important element of lake paleoclimate studies. The lake-level studies of Street-Perrott and Harrison (1984, 1985) used geomorphic information (e.g., ancient shorelines), stratigraphic, geochemical, paleoecological, and archeological information for lake-level reconstruction. The geomorphological information included topographic information that permitted them to focus studies on potential outflows. Workers on pre-Quaternary lake deposits must base the interpretation of lake hydrologic status on much sparser information.

Smoot (1985) listed the criteria by which closed lakes might be recognized in the older geologic record. These include (1) a systematic increase in grain size towards all lake boundaries, (2) paleocurrent patterns that indicate flow toward all lake boundaries, (3) local provenance of coarse-grained sediments, (4) occurrence of evaporites or traces of evaporites, and (5) cyclicity of fine-grained sediments in the center of the basin. However, lack of these features does not mean the lake was open. The first through third criteria depend on excellent preservation of the lateral extent of the lake. Evaporites are good indicators, but not all closed lakes are saline (Gore 1989). In addition, the sedimentary record of a change in the level of a closed lake can resemble a change between an open and a closed lake, a common occurrence. Thus cyclicity appears to be the best indicator that a lake was closed and therefore sensitive to climate, and that the cyclicity indicates climatic change.

The lacustrine deposits of the Triassic rift basins of eastern North America, which extend from North Carolina to Nova Scotia, are among the rare cases in the pre-Quaternary record in which several coeval lake systems can be used to reveal regional climatic patterns. In general, sediments from the northern basins indicate drier climates, and sediments from the southern basins indicate wetter climates (figure 5.35; Olsen 1990). The overall evolution of the basins from shallow to deep to shallow was regarded by Olsen (1990) to be principally the result of tectonic control, based on basin subsidence and fill models of Schlische and Olsen (1990). Differences among the eastern North American rift basins, however, are probably attributable to climate because the tectonics were similar.

Olsen (1990) classified the eastern North American rift basins into three types. In the Newark-type basins, water influx approximately equalled water loss. Changes in precipitation resulted in dramatic changes in the sediments deposited (figure 5.36b), which Olsen (1990) attributed to changes in lake level. The Richmond-type basin category (figure 5.36a) comprises only two basins, the Richmond and Taylorsville Basins and, possibly, the Deep River and Farmville Basins. The basins contain coal beds and lacustrine sediments that show no evidence of subaerial exposure; water influx must have exceeded water loss from evapotranspiration. The Fundy-type basins (figure 5.36c) alternated between shallow, perennial lakes and playas with salt crusts. Evaporites and eolian deposits are associated with the lacustrine deposits. Water influx in these basins only rarely matched water loss.

Olsen (1990) attributed the cyclicity in the Newark Supergroup (figure 5.36) to the strong monsoonal climate of Pangea coupled with Milankovitch forcing, to which the Newark-type basins would be expected to have been particularly sensitive. He assigned depth ranks to each of the lacustrine facies, based on sedimentary structures and fossils, and then performed spectral (in this case, Fourier) analysis on the depth-rank logs. Within the limitations of the dating methods, which included paleontological, radiometric, and varve-count dates, the power spectra from the spectral analysis conformed to expected frequencies and harmonics of Milankovitch cycles (see also Van Houten 1964; Olsen and Kent 1996).

Olsen's (1986) depth-ranking system of the lacustrine facies is notable for two reasons. First, it

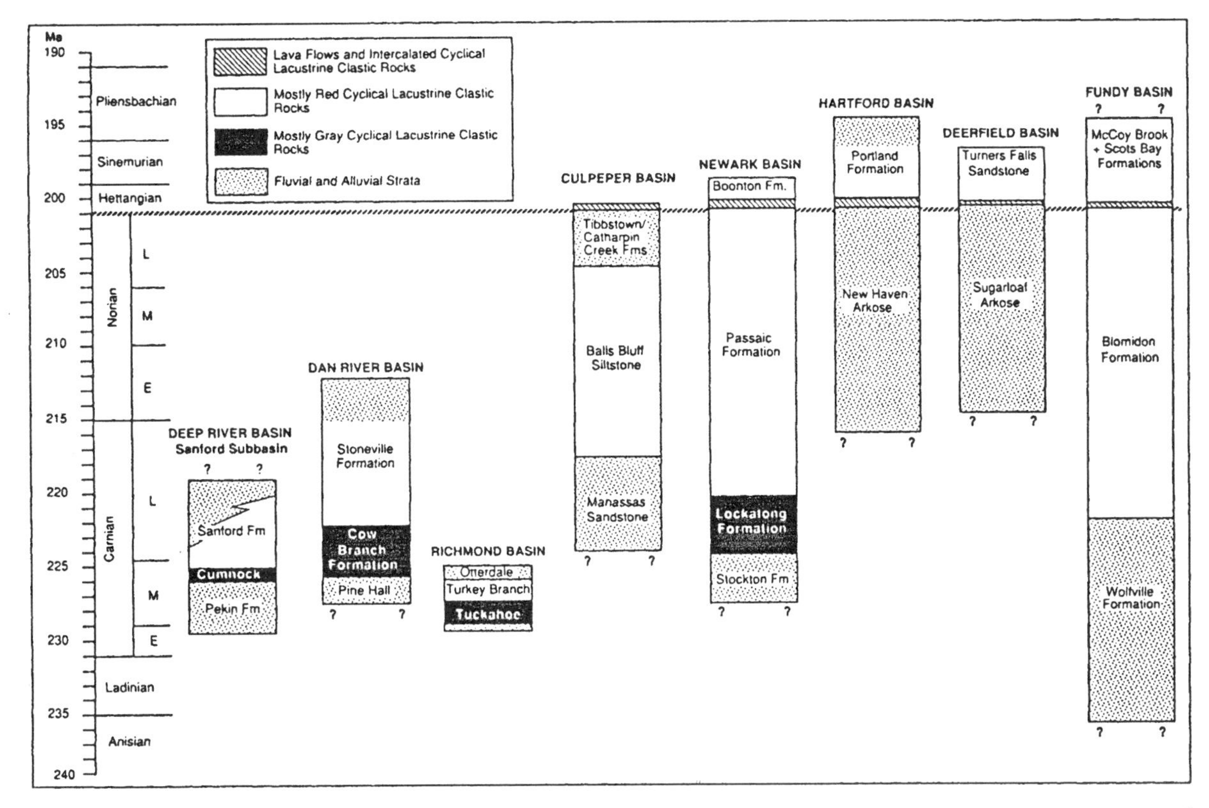

Figure 5.35. Correlation and facies from basins containing the Newark Supergroup (Triassic-Jurassic; eastern North America), arranged generally south *(left)* to north *(right)*. From Olsen, Schlische, and Gore (1989) in *Tectonic, Depositional, and Paleoecological History of Early Mesozoic Rift Basins, Eastern North America.* International Geological Congress Field Trip T351, copyright by the American Geophysical Union.

established a method for estimating changes in lake level that could potentially be used anywhere in a lake basin and that allows comparison of lake-center and lake-margin changes if the appropriate sites are available. The method is only semiquantitative but is consistent in its relative estimates because it depends partly on the principle of superposition. Moreover, the method does not depend on specific lithologies. Rather, similar rankings could be established for any lacustrine deposit, regardless of the lithology, and changes in the depth ranks could be compared among different basins. Second, although the method is semiquantitative with respect to actual depth, the rank values permit quantitative analysis of changes in lake depth, such as the spectral analyses of Olsen (1986, 1990) or the quantitative stratigraphic analyses of Kominz and Bond (1990).

Smoot and Olsen (1985, 1988) provided detailed analyses of the information about lakes that can be derived from massive mudstones, which are common in lacustrine rocks. They showed the large amount of information that can be derived from such rocks and, although they did not specifically address climate in the basins they examined, their descriptions clearly could play an important role in paleoclimate interpretation from lacustrine sediments. Smoot and Olsen (1985) divided massive mudstones into four types—mud-cracked, root-disrupted, burrowed, and sand-patch—that they regarded as end members (table 5.10). *Mud-cracked* massive mudstone indicates alternating wetting and drying of the sediment. The "crumb" fabrics in this type of mudstone are characteristic of an aggrading playa mudflat, and the breccia fabric is probably more characteristic of slightly wetter conditions. *Burrowed* massive mudstone contains a variety of ichnofossils, of which *Scoyenia* is the most common; this trace has convex laminae (meniscus structures) and textured outer walls. This facies can occur in several different environments. *Root-disrupted* massive mudstone is characteristic of vegetated floodplains, river overbank, or lake margin. *Sand-patch* massive

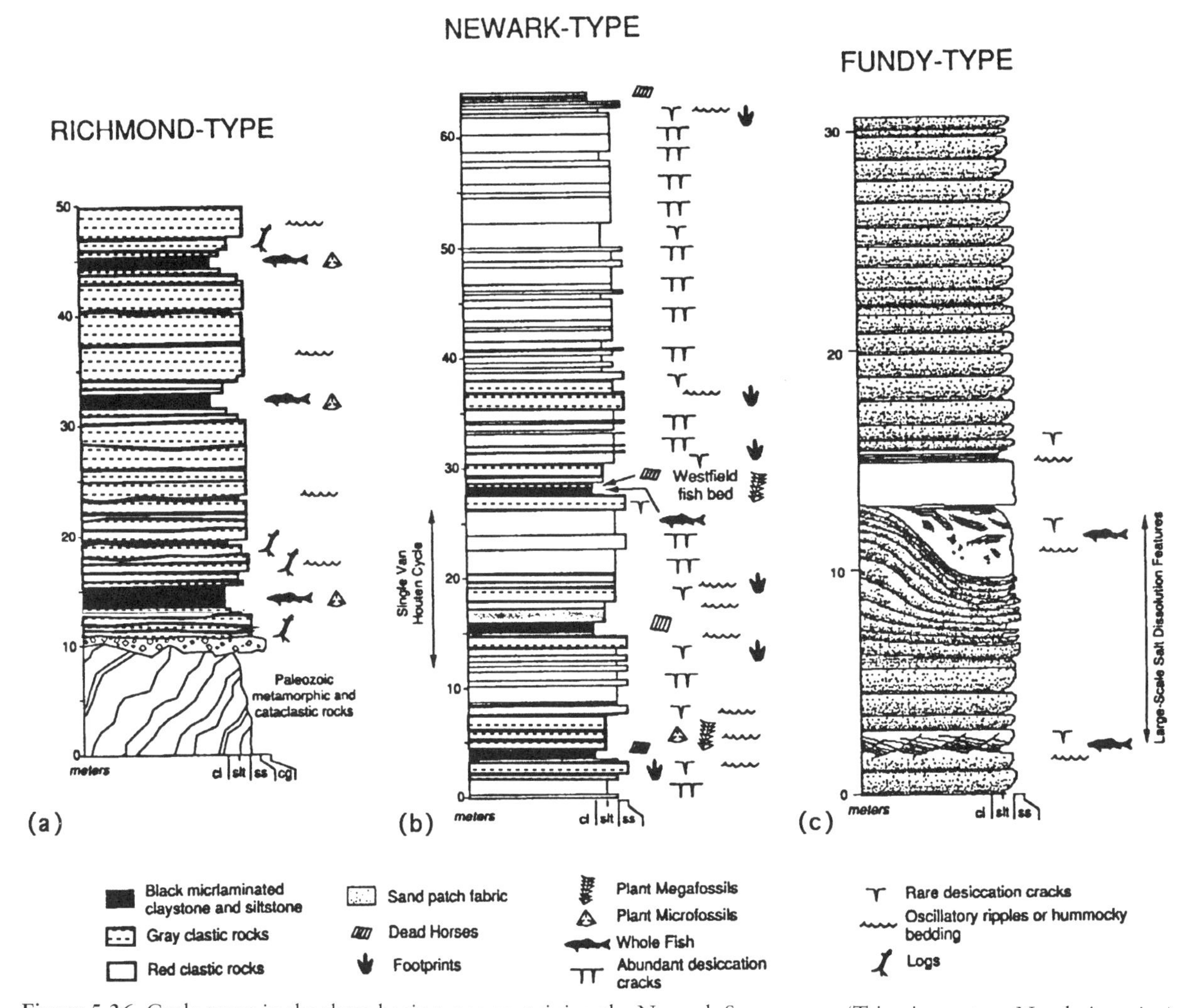

Figure 5.36. Cycle types in the three basin types containing the Newark Supergroup (Triassic; eastern North America). See text for explanation. From Olsen, Schlische, and Gore (1989) in *Tectonic, Depositional, and Paleoecological History of Early Mesozoic Rift Basins, Eastern North America*. International Geological Congress Field Trip T351, copyright by the American Geophysical Union.

mudstone results from wind-blown sand that becomes trapped in the surficial irregularities of an efflorescent salt crust.

One of the detailed analyses performed by Smoot and Olsen (1988) is illustrated in figure 5.37. The entire section illustrated would, in many studies, be described simply as "red and gray silty, massive mudstone." Their analysis of the massive mudstones permitted Smoot and Olsen (1988) to define 30 cycles in the lake sediments.

Types of Lacustrine Deposits

The very presence of lacustrine deposits in a sequence tends to encourage a perception of a humid and moderately seasonal to aseasonal climate unless the deposits have some notable component, such as evaporites, that gives more specific information. However, the mere existence of lakes is not necessarily related to climate (Lambiase 1990 and others). Rather, it is the composition of the lake and/or lake-margin sedimentary rocks that can contribute to interpretations of paleoclimate. Three types of lake deposits in particular have received much attention with regard to their potential as paleoclimatic indicators. These are glacial-lake deposits, deposits of saline and/or alkaline lakes, and organic-rich lake deposits. Saline and saline, alkaline lakes commonly include organic-rich intervals, so the following division should be recognized as somewhat arbitrary. Glacial lake deposits are discussed in that section later in this chapter.

Table 5.10.
Descriptions of four types of massive, lacustrine mudstones

Mud-cracked massive mudstone
- Jagged cracks filled with sandstone, mudstone, or cement, polygonal in plan view
- Breccia and crumb fabrics
  - Breccia fabric: typical mud-cracked fabric, with angular intraformational clasts
  - Crumb fabric: smaller cracks with smaller polygons, mm-scale mud clumps, laminoid and ovoid cement-filled vesicles

Burrowed massive mudstone
- 0.5–1 cm wide tubes, filled with sandstone or mudstone, with constant diameters and randomly oriented

Root-disrupted massive mudstone
- Bifurcating and tapering tubes, ranging in size from 1 to several mm, associated with small, spherical nodules of calcite
- Large tubes generally filled with sandstone or mudstone, smaller ones clay lined and filled with carbonate mineral cement

Sand-patch massive mudstone
- 1–5 cm long irregular pods of sandstone or siltstone with internal jagged cracks, varying grain size, cuspate contacts with surrounding mudstone

Descriptions from Smoot and Olsen (1985).
See text for environmental significance of each type.

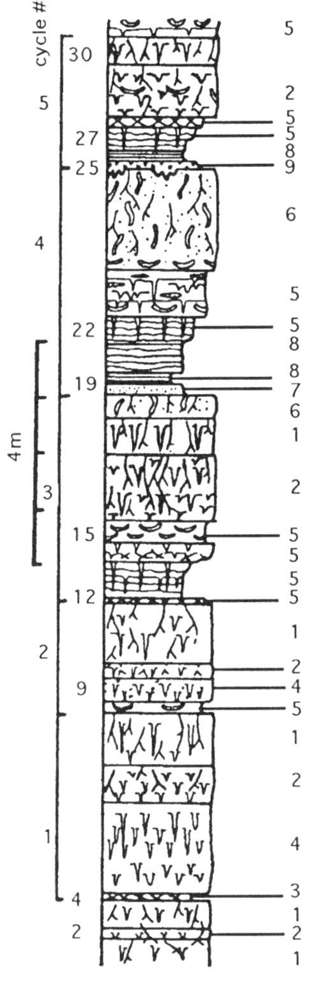

Figure 5.37. Stratigraphic section of the Balls Bluff Siltstone (Triassic; Virginia, U.S.). Lithologic descriptions (greatly simplified from those in Smoot and Olsen 1988): *1*, Red or gray silty mudstone, abundant root casts, large narrow cracks; *2*, red or gray silty mudstone, root casts filled with carbonate and following polygons; *3*, greenish-gray mudstone, brecciated; *4*, red silty mudstone with crumb fabric and vesicles; *5*, thin beds of tan-weathering, dolomitic siltstone broken into polygonal curls, scoured basal contacts and internal oscillation ripple cross-lamination, vesicles and crumb fabric in some beds; *6*, black sandy mudstone, root casts, mudcracks; *7*, sandstone with black shale intraclasts; *8*, black organic-rich shale; *9*, oolitic calcarenite. Reprinted from *Triassic-Jurassic Rifting, Part A,* Developments in Geotectonics 22 (W. Manspeizer, ed.), Smoot, J.P. and P.E. Olsen, "Massive mudstones in basin analysis and paleoclimatic interpretation of the Newark Supergroup," pp. 249–74 (1988) with kind permission of Elsevier Science-NL, Sara Burgerhartstraat 25, 1055 KV Amsterdam, The Netherlands.

Saline and/or alkaline lakes

Not all saline or saline, alkaline lakes form evaporites. If they can be distinguished from freshwater lake deposits, the deposits of saline, alkaline lakes could have paleoclimatic significance. Smoot and Lowenstein (1991) addressed this issue and, summarizing observations from modern saline lakes and playas, established criteria by which ancient perennial, saline lake deposits might be identified. These include (1) organic-matter-rich laminations, (2) absence or paucity of fossils, (3) evidence of intrasediment evaporites, most commonly pseudomorphs of gypsum and halite or salt deformation of the sediment (Smoot and Castens-Seidell 1994), and (4) occurrence of alkaline minerals such as dolomite, zeolites, diagenetic feldspars, and Mg-rich clays. In the examples they cited, mudcracks also were an important criterion. Another criterion is associated alluvial fans.

An example of an ancient lacustrine deposit that is interpreted as a saline lake but that does not have evaporites is the Lower Jurassic East Berlin Formation, part of the Newark Supergroup in the Hartford Basin (Connecticut and Massachusetts, U.S.; Gierlowski-Kordesch and Rust 1994). This unit meets many of the criteria established by Smoot and Lowenstein (1991) for the recognition of such lakes. The unit consists of sandstone, red mudstone and shale containing minor sandstone, gray-green mudstone and shale, and black shale. The sandstones are interpreted as sheet-flood deposits on a distal braidplain or sand flat. Disrupted (referring to the discontinuous nature of the laminae) red mudstones, some mottled, have clay slickensides, small fenestrae, and mudcracks as deep as 1 m. The mudcracks are sometimes filled with sand even in the absence of an overlying sandstone bed. These are interpreted as alluvial plain or sand flat mudstones altered to vertisols. The sand-filled mudcracks resulted either from sand trapped during eolian transport, removal or lack of development of an overlying source sand layer, or dilution of the overlying sand by subsequent transport of sand-sized mud aggregates. Black shales are either irregularly laminated carbonate or siltstone, with poorly preserved fish, rare conchostracans, or magnesite micronodules, or flat-laminated clay and carbonate siltstone with well-preserved fish, conchostracans, rare ostracodes, and coprolites. Gray or black mudrocks ("stratified mudrocks" in their terminology) have fine-grained, current-laminated sandstone and siltstone and mudcracks that are generally a few centimeters long, tend to increase in density upward in the units, and are most common where this facies overlies, rather than underlies, the black shales. Fossils include plants, dinosaur trackways, and rare burrows and conchostracans. Traces of analcime, gypsum, and halite crystal molds also occur.

Other deposits that have many of the features described by Smoot and Castens-Seidell 1994 and Smoot and Lowenstein (1991) are the Lower Triassic "Bundsandstein" in Helgoland (Denmark) and Upper Triassic Malmros Klint "Member" in Jameson Land, Greenland (Clemmensen 1979) and Jurassic units in the Culpeper Basin in northern Virginia, U.S. (Gore 1988).

Lakes with organic-rich bottom sediments

Permanently stratified lakes are called meromictic (e.g., Anderson et al. 1985) and are characterized by oxygen-poor bottom waters and organic-rich sediments. Because meromixis can be brought about by salinity stratification, many meromictic lakes are also saline, and thus the presence of organic-rich sediments is sometimes used as a criterion for the recognition of ancient saline lakes. However, permanent stratification can also be related to temperature, and some meromictic lakes are stratified by temperature, not salinity, or have bottom waters that are both colder and slightly more saline than the surface waters. These lakes can also accumulate organic-rich sediments. The issue is further confused by salinity stratification that might be the result of an influx of saline groundwaters that has nothing to do with climate (e.g., Anderson et al. 1985; Renaut and Tiercelin 1994). Meromictic lakes in North America are concentrated in high mid-latitudes, between about 42° and 55° latitude (Anderson et al. 1985) and in Africa close to the equator (Talbot 1988), two very different climate zones that have in common a generally more humid climate than the surrounding regions. Finally, it should be noted that not all meromictic lakes have organic-rich bottom sediments.

The origin of organic-rich lacustrine sediments has been the subject of some controversy. Talbot (1988), using the East African lakes as examples, argued that the sediments richest in organic matter are in meromictic (stratified) or monomictic (unstratified) lakes in humid climates with gentle winds. He emphasized that the organic-matter accumulation in saline lakes occurs during lake highstands when the surface water is fresh. Shape and orientation of the lake relative to wind also are important in limiting vertical mixing. Talbot (1988) observed that high lake levels in Africa in the Holocene corresponded to times of lower wind strength as determined from eolian dust.

The importance of changes in productivity, as opposed to stratification, in the accumulation of organic matter in lakes has not been widely discussed. Kelts (1988) noted that alkaline lakes are particularly productive because of the pH dependence of nutrient availability and CO_2 solubility, both of which are higher in alkaline lakes. Relatively little debate exists so far on the question of preservation versus production of organic matter, which has been so important in studies of marine organic-rich rocks.

It is commonly assumed that laminated sediments must indicate anoxic bottom waters, as otherwise the sediments would be bioturbated. Olsen (1985) presented some evidence to show that the presence of laminated sediments in lakes is not necessarily dependent on anoxic bottom waters. In Olsen's (1985) data set (figure 5.38), the oxygen content of the bottom waters is inversely related to primary productivity. The preservation of laminated sediments in such lakes results from the exclusion of benthic organisms by anoxic bottom waters. However, some lakes can be virtually sterile from lack of nutrients and, in these cases, benthic organisms are excluded because of the lack of food. Sediments in such lakes also can be laminated (figure 5.38), as can glaciolacustrine lakes (Ashley, Shaw, and Smith 1985), which might be low in nutrients as well as cold.

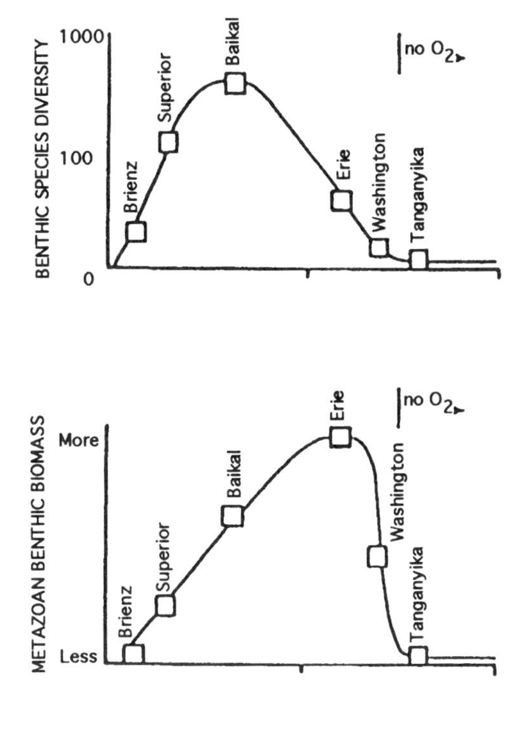

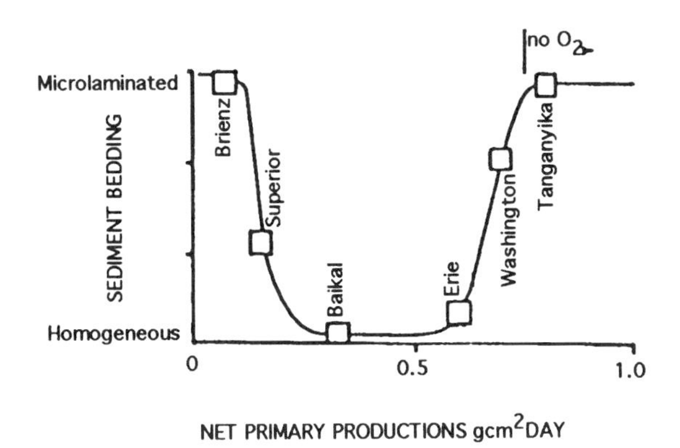

Figure 5.38. Relationship between the preservation of microlaminated sediments and net primary production and density and biomass of benthic organisms. From Olsen (1985).

These observations have important implications for the interpretation of the paleoecology of ancient lakes with laminated sediments.

The potential paleoclimatic information contained in lamination types remains to be exploited in the lacustrine rocks in the older geologic record except in evaporites (however, see Anderson and Dean 1995). The major use to date has been varve-counting as a method of dating lacustrine sequences with the intent of linking lacustrine cyclicity to Milankovitch cycles (Olsen 1986, 1990).

Geochemistry of Lacustrine Sediments

Whether a lake is a closed or open system can influence its sensitivity as a recorder of climate. Oxygen and carbon isotope analyses of lacustrine carbonates can provide some information that might permit distinguishing closed and open lakes in the geologic record where paleogeographic information is incomplete (Talbot and Kelts 1990). As in marine carbonates, the oxygen isotopic signature of primary lacustrine carbonates is a function of temperature and the isotopic composition of the water. The carbon isotopic signature of primary lacustrine carbonates is a function of the isotopic composition of the dissolved inorganic carbon at the time of precipitation, which is influenced by a complex interplay of numerous factors (McKenzie 1985). Generally, little or no correlation exists between $\delta^{18}O$ and $\delta^{13}C$ in open lakes, especially if the residence time of water in the lake is short; in those instances, $\delta^{18}O$ is invariant and related to the composition of inflow (Talbot 1990). In contrast, $\delta^{18}O$ and $\delta^{13}C$ usually covary in closed lakes; if the correlation is ≥ 0.7, a closed lake is indicated (Talbot 1990). In addition, the range of $\delta^{18}O$ values will be several per mil in a closed lake. The water balance in a closed lake influences $\delta^{18}O$, with heavier values resulting from the fact that evaporation, which preferentially removes ^{16}O, is the only outflow mechanism. However, under certain circumstances, an open lake might have covariance, possibly indicating long-term (Ka) climatic change (Drummond, Patterson, and Walker 1995).

Figure 5.39 is a summary of the possible factors affecting the position of the $\delta^{18}O$-$\delta^{13}C$ correlation line along the respective axes and the slope of the line (Talbot 1990). Each closed lake seems to have a unique signature that results in a conservative correlation line, and changes in the

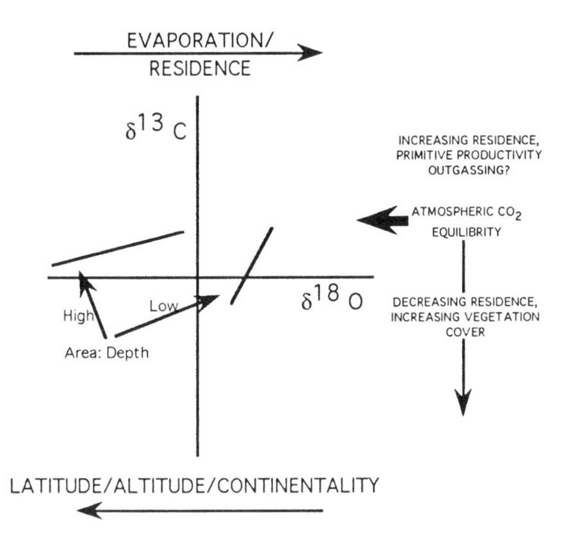

Figure 5.39. Summary of the controls on the relationship between $\delta^{18}O$ and $\delta^{13}C$ in closed lakes. Reprinted from *Chemical Geology* 80, Talbot, M.R., "A review of the palaeohydrological interpretation of carbon and oxygen isotopic ratios in primary lacustrine carbonates," pp. 261–79 (1990) with kind permission of Elsevier Science-NL, Sara Burgerhartstraat 25, 1055 KV Amsterdam, The Netherlands.

position and/or slope of the line might indicate some fundamental change in the hydrography of the basin. Talbot (1990) cited the Great Salt Lake as an example. Although a major climatic change caused the lake to shrink in size around 12 Ka, the $\delta^{18}O$ and $\delta^{13}C$ of the aragonitic muds maintained the same relationship.

Not all closed lakes show covariation of $\delta^{18}O$ and $\delta^{13}C$. For example, Pleistocene Lake Lisan (Israel) shows no such relationship, although the lake was almost certainly closed (Talbot 1990). McKenzie (1985) attributed this to a lack of biologic activity in this highly saline lake.

In one of the few attempts to apply $\delta^{18}O$-$\delta^{13}C$ covariance to ancient deposits, Suchecki, Hubert, and Birney de Wet (1988) compared the $\delta^{13}C$ and $\delta^{18}O$ signatures in carbonates from Lower Jurassic lacustrine rocks of the East Berlin and Portland formations in the Hartford Basin (Massachussetts and Connecticut, U.S.) and the Scots Bay Formation in the Fundy Basin (Nova Scotia, Canada). The East Berlin and Portland formations include black and grey mudstones that are considered to be the deposits of lake-level highstands (Gierlowski-Kordesch and Rust 1994). In both units, the carbonates near the black or grey mudstones show a pronounced shift toward heavier $\delta^{13}C$ values; Suchecki, Hubert, and Birney de Wet (1988) interpreted this as reflecting productivity in the lake, with ^{12}C sequestered in the organic matter now preserved in the mudstones (McKenzie 1985). The $\delta^{18}O$ values are highly variable and not related to lithology in these units; these results are consistent with expected relationships for diagenetic carbonates (Talbot and Kelts 1986; Talbot 1990). In contrast, both $\delta^{13}C$ and $\delta^{18}O$ in the Scots Bay Formation show variation with lithology, and the isotopes also show a loose covariation, although Suchecki, Hubert, and Birney de Wet (1988) noted that the most enriched values of both isotopes were near bedding surfaces that showed some evidence of exposure, alteration, or erosion. The trend of the covariation had a moderate slope, which, according to Talbot (1990) would indicate a lake of moderate area/depth ratio. Suchecki, Hubert, and Birney de Wet (1988) interpreted the subparallel trends in the $\delta^{18}O$ and $\delta^{13}C$ curves as indicative of the establishment of salinity stratification, leading to greater sequestration of ^{12}C in organic matter in the sediments during times of lake contraction and enrichment in ^{18}O. The trend of the covariation also lies in the negative quadrant for both $\delta^{18}O$ and $\delta^{13}C$ (see figure 5.39), which indicates higher latitude or altitude or greater continentality, low residence time of the water, and/or relatively dense vegetation cover. The Scots Bay Formation lake was likely deposited under conditions of extreme continentality (Parrish, Ziegler, and Scotese 1982) and at high altitudes (Hay et al. 1982). Associated sediments and temperature estimates from paleosols were inconsistent with dense vegetation, but the negative $\delta^{13}C$ values might indicate a lower residence time for the water.

Other applications of isotope geochemistry to lake sediments have concentrated on information that can be derived from oxygen isotopes about temperature, the source of the water, and the effects of evaporation. For these parameters, the interpretation of oxygen isotopes in lakes and rivers is far more difficult than in the oceans. Source and evaporation effects are negligible in the oceans and the relationship of isotopic composition of the water to temperature and ice-volume effects is relatively straightforward. In lakes, however, large variations in oxygen isotopic composition of the waters can be related with equal probability to temperature, evaporation, or variation in the original source or sources of the water whose composition is recorded in shells or lake carbonates. In addition to variations in the original sources for the water, there are strong effects of latitude, altitude, and mean annual temperature (Gregory et al. 1989; Drummond et al. 1993; Norris et al. 1996). An example is the upper Miocene Camp Davis Formation (Wyoming, U.S.; Drummond et al. 1993). The $\delta^{18}O$ values in the carbonates were −21.5‰ to −35.2‰ (PDB). These values are very light, and Drummond et al. (1993) concluded that the water must have originated as glacial meltwater because the only comparably depleted values found today are in snow. The only $\delta^{18}O$ values close to those from the Camp Davis Formation are from calcite cement in sandstone from Antarctica and from very high-latitude sites in Cretaceous rocks of Australia (Gregory et al. 1989).

If carbonate precipitation occurred at 20°C in the Camp Davis lake (most occurs at cooler temperatures in modern lakes), the isotopic composition of the water would have been −26.7‰; if carbonate precipitation occurred at 30°C (an estimate that is almost certainly too high for this region), the isotopic composition of the water would have been −24.6‰. Even this latter, "best-case" scenario puts the Camp Davis lacustrine limestones at the extreme lower edge of the range of values in lake carbonates at the same latitude today (Drummond et al. 1993). If modern relationships between air temperature and isotopic composition of the waters are extrapolated to fit the Camp Davis data, the derived air temperatures are inconsistent with paleobotanical information that puts temperatures about the same as today's. Drummond et al. (1993) also ruled out an altitude effect. The Camp Davis catchments would have to have been between 3.9–5.1 km altitude in the Miocene; today, the region is at an altitude of 2.8 km, and no evidence exists for a major change in altitude since the Miocene (see comments on Lake Baikal by Kolodny and Luz 1991). Another possibility is recycling of meteoric water. If nearby lakes were large enough to be a moisture source for precipitation, the water derived from those lakes would have been isotopically light because of the double fractionation, first when precipitation filled the lake and then again when the lake water evaporated to provide a moisture source for precipitation in the surrounding regions. Several large lakes in the region could have contributed significantly to the moisture available for the precipitation to the lake in which the Camp Davis limestones were deposited. If one of these was the source for precipitation to the Camp Davis lake, the paleoaltitude for the Camp Davis catchment could have been as low as 3.3–4.4 km. Drummond et al. (1993) favored this explanation as the only plausible one, so they went on to discuss the discrepancy between the estimated altitude and the current altitude of 2.8 km. They suggested a combination of erosion and subsidence since the Miocene to explain the discrepancy. Ancient lake deposits or freshwater fossils give isotopic signatures that are very dissimilar to modern values, and interpreting this information has been challenging.

FLUVIAL AND ALLUVIAL DEPOSITS AND PRE-QUATERNARY GEOMORPHOLOGY

Hardly anyone would include fluvial and alluvial deposits on a list of paleoclimatic indicators. Nevertheless, the citation of climate as a control on the style of fluvial and alluvial sedimentation is quite common in the literature. Methods of climate interpretation of modern and Quaternary fluvial/alluvial deposits commonly require information on geomorphology that is only very rarely available in the older geologic record. However, some aspects of these methods might be more broadly applicable to the older record than has been generally realized. In addition, a small body of literature exists that addresses pre-Quaternary geomorphologic features and their relationship to paleoclimate and the use of sandstone composition to interpret paleoclimate. Although the use of such information is limited, a few studies are summarized here.

For the purposes of this book, the discussion here is focused narrowly on only those aspects that have potential for interpretation of climate in the pre-Quaternary geologic record and is confined to interpretation of the coarse-grained deposits (sandstone and conglomerate) and their stratigraphic architecture. Fluvial deposits in the broadest sense include paleosols, which are discussed elsewhere in this chapter. Some cold-climate indicators are found in fluvial deposits, most specifically in the coarse-grained facies. Because these are not necessarily indicative of glaciation, but only of cold climate, they are included here rather than in the section on glacial deposits. Finally, examples of pre-Quaternary geomorphology and sandstone composition and their applications to paleoclimate interpretation are presented.

Fluvial and Alluvial Processes

Rivers

Schumm and Brakenridge (1987) divided the issue of using river deposits to interpret climate into four categories—location, convergence, divergence, and sensitivity; a fifth category, important to ancient deposits, might be termed context. With respect to *location,* how one interprets river deposits can depend on the part of the river they were deposited in. Schumm and Brakenridge (1987) pointed out that one part of a river may be aggrading only in response to degradation upstream; the condition for the river overall may be degradation. This type of problem is relatively unimportant for fluvial deposits in the older geologic record, where the geomorphology is not preserved, because preserved fluvial deposits are by definition representative of an overall aggrading system—otherwise the rocks would not be there to study. Another problem of location is that interpretation of fluvial deposits might also vary when the rivers crossed more than one climate zone, as does, for example, the Nile River today. *Convergence* is different processes having the same effect and *divergence* is one process having different effects at different times (Schumm and Brakenridge 1987). For pre-Quaternary paleoclimatology, an important example of convergence includes the processes than can cause a stream to become braided, that is, high sediment load, rapid aggradation, and/or flashy or seasonal runoff. *Sensitivity* refers to the fact that a slight change in runoff, sediment yield, or stream power can cause either a slight change in fluvial morphology or a complete reorganization of the fluvial system, depending on how close the river is to certain thresholds when the change occurs. *Context* refers to the large-scale setting in time and place for a river. An excellent example of the importance of context was given by Bull (1991) in discussing the effects of glaciation on the Colorado River, U.S., which has always been in an arid or semi-arid climate. During the last glacial episode, glaciation and associated processes in the Rocky Mountains added more sediment to the system than the river could transport, so despite the lower sea level, the river aggraded. With the decrease in glaciation at the end of the episode, sediment supply dropped and degradation began, despite the fact that eustatic sea level was rising.

Vegetation is very important to fluvial processes. The relationships among vegetation, climate (effective precipitation), and sediment yield are illustrated in figure 5.40. An important parameter not expressed well in such diagrams is seasonality of precipitation, not only how variable rain is throughout the year but also the time of year the rain falls. For example, a Mediterranean-type climate, in which most of the annual rainfall is in the winter, yields more runoff per unit of precipitation than does a monsoonal climate, in which most rain falls in the summer, because evaporation loss is less in the cooler climate (Schumm and Brakenridge 1987). In addition, fluvial systems are significantly affected by the patchiness of rain in short time and small spatial scales. Storms have a disproportionate effect on fluvial erosion and sedimentation compared with less-concentrated precipitation (Bryan, Campbell, and Sutherland 1988; Hirschboeck 1988). In some environments, especially arid or semi-arid environments but even in some humid environments, the fluvial/alluvial record can consist entirely of storm-flood events. That storms and floods might increase or decrease dramatically in magnitude with only a modest change in average climate (Knox 1993) must also be considered in studies of fluvial/alluvial deposits.

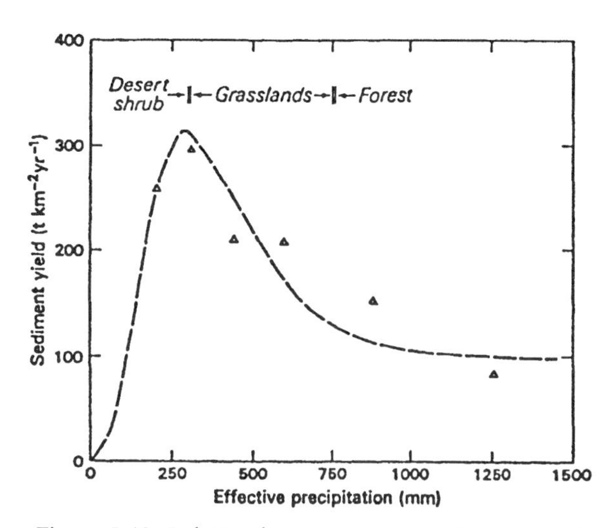

Figure 5.40. Relationships among vegetation type, sediment yield, and climate, expressed as effective precipitation. Effective precipitation is the amount of water available to support plants, weather rocks, form soils, and supply water for runoff, and is determined by precipitation, temperature, and windiness (Bull 1991). From Langbein and Schumm (1958) in *Eos* 39:1076–1084, copyright by the American Geophysical Union.

Landscape features are generally lost or only poorly represented in pre-Quaternary fluvial/alluvial deposits, so geologists have focused on variations in the stratigraphic architecture. Architecture in this sense refers to the geometry and lateral and vertical distribution of rock units, commonly emphasizing the sandstones (Miall 1990). Added to the architectural descriptions are numerous sedimentological features that can be useful for interpreting the functional morphology of the fluvial/alluvial system.

Geologists who have focused on climate interpretation of fluvial deposits have tended to emphasize braided-river deposits because such deposits seem to be confined to a narrower range of possible climatic conditions than meandering-river deposits, although braided streams themselves occur in a wide range of climates (Miall 1977). In sections with meandering-river deposits, the useful paleoclimatic information tends to come more from the fossils and fine-grained lithologies, such as paleosols and evaporites (e.g., Rust 1981). Characteristics of channel sandstones that are regarded as important for the interpretation of fluvial deposits are the width-to-height (-depth) ratio of individual channel sandstone bodies and their relationships to other sandstone bodies within a sequence, that is, whether several sandstone bodies are stacked vertically, offset from one another, isolated within fine-grained rocks, and so on. Braided-river deposits are multilateral (i.e., numerous channels along the lateral extent of the sandstone) and multistoried, with abundant internal scoured surfaces, coarse grain size, rapid textural changes, low ratio of shale to sandstone, unimodal paleocurrent trends, and presence of planar and trough crossbed sets low in the sequence. Channel-lag deposits, erosive bases, absence of sorting, and typically large size of extraformational clasts suggest single, high energy, short-lived events with a high sediment load (Miall 1977). These features, contrasted with typical features of meandering-stream deposits, are summarized in table 5.11.

Distinguishing between the deposits of ephemeral and perennial braided streams also has been an important exercise for interpreting paleoclimate, and criteria that can be used are listed in table 5.11. Ephemeral-stream deposits are more likely than perennial-stream deposits to have abundant mudcracked mud drapes, early cementation or case hardening reflected in steep scours and sand intraclasts, intercalations of wind-deflation gravels or eolian sand beds or dunes, abundant mud clasts or peloids, thick caliche or silcrete in the overbank, and root traces that disrupt the channel deposits because channels are the preferred sites for plant growth in arid environments (Smoot and Lowenstein 1991). Perennial-stream sandstones have more low-flow-regime structures, indicating deeper, more sustained flow; they may be meandering-river deposits (Dubiel and Smoot 1994). Dubiel and Smoot (1994) regarded the fine-grained overbank facies as most diagnostic in distinguishing between ephemeral and perennial streams. Thick floodplain deposits imply at least episodic flooding and high precipitation, although the precipitation can be distant from the river itself. Distribution and size of mudcracks, invertebrate trace fossils, and root traces all can be indicative of the frequency and size of floods.

The abundance of fine-grained facies is key to distinguishing between ephemeral and perennial streams and and for interpreting paleoclimate (Dubiel and Smoot 1994). Dubiel and Smoot (1994) included a lengthy discussion of mudcracks, which are commonly cited as evidence for drying. In their opinion, although mudcracks clearly indicate a drying episode, it is very difficult to infer the length or intensity of the drying event from these structures. In general, complex cracking is more consistent with more prolonged drying; narrow, jagged cracks with mostly dry conditions with intermittent wetting; and sinuous, branching mudcracks with repeated wetting and drying. Vesicular features indicate very dry conditions, and slickensides indicate repeated wetting and drying in clay-rich soils. In practice, comparative studies of mudcracks within or between formations might be useful, but an observation of mudcracks in scattered horizons is probably of limited use in paleoclimate interpretation.

Because fluvial deposition as studied in the pre-Quaternary geologic record encompasses much longer time intervals than more recent deposits, sorting out the effects of tectonics from the effects of climate becomes particularly crucial. Information on fluvial style is insufficient to decide among various climate scenarios (see, for example, Behrensmeyer and Tauxe 1982; Glover and Powell 1996). Paleobotanical, paleosol, and regional tectonic studies might provide enough additional information to make a decision. For example, Smith (1994) argued that channel geometry and abundance (that is, the sand-to-shale ratio) in the St. David Formation (Plio-

Table 5.11.
Features of braided- and meandering-stream deposits and distinguishing features of braided ephemeral and perennial streams

		Braided stream	
Braided stream	Meandering stream	Ephemeral	Perennial
Multilateral sand bodies	Isolated sand bodies	Syndepositional caliche-clast conglomerates present	Intraformational conglomerates common
Multistoried	Single-storied	Multistoried	Single-storied
Abundant internal scoured surfaces	Relatively few internal scours	Poorly sorted channel sandstones	Better-sorted channel sandstones
Coarse grain size	Fine grain size	Low aspect-ratio sandstone bodies	High aspect-ratio sandstone bodies
Rapid textural changes	Gradual textural changes	Broad, shallow, and flat-bottomed sand bodies, filled with braid-bar deposits	Sheet-like, coarse-grained sand bodies (15–20 m thick, 10s km wide)
Low ratio of shale to sandstone	High ratio of shale to sandstone	Critical to super-critical flow-regime structures	More low-flow-regime structures
Narrow range of paleocurrent directions	Wide range of paleocurrent directions	Fine-grained facies constitute < 5% of the rock	Fine-grained facies constitute 10–35% of the rock
Presence of planar and trough crossbed sets in the sequence	Absence of planar and trough crossbed sets	Aggradation	Lateral accretion more common
Abundant channel-lag deposits	Fewer, more confined channel-lag deposits	Abundant exposure features, e.g., early cementation or case hardening resulting in steep scours and sand intraclasts	Fewer exposure features
Fining-upward channel-abandonment deposits, rarely >3 m	Fine-grained fill of abandoned channels, commonly >3 m	Abundant flat mud-pebble intraclasts	Flad mud-pebble intraclasts less common
Point and linguloid bars, low-water bar accretion	Fining-upward point bars, epsilon crossbeds	Desiccation cracks, caliche, trace fossils, root traces in channels	Root traces in overbank
Erosive bases always present	Erosive bases not necessarily present	Episodic deposition	Continual deposition
Poorly sorted sediments	Well-sorted sediments	Intercalations of wind deflation gravels, eolian deposits	Wind-related features not intercalated with channel deposits
Large extraformational clasts	Small extraformational clasts		

From information in various sources (Miall 1977; DeLuca and Eriksson 1989; Smoot and Lowenstein 1991; Dubiel and Smoot 1994).

Pleistocene; Arizona, U.S.) were related to climate. A change from sheet channels to ribbon channels at 3.4 Ma was accompanied by increases in sediment-accumulation rate, the abundance of hydromorphic calcic paleosols, and the abundance of pond carbonates. Paleosol-carbonate isotopes indicated an increase in total precipitation and a decrease in the seasonality of precipitation at the same time (Smith et al. 1993). At about 2.7 Ma, sediment accumulation rates decreased, pond and hydromorphic paleosol carbonates became much less abundant, and sheet channels again predominated. Stable-isotope data indicated a decrease in precipitation at that time, which continued to about 1.6 Ma, when alluvial fan sedimentation commenced in the St. David Formation. The St.

David Formation was deposited over basin-margin faults and aggraded very slowly [evidence cited by Smith et al. (1993) for tectonic quiescence during St. David deposition]. Thus Smith (1994) concluded that variations in fluvial morphology were related to climatic change, a conclusion supported by information from the paleosols.

Alluvial fans

Until recently, many workers tended to associate alluvial fans with arid and semi-arid climates, but it is now recognized that fan landforms can occur in any climate. Stanistreet and McCarthy (1993) classified fans as debris-flow-dominated, braided-fluvial, and low-sinuosity, meandering-fluvial. Debris-flow-dominated fans are small scale (Dubiel and Smoot 1994) and have steep slopes and little vegetation (Stanistreet and McCarthy 1993; Dubiel and Smoot 1994). Deposition at the fan head may be channelized and comprises debris flows, whereas in the distal portions of the fan, deposition may occur in sheet flows and shallow, braided streams (Miall 1992a). Fluvial deposits are patchy, thin, coarse grained, and poorly sorted, with upper-flow-regime sedimentary structures indicating flashy (that is, short-lived, high-flow-rate) discharge (Dubiel and Smoot 1994). Lack of channel confinement beyond the proximal portions and flashy sedimentation characterize these landforms. These fans are common in arid climates; tectonics plays an important role in fan formation because fan formation requires a change from confined to unconfined flow within a short distance (Schumm, Mosley, and Weaver 1987 and references therein). With increasing runoff, fan deposits become increasingly fluvial in nature and, usually, larger. Because the fan landform is usually not preserved or is difficult to recognize in the geologic record, the deposits of braided-fluvial and low-sinuosity, meandering-fluvial fans (Stanistreet and McCarthy 1993) may be indistinguishable from fluvial deposits. Some workers would limit the definition of alluvial fans to only those that are debris- or sheet-flow dominated and demonstrably fan shaped (Blair and McPherson 1994), but this view is not widely accepted. Although debate about terminology has been intense, most geologists seem to take a pragmatic approach that recognizes wide variation in alluvial-fan deposits but draws a line between alluvial-fan deposits that can be recognized as such in the geologic record and those that are indistinguishable from fluvial deposits.

Alluvial fan deposits have been used for paleoclimate interpretation only to the extent that the fans are interpreted either as "arid" or "wet" alluvial fans. In both cases, the climate interpretation is generally made on the basis of other types of evidence, such as the presence of evaporites, eolian deposits, or coal beds, and the alluvial fan deposits themselves are subsidiary. Alluvial-fan sedimentation has received much more attention for the light it can throw on tectonic, rather than paleoclimatic, changes, although climate is commonly an element of most interpretations (e.g., Ridgway and DeCelles 1993 as one of many examples). The existence of fans requires tectonics to establish the geomorphic setting for fan formation, and the larger-scale features of fan architecture are most likely related to tectonics and intrinsic fan-depositional processes (DeCelles et al. 1991). Smaller elements, however (the first- to fourth-order lithosomes of DeCelles et al. 1991), may be related to climate, but not necessarily.

Indicators of cold climates in fluvial/alluvial deposits

Martini, Kwong, and Sadura (1993) described diagnostic features in fluvial deposits for ice-rafting; such features might be preserved in glacial outwash fan deposits, although they are much more likely to be preserved in valley-floor fluvial deposits, and ice-rafting is not necessarily related to glaciation per se. Their description of fluvial, ice-rafted sediments was "patchy silt drapes over grassy banks, lenses of unsorted coarse sand and pebbles in fine overbank deposits, lenses of poorly sorted gravel and isolated pebbles and boulders strewn on levees and the treeless deltaic floodplain." To my knowledge, these criteria have not yet been applied to ancient deposits.

Sand wedges and ice-wedge casts are vertical, sand-filled, wedge-shaped structures in gravel or finer-grained sediments. They form by incremental deformation by ice that freezes from water in the atmosphere. They are typically a few meters across in the widest portion at the top and as much as 10 m deep (Black 1976). Sand wedges are the vertical expression of thermal-contraction polygons, also called patterned ground. They indicate progressive drying in a permafrost-dominated environment and the sand is commonly eolian (Carter 1983). Sand wedges can be mistaken for other types of structures, such as those formed by simple frost-wedging or tension cracks (Black 1976). Diagnostic features are deformation of the surrounding host sediments and polygonal features in plan view (Black 1976).

Pre-Quaternary Geomorphology

Geomorphology has played a key role in interpretation of Quaternary continental climates; much of the Earth's surface today inherits its form from climates in the recent past. Unfortunately, land-surface features are poorly preserved in the older record. Although valleys and their fill deposits can be recognized (e.g., the Shinarump Member of the Chinle Formation, Arizona, U.S.), the complex terrace landforms that are so important to Quaternary paleoclimate reconstruction are not preserved well enough to allow a detailed reconstruction of past river behavior, and other types of landforms are equally rare. One notable exception is a large-scale, late Eocene erosion surface in the Front Range of Colorado, U.S., the study of which has raised some provocative ideas about the importance of the variability in storminess.

Chase (1991) indirectly modeled the effects of small versus large storms on erosion and sedimentation. Storms were simulated by "precipitons," that is, increments of precipitation or (once the water hits the ground) floods; the precipitons simulate erosion in response to storm runoff. When precipitons first fall on the model topography, they cause diffusion, representing slope wash, slumping, talus formation, and soil creep, processes that change the landscape over short spatial scales. High values of diffusion are a proxy for humid, rapid-weathering climates (Chase 1991). A range of values for diffusion could represent all climates that have steady rainfall, whether that rainfall occurs in arid or humid climates. In arid climates, steady rainfall (if such a climate existed) would have virtually no effect on the landscape; the amount of water available for erosion and transport would be negligible. In humid climates, vegetation stablilizes the slopes such that the erosive power is negligible, although runoff might be significant because of soil saturation; storms must be either intense or long lasting to produce mass wasting on heavily vegetated slopes (e.g., Kochel 1990). The model landscape was only slowly modified by diffusion, whereas diffusion plus erosion significantly altered the landscape with the same amount of precipitation.

Gregory and Chase (1994) applied this model to a study of the late Eocene erosion surface in Colorado, U.S.. The surface is low relief and now stands at about 2.5 km altitude. Late Eocene unconformities are widespread in the western United States (Dickinson et al. 1988; Gregory and Chase 1994). Previous studies had suggested that the erosion surface was formed by peneplanation processes at low altitude and then uplifted and dissected by a later orogeny. However, paleobotanical data from rocks just overlying the surface were interpreted as indicating an altitude of 2.2 to 3.3 km (Gregory 1992; Gregory and Chase 1992, 1994). Gregory and Chase (1994) investigated various tectonic models as explanations for the late Eocene erosion surface; no reasonable tectonic model was sufficient (see also Chase et al., in press). The only possible alternative explanation for such a smooth surface was some influence of climate, leading to the application of Chase's (1991) model. In the model, the surface was generated by small storms, and present topography by a reduction in the number and increase in the size of storms, consistent with paleobotanical information and climate models.

When an erosion surface stops eroding, it starts to weather and form paleosols and other weathering features. At any point in time, the Earth's surface is covered by a mosaic of vegetation and soil types, depending on local climate, vegetation, drainage, and so on. It is possible to trace such surfaces over short distances (a few kilometers) in the geologic record, and local drainage conditions, especially, are thus relatively easily documented (Kraus 1997). More difficult to study are large-scale surfaces, but if such surfaces can be documented, they provide detailed information about the spatial heterogeneity of paleoclimate. Simon-Coinçon, Thiry, and Schmitt (1997) documented karst and other weathering features on an early Tertiary paleosurface in France, showing variability on a large spatial scale. Despite the spatial variability, the temporal evolution of the surface was similar across large portions of the region. They attributed some of this temporal similarity to changes in climate, which were reflected across the region, despite the local variations. Tectonics also overprinted the spatial heterogeneity.

Composition of Continental Sandstones

Sandstone composition has been used principally as a way of studying the provenance of sands. Recent studies of sandstone composition have focused on tectonic influences and particularly on using sandstone composition to determine plate-tectonic setting (Dickinson and Suczek 1979). Workers in the field of sandstone petrology have long acknowledged that climate could have an important impact on sandstone composition, and a number of studies have shown a correlation between climate and the composition of fluvial sands (see review by Johnsson 1992). In general, the warmer, wetter, and more vegetated a region is (note that vegetation is controlled by temperature and precipitation), the more intense is the chemical weathering and the more compositionally mature (quartz rich) is the resulting sand.

It is clear from many studies (e.g., Suttner and Dutta 1986; Grantham and Velbel 1988; Johnsson, Stallard, and Meade 1988; Johnsson 1990; Heins 1993) that sandstone composition can, at the most, provide support for paleoclimate interpretations that are based principally on other types of information. Interpretations are dependent on knowledge of variations in the source rock, information that is unlikely to be available in the older geologic record. In addition, the sands that most closely reflect provenance and climate must come from the proximal parts of the basin, just the parts that are least likely to be preserved in the geologic record. Although the compositional changes in the rocks studied by Suttner and Dutta (1986) might be interpreted as evidence for paleoclimatic change, in the absence of other types of information, they could as easily be interpreted as changes in source rock with unroofing, stream capture, or changes in tectonics. The independent types of information required would be other paleoclimatic indicators and/or good paleogeographic and paleogeologic information for the units studied.

GLACIAL DEPOSITS AND RELATED INDICATORS

More than anything else, the landforms and deposits left by the Pleistocene continental ice sheets have defined the development of paleoclimatology as a discipline of geology. Although the current glacial episode[2] is only the most recent of several in Earth history, the fact that it is still ongoing and the preserved record of the most recent glacial advances so extensive and dramatic meant that for many decades, the field of paleoclimatology was, for all intents and purposes, "Ice-Age paleoclimatology." One might argue that because of this recency, glacial deposits have received attention all out of proportion to their importance as paleoclimatic indicators, or to the importance of glacial episodes in the climate history of the Earth. Regardless, the fact that Earth's climate has apparently oscillated, regularly or otherwise, between times of "icehouse" climates, when continental ice sheets were extensive, and longer periods of "greenhouse" climates, when continental ice sheets were not in evidence, has captured the imaginations of numerous paleoclimatologists, who have attempted to fit significant evolutionary, chemical, and sedimentologic, as well as climatic, events, into the oscillatory framework (e.g., Williams 1975; Fischer and Arthur 1977; Weissert and Lini 1991; Frakes, Francis, and Syktus 1992), a practice not universally accepted (Eyles 1993).

Indicators of glaciation are both direct and indirect. Direct indicators include sediments created by glacial action, whether directly deposited by ice or reworked, and glacial landforms at various scales. Indirect indicators are many and are treated in other sections of this book. These include periglacial eolian deposits, isotope data from rocks and fossils, paleobiogeography, ikaite, sand wedges, and clay minerals; some of these may merely indicate very cold climate and not necessarily the presence of ice. An indirect indicator of glaciation discussed in this section is glacio-eustatic sea-level change. For the purposes of the following discussion, the principal indicators are summarized in table 5.12.

Much of the older literature on the Earth's glacial record assumed that all coarse-textured, poorly sorted deposits—diamictites—were glacial in origin (Eyles 1993). Although some earlier workers recognized the problems with this assumption, an influential paper by Schermerhorn (1974) forced the recognition that such deposits form in a variety of ways (Eyles 1993; see Chumakov and Frakes 1997 for a recent discussion of one controversial unit). The familiar term "tillite" is now restricted specifically to diamictite derived from sediments deposited directly by glaciers (Eyles, Eyles, and Miall 1983). The common terms for the unlithified counterparts are "till" and "diamict" (Eyles, Eyles, and Miall 1983). Hambrey, Ehrmann, and Larsen (1991)

2. The term "glacial episode" is used here for the long-term events, e.g., the Permo-Carboniferous glacial episode. The terms "glacial advances" and "glacial retreats" are used to refer to the shorter-term events, of which there were generally several during a glacial episode. "Ice Ages" is here restricted to Pleistocene glacial cycles.

Table 5.12.
Indicators of glaciation in glacial depositional systems

Glacioterrestrial depositional systems	
Subglacial	Moderate preservation potential; characterized by scoured, deeply dissected, and/or deformed bedrock and deposits of lodgement, deformation, and melt-out till
Glaciolacustrine	Moderate preservation potential; characterized by varves, dropstones, coarse fan-delta and gravity flow deposits, ice-deformed bottom deposits*
Supraglacial	Poor preservation potential
Glaciated valley	Poor preservation potential
Glaciofluvial	Poor preservation potential as recognizably glacial deposits
Glaciomarine depositional systems	
Ice-contact	Poor preservation potential; characterized by "rapid lateral and vertical facies changes, highly irregular sediment geometries and a great range of facies types" (Eyles 1993)
Shelf and slope	High preservation potential
Gravity flows	Various types of graded or massive deposits; abundance of coarse-grained gravity-flow deposits associated with debris flow deposits may be diagnostic
Ice-rafting	Dropstones in fine-grained sediments and diamictites interbedded with hemipelagic sediments

From information in Eyles (1993).

*These features can be indistinguishable from their marine counterparts, and thus are not discussed separately. See text for further explanation of indicators that have moderate or high preservation potential. See figure 5.41 for positions of the depositional systems and typical stratigraphic columns.

recommended "diamicton" for the unlithified deposits and "diamict" as an inclusive term for both the lithified and unlithified deposits, but the terminology suggested by Eyles, Eyles, and Miall (1983) will be used here.

Diamictites can be of glacial origin but reworked, in which case the term "tillite" is inappropriate. Instead, the term "glacially influenced" is used for deposits or basins whose formation and evolution substantially depended on sediment flux from glaciers, glacio-eustatic sea-level changes, or isostatic loading of the crust by ice (Eyles, Eyles, and Miall 1985). Although "glacially influenced marine depositional systems" is more correct, these systems will be referred to here by the less accurate but shorter "glaciomarine." The emphasis in recent years has been less on glacial sediments and more on glacial depositional systems (Brodzikowski and van Loon 1991; Eyles 1993). The principal glacial depositional systems are illustrated in figure 5.41 (Eyles and Eyles 1992).

Notwithstanding the spectacular continental deposits of the Pleistocene Ice Ages, by far the most abundant indicators of glaciation are glaciomarine deposits. Eyles (1993), in a comprehensive review of glacial deposits, estimated that >95% by volume of glacial deposits are glaciomarine. Overall, the sediments in glaciomarine systems have a higher preservation potential than do those in glacioterrestrial systems (see table 5.12; Eyles 1993). Among the glacioterrestrial depositional systems, the subglacial and glaciolacustrine systems probably have the highest preservation potential. The other depositional systems—supraglacial, glaciated valley, and glaciofluvial—have low preservation potential. The sediments deposited in these systems are likely to be reworked and resedimented to the extent that their glacial origin might be impossible to discern (Eyles 1993). These deposits will not be discussed further here, and the reader is referred to the review by Brodzikowski and van Loon (1991).

Indicators of Glaciation

A number of geologic features are regarded as diagnostic of the action of ice and are particularly important in the interpretation of diamictites for which contextual data are poor or lacking. In terrestrial environments, these features include abraded bedrock; in both terrestrial and aquatic environments, striated and faceted clasts; and in aquatic environments, dropstones and ice-rafted debris (Hambrey and Harland 1981a).

Abraded bedrock

Bedrock abraded by ice acquires characteristic surface features (Hambrey and Harland 1981b). These include polished surfaces, crescentic chattermarks, nailhead striae, intersecting sets of striations, and a hummocky character called "rouche moutonée" (all beautifully illustrated by Hambrey and Harland 1981b). Striations are

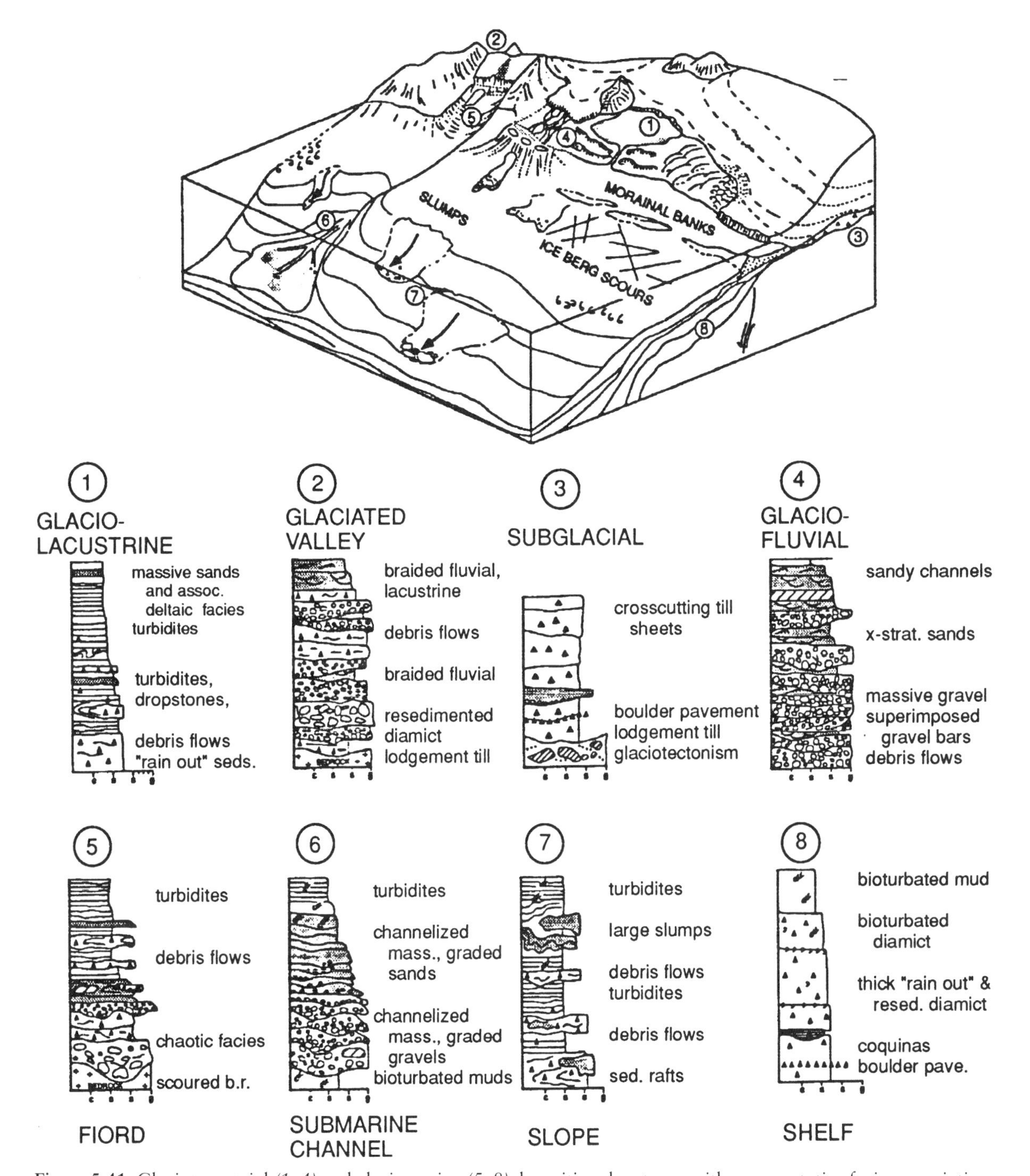

Figure 5.41. Glacioterrestrial *(1–4)* and glaciomarine *(5–8)* depositional systems, with representative facies associations. Numbers on the block diagram correspond to those with the stratigraphic columns. From Eyles and Eyles (1992).

formed by processes other than glacial action, but Hambrey and Harland (1981b) regarded nailhead striae, crescentic gouges, and several intersecting sets of striations as diagnostic. Polished surfaces in the absence of other features are also not unique to the glacial environment; rivers, for example, can leave highly polished bedrock surfaces (Baker 1988).

Surface features and composition of clasts

Like bedrock, cobble-sized or larger clasts subjected to glacial abrasion can be polished and striated, but these features can be formed in a variety of other ways (Eyles 1993). Eyles (1990, 1993) emphasized that, to be diagnostic of glacial action, the clasts must be faceted as well as striated, specifically into a bullet shape; such clasts might, in reworked glacial sediments, be the only indication of glaciation in the record. However, in most cases, workers do not emphasize the shapes of the clasts. Clasts can also have two intersecting sets of striations (Eyles and Eyles 1989). By no means will all clasts in a glacially derived diamictite show features caused by abrasion by ice. The percent of clasts showing striations in glacial sediments ranges from 0.1% to 28%, depending on clast size, composition, and possibly other factors not determined (Drewry 1986). Ice reshaping favors survival of bladed and spherical clasts (Drewry 1986).

In sand-sized clasts, numerous surface features are considered diagnostic of glacial action, particularly when they occur together (Margolis and Kennett 1971). These include large-scale (>1 μm) breakage blocks, large (>1 μm) conchoidal fractures, sharp angular outline, high relief (>1 μm), and smooth fracture planes. In addition, glacially influenced sand grains commonly have randomly oriented striations, semiparallel, step-like fractures, arc-shaped steps (chattermarks), and meandering ridges (>5 μm in length). Particularly diagnostic are chattermark trails, especially on garnets, and angular quartz grains with conchoidal fracture surfaces (Gravenor 1981; Hambrey and Harland 1981b). A wide range of composition of the sand grains and larger clasts is also considered important for an interpretation of glacial action (Hambrey and Harland 1981b; Krinsley and Trusty 1985). Eyles (1993) particularly emphasized "far-traveled," extrabasinal clasts.

Dropstones, ice-rafted debris, and sea-ice scour marks

Dropstones are outsized—coarse sand to boulder—clasts in finer-grained rock, usually siltstone to claystone. Dropstones are generally isolated (some writers call them "lonestones," e.g., Crowell 1995). Ice-rafted debris (IRD), on the other hand, consists of layers of coarse-grained sediment within a generally finer-grained sequence. IRD is interpreted to have fallen through the water column to the sea (or lake) floor from floating, sediment-laden masses of ice. IRD is informally distinguished from diamictite, which also can be deposited from melting ice, by being interbedded with fine-grained, aquatic sediments and by having been deposited at some distance from the glacier terminus. A gradation between pure diamictite and IRD as defined here is expected and observed.

Dropstones. Extensive deposits of dropstones that coincide in time with other evidence of glaciation were probably formed from clasts released from melting glacial ice. A particularly useful feature for diagnosis of glacially derived dropstones is clots of sediment that appear to adhere to larger clasts (Visser 1983; Eyles, Eyles, and Miall 1985). These originate as detritus frozen to a larger clast. They are most commonly observed in tillite but might occur in dropstones shed into shallow and/or near-freezing water.

Although dropstones commonly are assumed to be diagnostic of glaciation, they are not unequivocally so, except, possibly, for the Proterozoic. Even very large clasts are commonly and easily transported in the holdfasts of marine algae and in the root balls and stems of woody plants (Emery 1955, 1963; Gilbert 1990). This is an important mechanism of clast transport that possibly has been underestimated, with regard to both the size and the amount of clasts transported. For example, Frakes and Francis (1988) considered wood rafting for dropstones in the Cretaceous Bulldog Shale of Australia (Sheard 1990) because wood is present in the deposit, but they rejected that hypothesis because of quantity of material and size of clasts (to 3 m in diameter), despite the fact that clasts as large as 3 to 4 m have been reported from the root balls of modern trees (Emery 1955). The subject has received some attention in recent years (Gilbert 1990; Bennett, Doyle, and Mather 1996).

Perhaps the most telling evidence of the potential of vegetation as a significant source of dropstones is the observation that large erratics are occasionally noted in dredge samples from the modern ocean floor in regions where ice could not have been a factor (Emery 1955). Given the tiny area so sampled compared with the total area of the ocean floor, this observation implies that such erratics are actually quite common. In outcrop, such dropstones would be very prominent.

Ice-rafted debris. When clasts larger than the background aquatic sediments become abundant enough to form beds of diamictite, establishing a glacial origin becomes a matter of distinguishing the deposits from those of debris flows or turbidites, particularly if the diamictite has been reworked. This can be almost impossible under some circumstances. Six types of information are used to make the distinction. These are (1) presence of striated, faceted clasts, or surface features or shapes diagnostic of abrasion within ice (discussed in the previous section); (2) geometry of the deposits; (3) lack of graded bedding, current, or flow structures; (4) lack of a connected source; (5) grain-size distribution; and (6) stratigraphic and paleogeographic contextual information (Keany et al. 1976; Hambrey and Harland 1981a; Anderson 1983; Breza and Wise 1992; Eyles 1993).

Distal IRD rarely contains large clasts. Rather, the IRD is coarse sand or finer, and distinguishing it from distal turbidites or eolian sands can be difficult. Several workers have established various types of quantitative analyses for distinguishing IRD from other types of sediment. Two grain-size cutoffs have been used, medium sand (≥250 μm; Breza and Wise 1992) and very fine sand (>62 to 63 μm; Keany et al. 1976; Kennett and Barker 1990). The methods for quantifying the grains have also varied. Kent, Opdyke, and Ewing (1971) used dry weight percent of grains >250 μm, which included cobbles as large as 70 mm, for cores in the North Pacific. They noted grain-surface textures typical of glacially derived sands. Keany et al. (1976) used weight percent of the sand fraction (>62 to 250 μm) that had certain grain-surface textures, namely high degree of angularity, conchoidal surface fractures, and striations, to distinguish IRD. Margolis and Kennett (1971) took numerical percentages of glacially derived grains from point counts of 300 grains or all grains if they numbered fewer than 300. They also used grain-surface textures to identify glacially derived grains. In their study of ODP Site 748 (Kerguelen Plateau), Breza and Wise (1992) devised an IRD index, counting the entire suite of clastic grains in the medium to very coarse sand fraction (≥250 μm to <2 mm) to get the number of grains per gram of sediment.

The amount and character of the material transported by ice depends on a number of factors in addition to the variations in transport rate discussed previously. For example, Anderson (1983) noted that off Antarctica today, IRD is less than 10% of the sediment collected from the continental slope; that is, IRD decreases rapidly away from the shelf edge. However, this pattern could be unique to Antarctica because the currents keep icebergs close to and parallel to the coast for long periods. He also noted that till carried within and on top of glaciers is transported without grinding, and the clasts will thus lack the features regarded as diagnostic of glacial action. These clasts, though glacially derived, would not be useful for determining glacial origin of the diamictite in which they occur, and if they constitute a significant fraction of the sediment, correct interpretation of the diamictite might be very difficult.

Sea-ice scour marks. As icebergs traverse continental shelves on their way out to sea, they commonly leave grooves in the soft sediment at the sea floor. These ice-scour marks have been preserved in the geologic record (Rocha-Campos, Dos Santos, and Canuto 1994; Woodworth-Lynas and Dowdeswell 1994), sometimes interpreted as abraded bedrock (Woodworth-Lynas and Dowdeswell 1994). Criteria for their recognition are the presence of berms and evidence for soft-sediment deformation associated with the scours (Woodworth-Lynas and Dowdeswell 1994).

Glacial Deposits and Depositional Systems

Although the aforementioned features are regarded as diagnostic, their interpretation depends strongly on information about the deposits and depositional systems in which they occur, what Eyles (1993) called the contextual data. These are discussed in this section, divided into glacioterrestrial and glaciomarine depositional systems.

Glacioterrestrial deposits and depositional systems

Till in glacioterrestrial systems consists of three general types, lodgement till, deformation till, and melt-out till; many variations of these types have been described from the more recent record (Drewry 1986; Brodzikowski and van Loon 1991). Lodgement till forms as deposits at the base of the glacier over which the glacier slips, whereas deformation till is formed where some of the movement of the glaciers is accommodated within the sediments themselves. Lodgement till contains glacially shaped clasts, and deposits are generally <20 m thick. Deformation till is unstratified and is recognized by the presence of a deformed substrate of unconsolidated sediment and substrate sediments that are only partly assimilated into the till. Deformation till is probably the most common type; most of the Pleistocene till in North America and other regions is of this type (Eyles 1993). Glaciolacustrine deposits consist of laminated, coarse-grained gravity flow, and coarse- and fine-grained fan-delta deposits; bottom sediments in a glacially influenced lake may be deformed by ice movement. The laminated sediments (figure 5.42), commonly referred to as varves, are difficult to distin-

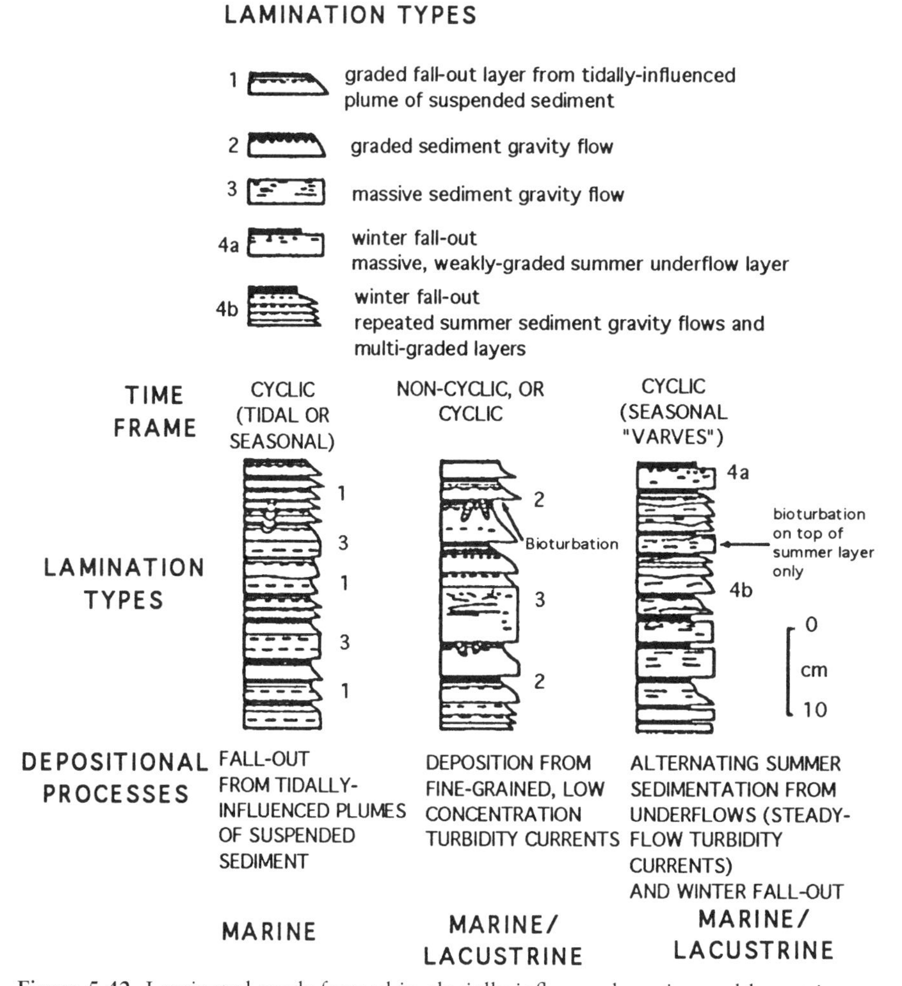

Figure 5.42. Laminated muds formed in glacially influenced marine and lacustrine environments. Reprinted from *Earth-Science Reviews* 35, Eyles, N., "Earth's glacial record and its tectonic setting," pp. 1–248 (1993) with kind permission of Elsevier Science-NL, Sara Burgerhartstraat 25, 1055 KV Amsterdam, The Netherlands.

guish from similar sediments formed in nonglacially influenced lakes and in many marine settings. Eyles (1993) emphasized that the recognition of laminated sediments as glacial varves, which record annual cycles of ice melt (but see Stihler, Stone, and Beget 1992), requires other contextual data. In addition, even glaciolacustrine laminated sediments can be formed by mechanisms other than seasonal melt (Ashley, Shaw, and Smith 1985). In any case, they are generally a minor part of the glaciolacustrine depositional system. Rain-out diamict and mass-flow deposits interbedded with fan-delta deposits are volumetrically much more important than laminated sediments (Eyles 1993). A descriptive classification system for these deposits was established by Eyles, Eyles, and Miall (1983).

Glaciomarine deposits and depositional systems

Eyles and Eyles (1992) and Eyles (1993) divided glaciomarine depositional systems into two parts, ice-contact depositional systems (proximal) and glacially influenced shelf and slope depositional systems (distal). Ice-contact depositional systems have very poor preservation potential because of subsequent scouring by ice or reworking and resedimentation of the deposits onto the shelf and slope (Eyles 1993). Ice-contact deposits are characterized by "rapid lateral and vertical facies changes, highly irregular sediment geometries and a great range of facies types" (Eyles 1993). Most sediment in the proximal system is brought in by meltwater and deposited within 10 km of the ice (Elverhøi 1984).

Glacially influenced deposits of the shelf and slope are formed by two processes, gravity flows of glacially derived sediment from shallow water and "rain-out" from the base of a floating ice sheet or from icebergs (IRD; Eyles 1993). These processes account for most glacial deposition, having created diamictites as thick as 3000 m in some ancient examples (Eyles 1993). The presence and extent of these deposits are controlled by the regional climate (Eyles 1993). Glaciers in temperate, humid climates produce large volumes of mud and meltwater that result in very high deposition rates of sediment if they discharge into marine basins (Anderson and Ashley 1991; Henrich 1991; Eyles 1993). In colder, drier climates, sedimentation rates in glaciomarine settings are an order of magnitude lower (Eyles 1993). Turbidites in glaciomarine settings are typically high-concentration flows, with a high proportion of coarse grains, particularly in proximal parts of the depositional system. Such deposits are commonly associated with debris flows (Eyles 1993). More distal deposits may be indistinguishable from fine-grained turbidites formed in other marine settings (Eyles 1993). Laminated sediments can be a greater component of glaciomarine than glaciolacustrine deposits (Eyles 1993), but laminations may be controlled by tidal or, possibly, solar influences (Williams 1981, 1988, 1989a,b; Sonett and Williams 1985; Williams and Sonett 1985; Sonett, Finney, and Williams 1988).

Eustatic Sea-Level Changes as Indicators of Glaciation

The one type of indirect evidence for glaciation discussed in this section is eustatic sea-level change. How to recognize eustatic sea-level fluctuations has been a source of major controversy best left to basin analysts (e.g., Vail, Mitchum, and Thompson 1977; Brandt 1986; Heckel 1986; Klein 1989; Miall 1992b). Like other types of evidence for glaciation, the interpretation of sea-level changes as glacially driven requires some contextual information. Sea-level changes have been best studied in the Quaternary, which will not be discussed further here, and in the Permo-Carboniferous. In addition, some work has been done on the short but apparently intense Late Ordovician glaciation.

Nearly everyone agrees that the stratigraphic cyclicity (cyclothems) that is so prominent in Permo-Carboniferous sections worldwide is the result of glacially driven eustatic sea-level changes. However, some controversy exists as to the explanation of the variation in stratigraphic expression of these cycles. The controversy has centered around whether climate or tectonics explains the variation. Early workers assumed that the stratigraphic expression of the cycles would somehow have to be consistent across an entire continent. Recent work, however, has shown that, just as continents are climatically heterogeneous today, they must have been in the past as well, and the expression of the cyclicity varied accordingly across the continent (Cecil 1990; Miller and West 1993). In addition, climatic change accompanying sea-level change introduces non-Waltherian elements to the stratigraphic expression of the cycles (Soreghan 1992a).

As has been emphasized throughout this chapter, the stratigraphic effects of changes in tectonic regime can mimic the effects of changes in climate. Sorting out the causes of relative sea-level change—eustasy versus local tectonics—is an even greater problem. With regard to the Permo-Carboniferous cyclothems in North America, the controversy has taken two forms: whether the primary cause of stratigraphic variation was tectonic or climatic, and whether the cyclothems have anything at all to do with eustatic sea-level change. The former controversy is expressed in papers by Heckel (1986) and Cecil (1990), who argued for a climatic control through lateral variations in climate, and by Klein and Willard (1989), who argued for tectonic control through differences in position on a flexed crustal plate. Both groups, however, accepted eustatic (i.e., glacially driven) sea-level change as the primary control on the cyclothems.

The second controversy has been addressed by Dickinson, Soreghan, and Giles (1994), who compared stratigraphic cycles in different basins in the southern Cordillera, U.S. Their criteria for attributing these cycles to glacio-eustasy were the following:

1. ubiquitous development of cycles in both the foreland basin of the Ouachita-Marathon orogenic system and in correlative non-foreland basins in Nevada and Utah;
2. absence of comparable cyclicity in adjacent older and younger sequences;
3. synchronous stratigraphic records across

foreland tectonic flexures and associated forebulges;
4. basinwide distribution of the cyclothems in selected basins;
5. interbasinal correlation of 17 individual cyclothems from the midcontinent region through the Ouachita-Marathon foreland (Boardman and Heckel 1989); and
6. apparent cyclothem duration of about 412 Ka, which resembles long-period Milankovitch cycles (Collier, Leeder, and Maynard 1990).

Dickinson, Soreghan, and Giles (1994) regarded (1) and (2) as the two key criteria for refuting the hypothesis that the cyclicity is the result of foreland basin tectonics.

In the absence of enough stable isotope data to estimate ice volume, Crowley and Baum (1991a) and Eyles (1993) used estimates of sea-level change from Permo-Carboniferous cyclothems to place constraints on ice volume. Crowley and Baum's (1991a) maximum estimates were 110×10^6 km^3 and a sea-level fall of 278 m, whereas Eyles' (1993) were 40×10^6 km^3 and 69 m, respectively. The difference was, according to Eyles (1993), the result of inclusion by Crowley and Baum (1991a) of glaciomarine sediments as indicators of ice-sheet extent, when such sediments were probably deposited well beyond the edge of the ice sheet (see also González-Bonorino and Eyles 1995).

Brenchley et al. (1994) used sea-level curves and stable isotope data (see figure 2.36) as evidence for Late Ordovician glaciation. These data indicate a single, strong event at the end of the Ashgill Age. This distinguishes the Late Ordovician glacial episode from the Pleistocene and Permo-Carboniferous episodes; the event was approximately 0.5 to 1 Ma in length, as opposed to 50 to 60 Ma for the Permo-Carboniferous glaciation and at least 35 Ma for Cenozoic glaciation. Brenchley et al. (1994) cited four criteria for recognition of a glacio-eustatic mechanism for sea-level change: (1) cycles should be of short duration, (2) they should be synchronous and globally identifiable on different continents, (3) they should coincide with periods of known glaciation, and (4) they should be reflected in the δ^{18}O stratigraphy. The Late Ordovician event fits all these criteria; note that criterion (3) is the contextual data whose importance was stressed by Eyles (1993).

Whether glaciation occurred at all during the Mesozoic is controversial. Some workers have tended to view all sea-level falls as glacio-eustatic; not least among the sources of controversy is the question of whether these falls are truly global (e.g., Miall 1992). The issue was reviewed extensively by Eyles (1993) and will not be repeated here, but two points are worth noting. First, the sea-level falls during the Mesozoic tend to be isolated in time, not repetitive as they were in the Quaternary and Permo-Carboniferous. Although the Late Ordovician sea-level drop also was isolated, substantial additional information exists for glaciation at that time. Second, supporting information for Mesozoic glaciations, in Siberia and Australia, is poor and controversial, and no supporting (or refuting) information for glaciation on Antarctica exists, although glaciation there is commonly presumed to have been possible. Eyles (1993) suggested that climatic conditions as known for polar latitudes at times during the Mesozoic (e.g., Spicer and Parrish 1986) could have been favorable to the formation of temperate glaciers, and that the absence of indisputable glacial debris could be because the glaciers never reached sea level. However, he also noted that, if ice sheets did exist, they would have to have been so small that they could not have significantly affected sea level.

Distribution of Glacial Deposits and Their Paleoclimatic Significance

Up to this point, glacial deposits have been discussed without regard to the size of the ice masses that formed them. The world may never have been completely ice free. But montane glaciers, regardless of the latitude at which they occur, do not have the same implications for global climate as do continental ice sheets. The "icehouse" times in Earth history are marked by evidence of extensive continental ice sheets, such as those that existed during the Pleistocene (Fischer and Arthur 1977), although recent workers have tried to broaden the definition of "icehouse" climates to include poorly defined cold climates (Frakes, Francis, and Syktus 1992). This problem has importance in the context of this book because a major problem for pre-Quaternary paleoclimatology is distinguishing the deposits of montane glaciers from those of continental ice sheets. Recall that almost all evidence of glaciation will be in glaciomarine sediments

(Eyles 1993). Deposits from montane glaciers provide information only about local climate—cool and wet enough to permit formation of glaciers at altitude and sufficiently low ablation rates to permit the glaciers to approach sea level. A small area of glacial deposits, then, cannot be taken as indicative of an "icehouse" state (Crowell 1995). Widespread mapping of changes in the paleosnowline, as has been done for the Pleistocene (Broecker and Denton 1990), is probably impossible in the older record.

Another important consideration is that glaciers in tectonically active montane terranes produce far more sedimentary debris per unit area than do those in less active terranes (Eyles 1993). For example, Eyles (1993) estimated that glacial erosion rates along the northern North American Cordillera have been on the order of 166 m/Ma since glaciation was initiated in the late Cenozoic, whereas Pleistocene glacial erosion rates along the passive northeastern margin of North America were 24 to 48 m/Ma, a third or less as rapid. Compounding the high erosion rates in the Cordillera is the fact that the climate there is temperate and humid, permitting high sediment-transport rates for the glaciers. Thus no simple relationship exists between the size of the area glaciated and the amount of material produced by the glaciers, and local climate is as important as global climate for determining glacial sediment production rates.

Of course, montane glaciation may be more extensive during "icehouse" times. For example, the Cordilleran ice sheet of North America began supplying glacially derived sediment to the northwest Pacific as early as the late Miocene (Eyles 1993), long before the establishment of the Laurentide ice sheet but well after initiation of the Antarctic ice sheet. Indeed, a continuum from small, isolated, montane glaciers, to glacier fields, to montane ice caps, to continental ice sheets would be expected as global climate cools, so temporal boundaries between "icehouse" and "greenhouse" states are rather arbitrary.

6

Paleoclimate Models and Data-Model Comparisons

Climate models play a very important role in pre-Quaternary paleoclimate studies. It is difficult to reconstruct pre-Quaternary paleoclimates on a global scale from the geologic record because the data for each time period tend to be scattered and unevenly distributed. Climate models provide a framework within which the data that do exist can be interpreted and provide testable hypotheses about the regions in which the data are poorly known. Climate models reinforce the need to remember the big picture while interpreting local paleoclimate records.

Three types of models have been important in paleoclimate studies—atmospheric circulation models, oceanic circulation models, and geochemical models. Atmospheric circulation models and geochemical models have both been treated in books, notably Crowley and North's (1991) book *Paleoclimatology,* and Holland's (1984) *The Chemical Evolution of the Atmosphere and Oceans.* Crowley and North's (1991) book highlights particular times in Earth history and focuses on atmospheric models of paleoclimate applied to those times. Their introductory chapter is a fine treatise on numerical atmospheric circulation models and provides the basis for much of the summary of atmospheric circulation models presented here; the interested reader might also obtain *A Climate Modelling Primer*, by Henderson-Sellers and McGuffie (1987). Holland's (1984) book preceded much of the work on modeling CO_2 levels that has dominated the literature for the past 10 years. Nevertheless, it provides excellent background for the understanding of atmospheric chemical processes and current modeling efforts. The reader should turn to those works for more complete treatments than are possible here. Oceanic circulation models are still relatively uncommon in studies of pre-Quaternary paleoclimatology.

In light of existing work on paleoclimate models and in accord with the theme of this book—study of paleoclimate from the geologic record—the principal purpose of this chapter is the discussion of data-model comparisons in the final section. However, to provide context for that discussion from the model side of the comparison, different types of models are first briefly discussed, and some of their applications are noted. Emphasis is on the last 20 years in the literature. The reason for this is that the study of paleoclimates has an important geographic component, and the study of pre-Quaternary paleoclimates was limited until geologists began to reach some sort of consensus on where the continents were at various times, which did not happen until the early 1980s.

Boundary conditions are specified states at the beginning of a model run. Although boundary conditions are commonly changed in successive model runs, boundary conditions are static, not dynamic, in the models. A simple example is the distribution of land and sea. Successive model runs might have the continents in different places, but for each model run, the paleogeography is fixed. Response is the change in the modeled system, for example, the atmosphere, as a result of the changes in boundary conditions. The term "forcing" is also sometimes used in paleoclimate modeling studies. Forcing is any change that might affect the model results. In the previous example, forcing is the paleogeographic change over several model runs. In climate modeling, forcing tends to be the parameter that is changed in different model runs, whereas boundary conditions are the parameters that remain the same throughout those model runs.

Sensitivity studies are designed to illustrate the behavior of a model to changes in boundary conditions. Typically, sensitivity studies involve a control run and a series of model runs, each of which has one parameter that is different from the control run. Because so many climate parameters are interdependent, a change in one parameter may involve several changes in the equations. A comparison between the results of the control run and the sensitivity run indicates the response of the model to the change. Sensitivity studies are not necessarily intended to illustrate the behavior of the climate system itself.

ATMOSPHERIC CIRCULATION MODELS

Atmospheric circulation models fall into two categories, conceptual and numerical, and there are several types within each category. Conceptual models result in non- or semiquantitative, nondynamic predictions of paleoclimate, whereas the predictions from numerical models are quantitative, although sometimes quite generalized; numerical models can be dynamic. The reader might want to review the sections on atmospheric circulation and climate in chapter 1 before reading this section.

Conceptual and Parametric Models of Atmospheric Circulation

Conceptual models of atmospheric circulation have been in the literature for some time. One of the earlier examples is from Nairn and Smithwick (1976), who modeled atmospheric circulation patterns over Pangea for the Permian. Their models were quite generalized, and they did not formally present a methodology for the construction of their models. Three types of conceptual models have been formally presented. These are the conceptual models of Parrish, the general procedures for which were established by K. Hansen in Ziegler et al. (1977) and presented by Parrish (1982), and the parametric models of Scotese and Summerhayes (1986) and Gyllenhaal et al. (1991; the so-called Fujita-Ziegler model).

Conceptual models

Parrish's models (Parrish 1982; Parrish, Ziegler, and Scotese 1982; and as applied in many subsequent papers) are simple in concept. They are based on observations about the effects of thermal heterogeneities of the Earth's surface on atmospheric circulation patterns and the distribution of moisture in the atmosphere. These thermal heterogeneities include the meridional temperature gradient, as modified by the rotation of the Earth, and land-sea thermal contrasts. In addition, the conceptual models take into account the influence of large mountain ranges on circulation, such as the production of rain shadows and the intensification of low-pressure zones by the high-altitude heat effect (Rowley et al. 1985; Parrish and Barron 1986; Parrish 1993).

The general algorithm for production of the models is, briefly, as follows:

1. subtropical high-pressure cells centered at about 30° latitude in winter and 20° in summer on eastern sides of ocean basins, most intense in summer;
2. subpolar low-pressure cells centered at 60° latitude, most intense in winter, especially if near a coastline;
3. continental low-pressure cells (summer) equatorward of 40°, toward the eastern sides of continents;
4. continental high-pressure cells (winter) poleward of 30°.

These general features are modified according to the specifics of the paleogeography.

The atmospheric circulation patterns predicted with the conceptual models are qualitative and dependent on an understanding of the geography of atmospheric general circulation today. This level of subjectivity raises the question of reproducibility of model results. However, in modeling exercises given to many students over the years, the conceptual models have proved to give fairly consistent results (A. M. Ziegler, personal communication, and personal observation). The models do well in explaining general paleoclimatic patterns observed in the geologic record.

Scotese-Summerhayes parametric model

The parametric models of Scotese and Summerhayes (1986) were an attempt to automate the process by which the conceptual models of Parrish (Parrish and Curtis 1982) were constructed. The parameters for the Scotese-Summerhayes model include generalized values for surface atmospheric pressure over the oceans and continents taken from modern climate. The zonal patterns that result are modified by several subroutines: seasonality, continentality, ocean effect, and western intensification. The seasonality subroutine seasonally adjusts the latitudes of the isobars to obtain what Scotese and Summerhayes (1986) called the effective latitude, which takes into account the tilt of the Earth's axis (obliquity) and the inertia of the atmosphere. The continentality subroutine models the azonal aspects of circulation over large continents by intensifying high (in the winter) or low (in the summer) pressure by an amount dependent on the distance of the continental grid cell from the ocean. The ocean effect subroutine averages the pressures of nearest latitudinal grid points except where mountains intervene. This subroutine was constructed mostly to smooth out sharp gradients between land and sea. The western intensification subroutine puts the most intense parts of the subtropical high-pressure cells off the west coasts of continents.

Scotese and Summerhayes (1986) were interested in predictions of the locations of upwelling zones, and they compared their results with the conceptual-model upwelling predictions of Parrish and Curtis (1982). The correspondence was good, with estimated similarities of 80% to 100%, so they regarded the attempt to generate the conceptual models by computer a success.

Fujita-Ziegler parametric model

The Fujita-Ziegler model (Gyllenhaal et al. 1991) is a combination of the purely conceptual approach of Parrish (1982; Parrish and Curtis 1982) and the computerized parametric approach of Scotese and Summerhayes (1986) in that it is semiquantitative but not computer drawn. The basis for the Fujita-Ziegler model is the observed distribution of average zonal and meridional sea-level pressure for each season (figures 6.1, 6.2). The charts in the figures are used to determine average zonal pressure as a function of season and the total amount of land at each latitude (figure 6.1) and deviations from the average zonal pressure as a function of the widths of oceans and continents (figure 6.2).

Several interesting features of these charts are worth noting. In figure 6.1, the wider the land area, the farther the thermal equator (lowest sea-level pressure in low latitudes) is moved into the summer hemisphere. This is seen in the 1008 mb isobar on the summer side of the equator line in the right-hand side of the diagram. This is the effect of continentality, and the most extreme expression of this today is in the vicinity of the Asian summer monsoon. Subpolar lows are centered over the oceans, so they become less intense over the continents as the continents get wider. This is seen by the concentration of the low-pressure isobars on the left side of the diagram. The subtropical high-pressure belt is more intense over the continents in the winter than in the summer and increases with increasing continental width (note the 1022 mb isobar in the winter half of the right-hand side of the diagram near 30°). The subtropical high pressure decreases with increasing continental width in the summer (summer half of the right-hand side of the diagram at 30° to 60°).

In figure 6.2, note that there is an optimal width for continents and oceans that will give the greatest deviation from the zonal mean pressure. The continental high pressure and oceanic low pressure are most intense at about 50° latitude, whereas the continental low pressure and oceanic high pressure are most intense at about 30° latitude. The reasons for this have to do with the thermal responses of the zonal circulation to the seasons in each realm. The continental low-pressure data in figure 6.2 may be skewed by the Asian monsoon, in that the optimal land width for creating the most intense low pressure might be greater than shown without the presence of the Himalaya Mountains and the latent heat release in the Asian summer monsoon. In other words, we do not know the limits of the effects of sensible heating in creating continental low-pressure cells. The anomaly at low latitudes for the winter ocean is caused by data from the southern Indian Ocean subtropical high-pressure cell, which is intensified in the winter by the Asian summer monsoon.

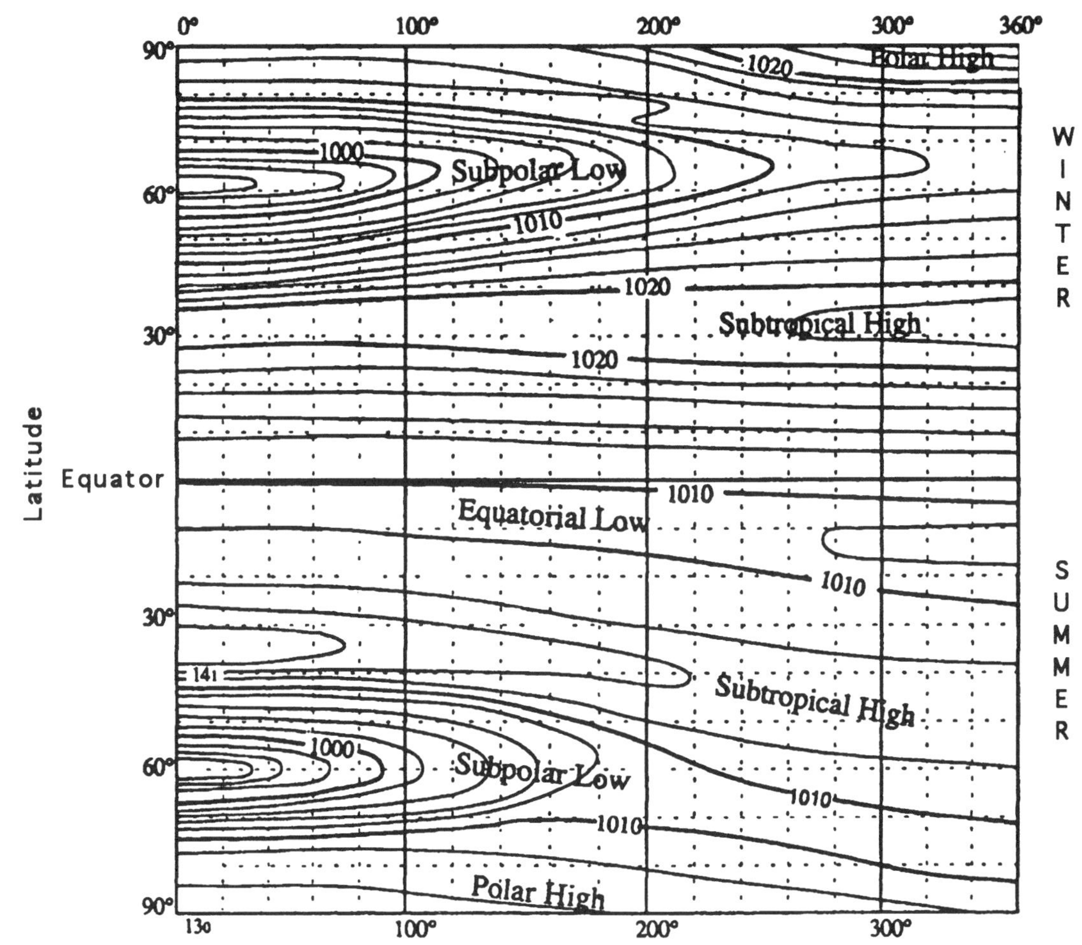

Figure 6.1. Diagram for predicting zonal mean air pressure. Horizontal axis: sum of widths of all continents (in degrees longitude) for each latitude. Note that latitude is divided into summer and winter hemispheres. Reprinted from *Palaeogeography, Palaeoclimatology, Palaeoecology* 86, Gyllenhaal, E.D., et al., "The Fujita-Ziegler model: a new semi-quantitative technique for estimating paleoclimate from paleogeographic maps," pp. 41–66 (1991) with kind permission of Elsevier Science-NL, Sara Burgerhartstraat 25, 1055 KV Amsterdam, The Netherlands.

The data that were used to compile figures 6.1 and 6.2 are taken from observations of modern climate, and therefore a number of the features are likely to be unique to today. The clearest examples are the pressure difference between the North and South Poles, seen as the strong increase in pressure across the top of figure 6.1, the anomalies created by the Asian summer monsoon, and the limits on land and ocean width (figure 6.2), which do not permit use of the information for supercontinents and superoceans wider than 250° longitude. Gyllenhaal et al. (1991) made no attempt to eliminate these peculiarities because they were interested in maintaining a realistic data set as the basis for the method. Regardless, in application, the Fujita-Ziegler model was successful in predicting the locations of evaporites and coal beds despite the fact that the parameters were specific to modern geography.

Applications of conceptual climate models

Of these models, the most extensively used in some form or another are those by Parrish (1982 et seq.). The models have been applied to prediction and interpretation of the distribution of marine upwelling indicators (Parrish 1982; Parrish and Curtis 1982; Parrish, Ziegler, and Humphreville 1983; Scotese and Summerhayes 1986; Parrish 1987b; Parrish 1995), coal beds and evaporites (Parrish, Ziegler, and Scotese 1982; Patzkowsky et al. 1991), and biogeography (see also Ziegler et al. 1977; Ziegler et al. 1981).

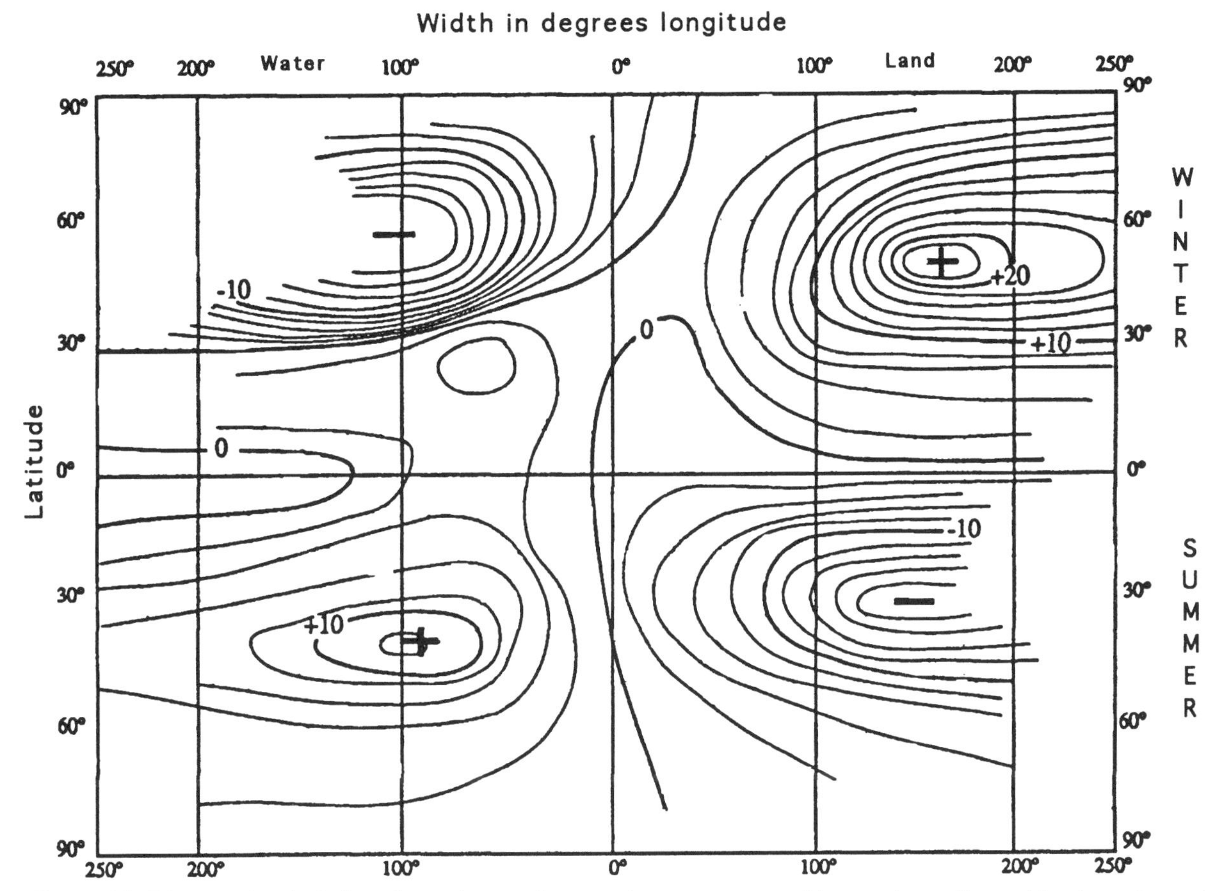

Figure 6.2. Diagram for deviations from the zonal mean air pressure caused by continentality and land-sea thermal contrast. The diagram is divided vertically into summer and winter hemispheres and horizontally into water (ocean) and land (continent). The deviation is determined by the season and the maximum width of the ocean or continent at a particular latitude and is subtracted from the zonal mean pressure determined from figure 6.1. The resulting value represents the maximum (for high pressure) or minimum (for low pressure) of the atmospheric circulation cell at that latitude and season. Blank areas are regions that have no analog in modern geography. Reprinted from *Palaeogeography, Palaeoclimatology, Palaeoecology* 86, Gyllenhaal, E.D., et al., "The Fujita-Ziegler model: a new semi-quantitative technique for estimating paleoclimate from paleogeographic maps," pp. 41–66 (1991) with kind permission of Elsevier Science-NL, Sara Burgerhartstraat 25, 1055 KV Amsterdam, The Netherlands.

Numerical Models of Atmospheric Circulation

Most numerical models used in paleoclimate studies fall into two general categories, energy balance models (EBMs) and global circulation models (GCMs). Both give numerical results that facilitate quantitative comparison of model results and paleoclimate data. EBMs are simpler and are based on conservation of energy, that is, the balance of incoming and outgoing radiation at the top of the atmosphere, with simple formulas for various fluxes (Crowley and North 1991). EBMs consider only temperature and derive other climatic processes from thermal fluxes. They make calculations over time steps of days or weeks, averaging changes over those time scales, and model the steady state, so they have no year-to-year variability. A schematic diagram of the energy inputs and outputs that might be considered in an EBM is shown in figure 6.3. EBMs are typically low resolution; indeed, many EBM results are presented as simple plots of calculation results by latitude, latitude being the principal dimension (Henderson-Sellers and McGuffie 1987). Nevertheless, the simplicity of EBMs has allowed investigation of problems that are appropriately treated with them or that would otherwise be prohibitively time consuming, for example, coupled ice-ocean-atmosphere models (e.g., Schmidt and Mysak 1996).

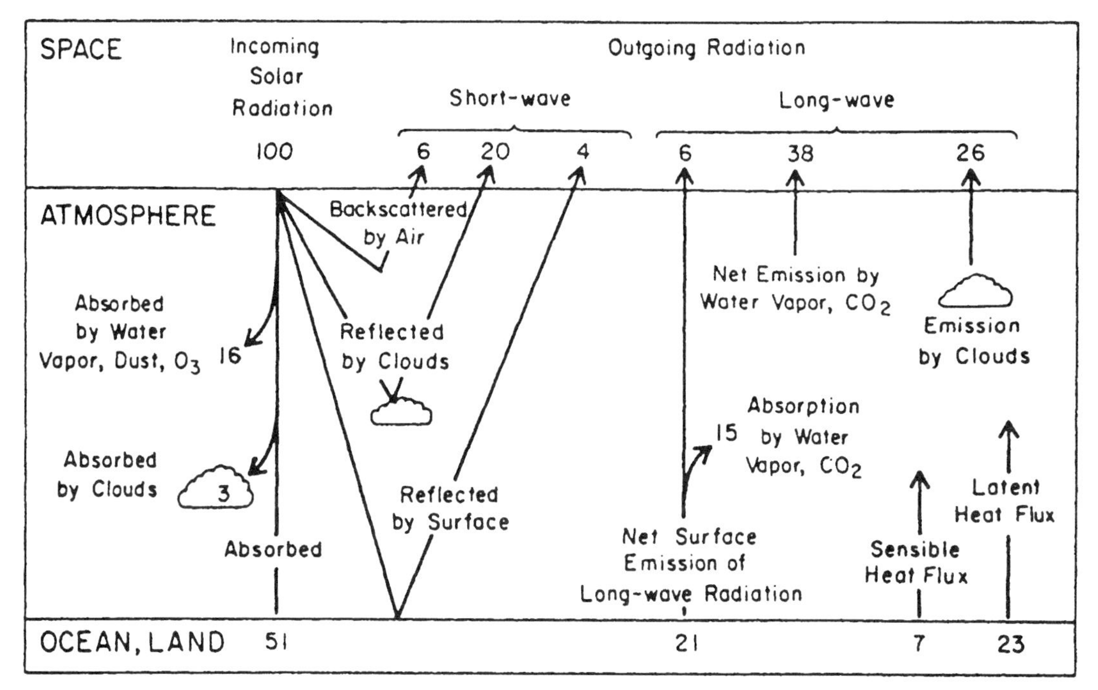

Figure 6.3. Schematic diagram of energy inputs and outputs that might be included in an energy balance model. Reprinted with permission from *Understanding Climatic Change: A Program for Action.* Copyright 1975 by the National Academy of Sciences. Courtesy of the National Academy Press, Washington, D.C.

Global circulation models, in contrast to EBMs, are complex and explicitly calculate climatic variables such as temperature, wind, and precipitation. Unlike EBMs, which present information as statistical averages, GCMs explicitly calculate and resolve the so-called synoptic-scale motions, that is, the 1000 km scale high- and low-pressure cells. The models are three-dimensional and time dependent (Barron and Moore 1994).

Global circulation models typically are higher resolution than EBMs. A typical GCM might calculate climatic variables on, say, a 4°-by-4° latitude-longitude grid and have 18 layers within the atmosphere. Latitude is commonly more finely divided because today, at least, more heat is transported latitudinally than longitudinally, so that changes occur over smaller spatial scales by latitude than by longitude. Time steps are typically 30 minutes over 5 to 20 model years. Some models use a spectral method based on spherical harmonics rather than a grid system, but the problems of spatial resolution are the same. See Crowley and North (1991:31–32) for an explanation of this method. Further discussion of gridding can be found in Barron and Moore (1994). The computational intensity of GCMs is great, involving 10^{10} or more calculations.

All climate models are governed by the same basic equations (Gates 1982; Henderson-Sellers and McGuffie 1987); for conservation of horizontal momentum (Newton's second law of motion):

$$\frac{\delta u}{\delta t} = -V_H \cdot \nabla u - \omega\frac{\delta u}{\delta p} + \frac{uv \tan \phi}{a} + 2v\Omega \sin \phi - \frac{1}{a \cos \phi}\frac{\delta \Phi}{\delta \lambda} + F_\lambda \tag{6.1}$$

and

$$\frac{\delta v}{\delta t} = -V_H \cdot \nabla V - \omega\frac{\delta v}{\delta p} - \frac{u^2 \tan \phi}{a} - 2u\Omega \sin \phi - \frac{1}{a}\frac{\delta \Phi}{\delta \phi} + F_\phi \tag{6.2}$$

for conservation of energy (heat, first law of thermodynamics):

$$\frac{\delta T}{\delta t} = -V_H \nabla T - \omega\frac{\delta T}{\delta p} - \frac{\alpha\omega}{C_p} + \frac{Q}{C_p}, \tag{6.3}$$

and for conservation of mass (continuity):

$$\frac{\delta q}{\delta t} = -V_H \cdot \nabla q - \omega\frac{\delta q}{\delta p} + (E - C). \tag{6.4}$$

Table 6.1.
Definition of terms in the fundamental equations of climate models

a	radius of the Earth (assumed to be spherical)
ϕ	latitude
λ	longitude
Ω	rotation rate of the Earth
∇	gradient on an isobaric surface
u	eastward component of the horizontal wind
v	northward component of the horizontal wind
$\mathbf{v}_H$	horizontal wind
Φ	geopotential of an isobaric surface
T	temperature
q	atmospheric specific humidity
p	pressure
F_λ, F_ϕ	friction terms
Q	diabatic heating
E	evaporation
C	condensation
α	specific volume
R	gas constant
ω	vertical mass flux

From Gates (1982).

These equations are prognostic; that is, they predict variations of horizontal velocity, temperature, and moisture with time (Gates 1982). They are used along with the diagnostic equation (a general equation for conservation of mass),

$$\frac{\delta\omega}{\delta p} + \nabla v_H = 0,$$

which describes vertical motion with respect to horizontal velocity and related variations in atmospheric pressure; the hydrostatic equation

$$\frac{\delta\Phi}{\delta p} = -\alpha;$$

and the atmospheric equation of state (sometimes known as the gas law),

$$p\alpha = RT.$$

The terms are defined in table 6.1. Dependent variables are calculated as a function of time and position and are temperature (T), the three components of air velocity (u, v, and vertical velocity), local pressure (p), and the concentration of water vapor (q) (Crowley and North 1991). The equations are written for a spherical coordinate system (Barron and Moore 1994).

Table 6.2.
Sub-grid-scale processes that are parameterized in atmospheric global circulation models

Turbulent transfer of heat, moisture, and momentum between the Earth's suface and the atmosphere
Turbulent transfer of heat, moisture, and momentum within the atmosphere by dry and moist (cumulus) convection
Condensation of water vapor
Transfer of solar and terrestrial radiation
Formation of clouds and their radiative interaction
Formation and dissipation of snow
Soil heat and moisture physics

From Schlesinger (1984).

Global circulation models were developed initially for weather prediction (Kutzbach 1985; Henderson-Sellers and McGuffie 1987), and more than a dozen models are in widespread use for weather and climate prediction (e.g., Cess et al. 1989). Of these, about eight are in widespread use for paleoclimate modeling (Crowley and North 1991; table 2.3). All these models use the same fundamental equations listed here, yet they give very different results, even in simulations of modern climate. Explicit model-model comparisons for any past time in Earth history have yet to be performed, so the differences among some of the models are illustrated for modern climates in North America in figure 6.4. In this experiment, CO_2 was doubled and all other specifications in each model remained the same. This is similar to comparing model results for some past climate except that geography was held the same as modern geography in all models.

All the models used the same boundary conditions of Earth radius, rotational speed, gravity, and geography. Why, then, did the models give such different results? The two principal reasons are lack of computational power and lack of knowledge. Many of the variables in the fundamental equations cannot be solved at the scales at which they operate in the atmosphere because those scales are much smaller than the grid cells specified for the models, even in higher-resolution models that have not yet been used for paleoclimate studies. Consider that one of the most important energy-transfer processes is convection in clouds and thunderstorms. These operate on horizontal scales of 10^0 to 10^1 km^2, whereas a grid cell in a typical GCM covers 10^3 to 10^4 km^2. An excellent illustration of an effect of grid size is

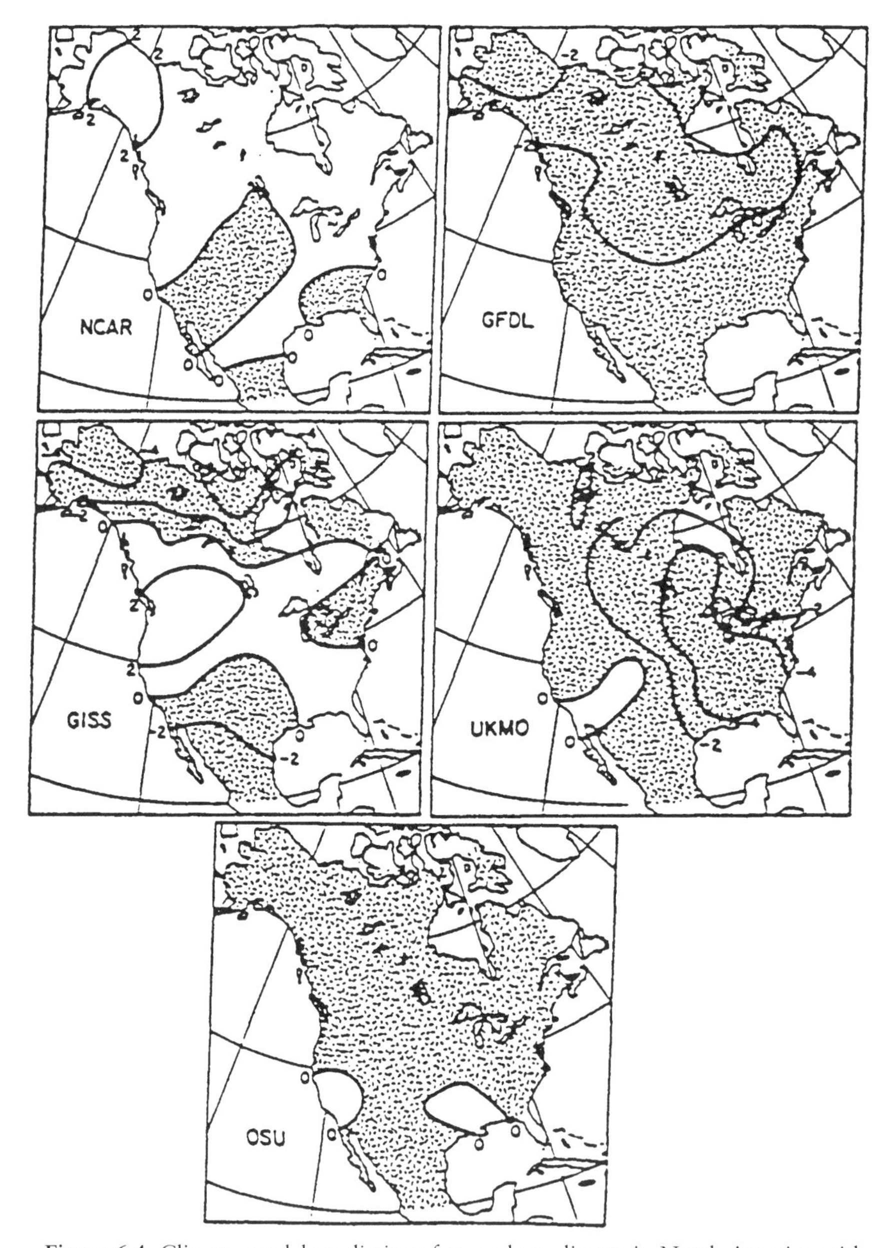

Figure 6.4. Climate-model predictions for modern climate in North America with a doubling of CO_2. The climate parameter illustrated is soil moisture. *Shaded areas* are those with less soil moisture than in simulations using present CO_2. The models are OSU, Oregon State University; GFDL, Geophysical Fluid Dynamics Laboratory (Princeton University); GISS, Goddard Institute of Space Sciences (NASA); UKMO, United Kingdom Meteorological Office; and NCAR, (U.S.) National Center for Atmospheric Research. From Kellogg and Zhao (1988). Permission granted by the American Meteorological Society.

illustrated in figure 6.5 (Pitman, Henderson-Sellers, and Yang 1990). GCMs predict rainfall as averaged over the total area of a grid cell (figure 6.5a). In these simulations, the authors kept everything the same, including the amount of rain per grid cell, except the area over which rainfall was distributed within each cell. Figure 6.5b shows evaporation and runoff with rainfall distributed over the entire area of each cell. Figure 6.5c and 6.5d show the same amount of rainfall distributed over half and one-tenth the area of each grid cell, respectively. The predictions of the relative importance of evaporation and runoff changed dramatically with decreasing area of rainfall. The difference is substantial because evaporated water is potentially recycled lo-

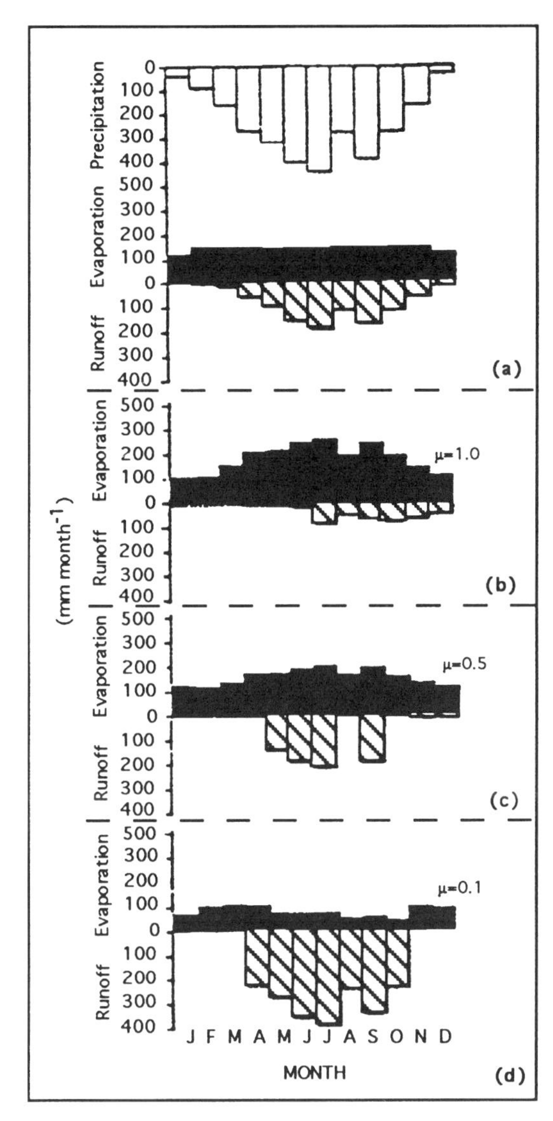

Figure 6.5. Simulations of seasonal precipitation, evaporation, and runoff in the Amazon Basin using different grid cell sizes in a land-surface model. *(a)* Land-surface model coupled with GCM. Note that evaporation is nearly constant and runoff reflects precipitation. *(b)* As in (a), except the land-surface model is decoupled from GCM. *(c)* As in (b), but with precipitation concentrated in half the standard grid cell. *(d)* As in (b), but with precipitation concentrated in one-tenth the standard grid cell. Although the land-surface model was decoupled from the GCM in experiments b to d, the results illustrate the effects of simulating more realistic distribution of precipitation (see text). Precipitation, winds, and annual and diurnal temperature variations are constant in all three cases. More recent models are less sensitive (A. J. Pitman, personal communication). From Pitman, Henderson-Sellers, and Yang (1990). Reprinted with permission from *Nature* 346:734–7. Copyright 1990 Macmillan Magazines Limited.

cally, whereas runoff is lost to the local area, with clear implications for the effects of climatic change on vegetation (Pitman, Henderson-Sellers, and Yang 1990).

Another problem with the low resolution of GCMs is that paleogeography cannot be accurately represented—many mountain ranges and epicontinental seaways were so narrow that they fall beneath the resolution of the model and are excluded—yet subgrid-scale topography can have a large effect on climate at the scale at which data on paleoclimatic indicators are collected (Parrish and Peterson 1988; Pollard and Schulz 1994). Examples are the Cretaceous Western Interior seaway of North America and late Cenozoic Paratethys in Asia (Ramstein et al. 1997). A continuous problem in paleoclimate modeling is continental interiors that are predicted to be far more seasonal and, most problematically, too cold in the winter, than indicated by the data (debated for the Eocene by Sloan and Barron 1990, 1991; Wing 1991; Wing and Greenwood 1993; Markwick 1994; Greenwood and Wing 1995). Valdes, Sellwood, and Price (1996) used the United Kingdom Universities Global Atmospheric Modelling Program (UGAMP) model, which has a higher spatial resolution than the versions of the NCAR model that were used by other workers, particularly with respect to longitude. This resolution allowed the seaway, which was narrow parallel to longitude, to be included in the paleogeographic boundary conditions. Valdes, Sellwood, and Price (1996) found that the Western Interior seaway had a significant ameliorating effect on the climate of paleo-North America. However, they also cautioned that so many differences exist between the NCAR and UGAMP models that the warmer results cannot be confidently attributed to just the difference in the treatment of the Western Interior seaway. Among other possible explanations for the different results are differences in the treatment of mountains, clouds, and oceans.

To bring the scale of calculation down to the scale of the processes would require a prohibitive number of calculations, even with modern supercomputers. However, even if we had infinitely powerful computers, a scale reduction would suppose that we know enough about how these processes work to quantify them, and the second major reason for different model results is that we do not clearly know how small-scale (and other) processes work. Both of these reasons—the need to keep the resolution of the models low and a lack of knowledge of fundamental processes—require climate modelers to make some mathematical assumptions about how climate works and to incorporate those assumptions into their calculations. For example, the variable q, specific humidity (see table 6.1), incorporates a host of assumptions about how heat and moisture are transported within the atmosphere. The frictional term includes all small-scale convective and turbulent transfers of momentum within the atmosphere and between the atmosphere and the Earth's surface. The differences among the models, then, comes primarily from how different groups of modelers specify those processes that are small scale and/or poorly known in their models. This is called parameterization (Henderson-Sellers and McGuffie 1987). The state of knowledge of some climatic processes is so poor that a wide range of parameterizations might all be considered reasonable. A summary of the subgrid-scale processes that are parameterized in GCMs is shown in table 6.2.

Among the most variably parameterized climate factors are clouds, and the results of large-scale model-model comparison experiments serve very well to illustrate the problem. Cess et al. (1989, 1996) compared the cloud feedback sensitivities and parameterizations of 19 global circulation models. Without cloud feedback, all models would tend to give similar results with any kind of climatic forcing (Cess et al. 1989). In their initial comparison (Cess et al. 1989), the model results diverged over a large range. In a later comparison (Cess et al. 1996), after atmospheric GCMs had undergone further development, the divergence in cloud-feedback results was much less, although they still ranged from positive to moderately negative. (Similarly, the rather old experiment illustrated in figure 6.4 might not give such divergent results if it were run today.) In addition, Cess et al. (1996) noted that similar results were achieved differently in different models. They used differences in shortwave radiation (cooling, SW) and longwave warming (LW) among the models to illustrate this. For example, those showing near-zero cloud feedback did so because the two radiation terms were negligible, the terms were modest but compensatory, or the terms were large but compensatory. Cess et al. (1996) concluded that the convergence in results, then, does not necessarily reflect increased accuracy of the models or of our understanding of atmospheric processes.

In paleoclimate models, parameters and boundary conditions that are prescribed—in addition to rotational speed, geography, and clouds—and that might vary, are the solar constant, which has changed over long time scales; energy input from the Sun, which varies according to the Earth's orbital parameters; composition of the atmosphere and the thermodynamic and radiation constants of the various atmospheric gasses; and albedo, that is, the reflectivity of the Earth's surface. It is not difficult to imagine that all of these parameters might be viewed—and estimated—quite differently by different modelers.

Climate-model results tend to be presented as means, which is useful for studying long-term, large-scale climatic change but which may be unrealistic even at those scales under certain circumstances. Some climatic changes may occur as threshold events or the same starting conditions might give rise to multiple equilibria (North and Crowley 1985; Barron, Boyle, and Leinen 1991; Hoskins 1993).

An additional limitation of numerical atmospheric climate models is that most have not been fully coupled to oceanic general circulation models for paleoclimate studies (a recent exception is Bush and Philander 1997), although fully coupled models are rapidly being developed (e.g., Boville and Gent 1998). The development of coupled models has lagged because the atmosphere and the ocean operate at such different rates that coupling them is a mathematical challenge. The oceans are treated in one of two ways in many atmospheric circulation models: as "swamps" or as "slabs." A swamp ocean is a source of moisture for the model, and the mixed-layer (upper 50 m) temperature is calculated on the basis of surface energy balance (Barron and Moore 1994). For a slab ocean, the mixed-layer temperature is calculated as in a swamp ocean, but annual cycles that require ocean heat capacity for full characterization are possible and ocean heat transport can be specified.

One of the most common uses for EBMs in paleoclimate modeling is to examine the sensitivity of the climate (energy balance) to various perturbations (Kutzbach 1985; Crowley and North 1991). For example, Barron (1983) used an EBM to investigate climatic response to various assumptions about cloud cover and land-surface albedo on a Cretaceous Earth (figure 6.6). The latitudinal plots show that Cretaceous climate was likely to have been sensitive to both albedo and cloud cover changes compared with today's climate, especially at the poles. Another useful

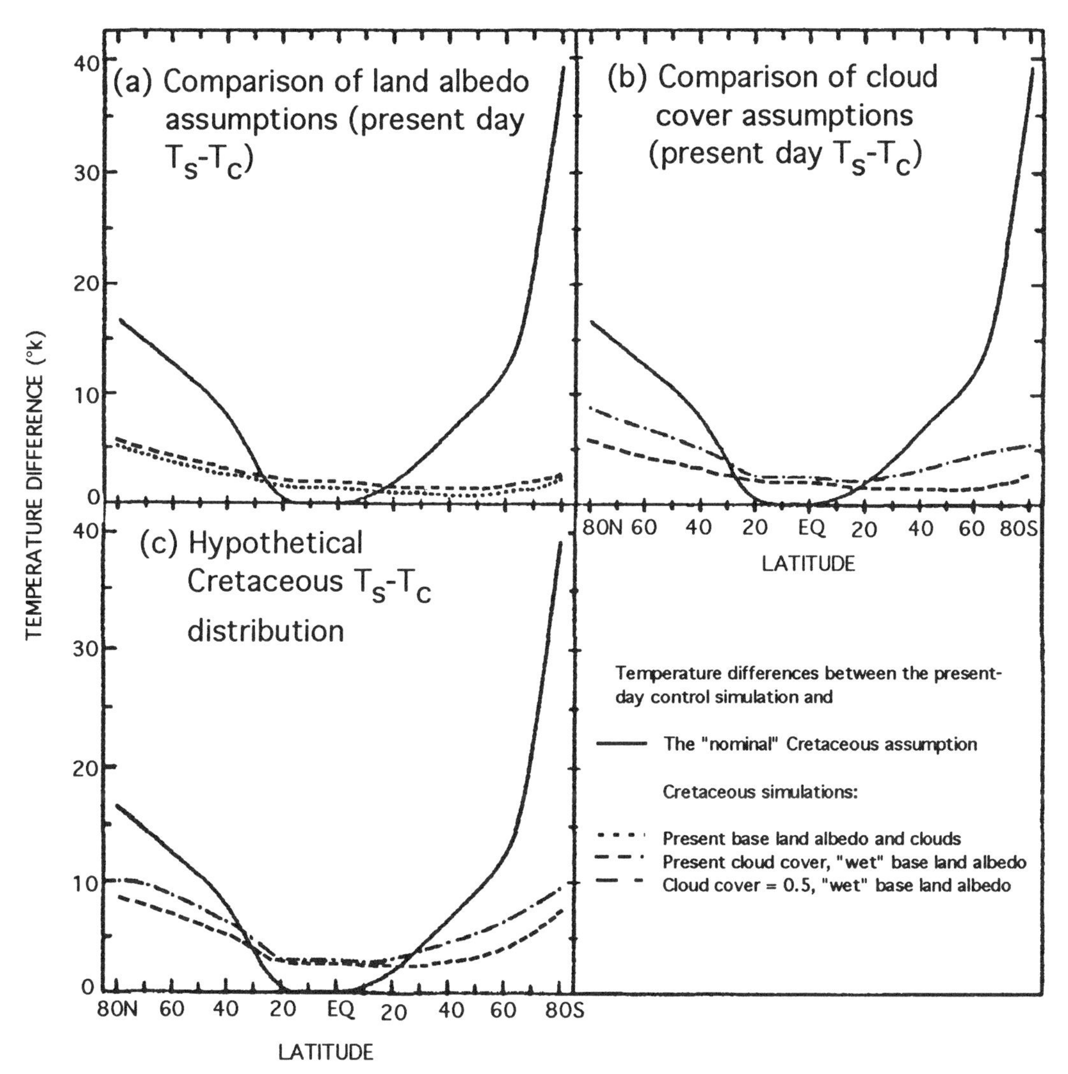

Figure 6.6. Predicted temperature differences between present-day climate and Cretaceous climate with various assumptions about land albedo and cloud cover. The "nominal" Cretaceous (*solid line*) is nearly the inverse of the modern latitudinal temperature profile, in that Barron (1983) assumed that modern and Cretaceous equatorial temperatures were the same but Cretaceous polar temperatures were much warmer, especially the South Pole; this resulted in the large differences for those regions indicated by the solid line. "Wet" base land albedo means that the simulations were run with low land albedos, consistent with a vegetated surface. Reprinted from *Earth-Science Reviews* 19, Barron, E.J., "A warm, equable Cretaceous: the nature of the problem," pp. 305-38 (1983) with kind permission of Elsevier Science-NL, Sara Burgerhartstraat 25, 1055 KV Amsterdam, The Netherlands.

application of EBMs is in coupling processes with vastly different response times, for example the atmosphere and oceans or atmosphere and ice (Kutzbach 1985). Comparison of model results to paleoclimate data is limited to surface temperature distribution (Barron and Moore 1994). EBMs have also been used to simulate surface-temperature patterns on global paleogeographic maps (Crowley, Hyde, and Short 1989; Crowley, Baum, and Kim 1993).

Global circulation models have been applied to a wide variety of problems, particularly since Barron and Washington started their pioneering work applying an NCAR global circulation model to the problem of mid-Cretaceous global warmth (Barron and Washington 1981, 1982a,b, 1984, 1985; see also Barron et al. 1995 and Barron 1990). Barron and his students have gone on to investigate the Eocene megathermal (Sloan and Barron 1990, 1992; Barron and Moore 1994; Sloan 1994; Sloan and Rea 1995; Sloan, Walker, and Moore 1995), the distribution of marine upwelling zones (Barron 1985a; Kruijs and Barron 1990), and many aspects of mid-

Cretaceous climate (e.g., Barron 1984; Barron, Arthur, and Kauffman 1985; Barron and Washington 1985; Glancy et al. 1993; Barron et al. 1995). Other applications of numerical atmospheric circulation models to the deep past have been to the problem of supercontinents and their climatic consequences (Crowley, Mengel, and Short 1987; Crowley, Hyde, and Short 1989; Kutzbach and Gallimore 1989; Chandler, Rind, and Ruedy 1992; Crowley, Baum, and Hyde 1992; Valdes and Sellwood 1992; Crowley, Baum, and Kim 1993; Kutzbach and Ziegler 1993; Valdes 1993), the effects of seasonality on climatic change (North and Crowley 1985; Crowley et al. 1986; Crowley, Hyde, and Short 1989), pre-Pleistocene glaciations (Hyde, Kim, and Crowley 1990; Baum and Crowley 1991; Crowley and Baum 1991a, 1992; Crowley, Baum, and Hyde 1991; Crowley, Yip, and Baum 1994; Gibbs, Barron, and Kump 1997), and the paleohydrological cycle (Hay, Barron, and Thompson 1990a,b), particularly as it relates to the distribution of coal beds (Otto-Bliesner 1993, 1995), evaporites (Pollard and Schulz 1994), and bauxites (Price, Valdes, and Sellwood 1997). These studies are all in addition to those that focus on the general effects of simply moving the continents from their present positions (Barron 1985b; Hauglustaine and Gérard 1992) and orographic effects (Ruddiman, Prell, and Raymo 1989; Prell and Kutzbach 1992; Ramstein et al. 1997).

Numerical climate models have been used extensively in the study of Pleistocene climates, particularly with regard to variations in solar energy (Milankovitch cycles; e.g., Kutzbach and Guetter 1986; Harrison, Kutzbach, and Behling 1991; Kutzbach, Gallimore, and Guetter 1991), monsoons (Kutzbach and Otto-Bliesner 1982; Kutzbach 1983; Kutzbach and Street-Perrott 1985; Prell and Kutzbach 1992), and paleogeography (Ruddiman and Kutzbach 1991; Kutzbach, Prell, and Ruddiman 1993). Inherent or explicit in many of these studies of both pre-Pleistocene and Pleistocene climates is the effect of CO_2. High-resolution regional climate models also are being applied to the more recent past (Hostetler et al. 1994).

OCEANIC CIRCULATION MODELS

Oceans present both relative advantages and relative disadvantages for modeling compared with the atmosphere (Crowley and North 1991). Advantages to modeling the oceans are the absence of equivalent complex radiative processes such as clouds and the fact that water can be considered incompressible (Crowley and North 1991). The most important disadvantage is the horizontal distance greater than which the rotation of the Earth becomes important in affecting flow. This distance in the atmosphere is about 1000 km, whereas in the oceans, it is about 100 km (Crowley and North 1991). The effect of this difference is that, to achieve the same accuracy as atmospheric models, oceanic GCMs must have a resolution greater than atmospheric GCMs (Crowley and North 1991). Another significant disadvantage is the complexity of the ocean basins in size and shape and the much larger frictional effects. Although surface characteristics strongly control atmospheric circulation, the atmosphere is not confined except by gravity.

Atmospheric circulation models do not ignore the oceans; at the very least, they treat the oceans as sources of heat and moisture. Oceanic circulation models likewise do not ignore the atmosphere. Rather, the atmosphere serves as a source of kinetic energy through winds, evaporation, or precipitation and also as a source and sink for heat. Evaporation and precipitation affect the salinity of the surface layer, which can result in greater or lesser overturn. Clearly, the ideal situation would be to run fully coupled ocean-atmosphere models, as the interactions among the two realms are so important in determining climate. However, coupled models are few and, until recently (Bush and Philander 1997), limited to studies of modern and Quaternary climates (an example is Schlesinger and Verbitsky 1996). The reason is that the oceans and atmosphere operate on completely different time scales. Large-scale atmospheric waves (as expressed in polar fronts or hurricanes, for example) traverse the entire circuit of the globe in a few days. In contrast, ocean currents require weeks or months to traverse the ocean basins, and the response time of the deep ocean is decades to millenia. Processes that result from coupling of the atmosphere and ocean, such as El Niño, recur on time scales of a few years, but these can affect even local and short-term processes (Luther 1995).

Oceanic circulation models, like atmospheric circulation models, can be divided into conceptual and numerical models. Conceptual oceanic circulation models are very much like conceptual atmospheric circulation models and are usually used to elucidate surface circulation patterns. Ex-

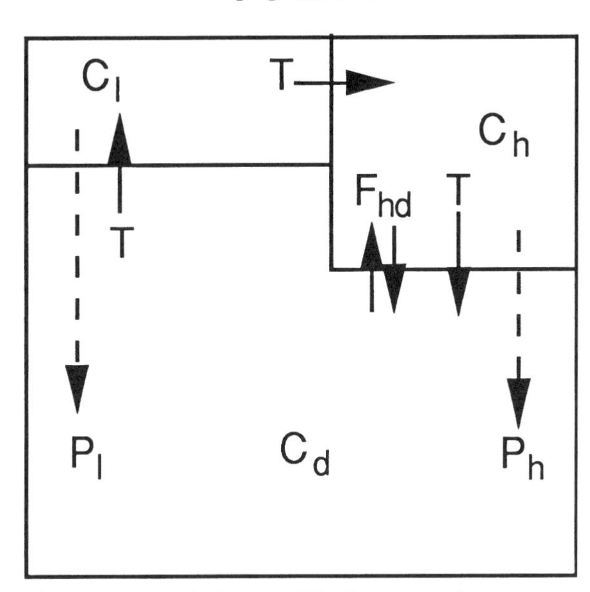

Figure 6.7. Simple box model of nutrient flux, used to assess the development of anoxic water masses. This could also be regarded as a geochemical model, but ocean circulation is an important element. l, h, and d refer to low- and high-latitude surface and deep water, respectively; T and F_{hd} are circulation parameters, analogous to large-scale thermohaline (e.g., bottom-water generation in the North Atlantic) and local convective overturn (e.g., deep-water generation in the Antarctic convergence), respectively; P is the particulate rain of phosphate from high-latitude deep water (P_h) and low-latitude surface water (P_l), respectively. From Sarmiento, Herbert, and Toggweiler (1988a) in *Global Biogeochemical Cycles* 2:115–28, copyright by the American Geophysical Union.

amples are predictions of warm and cold surface currents for the Paleozoic (Ziegler et al. 1981), large-scale oceanic circulation in the paleo-Pacific Ocean in the Jurassic (Parrish 1992), and predictions of upwelling zones (Parrish 1995). Jewell (1995) provided a simple extrapolation of equatorial circulation processes to predict the effects of wide oceans on equatorial productivity.

Numerical oceanic circulation models can be categorized into three types: box models, atmosphere-forced oceanic GCMs, and "true" oceanic GCMs (OGCMs). Box models are loosely equivalent to atmospheric energy-balance models in that they are not dynamic but rather calculate long-term statistical averages. Atmosphere-forced OGCMs are simply models of ocean-surface circulation derived from wind forcing generated by atmospheric GCMs and also are not dynamic (with respect to the ocean; e.g., Kutzbach, Guetter, and Washington 1990). "True" OGCMs include ocean-basin geometry and have at least two and usually more layers within the oceans, so that forcing in addition to wind friction can be modeled.

Numerical oceanic circulation models have been applied to many problems, including surface circulation patterns and tides (Ericksen and Slingerland 1990, using a confined circulation model), development of dysoxic or anoxic water masses (Sarmiento, Herbert, and Toggweiler 1988a,b, using box models), changes in deep-water circulation with variations in temperature and salinity at the surface (Wilde and Berry 1982, density mixing model), and stratification patterns in epicontinental seaways (Hay, Eicher, and Diner 1993, density mixing model similar to that of Wilde and Berry 1982).

An example of a box model of nutrient fluxes designed to estimate the development of anoxia is shown in figure 6.7; an example of a stratification model based on balances of water densities is shown in figure 6.8. The density-mixing models of Wilde and Berry (1982) and Hay, Eicher, and Diner (1993) relied on the effects of temperature, runoff, and evaporation on water density and how waters of different densities mix to determine how water masses would behave at depth.

Surface-circulation patterns in the oceans have also been simulated from wind stress, temperature, and precipitation-evaporation patterns from atmospheric GCMs (e.g., Ericksen and Slingerland 1990; Kutzbach, Guetter, and Washington 1990). One of the more sophisticated applications of such an approach was the pseudo-coupled modeling of the Western Interior seaway (Cretaceous, North America) by Slingerland et al. (1996). Using an atmospheric circulation model, they predicted atmospheric climate, including wind stress, precipitation, evaporation, and temperature, for the region and then used the output to set the boundary conditions for a numerical coastal ocean model. They found that the seaway experienced estuarine circulation with both the paleo-Arctic and Tethyan seaways and were able to explain several aspects of the faunal patterns that had resisted interpretation. Application of true OGCMs in paleoceanography has been relatively sparse and mostly by German workers using the so-called Hamburg model (Maier-Reimer, Mikolajewicz, and Hasselmann 1993). Notable studies are those by Heinze, Maier-Reimer, and Winn (1991) on oceanic modulation of atmospheric pCO_2, and by Maier-Reimer, Mikolajewicz, and Crowley (1990) and

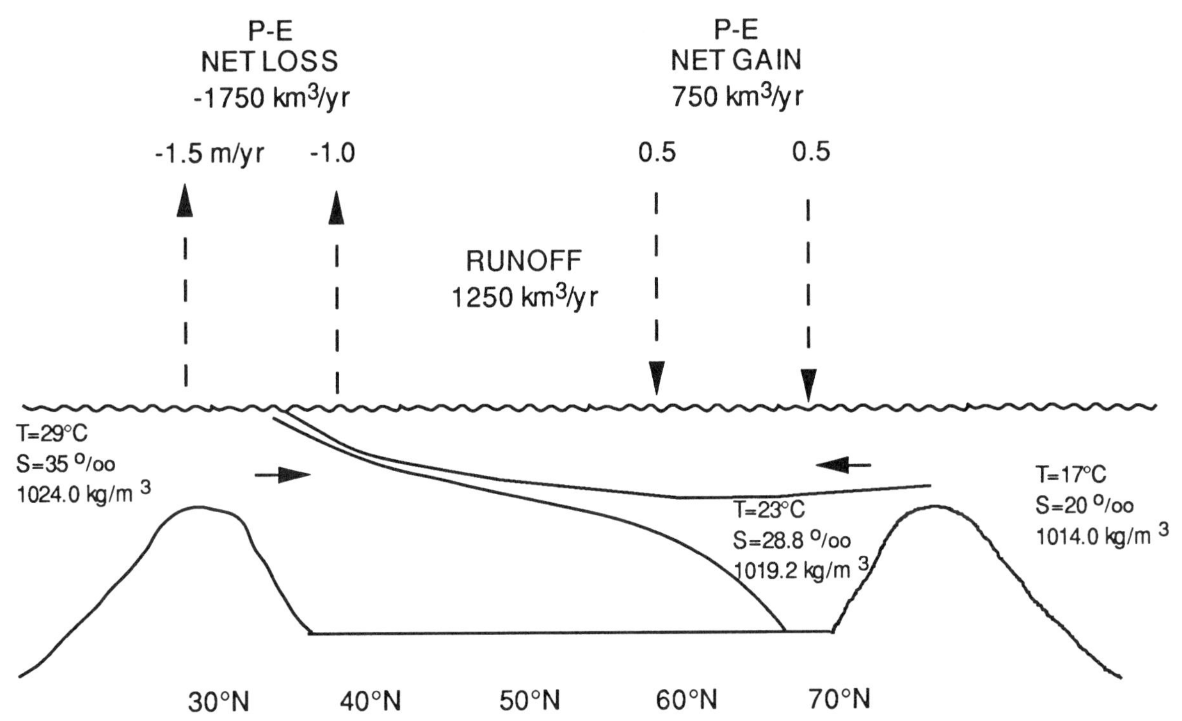

Figure 6.8. Water-balance model of the Cretaceous Western Interior seaway assuming a low latitudinal temperature gradient and a high salinity gradient between the Cretaceous Arctic and Gulf of Mexico water masses. This was one of several water-balance scenarios presented by the authors. From Hay, Eicher, and Diner (1993).

Mikolajewicz et al. (1993) on the effects on ocean circulation of oceanic gateways such as the Drake Passage. These studies used modern continental positions, closing or opening certain connections between the continents.

GEOCHEMICAL MODELS

Geochemical models seek to quantify and illuminate the exchange of elements and molecules among different parts of the Earth system. The key parts of geochemical models are mass conservation (for example, if a carbon atom disappears from one reservoir, it must appear in another), time, and flux (Van Cappellen and Wang 1995). Indeed, the mathematical basis for geochemical models (and geochemisty in general) is comprehensively and fully described (W. S. Broecker, cited in Delaney and Boyle 1988) as

$$\text{inputs} = \text{outputs}$$

for a given reservoir at steady state. The plural indicates that the reservoir can change; however, mass is conserved for each component in the system as a whole.

Many geochemical models exist, but not all are related to paleoclimate. Climate-related geochemical models almost always have carbon in its various forms as a primary or collateral focus. This is because the cycling of carbon is strongly dependent on climatic processes; some of the gaseous forms of carbon, that is, CO_2 and CH_4, are greenhouse gasses; and carbon, in the form of petroleum and coal, is of great economic interest. Where carbon is not the main focus, it still plays a strong role in the models, as in the cycling of phosphorus (e.g., Kump 1990; Van Cappellen and Ingall 1994).

In this book, some of the ways geochemical cycles are modeled have already been mentioned, namely interpretations of carbon isotopic signatures of organisms in the oceans and of the distribution of phosphate through time. In those cases, the models were in the context of the interpretation of specific paleoclimatic indicators. Geochemical models are implicitly or explicitly elements of some of the ocean models discussed in the previous section, that is, those that were at least partly directed toward understanding the distribution of dissolved oxygen in the oceans (e.g., Wilde and Berry 1982; Sarmiento, Herbert, and Toggweiler 1988a).

Conceptual geochemical models are mostly useful for illustrating geochemical cycles and have only general predictive value. Almost all geochemical models are numerical to some extent, and most are box models. Perhaps the most famous such model is the so-called BLAG model (taken from the names of its originators, R. A. Berner, A. Lasaga, and R. Garrels), a version of which is shown in figure 6.9. Shown in the figure are estimates of the sizes of various sources and sinks and the magnitudes of the fluxes among them. Perturbation of these parameters according to estimates of rates of sea-floor spreading and rates of weathering (related to temperature and continental rainfall) results in changes in the sizes of the sources and sinks and magnitudes of the fluxes. Of particular interest to paleoclimatology are predictions that atmospheric CO_2 was higher in the Eocene and much higher in the Late Cretaceous than at present. Berner (1990b, 1991, 1993, 1994) has gone on to refine the model by revising estimates and adding, for example, the effects of plants.

Geochemical models have two principal limitations. First, like all models, they are only as good as the estimates that go into them. Geochemical cycles are highly complex. Simple geochemical models that incorporate a few components may produce internally consistent results but may produce unexpected and unrealistic results when other components are added (Delaney and Boyle 1988). Many important parameters, for example, rate laws and diffusion and reaction coefficients for various molecules, are very poorly known (Van Cappellen and Wang 1995). Second, mass and isotope balances can be modeled relatively easily over long time periods ($\geq 10^5$ years) because one can assume a steady state (Kump 1993), and such applications are commonly used in paleoclimatology. However,

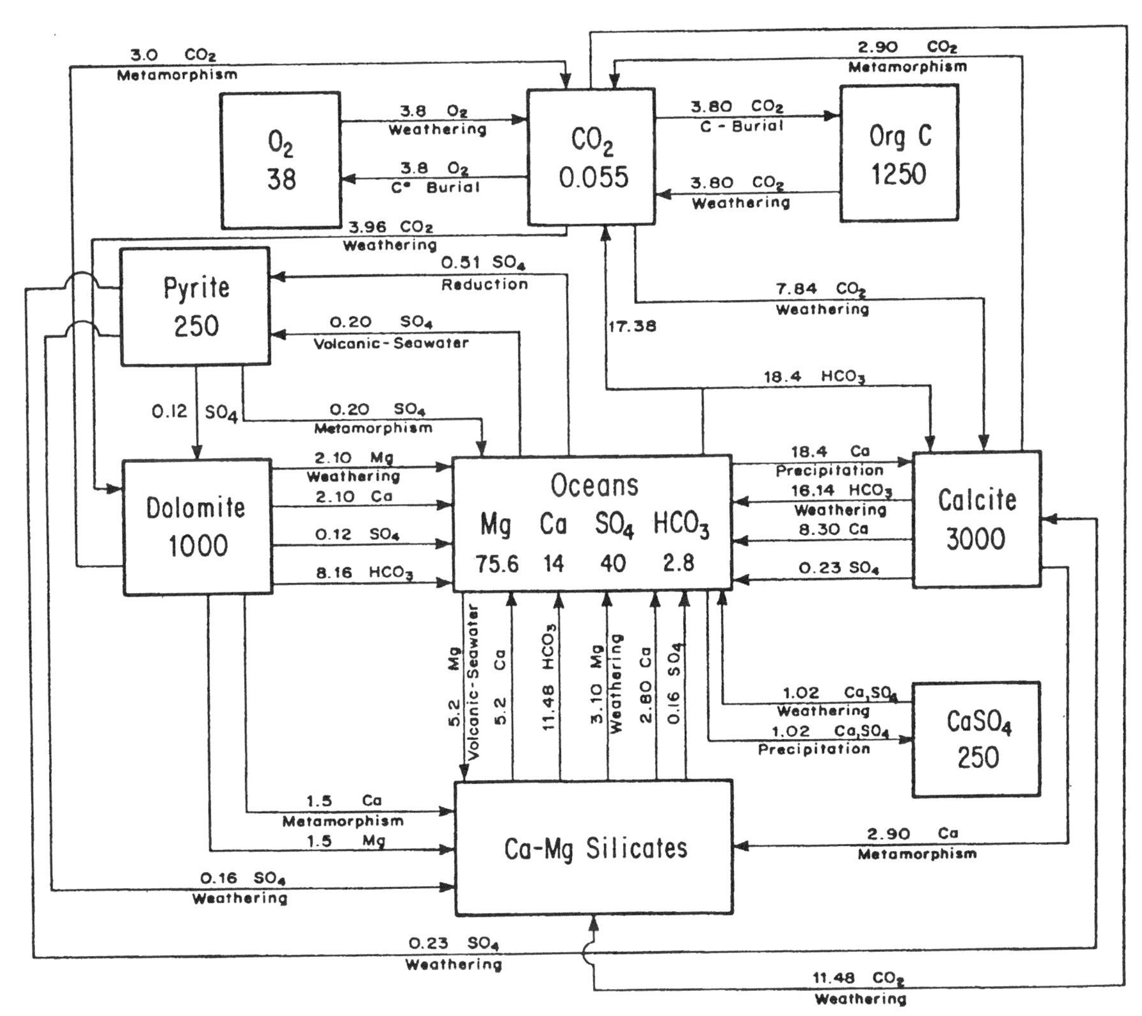

Figure 6.9. The BLAG geochemical model. From Lasaga, Berner, and Garrels (1985) in *The Carbon Cycle and Atmospheric CO_2: Natural Variations Archean to Present,* pp. 397–411. Geophysical Monograph 32, copyright by the American Geophysical Union.

the observable processes, from which flux estimates are derived, are short term. Therefore the assumption of steady state relies on short residence times (defined as the ratio of the inventory of a component in a reservoir to its total input or output flux) relative to the time scale of the modeled changes; if residence times are substantially longer, trends would reflect the component with the longer response time (see also Berner et al. 1993; Kump 1993).

USE OF MODELS IN PALEOCLIMATE STUDIES

Notwithstanding their limitations, models are very helpful for gaining insight into paleoclimatic processes. Models cannot be verified, validated, or otherwise shown to be "true" in any sense of being an accurate representation of the physical world (Greenwood 1989; Konikow and Bredehoeft 1992; Oreskes, Shrader-Frechette, and Belitz 1994; papers in Molnia 1996). This does not diminish, however, the heuristic value of models and, as shown very well by Crowley and North (1991), Barron and Moore (1994), and many others, paleoclimatology has advanced significantly in the two decades or so since climate models came into widespread use.

The word "test" is commonly used to refer to the process of comparing model results with geologic data, including in this book, but that word is misleading because models are not tested in the sense of the scientific method. Rather, model results can suggest rejectable hypotheses that are testable, and what paleoclimatologists do is compare model results with the geologic data and thereby learn something about how the Earth's climate system works. Following are some examples of model-data comparisons, presented with the intention of showing how such comparisons have benefited the understanding and investigation of problems in paleoclimatology. The ideas about paleoclimate outlined in these examples often developed over a series of papers by people working on the same problem using very different approaches.

Rowley et al. (1985) addressed the presence of abundant coal beds in tropical Laurasia and argued that the presence of a large mountain range on the equator—raised by the collision of Gondwana and Laurasia—enhanced the tropical Hadley circulation and temporarily countered the tendency for a large-scale monsoonal climate to develop on the supercontinent of Pangea (see Parrish 1993). Subsequently, Otto-Bliesner (1993) applied a GCM to the problem of tropical circulation (40°N to 40°S) for the same time period, performing model runs for low-latitude Pangea with and without mountains. She found that, with mountains, the tropical coal beds nearly all coincided with regions of year-round high precipitation and soil moisture, whereas without mountains, the coal-bed region was mostly dry at least part of the year (July), a condition not conducive to the formation of peats. In this example, GCM results provided support for a previous, qualitative hypothesis and matched the geologic data relatively well. Concurrence between data and models provides part of the basis for more detailed investigation of the studied paleoclimatic patterns and processes, as suggested by Otto-Bliesner (1993).

A clearly beneficial interaction between models and data has been in debate about the tendency for GCMs to predict continental interiors that are too cold, as determined from paleontological information. The model-data comparisons revealed this problem early on, and numerous attempts, starting with Barron and Washington (1984), were made to explain this discrepancy. These attempts have culminated in exploration of the combined effects of different influences on global warmth (Barron et al. 1995) and investigation of sub-grid-scale variations in paleogeography (Valdes, Sellwood, and Price 1996).

Barron et al. (1995) experimented with geography, CO_2, and oceanic heat flux and found that a combination of Cretaceous geography, CO_2 four times (4×) present levels, and poleward heat transport of 1.2×10^{15} W (watts) greater than at present produced the closest match with Cretaceous paleoclimate data. None of these factors by itself significantly altered the model results. No mechanism was known for increased oceanic heat transport and thus it was reasonable to look for independent evidence of such an increase before attempting to explain how it happened. Johnson et al. (1996) suggested that such evidence lies in the poleward expansion of tropical marine invertebrates coupled with the collapse of equatorial reef ecosystems and expansion of anoxic water masses. Although their model was speculative (Johnson 1997; Skelton and Donovan 1997), it is rich with testable hypotheses, and, important in the present context, suggests that seeking a mechanism for increased oceanic heat transport would be a fruitful endeavor. Since Barron et al.'s (1995) simulations, coupled atmosphere-ocean EBMs (Schmidt and Mysak 1996) and GCMs have given results indicating increased heat transport by the oceans for the Late Cretaceous (Bush and Philander 1997).

Another example along similar lines is the study of Late Permian climates incorporating large inland lakes and seas (Kutzbach and Ziegler 1993) interpreted from the geologic record in the Southern Hemisphere. Yemane (1993) had been critical of model results showing extremely cold climates for the continental interior of Pangea during the southern winter (e.g., Crowley, Hyde, and Short 1989; Kutzbach and Gallimore 1989), despite evidence of a flourishing biota. The models had been run on paleogeographic reconstructions that did not include the large lakes that Yemane (Yemane, Siegenthaler, and Kelts 1989; Yemane 1993) had interpreted from the geologic record. Kutzbach and Ziegler (1993) responded to this criticism and found that inclusion of the lakes and seas in the models reduced the mean annual range of temperature at 45°S from 50°C to 20°C and raised the mean winter temperature from −5°C to 10°C, values more consistent with the geologic and paleontologic data from the region. They also found the same control on precipitation by near-equatorial mountains that was seen in simulations of the Late Carboniferous tropics by Otto-Bliesner (1993; also Wilson et al. 1994).

Kutzbach and Ziegler (1993) went on to use the GCM to simulate the distribution of plant biomes. They did this by having the model predict the distribution of the combinations of temperature, precipitation, and seasonality that are the climatic constraints on modern biomes (see table 4.2). Their purpose was to see whether the model could simulate Late Permian biomes, defined previously from sedimentological and paleobotanical data (Ziegler 1990). The biomes in the region of the lakes had been interpreted as cool temperate with moderately to nonseasonal rainfall (biome 6, table 4.2), inconsistent with the continental interior setting of previous paleogeographic reconstructions (Ziegler 1990). The model predicted a mixture of the previously interpreted biome (6) and a biome (5 in table 4.2) that is characteristic of slightly warmer climates.

A data-model comparison study by Greenwood and Wing (1995), which, as of this writing, was the latest contribution to a debate about cold winters predicted for interior North America by climate models for the Eocene, is an example where model results prompted a compilation of relevant paleobotanical data. This debate was similar to that for the Cretaceous, centering around climate-model predictions that produced winters colder than interpreted from the paleontologic record. Unlike the Cretaceous, no subgrid-scale interior seaway existed in the Eocene, but there were large lakes, and Sloan (1994) performed a model run that included a large lake, coupled with 2× present CO_2. Unlike the much larger Southern Hemisphere lakes and seas incorporated by Kutzbach and Ziegler (1993), the Eocene lake affected climate only in the immediate vicinity and did not solve the discrepancy between warm-climate floras and cold-continent model results farther north (Greenwood and Wing 1995). A model run with 6× present CO_2, at the high range for Eocene CO_2 estimates, gave only slightly better results (Sloan 1994; Sloan did not couple 6× present CO_2 with the lake effect). It is instructive to note the authors' diverging conclusions: Sloan (1994) regarded the results (figure 6.10) as rather good, whereas Greenwood and Wing (1995) were of the opinion that the results indicate that significant flaws remain in the climate models. However, in a way, their respective opinions are beside the point. The benefit of the debate has been less in getting a "right" answer than in the efforts that have gone into the research. Following the lead of Barron et al. (1995) for the Cretaceous, Sloan, Walker, and Moore (1995) considered increased ocean heat transport for the Eocene (Sloan and Rea 1995). They did not regard increased oceanic heat transport as more likely over other processes as the primary cause of the early Eocene warm climate.

Data-model comparisons have likewise been fruitful for studies using conceptual models. Conceptual global rainfall predictions for the Early Triassic and Early Jurassic (Parrish, Ziegler, and Scotese 1982) indicated year-round aridity for the Colorado Plateau region (U.S.), partly because the region was on the downwind side of the continent. However, Late Triassic plants and sediments there, in the Chinle Formation, indicated at least a seasonally wet climate (e.g., Ash 1978). When the atmospheric circulation models (from which rainfall had been predicted) were extrapolated to the scale of the Colorado Plateau and compared with paleocurrent directions from

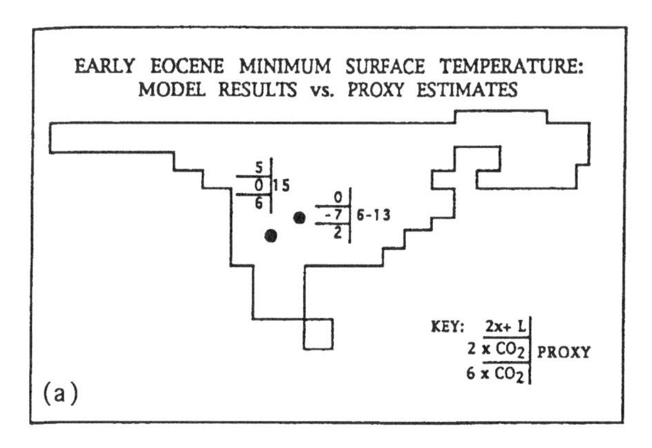

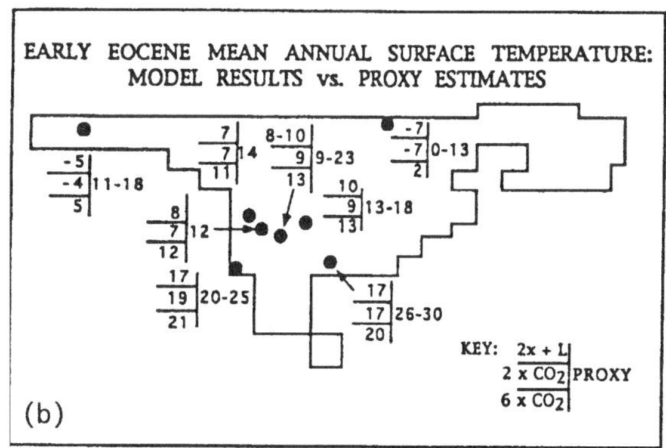

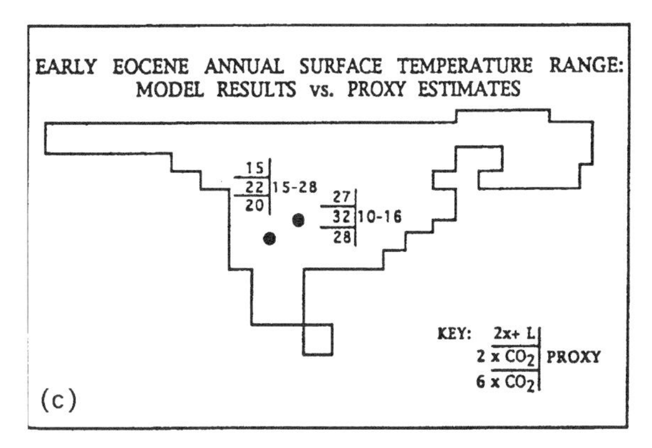

Figure 6.10. GCM temperature estimates for the early Eocene of North America compared with temperature estimates from plants. At each data point, the values on the left are different model runs and those on the right estimates from paleobotanical information. Model runs: 2× + L, 2 times modern CO_2 levels plus lake effect (see text); 2× CO_2, 2 times modern CO_2 alone; 6× CO_2, 6× modern CO_2 alone. From Sloan (1994) in *Geology.* Reproduced with permission of the publisher, the Geological Society of America, Boulder, Colorado, USA. Copyright © 1994 Geological Society of America, Inc.

eolian sandstones, it became apparent that the strongest winds underwent a 90° shift during that interval (Parrish and Peterson 1988). This 90° shift would have permitted the transport of moisture from the paleo-Pacific, a much closer source of moisture. It was the characteristics of the Chinle Formation that drew attention to the deficiency in the original models and the eolian sandstones that provided the critical data for that interpretation.

In some of these examples, conceptual model results were later supported by numerical model results. Conceptual models still have value, therefore, as a first pass at a global paleoclimate problem. However, numerical models allow workers to ask a variety of detailed questions, for example, the effects of varying CO_2 levels, that are out of the reach of conceptual models. In addition, as geologists and paleontologists accumulate greater amounts of quantitative data on paleoclimates, the attraction of numerical-model results against which to compare those data increases.

7

Integrative Paleoclimatology

Many of the strongest studies cited in this book have been those that have used a variety of methods and types of information. The strength of the conclusions of these studies does not mean they are not controversial. As refinements to previously developed methods are made and new methods developed, the conclusions might change. Or, workers might disagree on how much weight to give the various kinds of evidence and thus arrive at different conclusions. Regardless, the more data used, the more the interpretations are constrained.

One example of how conclusions are strengthened and modified as new methods are developed is the evolution of the interpretation of Cretaceous climate on the North Slope of Alaska. Early work, using plant megafossils, concluded only that the climate was warm-temperate and then became cooler (Smiley 1967); these designations were entirely qualitative. In addition, the vegetation was interpreted to have alternated between "upland" and "lowland" floras, with the implication that this was controlled by climate (May and Shane 1985).

Using leaf-margin analysis on angiosperms, R. A. Spicer and I concluded that mean annual temperature (MAT) in the early Late Cretaceous was about 10° ± 3°C (Spicer and Parrish 1986). Although too few angiosperms were found in the upper Upper Cretaceous flora to use the leaf-margin analysis method, we were able to estimate the temperature drop by that time, to between 2° and 6°C, on the basis of a drastic drop in diversity of plants, from >100 forms to about 10. A lower limit was placed on MAT by the fact that the forest type—mixed coniferous—was essentially the same as earlier in the Cretaceous, but with much lower diversity. Wolfe's (1979) nomogram, shown in figure 4.5, indicates that 2° to 6°C is about the lower limit for MAT for that forest type.

One limitation on vegetational analysis is that forest types are also influenced by precipitation; Wolfe (1979) took pains to keep that variable more or less constant in compiling his original data set. Thus one possibility for the change in North Slope vegetation through the Upper Cretaceous was that climate had become drier, which would also diminish diversity. Sedimentological analysis, however, showed that the rivers were large, meandering, and perennial (Phillips 1987, 1990). In addition, coal beds occurred throughout the section. Although they were thinner and less common in the upper Upper Cretaceous, no evidence existed that the reason for this thinning was a drier climate. Rather, the stratigraphic relationships indicated that sediment influx was probably too high to allow development of peat swamps.

The drop in temperature was also supported by a slight decrease in average growth-ring width and diameter in the fossil logs and by an increase in the number of false rings (Parrish and Spicer 1988a; Spicer and Parrish 1990). In addition, although the diversity of leaf taxa decreased, the diversity of pollen taxa increased, suggesting a larger component of herbaceous taxa. Herbs are typically opportunistic and well suited to severe conditions because they simply regenerate each growing season.

The presence of dinosaurs and a turtle suggested that we might put a lower limit on the cold-month mean temperature, but it turned out that the probable cold tolerances for those organisms were much lower than for the plants (Parrish et al. 1987). Likewise, we sampled numerous shell fragments of inoceramid bivalves, but all turned out to be too altered to provide paleotemperature data (J. T. Parrish and R. A. Spicer, unpublished). In summary, with new developments in paleobotanical analysis and better understanding of the sedimentological regime on the North Slope, we were able to bring different lines of evidence to bear on the problem, each of which offered slightly different information and augmented all the others. The strength of the interpretation provided concrete information about, and constraints on, the climate that helped inspire examination of previous modeling studies of the Cretaceous and led to investigations on the role of heat transport and vegetation in creating warm polar climates (Barron et al. 1995; Johnson et al. 1996; Bush and Philander 1997; Otto-Bliesner and Upchurch 1997).

Sometimes much evidence weighs in on both sides of a paleoclimate controversy. For example, one of the debates that has emerged in Tertiary climatology is the timing of the onset of glaciation in Antarctica. Two recent Ocean Drilling Program legs (113 and 120) were put together to address just this problem. In the Leg 120 report, Wise et al. (1992) summarized the evidence and arguments for and against an early Oligocene ice sheet (table 7.1). Note that the evidence and arguments on either side are usually not directly contradictory. Rather, different types of information are emphasized, for example, the strong positive $\delta^{18}O$ shift, interpreted to indicate cooling, versus lack of geologic evidence for a sea-level drop that would be expected with formation of a significant ice sheet. On balance, Wise et al. (1992) favored the presence of an ice sheet in the early Oligocene but were unsure of its persistence. They postulated that, despite the sharp isotope shift, the earliest Oligocene cooling was nevertheless small in comparison with the overall cooling trend for the Eocene. They placed particular weight on evidence such as that in figure 7.1, which shows a correlation of the single bed of ice-rafted debris (IRD) in the southern Indian Ocean with a strong shift in both the $\delta^{18}O$ and $\delta^{13}C$ records (Zachos et al. 1992). Although either the IRD or the isotope records could record local events, it is unlikely, in Wise et al.'s (1992) view, that the relationship between these two disparate types of evidence would reflect a solely local event.

For the most part, the use of marine fossils for paleoclimate interpretation falls into two general approaches, biogeographic and isotope analyses, although in some cases, morphologic data are also useful. Biogeographic analyses provide excellent general information about climate, particularly at large scales. Although biogeography does not give quantitative results unless calibrated with isotope data, the biogeographic trends that are seen in time and space can nevertheless provide an excellent qualitative picture of paleoclimatic trends. Despite the problems with many studies of paleobiogeography raised by Rosen (1992), biogeography has contributed substantially to our understanding of large-scale evolution of climate. For example, interpretation of the warmth of the Silurian Period, which is now noted in any number of graphs and books as a "greenhouse" time (e.g., Frakes, Francis, and Syktus 1992), is based in large part on the fact that numerous groups of organisms were highly cosmopolitan at that time. That interpretation of Silurian climate is bolstered by the presence of widespread carbonate reefs and platforms.

Isotopes of oxygen are useful in attempts to quantify climate, particularly temperature, and carbon isotopes are useful for understanding variations in biologic productivity in the oceans and their causes. Where interpreted as indicating high biologic productivity, $\delta^{13}C$ data are bolstered by the presence of other indicators of productivity, such as productivity-related faunas determined from factor analysis, abundant organisms, especially those high in the food chain, and organic- and phosphate-rich sediments. This approach was illustrated very nicely in studies of the Peterborough Member of the Oxford Clay (Jurassic; United Kingdom). A multinational group of scientists tackled the details of sedimentation of this unit, which is organic rich and notable for its abundant vertebrate fossils and ammonites (Hudson 1994). The purpose was to reconstruct the depositional environment and perhaps solve controversy about the formation's genesis, specifically the mechanisms of organic-matter sedimentation and preservation. This unit does not contain the full suite of sediments con-

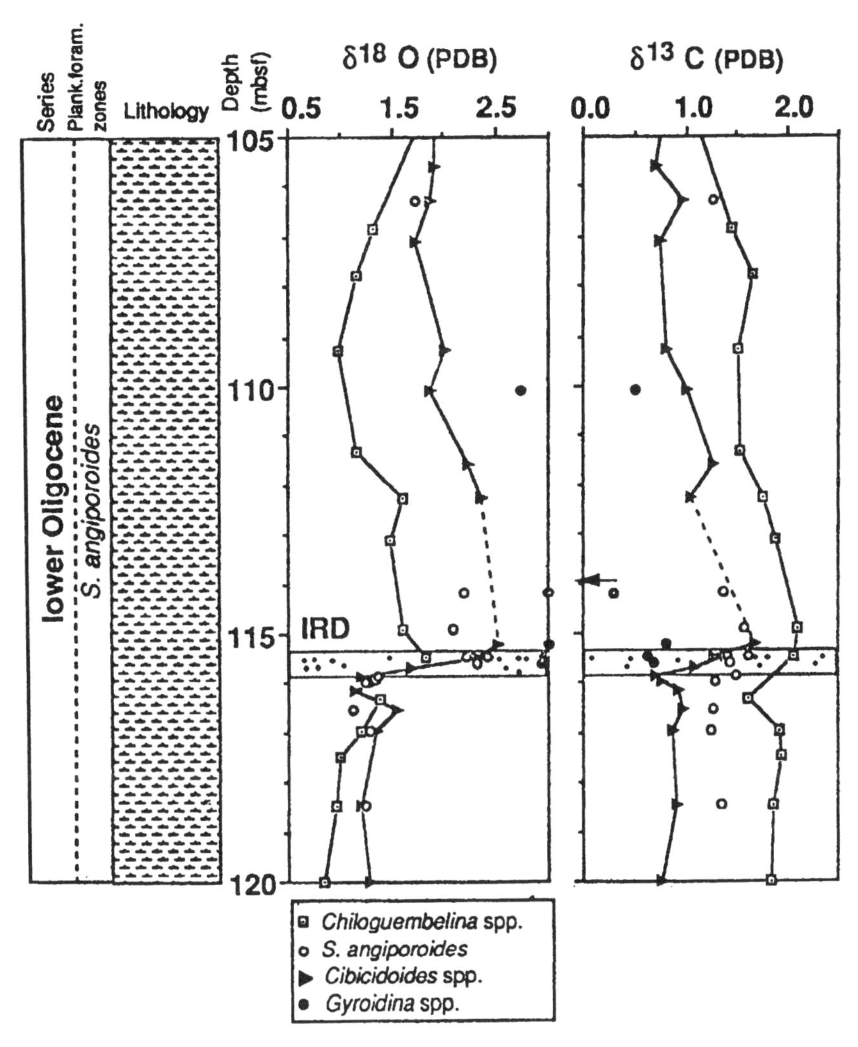

Figure 7.1. Correlation of ice-rafted debris and shifts in $\delta^{18}O$ and $\delta^{13}C$ at ODP Site 748 (southern Indian Ocean). From Zachos et al. (1992).

sidered characteristic of a well-developed upwelling zone. Although there does not seem to have been a collective conclusion about the overall depositional and chemical environment (Hudson 1994), the evidence shows that the Oxford Clay was deposited in shelf waters that were shallow enough to permit significant organic accumulation but deep enough to allow maintenance of a substantial thermal gradient from surface to bottom (Anderson et al. 1994). The climate was warm, with a mean annual temperature ≥20°C, and the marine system highly productive, that is, the organic matter was derived from plankton that were abundant enough to support a large population of large organisms.

Conclusions about the depth, as general as they were, were derived from reconstruction of the local paleogeography, sedimentology, and the variation in oxygen isotopes from animals that lived at different depths (Anderson et al. 1994; Hudson and Martill 1994; Macquaker 1994). Conclusions about the warmth of the climate came from oxygen isotopes in surface-dwelling organisms, which presumably recorded water temperatures closest to atmospheric temperatures (Anderson et al. 1994). Conclusions about the origin of the organic matter came from organic geochemical and carbon isotope data (Kenig et al. 1994). High productivity is indicated by the amount of organic matter (as high as 16.6% in the Peterborough Member), the abundance and large size of fecal pellets, the elaborate trophic structure coupled with relatively short paths from primary producers to top carnivores, and the presence of giant planktivores (Belin and Kenig 1994; Kenig et al. 1994; Martill et al.

Table 7.1.
Evidence for and against an early Oligocene ice sheet on Antarctica

Evidence and arguments for	Evidence and arguments against
Glaciomarine sediments of early Oligocene age have been discovered at three widely spaced localities around Antarctica and such widespread glaciation must be an ice sheet rather than montane glaciation.	No sedimentary evidence exists to suggest extensive formation of sea ice, implying that climate was not cold enough to support an ice sheet.
Ice-rafted debris (IRD) has been detected as far as 1000 km beyond the Antarctic margin, thus requiring large icebergs for transport.	IRD could have been transported by very dirty icebergs calved from montane glaciers.
Lack of evidence for sea ice close to the Antarctic margin indicates that cooling was the mechanism of psychrosphere development, rather than generation of saline, cold water with formation of sea ice.	IRD at Site 693 (Weddell Sea, close to the Antarctic margin) is volumetrically minor; IRD at Site 748 (southern Indian Ocean, distant from the Antarctic margin) represents a short-lived event that does not recur and could be of only local significance.
Grounded ice in the Ross Sea and Prydz Bay extended beyond the limits of the Recent ice sheet.	The continental shelves of Antarctica have been eroded and depressed by the Recent ice sheet and therefore are deeper than they would be otherwise. Therefore, they were likely much shallower in the early Oligocene and montane glaciers could have been grounded much farther out, so that no ice sheet is needed to explain the sediments in Prydz Bay and Ross Sea.
The benthic and/or planktonic foraminiferan $\delta^{18}O$ isotope records must reflect evidence of an ice-volume effect, otherwise the tropical planktonic values are too low, indicating unreasonably hot temperatures (see Miller, Fairbanks, and Mountain 1987).	The effect of the earliest Oligocene cooling on nannofossils was not mirrored in ostracodes and benthic foraminifera, as would be expected if high-latitude and bottom-water temperatures varied together.
Paleobiogeographic data on microfossils show a significant cooling and steepening of the latitudinal gradent during the earliest Ogliocene.	Nannofossil floras at high latitudes in the Ross Sea and Prydz Bay were high diversity during much of the early Oligocene.
The absence of recycled upper Eocene or lower Oligocene marine microfossils in outcrops of tills in the Pliocene Sirius Group indicates that interior seaways were covered by glacial ice at those times.	Beech forests persisted on Antarctica throughout the Oligocene, implying mild climates.
The positive $\delta^{18}O$ shift in the Oligocene is the sharpest one in the Cenozoic and is correlated with ice-rafting well beyond the margin of Antarctica.	No profound sea-level drop is recorded in earliest Oligocene sequences around the world.
	The stable isotope record is inherently ambiguous and available data can be interpreted without resorting to an ice sheet.

It should be noted that more recent studies have negated the hypothesis that the beech forests indicated mild climates (Francis and Hill 1996).
IRD, ice-rafted debris.

1994). The mechanism of high biologic productivity, that is, the source of nutrients and how they were delivered to surface waters, remains obscure, but the small amount of terrestrial organic matter, despite the proximity of land (Belin and Kenig 1994; Hudson and Martill 1994; Kenig et al. 1994), strongly suggests marine upwelling.

No one of the studies provided conclusive evidence about the environment of the Peterborough Member. Temperature was determined using isotopes from both invertebrates and vertebrates and productivity was understood from information from vertebrates, sediments, and organic and isotope geochemistry. By itself, each of these studies had a number of limitations, discussed at length by the authors. But the accumulation of data all pointing to certain conclusions gives a much clearer picture of the environment than was possible before. Notwithstanding accumulations of data such as those for the Peterborough Member and for the Antarctic glaciation problem, controversy that exists around some studies indicates that how one weighs various types of information can have a dramatic effect on how paleoclimate is interpreted.

Throughout this book, examples of the use of paleoclimatic indicators have shown that spatial or temporal trends are sometimes more helpful for interpreting paleoclimatic patterns than discrete data points. This is especially true of terrestrial lithologic indicators, even where stable isotope information is available. For example, the fascinating climate of Pangea, which has received much attention from both geologists and climate modelers (see review in Parrish 1993), has been studied mostly in continental sediments. However, as argued by Mutti and Weissert (1995), the effects of such a strong terrestrial climate system should be recorded even in Tethyan shallow-water carbonates, because the amount and seasonality of rainfall in low latitudes on the eastern side of Pangea would be enough to affect even the shallow marine environment.

Mutti and Weissert (1995) used oxygen isotope data from the carbonate platforms, along with sedimentologic information from paleosols, fluvial deposits, and evaporites from carbonate platform sequences in the Italian Alps and information on dissolution and cementation of the carbonates. The carbonate platform sequence is interrupted in the Carnian (Upper Triassic) by a paleokarst horizon (underlying the Calcare Rosso, which itself contains karst), siliciclastic deltaic deposits of the Arenaria di Val Sabbia and Gorno Formations, and evaporitic deposits of the San Giovanni Bianco.

The $\delta^{18}O$ values for host rock and cements average −5.8‰ to −6.0‰, depleted in ^{18}O with respect to Triassic seawater, which was about −2.3‰ based on analyses from aragonitic shells. The depletion could be the result of either deep burial diagenesis or the influence of meteoric waters. The analyses of the cements were from nonluminescent portions, which Mutti and Weissert (1995) regarded as evidence for meteoric influence as the principal control on isotopic composition. In addition, parts of the limestones that show the direct effects of burial, such as microfractures, have isotopic compositions markedly lower (−9‰ to −12‰). Assuming groundwater temperatures of 20° to 25°C, the isotopic composition of the meteoric waters would have been about −6‰, lower than in average low-latitude meteoric cements. This suggests a rainout effect from high altitudes, continentality, or amount. Because these cements were formed close to sea level at the continental margin, the amount effect must be the explanation for the low values if the cements reflect meteoric waters. Rainfall in megamonsoonal climates, though seasonal, is usually great, lending credence to the hypotheses of meteoric influence and amount effect. In this study, the absolute values of $\delta^{18}O$ were less important than the trends, that is, depletion relative to aragonite and in rocks showing evidence of burial.

The $\delta^{18}O$ data alone, however, could not provide much insight on the processes without other information. Mutti and Weissert (1995) interpreted the karst as indicative of high rainfall in this sequence. The karst penetrates deep (as much as 120 m) into the underlying carbonates (Esino Formation), and some of the dissolution cavities are filled with fibrous, isopachous, vadose cements. Teepee fabrics, which are deformation features characteristic of carbonates that have been subjected to desiccation, alternate with karstic horizons, which Mutti and Weissert (1995) interpreted to indicate alternation between wet and dry climates. Fluvial sediments are dominated by braided-stream deposits, which occur in regions with high but seasonal rainfall. The fluvial deposits are interbedded with reworked caliche nodules and anhydrite. The fluvial deposits are also red beds, with diagenetic features supporting the interpretation of reddening under a climate with fluctuating rainfall.

Overall, the data are consistent with a strongly monsoonal climate, evolving from wetter overall, but still seasonal, to arid, and back again through the section on long time scales (10^6 years), with superimposed shorter-term climatic fluctuations. Mutti and Weissert (1995) specifically noted that the arid interval was correlative with a wetter (but nevertheless seasonal) interval in western Pangea, which had been attributed to reversal of equatorial flow during the monsoon maximum (Parrish and Peterson 1988; Dubiel et al. 1991). Thus this study provides information that is critical to understanding the geography and evolution of Pangean climate.

Although individual indicators may have broad distributions along a paleoclimatic gradient, several lithologic indicators taken together can narrow the constraints on paleoclimate interpretation.

This was shown particularly well by Gyllenhaal (1991). His gradient analysis was presented in figure 5.33 and was based on data such as those in figures 5.1, 5.13, and 5.27, as well as similar histograms for vertisols and caliche. The spatial scale of trend analyses is usually large because of scatter in the data, but also because climate variations at the scale recorded in the geologic record on land are large. By tracing changes in lithologic indicators laterally, the effects of climate and tectonics can also be discerned (Cecil 1990). For example, in Pennsylvanian rocks of the eastern United States, lacustrine limestones grade vertically into siltstone and sandstone and eventually coal, after which the sequence is reversed (Cecil 1990). By themselves, these changes are not necessarily indicative of climatic change, but the cycle is mirrored laterally, in better-drained regions, by vertisols and aridisols that grade vertically into fluvial rocks and eventually into oxisols, with a reversal corresponding to the reversal in the lacustrine limestone-coal sequence (Cecil 1990). Given the strong paleoclimate signal in the paleosols and the correlation of the cycles, the changes are much more confidently attributed to paleoclimate as the primary control. Variations in cycles through time allowed not only refinement of previously established climate curves (Cecil et al. 1985), but addition of information about changes in seasonality (Cecil 1990; West, Archer, and Miller 1997).

One of the characteristics of many of the lithologic indicators of paleoclimate is that they would be classified as chemical sedimentary rocks. These include coal beds, evaporites, and different types of paleosols. Cecil et al. (1985, 1993) emphasized this characteristic as being particularly important to paleoclimate interpretation. They argued that the physical aspects of sedimentary rocks are controlled by tectonics and eustacy, whereas the chemical environment is controlled by climate. This argument is supported by compilations such as that in figure 5.33. Changes in the water table that are not related to climatic change but, rather, to tectonics or eustacy, cannot be easily ruled out, but documentation of widespread, correlative changes can help distinguish these influences (Cecil 1990; Dickinson, Soreghan, and Giles 1994).

Parrish, Demko, and Tanck (1993) provided a preliminary model linking many of the terrestrial paleoclimatic indicators to each other and to the important paleoclimatic and physical controls (figure 7.2). This model was partly empirical. The flowchart emphasizes the importance of climate in determining most of the chemical sedimentary indicators, as suggested by Cecil et al. (1993). A fundamental divide is in temperature. At low temperatures, the only chemical sediment that will form is peat unless temperatures are low enough that freezing can result in carbonate precipitation. Another fundamental divide is seasonality. Where overall temperatures are low, seasonality in neither temperature nor rainfall appears to have a strong effect on the sediments. At high temperatures, however, seasonality in temperature and/or rainfall can have a profound effect on the sedimentary record. With regard to seasonality, the model is oversimplified. It does not, for example, take into account the difference between winter-wet and summer-wet climates at high temperatures, although, unless the rainfall seasonality is extreme, it is not clear that it would be possible to distinguish between such environments in the geologic record. Tectonic activity is important in most cases for determining whether the sediments will be dominated by physical or chemical control. In most cases, high tectonic activity leads to sediments that are physically controlled, although this does not rule out the expression of some paleoclimatic indicators, such as eolian deposits. Where tectonic activity is high, only strongly seasonal climates are likely to leave a chemical signature in the rocks.

A growing trend in paleoclimatology is the integration of marine and terrestrial records, certainly an ultimate goal in understanding the Earth system, but one that has been hampered by traditional divisions in the field and by the different scales at which climate operates in the two realms. To confidently show a causal mechanism that affected both realms would require collecting data that might be expected to show a signal that changed as rapidly in the oceans as in the atmosphere and that recorded a single, strong event. Changes in the proportions of stable isotopes of oxygen and carbon in the global reservoir (water-bound oxygen and CO_2) meet the first criterion, and a single, strong event is recorded in those data near the Paleocene-Eocene boundary in both the marine and terrestrial records (Koch, Zachos, and Gingerich 1992; Koch, Zachos, and Dettman 1995). In the marine realm, strong negative shifts in both $\delta^{18}O$ and $\delta^{13}C$ are found in surface-dwelling foraminifera just above the Paleocene-Eocene boundary

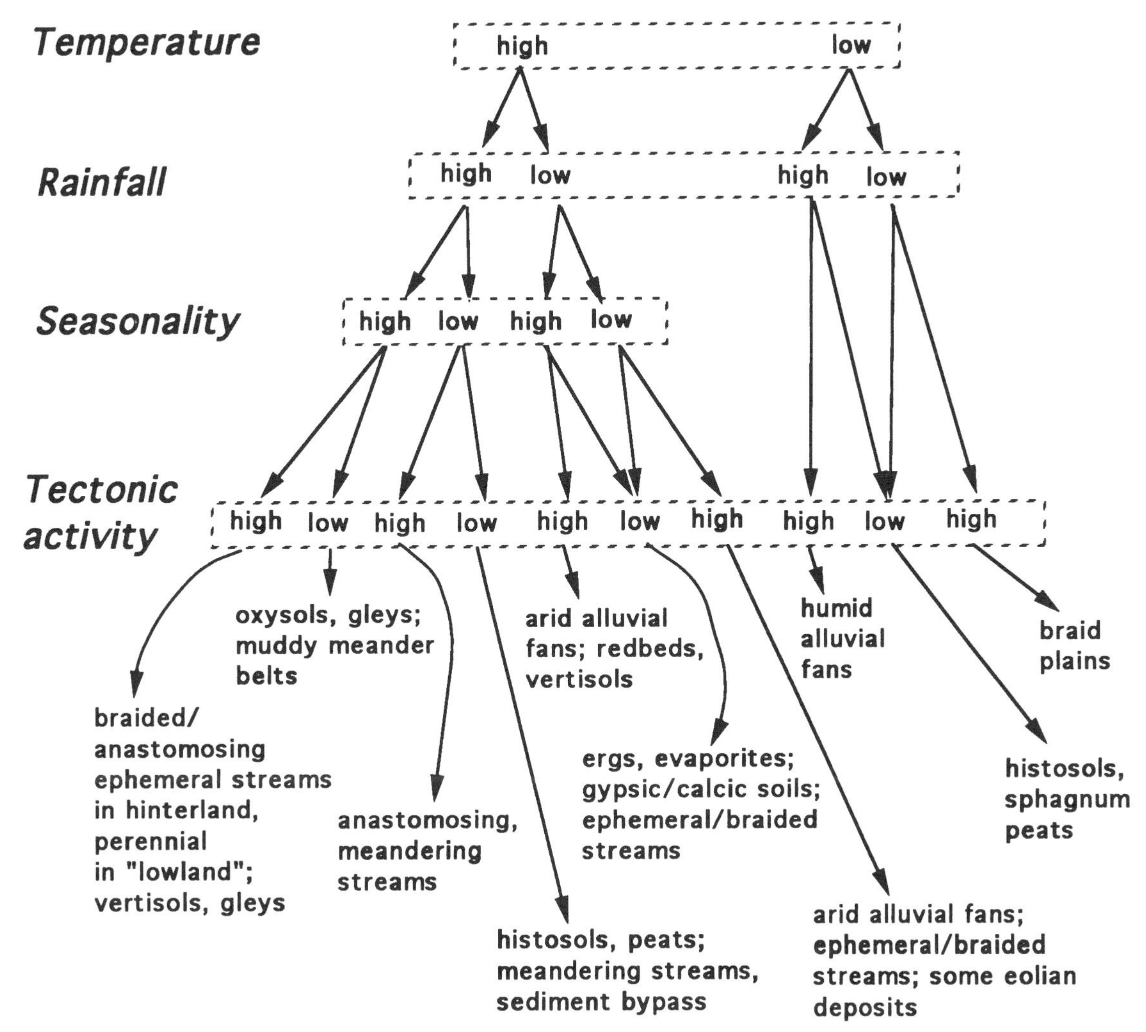

Figure 7.2. Model of climatic control on the formation of terrestrial paleoclimatic indicators. From Parrish, Demko, and Tanck (1993) in *Philosophical Transactions of the Royal Society of London,* Ser. A 344:21–5, with permission from The Royal Society.

as defined by magnetostratigraphy. In addition, $\delta^{18}O$ data converge between high- and low-latitude sites (Koch, Zachos, and Dettman 1995). On land, a strong negative shift in $\delta^{13}C$ is found in paleosol carbonates and the tooth enamel of vertebrates right at the boundary, as defined by on-land magnetostratigraphy and mammalian biostratigraphic zones. The questions are, first, is the apparent difference in relationship to the boundary significant and, second, what do the isotopic shifts mean?

The first problem, that of the exact correlation, shows one of the reasons why correlating climatic change in the marine and terrestrial realms is so difficult. First, the biostratigraphic time scales, which are commonly the basis for major chronostratigraphic boundaries, are based on different organisms. This is why the magnetostratigraphic time scale is so important; magnetic reversals are recorded simultaneously in terrestrial and oceanic sediments. Second, sedimentation styles are very different in the terrestrial and marine realms. Koch, Zachos, and Dettman (1995) noted that in both the marine and terrestrial records, the event occurred one third to one half of the way from the Chron 24–Chron 25 boundary. By assuming constant sedimentation rate, they dated the base of the event in the terrestrial section at 54.88 Ma, compared with 54.68 Ma estimated for the event in the marine (Kennett and Stott 1990). There are three possible interpretations for the 200,000 year difference. One is that it represents the remaining error between the two sections, possibly the result of an incorrect assumption of constant sedimentation rate. This was the interpretation favored by Koch, Zachos, and Dettman (1995), partly because the boundary has been the subject of constant revision. Indeed, they argued that the event would be a good chemostratigraphic marker for the boundary. Another is that the difference is real, representing a lag in response in the marine realm to whatever forcing caused the isotopic shifts. A third possibility is that the difference is real and that there is no connection between the shifts in the terrestrial and marine records. Neither of the latter two possibilities was considered by Koch, Zachos, and Dettman (1995), and the third is one that no paleoclimatologist is likely to consider unless further efforts to understand or dispel the difference proved futile.

The interpretation offered by Koch, Zachos, and Dettman (1995) for the isotopic shifts is abrupt and intense warming at high latitudes ($\delta^{18}O$ shift) accompanied by rapid erosion of carbon, possibly from terrestrial sediments (Rea et al. 1990; $\delta^{13}C$ shift). Dickens et al. (1995) suggested release of isotopically light methane from gas hydrates during oceanic warming. The event was accompanied by an increase in the cosmopolitanism of terrestrial mammals, which Koch, Zachos, and Dettman (1995) suggested might have resulted from high-latitude warming on land as well in the oceans. If large amounts of methane were released, that might have effected the warming on land because methane is a greenhouse gas (Sloan et al. 1992; Dickens et al. 1995). The links in the carbon record between marine surficial carbonate (foraminifera) and mammal teeth are through the atmosphere and plants.

The theme of this book is as follows: *Paleoclimate interpretations for every time and place are strengthened by the use of as many different indicators as possible and by consideration of those indicators in their regional and global contexts.* The studies cited in this chapter are just a few of those that have taken full advantage of this principle. The prospects for a previously unimaginable level of detail in our understanding of the climates of deep time are excellent as more integrated studies are produced and are themselves integrated.

SCOTT W. STARRATT

References

Abrantes, F. 1992. Palaeoproductivity oscillations during the last 130 ka along the Portuguese and NW African margins. In C. P. Summerhayes, W. L. Prell, and K. C. Emeis, eds., *Upwelling Systems: Evolution Since the Early Miocene,* pp. 499–510. Geological Society of London Special Publication 64.

Adams, C. G. 1973. Some Tertiary foraminifera. In A. Hallam, ed. *Atlas of Palaeobiogeography,* pp. 453–68. Amsterdam: Elsevier.

———. 1992. Larger foraminifera and the dating of Neogene events. In R. Tsuchi and J. C. Ingle, Jr., eds., *Pacific Neogene: Environment, Evolution, and Events,* pp. 221–35. Tokyo: University of Tokyo Press.

Adams, C. G., D. E. Lee, and B. R. Rosen. 1990. Conflicting isotopic and biotic evidence for tropical sea-surface temperatures during the Tertiary. *Palaeogeography, Palaeoclimatology, Palaeoecology* 77:289–313.

Addicott, W. O. 1969. Tertiary climatic change in the marginal northeastern Pacific Ocean. *Science* 165:583–6.

———. 1970. Latitudinal gradients in Tertiary molluscan faunas of the Pacific coast. *Palaeogeography, Palaeoclimatology, Palaeoecology* 8:287–312.

Adlis, D. S., E. L. Grossman, T. E. Yancey, and R. D. McLerran. 1988. Isotope stratigraphy and paleodepth changes of Pennsylvanian cyclical sedimentary deposits. *Palaios* 3:487–506.

Agassiz, L. 1828. On the erratic blocks of the Jura. *Edinburgh New Philosophical Journal* 24:176–9.

Aigner, T. 1982. Calcareous tempestites: Storm-dominated stratification in Upper Muschelkalk limestones (Middle Trias, SW-Germany). In G. Einsele and A. Seilacher, eds., *Cyclic and Event Stratification,* pp. 180–98. Berlin: Springer-Verlag.

Allen, J. R. L. 1996. Wind blown trees as a palaeoclimatic indicator: The character and role of gusts. *Palaeogeography, Palaeoclimatology, Palaeoecology* 121:1–12.

Allmon, W. D. 1988. Ecology of recent Turritelline gastropods (Prosobranchia, Turritellidae): Current knowledge and paleontological implications. *Palaios* 3:259–84.

———. 1993. Age, environment and mode of deposition of the densely fossiliferous Pinecrest Sand (Pliocene of Florida): Implications for the role of biological productivity in shell bed formation. *Palaios* 8:183–201.

Allmon, W. D., S. D. Emslie, D. S. Jones, and G. S. Morgan. 1996. Late Neogene oceanographic change along Florida's west coast: Evidence and mechanisms. *Journal of Geology* 104:143–62.

Almogi-Labin, A., A. Bein, and E. Sass. 1993. Late Cretaceous upwelling system along the southern Tethys margin (Israel): Interrelationship between productivity, bottom water environments, and organic matter preservation. *Paleoceanography* 8:671–90.

Altabet, M. A. and R. François. 1994. Sedimentary nitrogen isotopic ratio as a recorder for surface ocean nitrate utilization. *Global Biogeochemical Cycles* 8:103–16.

Alvin, K. L., C. J. Fraser, and R. A. Spicer. 1981. Anatomy and palaeoecology of *Pseudofrenelopsis* and associated conifers in the English Wealden. *Palaeontology* 24:759–78.

Alvin, K. L., R. A. Spicer, and J. Watson. 1978. A *Classopollis*-containing male cone associated with *Pseudofrenelopsis*. *Palaeontology* 21:847–56.

Ammons, R., W. J. Fritz, R. B. Ammons, and A. Ammons. 1987. Cross-identification of ring signatures in Eocene trees (*Sequoia magnifica*) from the Specimen Ridge locality of the Yellowstone fossil forests. *Palaeogeography, Palaeoclimatology, Palaeoecology* 60:97–108.

Amorosi, A. 1997. Detecting compositional, spatial, and temporal attributes of glaucony: A tool for provenance research. *Sedimentary Geology* 109: 135–53.

Anderson, D. M. and W. L. Prell. 1993. A 300 kyr record of upwelling off Oman during the late Quaternary: Evidence of the Asian southwest monsoon. *Paleoceanography* 8:193–208.

Anderson, J. B. 1983. Ancient glacial-marine deposits: Their spatial and temporal distribution. In B. F. Molnia, ed. *Glacial-Marine Sedimentation*, pp. 3–92. Mt. Kisco, NY: Plenum.

Anderson, J. B. and G. M. Ashley. 1991. Glacial marine sedimentation paleoclimatic signficance: A discussion. In J. B. Anderson and G. M. Ashley, eds., *Glacial Marine Sedimentation; Paleoclimatic Significance*, pp. 223–6. Geological Society of America Special Paper 261.

Anderson, R. Y. and W. E. Dean. 1995. Filling the Delaware Basin: Hydrologic and climatic controls on the Permian Castile Formation varved evaporite. In P. A. Scholle, T. M. Peryt, and D. S. Ulmer-Scholle, eds., *The Permian of Northern Pangea, v. 2*, pp. 61–78. Berlin: Springer-Verlag.

Anderson, R. Y., W. E. Dean, J. P. Bradbury, and D. Love. 1985. Meromictic lakes and varved lake sediments in North America. *U.S. Geological Survey Bulletin* 1607.

Anderson, R. Y., W. E. Dean, D. W. Kirkland, and H. I. Snider. 1972. Permian Castile varved evaporite sequence, West Texas and New Mexico. *Geological Society of America Bulletin* 83:59–86.

Anderson, R. Y. and D. W. Kirkland. 1960. Origin, varves, and cycles of Jurassic Todilto Formation, New Mexico. *American Association of Petroleum Geologists Bulletin* 44:37–52.

Anderson, T. F., B. N. Popp, A. C. Williams, L.-Z. Ho, and J. D. Hudson. 1994. The stable isotopic record of fossils from the Peterborough Member, Oxford Clay Formation (Jurassic), UK: palaeoenvironmental implications. *Journal of the Geological Society of London* 151:125–38.

Andreasson, F. P. and B. Schmitz. 1996. Winter and summer temperatures of the early middle Eocene of France from *Turritella* $\delta^{18}O$ profiles. *Geology* 24:1067–70.

Archer, D., M. Lyle, K. Rodgers, and P. Froelich. 1993. What controls opal preservation in tropical deep-sea sediments? *Paleoceanography* 8:7–21.

Arnold, E., M. Leinen, and J. King. 1995. Paleoenvironmental variation based on the mineralogy and rock-magnetic properties of sediment from Sites 885 and 886. In D. K. Rea, I. A. Basov, D. W. Scholl, and J. F. Allan eds., *Proceedings of the Ocean Drilling Project, Scientific Results, Leg 145*, pp. 231–45. College Station, TX: Ocean Drilling Program.

Arthur, M. A., T. F. Anderson, I. R. Kaplan, J. Veizer, and L. S. Land, eds. 1983. *Stable Isotopes in Sedimentary Geology*. SEPM Short Course 10.

Arthur, M. A., W. E. Dean, E. D. Neff, B. J. Hay, J. King, and G. Jones. 1994. Varve calibrated records of carbonate and organic carbon accumulation over the last 2000 years in the Black Sea. *Global Biogeochemical Cycles* 8:195–217.

Arthur, M. A., W. E. Dean, and L. M. Pratt. 1988. Geochemical and climatic effects of increased marine organic carbon burial at the Cenomanian/Turonian boundary. *Nature* 335:714–7.

Arthur, M. A., W. E. Dean, and D. A. V. Stow. 1984. Models for the deposition of Mesozoic-Cenozoic fine-grained organic-carbon-rich sediment in the deep sea. In D. A. V. Stow and D. J. W. Piper, eds., *Fine-Grained Sediments: Deep-Water Processes and Facies*, pp. 527–59. Geological Society of London Special Publication 15.

Arthur, M. A. and H. C. Jenkyns. 1981. Phosphorites and paleoceanography. Paper read at *Proceedings of the 26th International Geological Congress, Symposium C4, Geology of Oceans*, at Paris, 7–17 July 1980.

Arthur, M. A. and B. B. Sageman. 1994. Marine black shales: Depositional mechanisms and environments of ancient deposits. *Annual Reviews of Earth and Planetary Sciences* 22:499–551.

Arthur, M. A., S. O. Schlanger, and H. C. Jenkyns. 1987. The Cenomanian-Turonian oceanic anoxic event, II. Paleoceanographic controls on organic-matter production and preservation. In J. Brooks and A. J. Fleet, eds., *Marine Petroleum Source Rocks*, pp. 401–20. Geological Society of London Special Publication 26.

Ash, S. R. 1978. Geology, paleontology, and paleoecology of a Late Triassic lake, western New Mexico. *Brigham Young University Geology Studies*, 25, pt. 2.

Ashley, G. M., J. Shaw, and N. D. Smith. 1985.

Glacial Sedimentary Environments. Society of Economic Paleontologists and Mineralogists Short Course 16.

Atchley, S. C. and D. B. Loope. 1993. Low-stand aeolian influence on stratigraphic completeness: Upper member of the Hermosa Formation (latest Carboniferous), southeast Utah, USA. In K. Pye and N. Lancaster, eds., *Aeolian Sediments, Ancient and Modern*, pp. 127–49. Special Publications of the International Association of Sedimentologists 16. Oxford: Blackwell Scientific.

Ayliffe, L. K. and A. R. Chivas. 1990. Oxygen isotope composition of the bone phosphate of Australian kangaroos: Potential as a palaeoenvironmental indicator. *Geochimica et Cosmochimica Acta* 54:2603–9.

Babinot, J.-F. and J.-P. Colin. 1992. Marine ostracode provincialism in the Late Creatceous of the Tethyan realm and the Austral Province. *Palaeogeography, Palaeoclimatology, Palaeoecology* 92:283–93.

Bachhuber, F. W. 1992. A pre-late Wisconsin paleolimnologic record from the Estancia Valley, central New Mexico. In P. U. Clark and P. D. Lea, eds., *The Last Interglacial-Glacial Transition in North America*, pp. 289–307. Geological Society of America Special Paper 270.

Bagnold, R. A. 1941. *The Physics of Blown Sand and Desert Dunes*. London: Methuen.

Bailey, I. W. and E. W. Sinnott. 1915. A botanical index of Cretaceous and Tertiary climates. *Science* 41:831–4.

———. 1916. The climatic distribution of certain types of angiosperm leaves. *American Journal of Botany* 3:24–39.

Bailey, S. W. 1980. Summary of recommendations of AIPEA Nomenclature Committee. *Clay Minerals* 15:85–93.

Baird, G. C. and C. E. Brett. 1991. Submarine erosion on the anoxic sea floor: Stratinomic, palaeoenvironmental, and temporal significance of reworked pyrite-bone deposits. In R. V. Tyson and T. H. Pearson, eds., *Modern and Ancient Continental Shelf Anoxia*, pp. 233–57. Geological Society of London Special Publication 58.

Baker, V. R. 1988. Flood erosion. In V. R. Baker, R. C. Kochel, and P. C. Patton, eds., *Flood Geomorphology*, pp. 81–95. New York: John Wiley and Sons.

Bakker, R. T. 1971. Dinosaur physiology and the origin of mammals. *Evolution* 25:636–58.

———. 1975. Dinosaur rennaissance. *Scientific American* 232(4):58–78.

Baldauf, J. G. and J. A. Barron. 1990. Evolution of biosiliceous sedimentation patterns—Eocene through Quaternary: Paleoceanographic response to polar cooling. In U. Bleil and J. Thiede, eds., *Geological History of the Polar Oceans: Arctic Versus Antarctic*, pp. 575–607. Amsterdam: Kluwer Academic.

Ballantyne, C. K. and G. Whittington. 1987. Niveo-aeolian sand deposits on An Teallach, Wester Ross, Scotland. *Transactions of the Royal Society of Edinburgh: Earth Sciences* 78:51–63.

Bambach, R. K. 1990. Late Palaeozoic provinciality in the marine realm. In W. S. McKerrow and C. R. Scotese, eds., *Palaeozoic Palaeogeography and Biogeography*, pp. 307–23. Geological Society of London Memoir 12.

Bárdossy, G. 1982. *Karst Bauxites*. Developments in Economic Geology 14. Amsterdam and Budapest: Elsevier and Akadémiai Kiadó.

Bárdossy, G. and G. J. J. Aleva. 1990. *Lateritic Bauxites*. Developments in Economic Geology 27. Amsterdam: Elsevier.

Barnola, J. M., D. Raynaud, Y. S. Korotkevitch, and C. Lorius. 1987. Vostok ice core: A 160,000-year record of atmospheric CO_2. *Nature* 329:408–14.

Barrick, R. E., A. G. Fischer, Y. Kolodny, B. Luz, and D. Bohaska. 1992. Cetacean bone oxygen isotopes as proxies for Miocene ocean composition and glaciation. *Palaios* 7:521–31.

Barrick, R. E., W. J. Showers, and A. G. Fischer. 1996. Comparison of thermoregulation of four ornithischian dinosaurs and a varanid lizard from the Cretaceous Two Medicine Formation: Evidence from oxygen isotopes. *Palaios* 11:295–305.

Barron, E. J. 1983. A warm, equable Cretaceous: The nature of the problem. *Earth-Science Reviews* 19:305–38.

———. 1984. Climatic implications of the variable obliquity explanation of Cretaceous-Paleogene high-latitude floras. *Geology* 12:595–8.

———. 1985a. Numerical climate modeling, a frontier in petroleum source rock prediction: Results based on Cretaceous simulation. *American Association of Petroleum Geologists Bulletin* 69: 448–59.

———. 1985b. Explanations of the Tertiary global cooling trend. *Palaeogeography, Palaeoclimatology, Palaeoecology* 50:45–61.

———. 1990. *Atlas of Cretaceous Model Results*. Earth System Science Center Publication. University Park, PA: Pennsylvania State University.

Barron, E. J., M. A. Arthur, and E. G. Kauffman. 1985. Cretaceous rhythmic bedding sequences: A plausible link between orbital variations and climate. *Earth and Planetary Science Letters* 72: 327–40.

Barron, E. J., E. Boyle, and M. Leinen. 1991. Abrupt changes and major reorganizations in ocean-atmosphere circulation. In G. S. Mountain and M. E. Katz, eds., *Advisory Panel Report on Earth System History*, pp. 37–48. Washington, D.C.: Joint Oceanographic Institutions, Inc.

Barron, E. J., P. J. Fawcett, W. H. Peterson, D. Pollard, and S. L. Thompson. 1995. A "simulation" of mid-Cretaceous climate. *Paleoceanography* 10:953–62.

Barron, E. J. and G. T. Moore. 1994. *Climate Model Application in Paleoenvironmental Analysis.* SEPM (Society for Sedimentary Geology) Short Course 33.

Barron, E. J., E. Saltzman, and D. A. Price. 1984. Occurrence of *Inoceramus* in the South Atlantic and oxygen isotopic paleotemperatures in Hole 530A. In W. W. Hay, J. C. Sibuet, et al. *Initial Reports of the Deep-Sea Drilling Project, Leg 75*, pp. 893–904. Washington, D.C.: U.S. Government Printing Office.

Barron, E. J. and W. M. Washington. 1981. Modeling Cretaceous climate using realistic geography: Simulations with the NCAR Global Circulation Model. *Geological Society of America Abstracts with Program* 13:404.

———. 1982a. Atmospheric circulation during warm geologic periods: Is the equator-to-pole surface-temperature gradient the controlling factor? *Geology* 10:633–6.

———. 1982b. Cretaceous climate: A comparison of atmospheric simulations with the geologic record. *Palaeogeography, Palaeoclimatology, Palaeoecology* 40:103–33.

———. 1984. The role of geographic variables in explaining paleoclimates: Results from Cretaceous climate model sensitivity studies. *Journal of Geophysical Research* 89:1267–79.

———. 1985. Warm Cretaceous climates: High atmospheric CO_2 as a plausible mechanism. In E. T. Sundquist and W. S. Broecker, eds., *The Carbon Cycle and Atmospheric CO_2: Natural Variations Archean to Present*, pp. 546–53. Geophysical Monograph 32. Washington, D.C.: American Geophysical Union.

Barron, J. A. 1992. Pliocene paleoclimatic interpretation of DSDP SITE 580 (NW Pacific) using diatoms. *Marine Micropaleontology* 20:23–44.

———. 1993. Diatoms. In J. H. Lipps, ed. *Fossil Prokaryotes and Protists*, pp. 155–67. Oxford: Blackwell Scientific.

Barron, J. A. and J. G. Baldauf. 1989. Tertiary cooling steps and paleoproductivity as reflected by diatoms and biosiliceous sediments. In W. H. Berger, V. S. Smetacek, and G. Wefer, eds., *Productivity of the Ocean: Present and Past*, pp. 341–54. New York: John Wiley and Sons.

Barry, J. C., M. E. Morgan, L. J. Flynn, D. Pilbeam, L. L. Jacobs, E. H. Lindsay, S. M. Raza, and N. Solounias. 1995. Patterns of faunal turnover and diversity in the Neogene Siwaliks of northern Pakistan. *Palaeogeography, Palaeoclimatology, Palaeoecology* 115:209–26.

Barry, R. G. and R. J. Chorley. 1987. *Atmosphere, Weather, and Climate*, 5th ed. London: Methuen.

Bartlein, P. J. and C. Whitlock. 1993. Paleoclimatic interpretation of the Elk Lake pollen record. In J. P. Bradbury and W. E. Dean, eds., *Elk Lake, Minnesota: Evidence for Rapid Climate Change in the North-Central United States*, pp. 275–93. Geological Society of America Special Paper 276.

Bartram, K. M. 1987. Lycopod succession in coals: An example from the Low Barnsley Seam (Westphalian B), Yorkshire, England. In A. C. Scott, ed. *Coal and Coal-Bearing Strata*, pp. 187–99. Geological Society of London Special Publication 32.

Bates, R. L. and J. A. Jackson, eds. 1984. *Dictionary of Geological Terms.* New York: Anchor Books, Doubleday.

Baturin, G. N. 1983. Some unique sedimentological and geochemical features of deposits in coastal upwelling regions. In J. Thiede and E. Suess, eds., *Coastal Upwelling: Its Sediment Record, Part B*, pp. 11–27. Mt. Kisco, NY: Plenum.

Baum, S. K. and T. J. Crowley. 1991. Seasonal snowline instability in a climate model with realistic geography: Application to Carboniferous (~300 MA) glaciation. *Geophysical Research Letters* 18:1719–22.

Bé, A. W. H. 1977. An ecological, zoogeographic and taxonomic review of Recent planktonic foraminifera. In A. T. S. Ramsay, ed. *Oceanic Micropalaeontology Vol. 1*, pp. 1–102. London: Academic Press.

Beaufort, L. and M.-P. Aubry. 1992. Paleoceanographic implications of a 17-m.y.-long record of high-latitude Miocene calcareous nannoplankton fluctuations. In S. W. Wise, Jr., R. Schlich, et al. *Proceedings of the Ocean Drilling Program, Scientific Results, Leg 120*, pp. 539–49. College Station, TX: Ocean Drilling Program.

Beauvais, L. 1992. Palaeobiogeography of the Early Cretaceous corals. *Palaeogeography, Palaeoclimatology, Palaeoecology* 92:233–47.

Beerling, D. J. 1997. Interpreting environmental and

biological signals from the stable carbon isotope composition of fossilized organic and inorganic carbon. *Journal of the Geological Society of London* 154:303–9.

Beerling, D. J. and W. G. Chaloner. 1993. Evolutionary responses of stomatal density to global CO_2 change. *Biological Journal of the Linnean Society* 48:343–53.

———. 1994. Atmospheric CO_2 changes since the last glacial maximum: Evidence from the stomatal density record of fossil leaves. *Review of Palaeobotany and Palynology* 81:11–7.

Behrensmeyer, A. K. 1978. Taphonomic and ecologic information from bone weathering. *Paleobiology* 4:150–62.

Behrensmeyer, A. K. and L. Tauxe. 1982. Isochronous fluvial systems in Miocene deposits of northern Pakistan. *Sedimentology* 29:331–52.

Belin, S. and F. Kenig. 1994. Petrographic analyses of organo-mineral relationships: Depositional conditions of the Oxford Clay Formation (Jurassic), UK. *Journal of the Geological Society of London* 151:153–60.

Bennett, M. R., P. Doyle, and A. E. Mather. 1996. Dropstones: Their origin and significance. *Palaeogeography, Palaeoclimatology, Palaeoecology* 121:331–9.

Benson, L. V., D. R. Currey, R. I. Dorn, K. R. Lajoie, C. G. Oviatt, S. W. Robinson, G. I. Smith, and S. Stine. 1990. Chronology of expansion and contractions of four Great Basin lake systems during the past 35,000 years. *Palaeogeography, Palaeoclimatology, Palaeoecology* 78:241–86.

Berger, W. H. 1968. Planktonic foraminifera: Selective solution and paleoclimatic interpretation. *Deep-Sea Research* 15:31–43.

———. 1970. Biogenous deep-sea sediments: Fractionation by deep-sea circulation. *Geological Society of America Bulletin* 81:1385–402.

———. 1982. Increase of carbon dioxide in the atmosphere during deglaciation: The coral reef hypothesis. *Naturwissenschaften* 69:87–8.

Berger, W. H. and R. S. Keir. 1984. Glacial-Holocene changes in atmospheric CO_2 and the deep-sea record. In J. E. Hansen and T. Takahashi, eds., *Climate Processes and Climate Sensitivity*, pp. 337–51. Geophysical Monograph 29. Washington, D.C.: American Geophysical Union.

Berger, W. H., R. M. Leckie, T. R. Janecek, R. Stax, and T. Takayama. 1993. Neogene carbonate sedimentation on Ontong Java Plateau: Highlights and open questions. In W. H. Berger, L. W. Kroenke, L. A. Mayer, et al. *Proceedings of the Ocean Drilling Program, Scientific Results, Leg 130*, pp. 711–44. College Station, TX: Ocean Drilling Program.

Berger, W. H. and A. Spitzy. 1988. History of atmospheric CO_2: Constraints from the deep-sea record. *Paleoceanography* 3:401–12.

Berggren, W. A. 1977. Atlas of Palaeogene planktonic foraminifera. In A. T. S. Ramsay, ed. *Oceanic Micropalaeontology Vol. 1*, pp. 205–99. London: Academic Press.

———. 1978. Marine micropaleontology: An introduction. In B. U. Haq and A. Boersma, eds., *Introduction to Marine Micropaleontology*, pp. 1–17. New York: Elsevier.

Berggren, W. A., D. V. Kent, and J. J. Flynn. 1985. Paleogene geochronology and chronostratigraphy. In N. J. Snelling, ed. *The Chronology of the Geological Record*, pp. 141–95. Geological Society of London Memoir.

Berggren, W. A., D. V. Kent, and J. A. Van Couvering. 1985. Neogene geochronology and chronostratigraphy. In N. J. Snelling, ed. *The Chronology of the Geological Record*, pp. 211–59. Geological Society of London Memoir.

Bergström, S. M. 1973. Ordovician conodonts. In A. Hallam, ed. *Atlas of Palaeobiogeography*, pp. 47–58. Amsterdam: Elsevier Scientific.

———. 1990. Relations between conodont provincialism and the changing palaeogeography during the Early Palaeozoic. In W. S. McKerrow and C. R. Scotese, eds., *Palaeozoic Palaeogeography and Biogeography*, pp. 105–21. Geological Society of London Memoir 12.

Berner, R. A. 1981. A new geochemical classification of sedimentary environments. *Journal of Sedimentary Petrology* 51:359–65.

———. 1990a. Diagenesis of phosphorus in sediments from non-upwelling areas. In W. C. Burnett and S. R. Riggs, eds., *Phosphate Deposits of the World. Vol. 3, Neogene to Modern Phosphorites*, pp. 27–32. Cambridge: Cambridge University Press.

———. 1990b. Atmospheric carbon dioxide levels over Phanerozoic time. *Science* 249:1382–6.

———. 1991. A model for atmospheric CO_2 over Phanerozoic time. *American Journal of Science* 291:339–76.

———. 1992. Palaeo-CO_2 and climate. *Nature* 358:114.

———. 1993. Paleozoic atmospheric CO_2: Importance of solar radiation and plant evolution. *Science* 261:68–70.

———. 1994. GEOCARB II: A revised model of atmospheric CO_2 over Phanerozoic time. *American Journal of Science* 294:56–91.

Berner, R. A., K. C. Ruttenberg, E. D. Ingall, and J.-L. Rao. 1993. The nature of phosphorus burial in modern marine sediments. In R. Wollast, F. T. Mackenzie, and L. Chou, eds., *Interactions of C, N, P and S Biogeochemical Cycles and Global Change*, pp. 365–78. NATO ASI Series I, Global Environmental Change 4. Berlin: Springer-Verlag.

Berner, R. A. and D. M. Rye. 1992. Calculation of the Phanerozoic strontium isotope record of the oceans from a carbon cycle model. *American Journal of Science* 292:136–48.

Berry, W. B. N. 1974. Facies distribution patterns of some marine benthic faunas in Early Paleozoic platform environments. *Palaeogeography, Palaeoclimatology, Palaeoecology* 15:152–68.

Berry, W. B. N. and P. Wilde. 1978. Progressive ventilation of the oceans—An explanation for the distribution of the lower Paleozoic black shales. *American Journal of Science* 278:257–75.

———. 1990. Graptolite biogeography: Implications for palaeogeography and palaeoceanography. In W. S. McKerrow and C. R. Scotese, eds., *Palaeozoic Palaeogeography and Biogeography*, pp. 129–37. Geological Society of London Memoir 12.

Berry, W. B. N., P. Wilde, and M. S. Quinby-Hunt. 1987. The oceanic non-sulfidic oxygen-minimum zone: A habitat for graptolites? *Bulletin of the Geological Society of Denmark* 35:103–14.

Bertram, C. J., H. Elderfield, R. J. Aldridge, and S. Conway Morris. 1992. $^{87}Sr/^{86}Sr$, $^{143}Nd/^{144}Nd$ and REEs in Silurian phosphatic fossils. *Earth and Planetary Science Letters* 113:239–49.

Besly, B. M. and C. R. Fielding. 1989. Paleosols in Westphalian coal-bearing and red-bed sequences, central and northern England. *Palaeogeography, Palaeoclimatology, Palaeoecology* 70:303–30.

Bestland, E. A., G. J. Retallack, A. E. Rice, and A. Mindszenty. 1996. Late Eocene detrital laterites in central Oregon: Mass balance geochemistry, depositional setting, and landscape evolution. *Geological Society of America Bulletin* 108:285–302.

Birkeland, P. W. 1974. *Pedology, Weathering, and Geomorphological Research*. New York: Oxford University Press.

Bischoff, J. L., R. Julià, W. C. Shanks, III, and R. J. Rosenbauer. 1994. Karstification without carbonic acid: Bedrock dissolution by gypsum-driven dedolomitization. *Geology* 22:995–8.

Black, R. F. 1976. Periglacial features indicative of permafrost: Ice and soil wedges. *Quaternary Research* 6:3–26.

Blair, T. C. and J. G. McPherson. 1994. Alluvial fans and their natural distinction from rivers based on morphology, hydraulic processes, sedimentary processes, and facies assemblages. *Journal of Sedimentary Research* A64:450–89.

Blakey, R. C. and L. T. Middleton. 1983. Permian shoreline eolian complex in central Arizona: Dune changes in repsonse to cyclic sealevel changes. In M. E. Brookfield and T. S. Ahlbrandt, eds., *Eolian Sediments and Processes*, pp. 551–81. Developments in Sedimentology 38. Amsterdam: Elsevier.

Blodgett, R. B., D. M. Rohr, and A. J. Boucot. 1990. Early and Middle Devonian gastropod biogeography. In W. S. McKerrow and C. R. Scotese, eds., *Palaeozoic Palaeogeography and Biogeography*, pp. 277–84. Geological Society of London Memoir 12.

Blodgett, R. H. 1984. Nonmarine depositional environments and paleosol development in the Upper Triassic Dolores Formation, southwestern Colorado. In D. C. Brew, ed. *Four Corners Geological Society Field Trip Book*, pp. 46–92. Durango, Colorado.

———. 1985a. Paleovertisols as indicators of climate. *American Association of Petroleum Geologists Bulletin* 69:239.

———. 1985b. Paleovertisols—Their utility in reconstructing ancient fluvial floodplain sequences. *Third International Fluvial Sedimentology Conference Abstracts*, pp. 10.

Blodgett, R. H., J. P. Crabaugh, and E. F. McBride. 1993. The color of red beds—a geologic perspective. *Soil Color*, pp. 127–59. Soil Science Society of America Special Publication no. 31. Madison, WI: Soil Science Society of America.

Boardman, D. R., II and P. H. Heckel. 1989. Glacial-eustatic sea-level curve for early Late Pennsylvanian sequence in north-central Texas and biostratigraphic correlation with curve for midcontinent North America. *Geology* 17:802–5.

Boardman, E. L. 1989. Coal measures (Namurian and Westphalian) Blackband Iron Formations: Fossil bog iron ores. *Sedimentology* 36:621–33.

Bodén, P. and J. Backman. 1996. A laminated sediment sequence from the northern North Atlantic Ocean and its climatic record. *Geology* 24: 507–10.

Bodnar, D. A. 1984. Stratigraphy, age, depositional environments, and hydrocarbon source rock evaluation of the Otuk Formation, north-central Brooks Range, Alaska. MS, University of Alaska, Fairbanks.

Boersma, A. and I. Premoli Silva. 1991. Distribution of Paleogene planktonic foraminifera—analogies with the Recent? *Palaeogeography, Palaeoclimatology, Palaeoecology* 83:29–48.

———. 1989. Atlantic Paleogene biserial heterohelicid foraminifera and oxygen minima. *Paleoceanography* 4:271–86.

Boltovskoy, D. and W. R. Reidel. 1987. Polycystine radiolaria of the California Current region: Seasonal and geographic patterns. *Marine Micropaleontology* 12:65–104.

Boucot, A. J. 1975. *Evolution and Extinction Rate Controls*. Amsterdam: Elsevier.

———. 1990. Silurian biogeography. In W. S. McKerrow and C. R. Scotese, eds., *Palaeozoic Palaeogeography and Biogeography*, pp. 191–6. Geological Society of London Memoir 12.

Boucot, A. J. and J. Gray. 1982. Paleozoic data of climatological significance and their use for interpreting Silurian-Devonian climate. In W. H. Berger and J. C. Crowell, eds., *Climate in Earth History*, pp. 189–98. Washington, D.C.: National Academy Press.

Bourrouilh-Le Jan, F. G. 1980. *Phosphates, sols bauxitiques et karsts dolomitiques du Centre et SW Pacifique Comparaisons sédimentologiques et geochimiques*. Geologie Comparée des Gisements de Phosphates et de Pétrole. Orleans.

Bourrouilh-Le Jan, F. G. and L. C. Hottinger. 1988. Occurrence of rhodolites in the tropical Pacific—a consequence of mid-Miocene paleo-oceanographic change. *Sedimentary Geology* 60:355–67.

Boville, B. A. and P. R. Gent. 1998. The NCAR Climate System Model, version one. *Journal of Climate* 11:1115–1130.

Bowen, R. 1961. Paleotemperature analyses of Mesozoic Belemnoidea from Australia and New Guinea. *Geological Society of America Bulletin* 72:769–74.

Bowler, J. M. 1973. Clay dunes: Their occurrence, formation and environmental significance. *Earth-Science Reviews* 9:315–38.

Bown, T. M. and M. J. Kraus. 1983. Ichnofossils of the alluvial Willwood Formation (lower Eocene), Bighorn Basin, northwest Wyoming, U.S.A. *Palaeogeography, Palaeoclimatology, Palaeoecology* 43:95–128.

Boyle, E. 1993. Measures of productivity. *Nature* 362:21–2.

Boyle, E. A. 1986. Paired carbon isotope and cadmium data from benthic foraminifera: Implications for changes in oceanic phosphorus, oceanic circulation, and atmospheric carbon dioxide. *Geochimica et Cosmochimica Acta* 50:265–76.

———. 1988. Cadmium: Chemical tracer of deep-water paleoceanography. *Paleoceanography* 3: 471–90.

Brachert, T. C., C. Betzler, P. J. Davies, and D. A. Feary. 1993. Climatic change: Control of carbonate platform development (Eocene-Miocene, Leg 133, northeastern Australia). In J. A. McKenzie, P. J. Davies, A. Palmer-Julson, et al. *Proceedings of the Ocean Drilling Program, Scientific Results, Leg 83*, pp. 291–300. College Station, TX: Ocean Drilling Program.

Bradbury, J. P. 1988. Fossil diatoms and Neogene paleolimnology. *Palaeogeography, Palaeoclimatology, Palaeoecology* 62:299–316.

Bradley, R. S. 1985. *Quaternary Paleoclimatology*. Boston: Allen and Unwin.

Braitsch, O. 1971. Rhythmic bedding and influxes. *Salt Deposits: Their Origin and Composition*, pp. 251–8. Minerals, Rocks and Inorganic Materials 4. Berlin: Springer-Verlag.

Bralower, T. J. 1988. Calcareous nannofossil biostratigraphy and assemblages of the Cenomanian-Turonian boundary interval: Implications for the origin and timing of oceanic anoxia. *Paleoceanography* 3:275–316.

Bralower, T. J. and H. R. Thierstein. 1984. Low productivity and slow deep-water circulation in mid-Cretaceous oceans. *Geology* 12:614–8.

———. 1987. Organic carbon and metal accumulation rates in Holocene and mid-Cretaceous sediments: Palaeoceanographic significance. In J. Brooks and A. J. Fleet, eds., *Marine Petroleum Source Rocks*, pp. 345–69. Geological Society of London Special Publication 26.

Bralower, T. J., J. C. Zachos, E. Thomas, M. Parrow, C. K. Paull, D. C. Kelly, I. P. Silva, W. V. Sliter, and K. C. Lohmann. 1995. Late Paleocene to Eocene paleoceanography of the equatorial Pacific Ocean: Stable isotopes recorded at Ocean Drilling Program Site 865, Allison Guyot. *Paleoceanography* 10:841–65.

Brand, U. 1994. Continental hydrology and climatology of the Carboniferous Joggins Formation (lower Cumberland Group) at Joggins, Nova Scotia: Evidence from the geochemistry of bivalves. *Palaeogeography, Palaeoclimatology, Palaeoecology* 106:307–21.

Brandley, R. T. and F. F. Krause. 1994. Thinolite-type pseudomorphs after ikaite: Indicators of cold water on the subsequatorial western margin of Lower Carboniferous North America. In B. Beauchamp, A. F. Embry and D. Glass, eds., *Pangea: Global Environments and Resources*, pp. 333–44. Canadian Society of Petroleum Geologists Memoir 17.

Brandt, K. 1986. Glacieustatic cycles in the Early Jurassic? *Neues Jahrbuch für Geologie und Paläontologie Mh.* 1986(5):257–74.

Brass, G. W., J. R. Southam, and W. H. Peterson.

1982. Warm saline bottom water in the ancient ocean. *Nature* 296:620–3.

Brassell, S. C., G. Elginton, I. T. Marlowe, U. Pflaumann and M. Sarnthein. 1986. Molecular stratigraphy: A new tool for climatic assessment. *Nature* 320:129–33.

Breed, C. S. and W. J. Breed. 1972. Invertebrates of the Chinle Formation. In W. J. Breed and C. S. Breed, eds., *Investigations in the Chinle Formation*, pp. 19–22. Bulletin of the Museum of Northern Arizona 47.

Bremner, J. M. 1983. Biogenic sediments on the South West African (Namibian) continental margin. In J. Thiede and E. Suess, eds., *Coastal Upwelling: Its Sediment Record, Part B*, pp. 73–104. Mt. Kisco, NY: Plenum.

Bremner, J. M. and J. Rogers. 1981. Major lithofacies of the Namibian continental margin. *Program Abstracts of the Conference on Coastal Upwelling: Its Sediment Record.*

———. 1990. Phosphorite deposits on the Namibian continental shelf. In W. C. Burnett and S. R. Riggs, eds., *Phosphate Deposits of the World. Vol. 3, Neogene to Modern Phosphorites*, pp. 143–52. Cambridge: Cambridge University Press.

Brenchley, P. J., J. D. Marshall, G. A. F. Carden, D. B. R. Robertson, D. G. F. Long, T. Meidla, L. Hints, and T. F. Anderson. 1994. Bathymetric and isotopic evidence for a short-lived Late Ordovician glaciation in a greenhouse period. *Geology* 22:295–8.

Brewster, N. A. 1980. Cenozoic biogenic silica sedimentation in the Antarctic Ocean. *Geological Society of America Bulletin* 91:337–47.

Breza, J. R. and S. W. Wise, Jr. 1992. Lower Oligocene ice-rafted debris on the Kerguelen Plateau: Evidence for East Antarctic continental glaciation. In S. W. Wise, Jr., R. Schlich, et al. *Proceedings of the Ocean Drilling Program, Scientific Results, Part 1, Leg 120*, pp. 161–78. College Station, TX: Ocean Drilling Program.

Briden, J. C. 1968. Paleoclimatic evidence of a geocentric axial dipole field. In R. A. Phinney, ed. *The History of the Earth's Crust*, pp. 178–94. Princeton: Princeton University Press.

Briden, J. C. and E. Irving. 1964. Palaeolatitude spectra of sedimentary palaeoclimatic indicators. In A. E. M. Nairn, ed. *Problems in Palaeoclimatology*, pp. 199–224. London: Wiley.

Briggs, D. E. G., E. N. K. Clarkson, and R. J. Aldridge. 1983. The conodont animal. *Lethaia* 16:1–14.

Brinkhuis, H. and U. Biffi. 1993. Dinoflagellate cyst stratigraphy of the Eocene/Oligocene transition in central Italy. *Marine Micropaleontology* 22: 131–83.

Brodzikowski, K. and A. J. van Loon. 1991. *Glacigenic Sediments*. Developments in Sedimentology 49. Amsterdam: Elsevier.

Broecker, W. S. 1993. An oceanographic explanation for the apparent carbon isotope-cadmium discordancy in the glacial Antarctic? *Paleoceanography* 8:137–9.

Broecker, W. S. and G. H. Denton. 1990. What drives glacial cycles? *Scientific American* 262(1): 49–56.

Broecker, W. S. and T.-H. Peng. 1982. *Tracers in the Sea*. Palisades, New York: Columbia University.

Bromley, R. and U. Asgaard. 1979. Triassic freshwater ichnocoenoses from Carlsberg Fjord, East Greenland. *Palaeogeography, Palaeoclimatology, Palaeoecology* 28:39–80.

Bronger, A. and N. Bruhn. 1989. Relict and recent features in tropical alfisols from south India. In A. Bronger and J. A. Catt, eds., *Paleopedology—Nature and Application of Paleosols*, pp. 107–28. Catena Supplement 16. Cremlingen-Destedt, Germany: Catena Verlag.

Brookfield, M. E. 1977. The origin of bounding surfaces in ancient aeolian sandstones. *Sedimentology* 24:303–32.

Brooks, C. E. P. 1949. *Climate Through the Ages.* London: Ernest Benn.

Brouwers, E. M., N. O. Jørgensen, and T. M. Cronin. 1991. Climatic signficance of the ostracode fauna from the Pliocene Kap København Formation, north Greenland. *Micropaleontology* 37: 245–67.

Brumsack, H.-J. 1980. Geochemistry of Cretaceous black shales from the Atlantic Ocean (DSDP Legs 11, 14, 36, and 41). *Chemical Geology* 31: 1–25.

———. 1981. Trace element geochemistry of Cretaceous black shales and paleoenvironmental consequences. *Program of Abstracts, Conference on Coastal Upwelling: Its Sediment Record, Advance Research Institute.*

———. 1986. The inorganic geochemistry of Cretaceous black shales (DSDP Leg 41) in comparison to modern upwelling sediments from the Gulf of California. In C. P. Summerhayes and N. J. Shackleton, eds., *North Atlantic Palaeoceanography*, pp. 447–62. Geological Society of London Special Publication 21.

Brumsack, H. J. and J. Thurow. 1986. The geochemical facies of black shales from the Cenomanian/Turonian boundary event (CTBE). *Mitteilungen aus dem Geologisch-Paläontologischen Institut der Universität Hamburg* 60:247–65.

Bryan, R. B., I. A. Campbell, and R. A. Sutherland. 1988. Fluvial geomorphic processes in semi-arid ephemeral catchments in Kenya and Canada. In A. M. Harvey and M. Sala, eds., *Geomorphic Processes in Environments with Strong Seasonal Contrasts. Vol. II: Geomorphic Systems*, pp. 13–35. Catena Supplement 13. Cremlingen-Destedt, Germany: Catena Verlag.

Bryant, J. D., P. N. Froelich, W. J. Showers, and B. J. Genna. 1996a. A tale of two quarries: Biologic and taphonomic signatures in the oxygen isotope composition of tooth enamel phosphate from modern and Miocene equids. *Palaios* 11:397–408.

———. 1996b. Biologic and climatic signals in the oxygen isotopic composition of Eocene-Oligocene equid enamel phosphate. *Palaeogeography, Palaeoclimatology, Palaeoecology* 126:75–89.

Bryant, J. D., B. Luz, and P. N. Froelich. 1994. Oxygen isotopic composition of fossil horse tooth phosphate as a record of continental paleoclimate. *Palaeogeography, Palaeoclimatology, Palaeoecology* 107:303–16.

Budyko, M. I. and A. B. Ronov. 1979. Chemical evolution of the atmosphere in the Phanerozoic. *Geochemistry International* 16:1–9.

Bukry, D. 1981. Silicoflagellate stratigraphy of offshore California and Baja California, Deep Sea Drilling Project Leg 63. In R. S. Yeats, B. U. Haq, et al. *Initial Reports of the Deep Sea Drilling Project, Leg 63*, pp. 539–57. Washington, D.C.: U.S. Government Printing Office.

Bull, W. B. 1991. *Geomorphic Responses to Climatic Change*. New York: Oxford University Press.

Burckle, L. H. 1978. Marine diatoms. In B. U. Haq and A. Boersma, eds., *Introduction to Marine Micropaleontology*, pp. 245–66. New York: Elsevier.

Burckle, L. H., R. Gersonde, and N. Abrams. 1990. Late Pliocene-Pleistocene paleoclimate in the Jane Basin region: ODP Site 697. In P. F. Barker, J. P. Kennett, et al. *Proceedings of the Ocean Drilling Project, Scientific Results, Leg 113*, pp. 803–9. College Station, TX: Ocean Drilling Program.

Burnett, W. C. 1977. Geochemistry and origin of phosphorite deposits from off Peru and Chile. *Geological Society of America Bulletin* 88:813–23.

———. 1990. Phosphorite growth and sediment dynamics in the modern Peru shelf upwelling system. In W. C. Burnett and S. R. Riggs, eds., *Phophate Deposits of the World. Vol. 3, Neogene to Modern Phosphorites*, pp. 62–72. Cambridge: Cambridge University Press.

Burnett, W. C., K. K. Roe, and D. Z. Piper. 1983. Upwelling and phosphorite formation in the ocean. In E. Suess and J. Thiede, eds., *Coastal Upwelling, Part A*, pp. 377–97. Mt. Kisco, NY: Plenum.

Burnett, W. C. and H. H. Veeh. 1977. Uranium-series disequilibrium studies in phosphorite nodules from the west coast of South America. *Geochimica et Cosmochimica Acta* 41:755–64.

Burnham, R. J. 1989. Relationships between standing vegetation and leaf litter in a paratropical forest. *Review of Palaeobotany and Palynology* 58:5–32.

Bush, A. B. G. and S. G. H. Philander. 1997. The late Cretaceous: Simulation with a coupled atmosphere-ocean general circulation model. *Paleoceanography* 12:495–516.

Cadle, A. B., B. Cairncross, A. D. M. Christie, and D. L. Roberts. 1993. The Karoo Basin of South Africa: Type basin for the coal-bearing deposits of southern Africa. *International Journal of Coal Geology* 23:117–57.

Cairns, S. D. and G. D. Stanley, Jr. 1987. Ahermatypic coral banks: Living and fossil counterparts. *Proceedings of the Fourth International Coral Reef Symposium* 1:611–8.

Calder, J. H. 1993. The evolution of a ground-water-influenced (Westphalian B) peat-forming ecosystem in a piedmont setting: The No. 3 seam, Springhill coalfield, Cumberland Basin, Nova Scotia. In J. C. Cobb and C. B. Cecil, eds., *Modern and Ancient Coal-Forming Environments*, pp. 153–80. Geological Society of America Special Paper 286.

———. 1994. The impact of climate change, tectonism and hydrology on the formation of Carboniferous tropical intermontane mires: The Springhill coalfield, Cumberland Basin, Nova Scotia. *Palaeogeography, Palaeoclimatology, Palaeoecology* 106:323–51.

Calvert, S. E. 1974. Deposition and diagenesis of silica in marine sediments. In K. J. Hsü and H. C. Jenkyns, eds., *Pelagic Sediments: On Land and Under the Sea*, pp. 273–99. International Association of Sedimentologists Special Publication 1.

———. 1987. Oceanographic controls on the accumulation of organic matter in marine sediments. In J. Brooks and A. J. Fleet, eds., *Marine Petroleum Source Rocks*, pp. 137–52. Geological Society of London Special Publication 26.

———. 1990. Geochemistry and origin of the Holocene sapropel in the Black Sea. In V. Ittekkot, S. Kempe, W. Michaelis, and A. Spitzy, eds., *Facets of Modern Biogeochemistry*, pp. 327–53. Berlin: Springer-Verlag.

Calvert, S. E., R. M. Bustin, and T. F. Pedersen. 1992. Lack of evidence for enhanced preservation of sedimentary organic matter in the oxygen minimum of the Gulf of California. *Geology* 20: 757–60.

Calvert, S. E. and R. E. Karlin. 1991. Relationships between sulphur, organic carbon, and iron in the modern sediments of the Black Sea. *Geochimica et Cosmochimica Acta* 55:2483–90.

Calvert, S. E., R. E. Karlin, L. J. Toolin, D. J. Donahue, J. R. Southon, and J. S. Vogel. 1991. Low organic carbon accumulation rates in Black Sea sediments. *Nature* 350:692–5.

Calvert, S. E., B. Nielsen, and M. R. Fontugne. 1992. Evidence from nitrogen isotope ratios for enchanced productivity during formation of eastern Mediterranean sapropels. *Nature* 359:223–5.

Calvert, S. E. and T. F. Pedersen. 1992. Organic carbon accumulation and preservation in marine sediments: How important is anoxia? In J. K. Whelan and J. W. Farrington, eds., *Productivity, Accumulation and Preservation of Organic Matter in Recent and Ancient Sediments*, pp. 231–63. New York: Columbia University Press.

———. 1993. Geochemistry of Recent oxic and anoxic marine sediments: Implications for the geological record. *Marine Geology* 113:67–88.

———. 1996. Sedimentary geochemistry of manganese: Implications for the environment of formation of manganiferous black shales. *Economic Geology* 91:36–47.

Calvert, S. E., T. F. Pedersen, P. D. Naidu, and U. von Stackelberg. 1995. On the organic carbon maximum on the continental slope of the eastern Arabian Sea. *Journal of Marine Research* 53:269–96.

Calvert, S. E. and N. B. Price. 1983. Geochemistry of Namibian shelf sediments. In E. Suess and J. Thiede, eds., *Coastal Upwelling, Part A*, pp. 337–75. Mt. Kisco, NY: Plenum.

Cameron, C. C., J. S. Esterle, and C. A. Palmer. 1989. The geology, botany and chemistry of selected peat-forming environments from temperate and tropical latitudes. *International Journal of Coal Geology* 12:105–56.

Campbell, C. A. and J. W. Valentine. 1977. Comparability of modern and ancient marine faunal provinces. *Paleobiology* 3:49–57.

Cannon, W. F. and E. R. Force. 1983. Potential for high-grade shallow-marine manganese deposits in North America. In W. C. Shanks, III, ed. *Cameron Volume on Unconventional Mineral Deposits*, pp. 175–89. New York: American Institute of Mining Engineers, Society of Mining Engineers.

Carannante, G., M. Esteban, J. D. Milliman, and L. Simone. 1988. Carbonate lithofacies as paleolatitude indicators: Problems and limitations. *Sedimentary Geology* 60:333–46.

Carbonel, P., J.-P. Colin, D. L. Danielopol, H. Löffler, and I. Neustrueva. 1988. Paleoecology of limnic ostracodes: A review of some major topics. *Palaeogeography, Palaeoclimatology, Palaeoecology* 62:413–61.

Carlquist, S. 1977. Ecological factors in wood evolution: A floristic approach. *American Journal of Botany* 64:887–96.

———. 1988. *Comparative Wood Anatomy*. Springer Series in Wood Science. Berlin: Springer-Verlag.

Caron, M. and P. Homewood. 1983. Evolution of early planktic foraminifers. *Marine Micropaleontology* 7:453–62.

Carpenter, S. J. and K. C. Lohmann. 1995. $\delta^{18}O$ and $\delta^{18}C$ of modern brachiopod shells. *Geochimica et Cosmochimica Acta* 59:3749–64.

Carpentier, B., A.-Y. Huc, and G. Bessereau. 1991. Wireline logging and source rocks—Estimation of organic carbon content by the CARBOLOG method. *The Log Analyst* May-June:279–97.

Carter, L. D. 1983. Fossil sand wedges on the Alaskan Arctic Coastal Plain and their paleoenvironmental significance. *Permafrost: Fourth International Conference, Proceedings*, pp. 109–14.

Casey, R. E. 1971a. Radiolarians as indicators of past and present water masses. In B. M. Funnell and W. R. Riedel, eds., *The Micropalaeontology of Oceans*, pp. 331–42. Cambridge: Cambridge University Press.

———. 1971b. Distribution of polycystine radiolaria in the oceans in relation to physical and chemical conditions. In B. M. Funnell and W. R. Riedel, eds., *The Micropalaeontology of Oceans*, pp. 151–9. Cambridge: Cambridge University Press.

———. 1977. The ecology and distribution of recent radiolaria. In A. T. S. Ramsay, ed. *Oceanic Micropalaeontology Vol. 2*, pp. 809–45. London: Academic Press.

———. 1993. Radiolaria. In J. H. Lipps, ed. *Fossil Prokaryotes and Protists*, pp. 249–84. Oxford: Blackwell Scientific Publications.

Cassiliano, M. L. 1997. Crocodiles, tortoises, and climate: A shibboleth re-examined. *Palaeoclimates* 2:47–69.

Cecil, C. B. 1990. Paleoclimate controls on stratigraphic repetition of chemical and siliciclastic rocks. *Geology* 18:533–6.

Cecil, C. B., F. T. Dulong, J. C. Cobb, and Supardi. 1993. Allogenic and autogenic controls on sedimentation in the central Sumatran basin as an analogue for Pennsylvanian coal-bearing strata in the Appalachian basin. In J. C. Cobb and C. B. Cecil, eds., *Modern and Ancient Coal-Forming Environments*, pp. 3–22. Geological Society of America Special Paper 286.

Cecil, C. B. and K. J. Englund. 1989. Origin of coal deposits and associated rocks in the Carboniferous of the Appalachian Basin. *Coal and Hydrocarbon Resources of North America Vol. T143*, pp. 84–104. 28th International Geological Congress 2. Washington, D.C.: American Geophysical Union.

Cecil, C. B., R. W. Stanton, F. T. Dulong, and J. J. Renton. 1982. Geologic factors that control mineral matter in coal. *Atomic and Nuclear Methods in Fossil Energy Research*:323–35.

Cecil, C. B., R. W. Stanton, F. T. Dulong, L. F. Ruppert, and J. J. Renton. 1981. A geochemical model for the origin of low-ash and low-sulfur coal. In *Mississippian-Pennsylvanian Boundary in the Central Part of the Appalachian Basin* Geological Society of America Field Trip No. 4:175–7.

Cecil, C. B., R. W. Stanton, S. G. Neuzil, F. T. Dulong, L. F. Ruppert, and B. S. Pierce. 1985. Paleoclimate controls on Late Paleozoic sedimentation and peat formation in the central Appalachian basin (U.S.A.). *International Journal of Coal Geology* 5:195–230.

Cerling, T. E. 1984. The stable isotopic composition of modern soil carbonate and its relationship to climate. *Earth and Planetary Science Letters* 71:229–40.

———. 1991. Carbon dioxide in the atmosphere: Evidence from Cenozoic and Mesozoic paleosols. *American Journal of Science* 291:377–400.

———. 1992. Use of carbon isotopes in paleosols as an indicator of the P(CO_2) of the paleoatmosphere. *Global Biogeochemical Cycles* 6:307–14.

Cerling, T. E., J. R. Bowman, and J. R. O'Neil. 1988. An isotopic study of a fluvial-lacustrine sequence: The Plio-Pleistocene Koobi Fora sequence, East Africa. *Palaeogeography, Palaeoclimatology, Palaeoecology* 63:335–56.

Cerling, T. E. and R. L. Hay. 1986. An isotopic study of paleosol carbonates from Olduvai Gorge. *Quaternary Research* 25:63–78.

Cerling, T. E. and J. Quade. 1993. Stable carbon and oxygen isotopes in soil carbonates. In P. K. Swart, K. C. Lohmann, J. McKenzie, and S. Savin, eds., *Climate Change in Continental Isotopic Records*, pp. 217–31. Geophysical Monograph 78. Washington, D.C.: American Geophysical Union.

Cerling, T. E., J. Quade, Y. Wang, and J. R. Bowman. 1989. Carbon isotopes in soils and palaeosols as ecology and palaeoecology indicators. *Nature* 341:138–9.

Cerling, T. E., Y. Wang, and J. Quade. 1993. Expansion of C4 ecosystems as an indicator of global ecological change in the late Miocene. *Nature* 361:344–5.

Cerling, T. E., V. P. Wright, and S. D. Vanstone. 1992. Further comments on using carbon isotopes in palaeosols to estimate the CO_2 content of the palaeo-atmosphere. *Journal of the Geological Society of London* 149:673–6.

Cess, R. D., G. L. Potter, J. P. Blanchet, G. J. Boer, S. J. Ghan, J. T. Kiehl, H. Le Treut, Z.-X. Li, X.-Z. Liang, J. F. B. Mitchell, J.-J. Morcrette, D. A. Randall, M. R. Riches, E. Roeckner, U. Schlese, A. Slingo, K. E. Taylor, W. M. Washington, R. T. Wetherald, and I. Yagai. 1989. Interpretation of cloud-climate feedback as produced by 14 atmospheric general circulation models. *Science* 245: 513–6.

Cess, R. D., M. H. Zhang, W. J. Ingram, G. L. Potter, V. Alekseev, H. W. Barker, E. Cohen-Solal, R. A. Colman, D. A. Dazlich, A. D. Del Genio, M. R. Dix, V. Dymnikov, M. Esch, L. D. Fowler, J. R. Fraser, V. Galin, W. L. Gates, J. J. Hack, J. T. Kiehl, H. Le Treut, K. K.-W. Lo, B. J. McAvaney, V. P. Meleshko, J.-J. Morcrette, D. A. Randall, M. R. Riches, E. Roeckner, J.-F. Royer, M. E. Schlesinger, P. V. Sporyshev, B. Timbal, E. M. Volodin, K. E. Taylor, W. Wang, and R. T. Wetherald. 1996. Cloud feedback in atmospheric general circulation models: An update. *Journal of Geophysical Research* 101:12791–4.

Chandler, M. A., D. Rind, and R. Ruedy. 1992. Pangaean climate during the Early Jurassic: GCM simulations and the sedimentary record of paleoclimate. *Geological Society of America Bulletin* 104:543–59.

Chapman, J. L. 1994. Distinguishing internal developmental characteristics from external palaeoenvironmental effects in fossil wood. *Review of Palaeobotany and Palynology* 81:19–32.

Chappell, J. and N. J. Shackleton. 1986. Oxygen isotopes and sea level. *Nature* 324:137–40.

Charpentier, R. R. 1984. Conodonts through time and space: Studies in conodont provincialism. In D. L. Clark, ed. *Conodont Biofacies and Provincialism*, pp. 11–32. Geological Society of America Special Paper 196.

Chase, C. G. 1991. Fluvial landsculpting and the fractal dimension of topography. *Geomorphology*:1–19.

Chase, C. G., K. M. Gregory, J. T. Parrish, and P. G. DeCelles. (In press.) Topographic history of the western Cordillera of North America and controls on climate. In T. J. Crowley and K. C. Burke, eds., *Tectonic Boundary Conditions for Climate Reconstructions*. Oxford: Oxford University Press.

Chavez, F. P. and J. R. Toggweiler. 1995. Physical estimates of global new production: The upwelling contribution. In C. P. Summerhayes, K.-C. Emeis, M. V. Angel, R. L. Smith, and B. Zeitzschel, eds., *Upwelling in the Ocean: Modern Processes and Ancient Records*, pp. 313–20. Chichester: John Wiley and Sons.

Chesnut, D. R., Jr. 1993. Gigantic Ordovician volcanic ash fall in North America and Europe: Biological, tectonomagmatic, and event-stratigraphic significance: Comment. *Geology* 21:381–2.

Chivas, A. R., P. De Deckker, J. A. Cali, A. Chapman, E. Kiss, and J. M. G. Shelley. 1993. Coupled stable-isotope and trace-element measurements of lacustrine carbonates as paleoclimatic indicators. In P. K. Swart, K. C. Lohmann, J. McKenzie, and S. Savin, eds., *Climate Change in Continental Isotopic Records*, pp. 113–21. Geophysical Monograph 78. Washington, D.C.: American Geophysical Union.

Choquette, P. W. and N. P. James. 1988. Introduction. In N. P. James and P. W. Choquette, eds., *Paleokarst*, pp. 1–21. New York: Springer-Verlag.

Chrintz, T. and L. B. Clemmensen. 1993. Draa reconstruction, the Permian Yellow Sands, northeast England. In K. Pye and N. Lancaster, eds., *Aeolian Sediments, Ancient and Modern*, pp. 151–61. Special Publication of the International Association of Sedimentologists 16. Oxford: Blackwell Scientific.

Chumakov, N. M. and L. A. Frakes. 1997. Mode of origin of dispersed clasts in Jurassic shales, southern part of the Yana-Kolyma fold belt, North East Asia. *Palaeogeography, Palaeoclimatology, Palaeoecology* 128:77–85.

Ciesielski, P. F. and K. R. Bjørklund. 1995. Ecology, morphology, stratigraphy, and the paleoceanographic significance of *Cycladophora davisiana davisiana*. Part II, Stratigraphy in the North Atlantic (DSDP Site 609) and Labrador Sea (ODP Site 646B). *Marine Micropaleontology* 25:67–86.

Ciesielski, P. F. and S. M. Case. 1989. Neogene paleoceanography of the Norwegian Sea based upon silicoflagellate assemblage changes in ODP Leg 104 sedimentary sequences. In O. Eldholm, J. Thiede, E. Taylor, et al. *Proceedings of the Ocean Drilling Program, Scientific Results, Leg 104*, pp. 527–41. College Station, TX: Ocean Drilling Program.

Cifelli, R. L. 1980. A reassessment of labyrinthodont paleolatitudinal distributions. *Palaeogeography, Palaeoclimatology, Palaeoecology* 30:121–31.

Clauer, N., E. Keppens, and P. Stille. 1992. Sr isotopic constraints on the process of glauconitization. *Geology* 20:133–6.

Clauzon, G., J.-P. Suc, F. Gautier, A. Berger, and M.-F. Loutre. 1996. Alternate interpretation of the Messinian salinity crisis: Controversy resolved? *Geology* 24:363–4.

Clemens, S. C., J. W. Farrell, and L. P. Gromet. 1993. Synchronous changes in seawater strontium isotope composition and global climate. *Nature* 363:607–10.

Clemens, S. C. and W. L. Prell. 1990. Late Pleistocene variability of Arabian Sea summer monsoon winds and continental aridity: Eolian records from the lithogenic component of deep-sea sediments. *Paleoceanography* 5:109–46.

Clemmensen, L. B. 1979. Triassic lacustrine red-beds and palaeoclimate: The "Bundsandstein" of Helgoland and the Malmros Klint Member of East Greenland. *Geologische Rundschau* 68:748–74.

———. 1987. Complex star dunes and associated aeolian bedforms, Hopeman Sandstone (Permo-Triassic), Moray Firth Basin, Scotland. In L. Frostick and I. Reid, eds., *Desert Sediments: Ancient and Modern*, pp. 213–31. Geological Society of London Special Publication 35.

Clemmensen, L. B., I. E. I. Øxnevad, and P. L. de Boer. 1994. Climatic controls on ancient desert sedimentation: Some late Palaeozoic and Mesozoic examples from NW Europe and the Western Interior of the USA. In P. L. de Boer and D. G. Smith, eds., *Orbital Forcing and Cyclic Sequences*, pp. 439–57. International Association of Sedimentologists Special Publication 19. Oxford: Blackwell Scientific.

Clemmensen, L. B. and H. Tirsgaard. 1990. Sand-drift surfaces: A neglected type of bounding surface. *Geology* 18:1142–5.

Cocks, L. R. M. 1972. The origin of the Silurian *Clarkeia* shelly fauna of South America, and its extension to West Africa. *Palaeontology* 15:623–30.

Cocks, L. R. M. and R. A. Fortey. 1990. Biogeography of Ordovician and Silurian faunas. In W. S. McKerrow and C. R. Scotese, eds., *Palaeozoic Palaeogeography and Biogeography*, pp. 97–104. Geological Society of London Memoir 12.

Cocks, L. R. M. and C. R. Scotese. 1991. The global

biogeography of the Silurian Period. In M. G. Bassett, P. D. Lane, and D. Edwards, eds., *The Murchison Symposium: Proceedings of an International Conference on the Silurian System*, pp. 109–22. Special Papers in Palaeontology 44, Palaeontological Association.

Cohen, A. 1981. Paleolimnological research at Lake Turkana, Kenya. *Paleoecology of Africa* 13: 61–82.

Cohen, A. S. 1982. Paleoenvironments of root casts from the Koobi Fora Formation, Kenya. *Journal of Sedimentary Petrology* 52:401–14.

———. 1990. Tectono-stratigraphic model for sedimentation in Lake Tanganyika, Africa. In B. J. Katz, ed. *Lacustrine Basin Exploration: Case Studies and Modern Analogs*, pp. 137–50. American Association of Petroleum Geologists Memoir 50. Tulsa, OK: American Association of Petroleum Geologists.

Cohen, A. S., R. Dussinger, and J. Richardson. 1983. Lacustrine paleochemical interpretations based on eastern and southern African ostracodes. *Palaeogeography, Palaeoclimatology, Palaeoecology* 43: 129–51.

Cohen, A. S., K.-E. Lezzar, J.-J. Tiercelin, and M. Soreghan. 1997. New palaeogeographic and lake level reconstructions of Lake Tanganyika: Implications for tectonic, climatic and biologic evolution in a rift lake. *Basin Research* 9:107–32.

Colbath, G. K. 1990. Palaeobiogeography of Middle Palaeozoic organic-walled phyotplankton. In W. S. McKerrow and C. R. Scotese, eds., *Palaeozoic Palaeogeography and Biogeography*, pp. 207–13. Geological Society of London Memoir 12.

Coles, K. S. and R. J. Varga. 1988. Early to middle Paleozoic phosphogenic province in terranes of the southern Cordillera, western United States. *American Journal of Science* 288:891–924.

Collier, R. E. L., M. R. Leeder, and J. R. Maynard. 1990. Transgressions and regressions: A model for the influence of tectonic subsidence, deposition and eustasy, with application to Quaternary and Carboniferous examples. *Geological Magazine* 127:117–28.

Collinson, M. E. 1986. Use of modern generic names for plant fossils. In R. A. Spicer and B. A. Thomas, eds., *Systematic and Taxonomic Approaches in Palaeobotany*, pp. 91–104. Systematics Association Special Volume 31.

Compton, J. S., D. A. Hodell, J. R. Garrido, and D. J. Mallinson. 1993. Origin and age of phosphorite from the south-central Florida Platform: Relation of phosphogenesis to sea-level fluctuations and $\delta^{13}C$ excursions. *Geochimica et Cosmochimica Acta* 57:131–46.

Compton, J. S., S. W. Snyder, and D. A. Hodell. 1990. Phosphogenesis and weathering of shelf sediments from the southeastern United States: Implications for Miocene $\delta^{13}C$ excursions and global cooling. *Geology* 18:1227–30.

Coney, P. J., D. L. Jones, and J. W. H. Monger. 1980. Cordilleran suspect terranes. *Nature* 288:329–33.

Cook, P. J. 1976. Sedimentary phosphate deposits. In K. H. Wolfe, ed. *Handbook of Strata-Bound and Stratiform Ore Deposits*, pp. 505–35 7. Amsterdam: Elsevier Scientific.

Cook, P. J. and M. W. McElhinny. 1979. A reevaluation of the spatial and temporal distribution of sedimentary phosphate deposits in the light of plate tectonics. *Economic Geology* 74:315–30.

Cook, P. J. and J. H. Shergold. 1986. Proterozoic and Cambrian phosphorites—Nature and origin. In P. J. Cook and J. H. Shergold, eds., *Phosphate Deposits of the World. Vol. 1, Proterozoic and Cambrian Phosphorites*, pp. 369–86. Cambridge: Cambridge University Press.

Copper, P. 1977. Paleolatitudes in the Devonian of Brazil and the Frasnian-Famennian mass extinction. *Palaeogeography, Palaeoclimatology, Palaeoecology* 21:165–207.

Corfield, R. M., M. A. Hall, and M. D. Brasier. 1990. Stable isotope evidence for foraminiferal habitats during the development of the Cenomanian/Turonian oceanic anoxic event. *Geology* 18:175–8.

Corrège, T. 1993. The relationship between water masses and benthic ostracod assemblages in the western Coral Sea, Southwest Pacific. *Palaeogeography, Palaeoclimatology, Palaeoecology* 105: 245–66.

Crabaugh, M. and G. Kocurek. 1993. Entrada Sandstone: An example of a wet aeolian system. In K. Pye, ed. *The Dynamics and Environmental Context of Aeolian Sedimenty Systems*, pp. 103–26. Geological Society Special Publication 72.

Creber, G. T. 1977. Tree rings: A natural data-storage system. *Biological Reviews* 52:349–83.

Creber, G. T. and W. G. Chaloner. 1984. Influence of environmental factors on the wood structure of living and fossil trees. *Botanical Review* 50: 357–448.

Crick, R. E. 1990. Cambro-Devonian biogeography of nautiloid cephalopods. In W. S. McKerrow and C. R. Scotese, eds., *Palaeozoic Palaeogeography and Biogeography*, pp. 147–61. Geological Society of London Memoir 12.

Cronin, T. M. 1991. Pliocene shallow water paleoceanography of the North Atlantic Ocean based on marine ostracodes. *Quaternary Science Reviews* 10:175–88.

Cronin, T. M. and H. J. Dowsett. 1990. A quantitative micropaleontologic method for shallow marine paleoclimatology: Application to Pliocene deposits of the western North Atlantic Ocean. *Marine Micropaleontology* 16:117–47.

Cronin, T. M., A. Kitamura, N. Ikeya, M. Watanabe, and T. Kamiya. 1994. Late Pliocene climate change 3.4–2.3 Ma: Paleoceanographic record from the Yabuta Formation, Sea of Japan. *Palaeogeography, Palaeoclimatology, Palaeoecology* 108:437–55.

Crowell, J. C. 1995. The ending of the late Paleozoic ice age during the Permian Period. In P. A. Scholle, T. Peryt, and D. S. Ulmer-Scholle, eds., *The Permian of Northern Pangea Vol. 1*, pp. 62–74. Berlin: Springer-Verlag.

Crowley, T. J. and S. K. Baum. 1991a. Toward reconciliation of Late Ordovician (~440 Ma) glaciation with very high CO_2 levels. *Journal of Geophysical Research* 96:22,597–22,610.

———. 1991b. Estimating Carboniferous sea-level fluctuations from Gondwanan ice extent. *Geology* 19:975–7.

———. 1992. Modeling late Paleozoic glaciation. *Geology* 20:506–10.

Crowley, T. J., S. K. Baum, and W. T. Hyde. 1991. Climate model comparison of Gondwanan and Laurentide glaciations. *Journal of Geophysical Research* 96:9217–26.

———. 1992. Milankovitch fluctuations on supercontinents. *Geophysical Research Letters* 19: 793–6.

Crowley, T. J., S. K. Baum, and K.-Y. Kim. 1993. General circulation model sensitivity experiments with pole-centered supercontinents. *Journal of Geophysical Research* 98:8793–800.

Crowley, T. J., W. T. Hyde, and D. A. Short. 1989. Seasonal cycle variations on the supercontinent of Pangaea: Implications for Early Permian vertebrate extinctions. *Geology* 17:457–60.

Crowley, T. J., J. G. Mengel, and D. A. Short. 1987. Gondwanaland's seasonal cycle. *Nature* 329: 803–7.

Crowley, T. J. and G. R. North. 1991. *Paleoclimatology*. New York: Oxford University Press.

Crowley, T. J., D. A. Short, J. G. Mengel, and G. R. North. 1986. Role of seasonality in the evolution of climate during the last 100 million years. *Science* 231:579–84.

Crowley, T. J., K.-J. J. Yip, and S. K. Baum. 1994. Snowline instability in a general circulation model: Application to Carboniferous glaciation. *Climate Dynamics* 10:363–76.

Culver, S. J. 1993. Foraminifera. In J. H. Lipps, ed. *Fossil Prokaryotes and Protists*, pp. 203–47. Oxford: Blackwell Scientific.

Cuomo, M. C. and P. R. Bartholomew. 1991. Pelletal black shale fabrics: Their origin and significance. In R. V. Tyson and T. H. Pearson, eds., *Modern and Ancient Continental Shelf Anoxia*, pp. 221–32. Geological Society of London Special Publication 58.

Curtis, C. D. 1990. Aspects of climatic influence on the clay mineralogy and geochemistry of soils, palaeosols and clastic sedimentary rocks. *Journal of the Geological Society of London* 147:351–8.

D'Angela, D. and A. Longinelli. 1990. Oxygen isotopes in living mammal's bone phosphate: Further results. *Chemical Geology* 86:75–82.

Darwin, C. R. 1842. Notes on the effects produced by ancient glaciers of Caernarvonshire, and on the boulders transported by floating ice. *London, Edinburgh, and Dublin Philosophical Magazine and Journal of Science* 21:180–8.

Davies, T. A., W. W. Hay, J. R. Southam, and T. R. Worsley. 1977. Estimates of Cenozoic oceanic sedimentation rates. *Science* 197:53.

Davis, G. A., J. W. H. Monger, and B. C. Burchfield. 1978. Mesozoic construction of the cordilleran "collage" central British Columbia to central California. In D. G. Howell and K. A. McDougall, eds., *Mesozoic Paleogeography of the Western United States, Pacific Coast Paleogeography Symposium 2*, pp. 1–32. Society of Economic Paleontologists and Mineralogists.

De Deckker, P. and R. M. Forester. 1988. The use of ostracods to reconstruct continental palaeoenvironmental records. In P. De Deckker, J. P. Colin, and J. P. Peypouquet, eds., *Ostracoda in the Earth Sciences*, pp. 175–99. Amsterdam: Elsevier.

Dean, W. E. 1978. Theoretical versus observed successions from evaporation of seawater. In W. E. Dean and B. C. Schreiber, eds., *Marine Evaporites*, pp. 74–85. Society of Economica Paleontologists and Mineralogists Short Course 4. Tulsa: Society of Economic Paleontologists and Mineralogists.

Dean, W. E., M. A. Arthur, and D. A. V. Stow. 1984. Origin and geochemistry of Cretaceous deep-sea black shales and multicolored claystones, with emphasis on Deep Sea Drilling Project Site 530, southern Angola Basin. In W. W. Hay, J.-C. Sibuet, et al. *Initial Reports of the Deep Sea Drilling Project, Leg 75*, pp. 819–44. Washington, D.C.: U.S. Government Printing Office.

Dean, W. E., G. R. Davies, and R. Y. Anderson.

1975. Sedimentological significance of nodular and laminated anhydrite. *Geology* 3:367–72.

Dean, W. E., J. V. Gardner, and R. Y. Anderson. 1994. Geochemical evidence for enhanced preservation of organic matter in the oxygen minimum zone of the continental margin of northern California during the late Pleistocene. *Paleoceanography* 9:47–61.

deBoer, P. L. 1982. Cyclicity and the storage of organic matter in middle Cretaceous pelagic sediments. In G. Einsele and A. Seilacher, eds., *Cyclic and Event Stratification*, pp. 456–75. Berlin: Springer-Verlag.

DeCelles, P. G., M. B. Gray, K. D. Ridgway, R. B. Cole, D. A. Pivnik, N. Pequera, and P. Srivastava. 1991. Controls on synorogenic alluvial-fan architecture, Beartooth Conglomerate (Palaeocene), Wyoming and Montana. *Sedimentology* 38:567–90.

Delaney, M. L. 1990. Miocene benthic foraminiferal Cd/Ca records: South Atlantic and western equatorial Pacific. *Paleoceanography* 5:743–60.

Delaney, M. L. and E. A. Boyle. 1988. Tertiary paleoceanic chemical variability: Unintended consequences of simple geochemical models. *Paleoceanography* 3:137–56.

Delaney, M. L. and G. M. Filippelli. 1994. An apparent contradiction in the role of phosphorus in Cenozoic chemical mass balances for the world ocean. *Paleoceanography* 9:513–27.

Delaney, M. L., B. N. Popp, C. G. Lepzelter, and T. F. Anderson. 1989. Lithium-to-calcium ratios in modern, Cenozoic, and Paleozoic articulate brachiopod shells. *Paleoceanography* 4:681–91.

Delorme, L. D. 1969. Ostracodes as Quaternary paleoecological indicators. *Canadian Journal of Earth Sciences* 6:1471–6.

DeLuca, J. L. and K. A. Eriksson. 1989. Controls on synchronous ephemeral- and perennial-river sedimentation in the middle sandstone member of the Triassic Chinle Formation, northeastern New Mexico, U.S.A. *Sedimentary Geology* 61:155–75.

Demaison, G. 1991. Anoxia vs. productivity: What controls the formation of organic-carbon-rich sediments and sedimentary rocks?: Discussion. *American Association of Petroleum Geologists Bulletin* 75:499–500.

Demaison, G. J. and G. T. Moore. 1980. Anoxic environments and oil source bed genesis. *American Association of Petroleum Geologists Bulletin* 64:1179–209.

deMenocal, P. B., W. F. Ruddiman, and E. M. Pokras. 1993. Influences of high- and low-latitude processes on African terrestrial climate: Pleistocene eolian records from equatorial Atlantic, Ocean Drilling Program Site 663. *Paleoceanography* 8:209–42.

Demko, T. M. 1995. Taphonomy of Fossil Plants in the Upper Triassic Chinle Formation. *Ph.D. thesis*, University of Arizona.

Demko, T. M. and J. T. Parrish. (In press.) Paleoclimatic setting of the Morrison Formation. In K. Carpenter, ed. *The Upper Jurassic Morrison Formation: An Interdisciplinary Approach*. Modern Geology.

Denys, C., C. T. Williams, Y. Dauphin, P. Andrews, and Y. Fernandez-Jalvo. 1996. Diagenetical changes in Pleistocene small mammal bones from Olduvia Bed I. *Palaeogeography, Palaeoclimatology, Palaeoecology* 126:121–34.

DePaolo, D. J. and K. L. Finger. 1991. High-resolution strontium-isotope stratigraphy and biostratigraphy of the Miocene Monterey Formation, central California. *Geological Society of America Bulletin* 103:112–24.

DePaolo, D. J. and B. L. Ingram. 1985. High-resolution stratigraphy with strontium isotopes. *Science* 227:938–41.

Dettman, D. L. and K. C. Lohmann. 1993. Seasonal change in Paleogene surface water $\delta^{18}O$: Freshwater bivalves of western North America. In P. K. Swart, K. C. Lohmann, J. McKenzie, and S. Savin, eds., *Climate Change in Continental Isotopic Records*, pp. 153–63. Geophysical Monograph 78. Washington, D.C.: American Geophysical Union.

Dhondt, A. V. 1992. Cretaceous inoceramid biogeography: A review. *Palaeogeography, Palaeoclimatology, Palaeoecology* 92:217–32.

Dickens, G. R., J. R. O'Neil, D. K. Rea, and R. M. Owen. 1995. Dissociation of oceanic methane hydrate as a cause of the carbon isotope excursion at the end of the Paleocene. *Paleoceanography* 10:965–71.

Dickens, G. R. and R. M. Owen. 1994. Late Miocene-early Pliocene manganese redirection in the central Indian Ocean: Expansion of the intermediate water oxygen minimum zone. *Paleoceanography* 9:169–81.

Dickinson, W. R., M. A. Klute, M. J. Hayes, S. U. Janecke, E. R. Lundin, M. A. McKittrick, and M. D. Olivares. 1988. Paleogeographic and paleotectonic setting of Laramide sedimentary basins in the central Rocky Mountains region. *Geological Society of America Bulletin* 100:1023–39.

Dickinson, W. R., G. S. Soreghan, and K. A. Giles. 1994. Glacio-eustatic origin of Permo-Carboniferous stratigraphic cycles: Evidence from the southern Cordilleran foreland region. *Tectonic and Eustatic Controls on Sedimentary Cycles*, pp. 25–34. SEPM Concepts in Sedimentology and Paleontology 4.

Dickinson, W. R. and C. A. Suczek. 1979. Plate tectonics and sandstone compositions. *The American Association of Petroleum Geologists Bulletin* 63:2164–82.

Diessel, C. F. K. 1992. *Coal-Bearing Depositional Systems*. Berlin: Springer-Verlag.

Dijkmans, J. W. A., E. A. Koster, J. P. Galloway, and W. G. Mook. 1986. Characteristics and origin of calcretes in a subarctic environment, Great Kobuk Sand Dunes, northwestern Alaska, U.S.A. *Arctic and Alpine Research* 18:377–87.

Dilley, F. C. 1973. Cretaceous larger foraminifera. In A. Hallam, ed. *Atlas of Palaeobiogeography*, pp. 403–19. Amsterdam: Elsevier Scientific.

DiMichele, W. A. and T. L. Phillips. 1994. Paleobotanical and paleoecological constraints on models of peat formation in the Late Carboniferous of Euramerica. *Palaeogeography, Palaeoclimatology, Palaeoecology* 106:39–90.

Dingle, R. V. and J. Giraudeau. 1993. Benthic Ostracoda in the Benguela System (SE Atlantic): A multivariate analysis. *Marine Micropaleontology* 22:71–92.

Dingus, A. S. 1984. Paleoenvironmental reconstruction of the Shublik Formation on the North Slope of Alaska.

Ditchfield, P. and J. D. Marshall. 1989. Isotopic variation in rhythmically bedded chalks: Paleotemperature variation in the Upper Cretaceous. *Geology* 17:842–5.

Ditchfield, P. W. 1997. High northern palaeolatitude Jurassic-Cretaceous palaeotemperature variation: New data from Kong Karls Land, Svalbard. *Palaeogeography, Palaeoclimatology, Palaeoecology* 130:163–81.

Dodd, J. R. and R. J. Stanton, Jr. 1975. Paleosalinities with a Pliocene Bay, Kettleman Hills, California: A study of the resolving power of isotopic and faunal techniques. *Geological Society of America Bulletin* 86:51–64.

Dodd, J. R. and R. J. Stanton. 1981. *Paleoecology, Concepts and Applications*. New York: John Wiley and Sons.

Domack, E. W. 1988. Biogenic facies in the Antarctic glacimarine environment: Basis for a polar glacimarine summary. *Palaeogeography, Palaeoclimatology, Palaeoecology* 63:357–72.

Donaldson, A. C., J. J. Renton, and M. W. Presley. 1985. Pennsylvanian deposystems and paleoclimates of the Appalachians. *International Journal of Coal Geology* 5:167–94.

Dorf, E. 1970. Paleobotanical evidence of Mesozoic and Cenozoic climatic changes. *Proceedings of the North American Paleontological Convention: Symposium on Paleoclimatology*, pp. 323–46. Lawrence, KS: Allen Press.

Douglas, R. G. 1979. Benthic foraminiferal ecology and paleoecology: A review of concepts and methods. *Foraminiferal Ecology and Paleoecology*, pp. 21–53. Society of Economic Paleontologists and Mineralogists Short Course 6.

———. 1981. Paleoecology of continental margin basins: A modern case history from the borderland of southern California. *Depositional Systems of Active Continental Margin Basins*, pp. 121–56. Short Course Notes.: Pacific Section, Society of Economic Paleontologists and Mineralogists.

Doyle, J. A., P. Biens, A. Doerenkamp, and S. Jardiné. 1977. Angiosperm pollen from the pre-Albian Lower Cretaceous of equatorial Africa. *Bulletin de Centre de Recherche Exploration-Production Elf-Aquitaine* 1:451–73.

Doyle, P. 1992. A review of the biogeography of Cretaceous belemnites. *Palaeogeography, Palaeoclimatology, Palaeoecology* 92:207–16.

Drewry, D. 1986. *Glacial Geologic Processes*. London: Edward Arnold.

Drewry, G. E., A. T. S. Ramsay, and A. G. Smith. 1974. Climatically controlled sediments, the geomagnetic field, and trade wind belts in Phanerozoic time. *Journal of Geology* 82:531–53.

Droser, M. L. and D. J. Bottjer. 1988. Trends in depth and extent of bioturbation in Cambrian carbonate marine environments, western United States. *Geology* 16:233–6.

Drummond, C. N., W. P. Patterson, and J. C. G. Walker. 1995. Climatic forcing of carbon-oxygen isotopic covariance in temperate-region marl lakes. *Geology* 23:1031–4.

Drummond, C. N., B. H. Wilkinson, and K. C. Lohmann. 1996. Climatic control of fluvial-lacustrine cyclicity in the Cretaceous Cordilleran Foreland Bain, western United States. *Sedimentology* 43:677–89.

Drummond, C. N., B. H. Wilkinson, K. C. Lohmann, and G. R. Smith. 1993. Effect of regional topography and hydrology on the lacustrine isotopic record of Miocene paleoclimate in the Rocky Mountains. *Palaeogeography, Palaeoclimatology, Palaeoecology* 101:67–79.

Dubiel, R. F. 1987. Sedimentology of the Upper Triassic Chinle Formation, southeastern Utah. *Ph.D. thesis*, University of Colorado.

———. 1992. Sedimentology and depositional history of the Upper Triassic Chinle Formation in the Uinta, Piceance, and Eagle Basins, northwestern Colorado and northeastern Utah. *U.S. Geological Survey Bulletin* 1787-W, pp. W1-W25.

Dubiel, R. F., J. T. Parrish, J. M. Parrish, and S. C. Good. 1991. The Pangaean megamonsoon—Evidence from the Upper Triassic Chinle Formation, Colorado Plateau. *Palaios* 6:347–70.

Dubiel, R. F., G. Skipp, and S. T. Hasiotis. 1992. Continental depositional environments and tropical paleosols in the Upper Triassic Chinle Formation, Eagle Basin, western Colorado. In R. M. Flores, ed. *Mesozoic of the Western Interior, Field Guidebook*, pp. 21–37. Denver: Rocky Mountain Section, SEPM.

Dubiel, R. F. and J. P. Smoot. 1994. Criteria for interpreting paleoclimate from red beds—a tool for Pangean reconstructions. In B. Beauchamp, A. F. Embry, and D. Glass, eds., *Pangea: Global Environments and Resources*, pp. 295–309. Canadian Society of Petroleum Geologists Memoir 17.

Duce, R. A., P. S. Liss, J. T. Merrill, E. L. Atlas, P. Buat-Menard, B. B. Hicks, J. M. Miller, J. M. Prospero, R. Arimoto, T. M. Church, W. Ellis, J. N. Galloway, L. Hansen, T. D. Jickells, A. H. Knap, K. H. Reinhardt, B. Schneider, A. Soudine, J. J. Tokos, S. Tsunogai, R. Wollast, and M. Zhou. 1991. The atmospheric input of trace species to the World Ocean. *Global Biogeochemical Cycles* 5:193–259.

Duke, W. L. 1985. Hummocky cross-stratification, tropical hurricanes, and intense winter storms. *Sedimentology* 32:167–94.

———. 1987. Hummocky cross-stratification, tropical hurricanes, and intense winter storms—Reply. *Sedimentology* 34:344–59.

Duplessy, J. C., A. W. H. Bé, and P. L. Blanc. 1981. Oxygen and carbon isotopic composition and biogeographic distribution of planktonic foraminifera in the Indian Ocean. *Palaeogeography, Palaeoclimatology, Palaeoecology* 33:9–46.

Dymond, J., E. Suess, and M. Lyle. 1992. Barium in deep-sea sediment: A geochemical proxy for paleoproductivity. *Paleoceanography* 7:163–81.

Dymond, J. and H. Veeh. 1975. Metal accumulation rates in the southeast Pacific and the origin of metalliferous sediments. *Earth and Planetary Science Letters* 28:13–22.

Eble, C. F. and W. C. Grady. 1993. Palynologic and petrographic characteristics of two Middle Pennsylvanian coal beds and a probable modern analogue. In J. C. Cobb and C. B. Cecil, eds., *Modern and Ancient Coal-Forming Environments*, pp. 119–38. Geological Society of America Special Paper 286.

Eder, W. 1982. Diagenetic redistribution of carbonate, a process in forming limestone-marl alternations (Devonian and Carboniferous, Rheinisches Schiefergebirge, W. Germany). In G. Einsele and A. Seilacher, eds., *Cyclic and Event Stratification*, pp. 98–112. Berlin: Springer-Verlag.

Edwards, L. E., P. J. Mudie, and A. de Vernal. 1991. Pliocene paleoclimatic reconstruction using dinoflagellate cysts: Comparison of methods. *Quaternary Science Reviews* 10:259–74.

Ehleringer, J. R., R. F. Sage, L. B. Flanagan, and R. W. Pearcy. 1991. Climate change and the evolution of C_4 photosynthesis. *Trends in Ecology and Evolution* 6:95–9.

Eicher, D. L. and R. Diner. 1989. Origin of the Cretaceous Bridge Creek cycles in the Western Interior, United States. *Palaeogeography, Palaeoclimatology, Palaeoecology* 74:127–46.

Einsele, G. 1982. Limestone-marl cycles (periodites): Diagnosis, significance, causes—A review. In G. Einsele and A. Seilacher, eds., *Cyclic and Event Stratification*, pp. 8–53. Berlin: Springer-Verlag.

Ekdale, A. A. and W. H. Berger. 1978. Deep-sea ichnofacies: Modern organism traces on and in pelagic carbonates of the western equatorial Pacific. *Palaeogeography, Palaeoclimatology, Palaeoecology* 23:263–78.

Ekdale, A. A., R. G. Bromley, and S. G. Pemberton. 1984. *Ichnology—the use of trace fossils in sedimentology and stratigraphy*. Society of Economic Paleontologists and Mineralogists, Short Course Notes 15.

Elder, R. L. and G. R. Smith. 1988. Fish taphonomy and environmental inference in paleolimnology. *Palaeogeography, Palaeoclimatology, Palaeoecology* 62:577–92.

Elder, W. P. 1987. The paleoecology of the Cenomanian-Turonian (Cretaceous) Stage boundary extinctions at Black Mesa, Arizona. *Palaios* 2:24–40.

———. 1988. Geometry of Upper Cretaceous bentonite beds: Implications about volcanic source areas and paleowind patterns, western interior, United States. *Geology* 16:835–8.

Elverhøi, A. 1984. Glacigenic and associated marine sediments in the Weddell Sea, fjords of Spitsbergen and the Barents Sea: A review. *Marine Geology* 57:53–88.

Emery, K. O. 1955. Transportation of rocks by driftwood. *Journal of Sedimentary Petrology* 25:51–7.

———. 1963. Organic transportation of marine sediments. In M. N. Hill, ed. *The Sea*, pp. 776–93. New York: Wiley Interscience.

Emiliani, C. 1955. Pleistocene temperatures. *Journal of Geology* 63:538–78.

Emiliani, C. and D. B. Ericson. 1991. The glacial/interglacial temperature range of the surface water of the oceans at low latitudes. In H. P. Taylor, Jr., J. R. O'Neil, and I. R. Kaplan, eds., *Stable Isotope Geochemistry: A Tribute to Samuel Epstein*, pp. 223–8. Geochemical Society Special Publication 3.

Erez, J. and B. Luz. 1983. Experimental paleotemperature equation for planktonic foraminifera. *Geochimica et Cosmochimica Acta* 47:1025–31.

Ericksen, M. C. and R. Slingerland. 1990. Numerical simulations of tidal and wind-driven circulation in the Cretaceous Interior Seaway of North America. *Geological Society of America Bulletin* 102: 1499–516.

Eshet, Y., A. Almogi-Labin, and A. Bein. 1994. Dinoflagellate cysts, paleoproductivity and upwelling systems: A Late Cretaceous example from Israel. *Marine Micropaleontology* 23:231–40.

Esteban, M. and C. F. Klappa. 1983. Subaerial exposure environment. In P. A. Scholle, D. G. Bebout, and C. H. Moore, eds., *Carbonate Depositional Environments*, pp. 2–54. American Association of Petroleum Geologists Memoir 33.

Esterle, J. S. and J. C. Ferm. 1990. On the use of modern tropical domed peats as analogues for petrographic variation in Carboniferous coal beds. *International Journal of Coal Geology* 16:131–6.

Estes, R. and J. H. Hutchison. 1980. Eocene Lower vertebrates from Ellesmere Island, Canadian Arctic Archipelago. *Palaeogeography, Palaeoclimatology, Palaeoecology* 30:325–47.

Eugster, H. P. and L. A. Hardie. 1978. Saline lakes. In A. Lerman, ed. *Lakes—Chemisty, Geology, Physics*, pp. 237–93. Heidelberg: Springer-Verlag.

Evans, R. 1978. Origin and significance of evaporites in basins around Atlantic margin. *American Association of Petroleum Geologists Bulletin* 62: 223–34.

Eyles, C. H., N. Eyles, and A. D. Miall. 1985. Models of glaciomarine sedimentation and their application to the interpretation of ancient glacial sequences. *Palaeogeography, Palaeoclimatology, Palaeoecology* 51:15–84.

Eyles, N. 1990. Marine debris flows: Late Precambrian "tillites" of the Avalonian-Cadomian orogenic belt. *Palaeogeography, Palaeoclimatology, Palaeoecology* 79:73–98.

———. 1993. Earth's glacial record and its tectonic setting. *Earth-Science Reviews* 35:1–248.

Eyles, N. and C. H. Eyles. 1989. Glacially-influenced deep-marine sedimentation of the Late Precambrian Gaskiers Formation, Newfoundland, Canada. *Sedimentology* 36:601–20.

———. 1992. Glacial depositional systems. In R. G. Walker and N. P. James, eds., *Facies Models: Response to Sea-level Change*, pp. 73–100. St. John's, Newfoundland: Geological Association of Canada.

Eyles, N., C. H. Eyles, and A. D. Miall. 1983. Lithofacies types and vertical profile modes; an alternative approach to the description and environmental interpretation of glacial diamict and diamictite sequences. *Sedimentology* 30:393–410.

Eyles, N. and M. B. Lagoe. 1989. Sedimentology of shell-rich deposits (coquinas) in the glaciomarine upper Cenozoic Yakataga Formation, Middleton Island, Alaska. *Geological Society of America Bulletin* 101:129–42.

Fagerstrom, J. A. 1987. *The Evolution of Reef Communities*. New York: John Wiley and Sons.

Fairbanks, R. G., M. Sverdlove, R. Free, P. H. Wiebe, and A. W. H. Bé. 1982. Vertical distribution and isotopic fractionation of living planktonic foraminifera from the Panama Basin. *Nature* 298:841–4.

Fairbanks, R. G., P. H. Wiebe, and A. W. H. Bé. 1980. Vertical distribution and isotopic composition of living planktonic foraminifera in the western North Atlantic. *Science* 207:61–3.

Falcon, R. M. S. 1989. Macro- and micro-factors affecting coal-seam quality and distribution in southern Africa with particular reference to the No. 2 seam, Witbank coalfield, South Africa. *International Journal of Coal Geology* 12:681–731.

Farlow, J. O. 1990. Dinosaur paleobiology. Part II. Dinosaur energetics and thermal biology. In D. B. Weishampel, P. Dodson, and H. Osm—lska, eds., *The Dinosauria*, pp. 43–55. Berkeley: University of California Press.

Farrell, J. W., T. F. Pedersen, S. E. Calvert, and B. Nielsen. 1995. Glacial-interglacial changes in nutrient utilization in the equatorial Pacific Ocean. *Nature* 377:514–7.

Faure, G. 1986. *Principles of Isotope Geochemistry*. 2nd ed. New York: Wiley.

Ferguson, D. K. 1985. The origin of leaf assemblages—new light on an old problem. *Review of Palaeobotany and Palynology* 46:117–88.

Filippelli, G. M. 1997. Intensification of the Asian monsoon and a chemical weathering event in the late Miocene-early Pliocene: Implications for late Neogene climate change. *Geology* 25:27–30.

Filippelli, G. M. and M. L. Delaney. 1992. Similar phosphorus fluxes in ancient phosphorite deposits and a modern phosphogenic environment. *Geology* 20:709–12.

———. 1994. The oceanic phosphorus cycle and continental weathering during the Neogene. *Paleoceanography* 9:643–52.

Finney, S. C. and X. Chen. 1990. The relationship of Ordovician graptolite provincialism to palaeogeography. In W. S. McKerrow and C. R. Scotese, eds., *Palaeozoic Palaeogeography and Biogeography*, pp. 123–8. Geological Society of London Memoir 12.

Fischer, A. G. 1986. Climatic rhythms recorded in strata. *Annual Review of Earth and Planetary Sciences* 14:351–76.

———. 1993. Cyclostratigraphy of Cretaceous chalk-marl sequences. In W. G. E. Caldwell and E. G. Kauffman, eds., *Evolution of the Western Interior Basin*, pp. 283–95. Geological Association of Canada Special Paper 39.

Fischer, A. G. and M. A. Arthur. 1977. Secular variations in the pelagic realm. In H. E. Cook and P. Enos, eds., *Deep-Water Carbonate Environments*, pp. 19–50. Society of Economic Paleontologists and Mineralogists Special Publication 25.

Fischer, A. G. and M. Sarnthein. 1988. Airborne silts and dune-derived sands in the Permian of the Delaware Basin. *Journal of Sedimentary Petrology* 58:637–43.

Flessa, K. W. 1975. Area, continental drift, and mammalian diversity. *Paleobiology* 1:189–94.

Flessa, K. W., S. G. Barnett, D. B. Cornue, M. A. Lomaga, N. Lombardi, J. M. Miyazaki, and A. S. Murer. 1979. Geologic implications of the relationship between mammalian faunal similarity and geographic distance. *Geology* 7:15–8.

Flessa, K. W. and J. J. Sepkoski. 1978. On the relationship between Phanerozoic diversity and changes in habitable area. *Paleobiology* 4:359–66.

Flores, R. M. 1993. Geologic and geomorphic controls of coal development in some Tertiary Rocky Mountain basins, USA. *International Journal of Coal Geology* 23:43–73.

Flower, B. P. and J. P. Kennett. 1993. Relations between Monterey Formation deposition and middle Miocene global cooling: Naples Beach section, California. *Geology* 21:877–80.

———. 1994. Oxygen and carbon isotopic stratigraphy of the Monterey Formation at Naples Beach, California. In J. S. Hornafius, ed. *Field Guide to the Monterey Formation Between Santa Barbara and Gaviota, California Vol. GB72*, pp. 59–66.: Pacific Section, American Association of Petroleum Geologists.

———. 1995. Middle Miocene deepwater paleoceanography in the southwest Pacific: Relations with East Antarctic ice sheet development. *Paleoceanography* 10:1095–112.

Flügel, E. and E. Flügel-Kahler. 1992. Phanerozoic reef evolution: Basic questions and data base. *Facies* 26:167–278.

Föllmi, K. B. 1993. Phosphorus and phosphate-rich sediments, an environmental approach. *Chemical Geology* 107:375–8.

———. 1995. 160 m.y. record of marine sedimentary phosphorus burial: Coupling of climate and continental weathering under greenhouse and icehouse conditions. *Geology* 23:859–62.

———. 1996. The phosphorus cycle, phosphogenesis, and marine phosphate-rich deposits. *Earth-Science Reviews* 40:55–124.

Föllmi, K. B., H. Weissert, M. Bisping, and H. Funk. 1994. Phosphogenesis, carbon-isotope stratigraphy, and carbonate-platform evolution along the Lower Cretaceous northern Tethyan margin. *Geological Society of America Bulletin* 106: 729–46.

Force, E. R. 1991. Geology of titanium-mineral deposits. *Geological Society of America Special Paper* 259.

Force, E. R., W. F. Cannon, R. A. Koski, K. T. Passmore, and B. R. Doe. 1983. Influences of ocean anoxic events on manganese deposition and ophiolite-hosted sulfide preservation. In T. M. Cronin, W. F. Cannon, and R. Z. Poore, eds., *Paleoclimate and Mineral Deposits*, pp. 26–9. U.S. Geological Survey Circular 822.

Fordyce, R. E. 1980. Whale evolution and Oligocene Southern Ocean environments. *Palaeogeography, Palaeoclimatology, Palaeoecology* 31:319–36.

———. 1982. The Austrasian marine vertebrate record and its climatic and geographic implications. In P. V. Rich and E. M. Thompson, eds., *The Fossil Vertebrate Record of Australia*, pp. 596–627. Melbourne: Monash University Press.

Forest, C. E., P. Molnar, and K. A. Emanuel. 1995. Palaeoaltimetry from energy conservation principles. *Nature* 374:347–50.

Forester, R. M. 1986. Determination of the dissolved anion composition of ancient lakes from fossil ostracodes. *Geology* 14:796–8.

Forester, R. M. 1991. Pliocene-climate history of the

western United States dervied from lacustrine ostracodes. *Quaternary Science Reviews* 10:133–46.

Forester, R. M. and E. M. Brouwers. 1985. Hydrochemical parameters governing the occurrence of estuarine and marginal estuarine ostracodes: An example from south-central Alaska. *Journal of Paleontology* 59:344–69.

Fortey, R. A. and L. R. M. Cocks. 1986. Marginal faunal belts and their structural implications, with examples from the Lower Palaeozoic. *Journal of the Geological Society of London* 143:151–60.

Frakes, L. A. 1979. *Climates Throughout Geologic Time*. New York: Elsevier.

Frakes, L. A. and B. R. Bolton. 1984. Origin of manganese giants: Sea-level change and anoxic-oxic history. *Geology* 12:83–6.

Frakes, L. A. and J. E. Francis. 1988. A guide to Phanerozoic cold polar climates from high-latitude ice-rafting in the Cretaceous. *Nature* 333:547–9.

Frakes, L. A., J. E. Francis, and J. I. Syktus. 1992. *Climate Modes of the Phanerozoic*. Cambridge: Cambridge University Press.

Francis, J. E. 1983. The dominant conifer of the Jurassic Purbeck Formation, England. *Palaeontology* 26:277–94.

———. 1984. The seasonal environment of the Purbeck (Upper Jurassic) fossil forests. *Palaeogeography, Palaeoclimatology, Palaeoecology* 48:285–307.

———. 1991. Palaeoclimatic significance of Cretaceous-early Tertiary fossil forests of the Antarctic Peninsula. In M. R. A. Thomson, J. A. Crame, and J. W. Thomson, eds., *Geological Evolution of Antarctica*, pp. 623–7. Cambridge: Cambridge University Press.

Francis, J. E. and R. S. Hill. 1996. Fossil plants from the Pliocene Sirius Group, Transantarctic Mountains: Evidence for climate from growth rings and fossil leaves. *Palaios* 11:389–96.

François, L. M. and J. C. G. Walker. 1992. Modelling the Phanerozoic carbon cycle and climate: Constraints from the $^{87}Sr/^{86}Sr$ isotopic ratio of seawater. *American Journal of Science* 292: 81–135.

François, R., M. A. Altabet, and L. H. Burckle. 1992. Glacial to interglacial changes in surface nitrate utilization in the Indian sector of the Southern Ocean as recorded by sediment $\delta^{15}N$. *Paleoceanography* 7:589–606.

François, R., S. Honjo, S. J. Manganini, and G. E. Ravizza. 1995. Biogenic barium fluxes to the deep sea: Implications for paleoproductivity reconstruction. *Global Biogeochemical Cycles* 9:289–303.

Frank, P. W. 1988. Conchostraca. *Palaeogeography, Palaeoclimatology, Palaeoecology* 62:399–403.

Freeman, K. H. and J. M. Hayes. 1992. Fractionation of carbon isotopes by phytoplankton and estimates of ancient CO_2 levels. *Global Biogeochemical Cycles* 6:185–98.

Fricke, H. C. and J. R. O'Neil. 1996. Inter- and intra-tooth variation in the oxygen isotope composition of mammalian tooth enamel phosphate: Implications for palaeoclimatological and palaeobiological research. *Palaeogeography, Palaeoclimatology, Palaeoecology* 126:91–9.

Fritts, H. C. 1976. *Tree Rings and Climate*. New York: Academic Press.

———. 1982. The climate-growth response. In M. K. Hughes, P. M. Kelly, J. R. Pilcher, and V. C. LaMarche, Jr., eds., *Climate from Tree Rings*, pp. 33–8. Cambridge: Cambridge University Press.

Froelich, P. N., M. L. Bender, N. A. Luedtke, G. R. Heath, and T. Devries. 1982. The marine phosphorus cycles. *American Journal of Science* 282:474–511.

Fryberger, S. G., T. S. Ahlbrandt, and S. Andrews. 1979. Origin, sedimentary features, and significance of low-angle eolian "sand sheet" deposits, Great Sand Dunes National Monument and vicinity, Colorado. *Journal of Sedimentary Petrology* 49:733–46.

Fulthorpe, C. S. and S. O. Schlanger. 1989. Paleo-oceanographic and tectonic settings of early Miocene reefs and associated carbonates of offshore Southeast Asia. *American Association of Petroleum Geologists Bulletin* 73:729–56.

Fulton, I. M. 1987. Genesis of the Warwickshire Thick Coal: A group of long-residence histosols. In A. C. Scott, ed. *Coal and Coal-Bearing Strata*, pp. 201–18. Geological Society of London Special Publication 32.

Gabbott, S. E., R. J. Aldridge, and J. N. Theron. 1995. A giant conodont with preserved muscle tissue from the Upper Ordovician of South Africa. *Nature* 374:800–3.

Gale, A. S., H. C. Jenkyns, W. J. Kennedy, and R. M. Corfield. 1993. Chemostratigraphy versus biostratigraphy: Data from around the Cenomanian-Turonian boundary. *Journal of the Geological Society of London* 150:29–32.

Gardner, J. V., W. E. Dean, and P. Dartnell. 1997. Biogenic sedimentation beneath the California Current system for the past 30 kyr and its paleographic implications. *Paleoceanography* 12:207–25.

Garrels, R. M. and A. Lerman. 1981. Phanerozoic cycles of sedimentary carbon and sulfur. *Proceedings of the National Academy of Sciences* 78:4652–6.

Gasperi, J. T. and J. P. Kennett. 1992. Isotopic evidence for depth stratification and paleoecology of Miocene planktonic foraminifera: Western equatorial Pacific DSDP Site 289. In R. Tsuchi and J. C. Ingle, Jr., eds., *Pacific Neogene: Environment, evolution, and events*, pp. 117–47. Tokyo: University of Tokyo Press.

Gasse, F. 1990. Tectonic and climatic controls on lake distribution and environments in Afar from Miocene to Present. In B. J. Katz, ed. *Lacustrine Basin Exploration: Case Studies and Modern Analogs*, pp. 19–41. American Association of Petroleum Geologists Memoir 50.

Gasse, F., R. Tehet, A. Durand, E. Gilbert, and J. C. Fontes. 1990. The arid-humid transition in the Sahara and the Sahel during the last deglaciation. *Nature* 346:141–6.

Gastaldo, R. A. 1986. Selected aspects of plant taphonomic processes in coastal deltaic regimes. In T. W. Broadhead, ed. *Land Plants*, pp. 27–44. Univerisity of Tennessee Studies in Geology 15.

Gastaldo, R. A., T. M. Demko, and Y. Liu. 1990. Carboniferous coastal environments and paleocommunities of the Mary Lee Coal Zone, Marion and Walker counties, Alabama. *A Guidebook for Field Trip VI, 39th Annual Meeting, Southeastern Section, Geological Society of America*: Geological Survey of Alabama.

Gates, W. L. 1982. Paleoclimate modeling—a review with reference to problems and prospects for the pre-Pleistocene. In W. H. Berger, and Crowell, J. C., ed. *Climate in Earth History*, pp. 26–42. Washington, D.C.: National Academy Press.

Geitgey, J. E. and T. R. Carr. 1987. Temperature as a factor affecting conodont diversity and distribution. In R. L. Austin, ed. *Conodonts: Investigative Techniques and Applications*, pp. 241–55. Chichester: Ellis Horwood, Ltd.

Geitzenauer, K. R., M. B. Roche, and A. McIntyre. 1977. Coccolith biogeography from North Atlantic and Pacific surface sediments. In A. T. S. Ramsay, ed. *Oceanic Micropalaeontology Vol. 2*, pp. 973–1008. London: Academic Press.

Gerson, R. and R. Amit. 1987. Rates and modes of dust accretion and deposition in an arid region—the Negev, Israel. In L. E. Frostick and I. Reid, eds., *Desert Sediments—Ancient and Modern*, pp. 157–68. Geological Sociey of London Special Publication 35.

Gersonde, R. 1986. Siliceous microorganisms in sea ice and their record in sediments in the southern Weddell Sea (Antarctica). In M. Richard, ed. *Proceedings of the 8th International Diatom Symposium*, pp. 549–66. Königstein, Germany: Koeltz Scientific.

Gibbs, M. T., E. J. Barron, and L. R. Kump. 1997. An atmospheric pCO_2 threshold for glaciation in the Late Ordovician. *Geology* 25:447–50.

Gierlowski-Kordesch, E. and B. R. Rust. 1994. The Jurassic East Berlin Formation, Hartford Basin, Newark Supergroup (Connecticut and Massachusetts): A saline lake-playa-alluvial plain system. In R. W. Renaut and W. M. Last, eds., *Sedimentology and Geochemistry of Modern and Ancient Saline Lakes*, pp. 249–65. SEPM (Society for Sedimentary Geology) Special Publication 50.

Gilbert, R. 1990. Rafting in glacimarine environments. In J. A. Dowdeswell and J. D. Scourse, eds., *Glacimarine Environments: Processes and Sediments*, pp. 105–20. Geological Society of London Special Publication 53.

Gili, E., J.-P. Masse, and P. W. Skelton. 1995. Rudists as gregarious sediment-dwellers, not reef-builders, on Cretaceous carbonate platforms. *Palaeogeography, Palaeoclimatology, Palaeoecology* 118:245–67.

Gingele, F. and A. Dahmke. 1994. Discrete barite particles in barium as tracers of paleoproductivity in South Atlantic sediments. *Paleoceanography* 9:151–68.

Giresse, P., J. Maley, and P. Brenac. 1994. Late Quaternary palaeoenvironments in the Lake Barombi Mbo (West Cameroon) deduced from pollen and carbon isotopes of organic matter. *Palaeogeography, Palaeoclimatology, Palaeoecology* 107:65–78.

Glancy, T. J., Jr., M. A. Arthur, E. J. Barron, and E. G. Kauffman. 1993. A paleoclimate model for the North American Cretaceous (Cenomanian-Turonian) epicontinental sea. In W. G. E. Caldwell and E. G. Kauffman, eds., *Evolution of the Western Interior Basin*, pp. 219–41. Geological Association of Canada Special Paper 39.

Glenn, C. R. 1990. Pore water, petrologic and stable carbon isotope data bearing on the origin of modern Peru margin phosphorites and associated authigenic phases. In W. C. Burnett and S. R. Riggs, eds., *Phosphate Deposits of the World. Vol. 3, Neogene to Modern Phosphorites*, pp. 46–61. Cambridge: Cambridge University Press.

Glenn, C. R. and M. A. Arthur. 1985. Sedimentary and geochemical indicators of productivity and oxygen contents in modern and ancient basins:

The Holocene Black Sea as the "type" anoxic basin. *Chemical Geology* 48:325–54.

Glenn, C. R., K. B. Föllmi, S. R. Riggs, G. N. Baturin, K. A. Grimm, J. Trappe, A. M. Abed, C. Galli-Olivier, R. E. Garrison, A. V. Ilyin, C. Jehl, V. Rohrlich, R. M. Y. Sadaqah, M. Schidlowski, R. E. Sheldon, and H. Siegmund. 1994a. Phosphorus and phosphorites: Sedimentology and environments of formation. *Eclogae Geologicae Helvetiae* 87:747–88.

Glenn, C. R., M. A. Arthur, J. M. Resig, W. C. Burnett, W. E. Dean, and R. A. Jahnke. 1994b. Are modern and ancient phosphorites really so different? In A. Iijima, A. M. Abed, and R. E. Garrison, eds., *Siliceous, Phosphatic and Glauconitic Sediments of the Tertiary and Mesozoic*, pp. 159–88. Proceedings of the 29th International Geological Congress, Part C. Utrecht: VSP Publishers.

Glenn, C. R. and J. D. Kronen. 1993. Origin and significance of late Pliocene phosphatic hardgrounds on the Queensland Plateau, northeastern Australian maring. In J. A. McKenzie, A. Palmer-Julson, et al. *Proceedings of the Ocean Drilling Program, Scientific Results, Leg 133*, pp. 525–34. College Station, TX: Ocean Drilling Program.

Glock, W. S. and S. Agerter. 1963. Anomalous patterns in tree rings. *Endeavour* 22:9–13.

Glover, B. W. and J. H. Powell. 1996. Interaction of climate and tectonics upon alluvial architecture: Late Carboniferous-Early Permian sequences at the southern margin of the Pennine Basin, UK. *Palaeogeography, Palaeoclimatology, Palaeoecology* 121:13–34.

Gobbett, D. J. 1973. Permian Fusulinacea. In A. Hallam, ed. *Atlas of Palaeobiogeography*, pp. 151–8. Amsterdam: Elsevier Scientific.

Goddéris, Y. and L. M. François. 1995. The Cenozoic evolution of the strontium and carbon cycles: Relative importance of continental erosion and mantle exchanges. *Chemical Geology* 126:169–90.

Goldberg, E. D. and G. O. S. Arrhenius. 1958. Chemistry of Pacific pelagic sediments. *Geochimica et Cosmochimica Acta* 13:153–212.

González-Bonorino, G. and N. Eyles. 1995. Inverse relation between ice extent and the late Paleozoic glacial record of Gondwana. *Geology* 23:1015–8.

Good, S. C., J. M. Parrish, and R. F. Dubiel. 1987. Paleoevironmental implications of sedimentology and paleontology of the Upper Triassic Chinle Formation, southeastern Utah. *Four Corners Geological Society Guidebook, 10th Field Conference*, pp. 117–8.

Gordon, W. A. 1975. Distribution by latitude of Phanerozoic evaporite deposits. *Journal of Geology* 83:671–84.

Gore, P. J. W. 1988. Lacustrine sequences in an early Mesozoic rift basin: Culpeper Basin, Virginia, USA. In A. J. Fleet, K. Kelts, and M. R. Talbot, eds., *Lacustrine Petroleum Source Rocks*, pp. 247–78. Geological Society of London Special Publication 40.

———. 1989. Toward a model for open- and closed-basin deposition in ancient lacustrine sequences: The Newark Supergroup (Triassic-Jurassic), eastern North America. *Palaeogeography, Palaeoclimatology, Palaeoecology* 70:29–51.

Goudie, A. S. 1983. Calcrete. In A. S. Goudie and K. Pye, eds., *Chemical Sediments and Geomorphology*, pp. 93–131. London: Academic Press.

Grantham, J. H. and M. A. Velbel. 1988. The influence of climate and topography on rock-fragment abundance in modern fluvial sands of the southern Blue Ridge Mountains, North Carolina. *Journal of Sedimentary Petrology* 58:219–27.

Gravenor, C. P. 1981. Appendix: Chattermark trails on garnets as an indicator of glaciation in ancient deposits. In M. J. Hambrey and W. B. Harland, eds., *Earth's Pre-Pleistocene Glacial Record*, pp. 17–21. Cambridge: Cambridge University Press.

Greenwood, D. R. and S. L. Wing. 1995. Eocene continental climates and latitudinal temperature gradients. *Geology* 23:1044–8.

Greenwood, H. J. 1989. On models and modeling. *The Canadian Mineralogist* 27:1–14.

Gregory, K. M. 1992. Late Eocene Paleolatitude, Paleoclimate, and Paleogeography of the Front Range Region, Colorado. *Ph.D. thesis,* University of Arizona.

———. 1994. Palaeoclimate and palaeovegetation of the 35 Ma Florissant flora, Front Range, Colorado. *Palaeoclimates—Data and Modelling* 1:23–57.

———. 1997. Are paleoclimate estimates biased by foliar physiognomic responses to increased atmospheric CO_2? *Palaeogeography, Palaeoclimatology, Palaeoecology* 124:39–52.

Gregory, K. M. and C. G. Chase. 1992. Tectonic significance of paleobotanically estimated climate and altitude of the late Eocene erosion surface, Colorado. *Geology* 20:581–5.

———. 1994. Tectonic and climatic significance of a late Eocene low-relief, high-level geomorphic surface, Colorado. *Journal of Geophysical Research* 99:20,141–20,60.

Gregory, K. M. and W. C. McIntosh. 1996. Paleoclimate and paleoelevation of the Oligocene Pitch-

Pinnacle flora, Sawatch Range, Colorado. *Geological Society of America Bulletin* 108:545–61.
Gregory, R. T. 1991. Oxygen isotope history of seawater revisited: Timescales for boundary event changes in the oxygen isotope composition of seawater. In H. P. Taylor, Jr., J. R. O'Neil, and I. R. Kaplan, eds., *Stable Isotope Geochemistry: A Tribute to Samuel Epstein*, pp. 65–76. Geochemical Society Special Publication 3.
Gregory, R. T., C. B. Douthitt, I. R. Duddy, P. V. Rich, and T. H. Rich. 1989. Oxygen isotopic composition of carbonate concretions form the Lower Cretaceous of Victoria, Australia: Implications for the evolution of meteoric waters on the Australian continent in a paleopolar environment. *Earth and Planetary Science Letters* 92:27–42.
Griffin, J. J., H. Windom, and E. D. Goldberg. 1968. The distribution of clay minerals in the World Ocean. *Deep-Sea Research* 15:433–59.
Grossman, E. L. 1992. Isotope studies of Paleozoic paleoceanography—Opportunities and pitfalls. *Palaios* 7:241–3.
———. 1994. The carbon and oxygen isotope record during the evolution of Pangea: Carboniferous to Triassic. In G. D. Klein, ed. *Pangea: Paleoclimate, Tectonics, and Sedimentation During Accretion, Zenith, and Breakup of a Supercontinent*, pp. 207–28. Geological Society of America Special Paper 288.
Grossman, E. L. and T.-L. Ku. 1986. Oxygen and carbon isotope fractionation of biogenic aragonite: Temperature effects. *Chemical Geology* 59:59–74.
Grossman, E. L., H.-S. Mii, and T. E. Yancey. 1993. Stable isotopes in Late Pennsylvanian brachiopods from the United States: Implications for Carboniferous paleoceanography. *Geological Society of America Bulletin* 105:1284–96.
Grossman, E. L., C. Zhang, and T. E. Yancey. 1991. Stable-isotope stratigraphy of brachiopods from Pennsylvanian shales in Texas. *Geological Society of America Bulletin* 103:953–65.
Gunatilaka, A. and S. Mwango. 1987. Continental sabkha pans and associated nebkhas in southern Kuwait, Arabian Gulf. In L. Frostick and I. Reid, eds., *Desert Sediments: Ancient and Modern*, pp. 187–203. Geological Society of London Special Publication 35.
Gunnell, G. F., M. E. Morgan, M. C. Maas, and P. D. Gingerich. 1995. Comparative paleoecology of Paleogene and Neogene mammalian faunas: Trophic structure and composition. *Palaeogeography, Palaeoclimatology, Palaeoecology* 115:265–86.
Gyllenhaal, E. D. 1991. How accurately can paleoprecipitation and paleoclimatic change be interpreted from subaerial disconformities? *Ph.D. thesis*, University of Chicago.
Gyllenhaal, E. D., C. J. Engberts, P. J. Markwick, L. H. Smith, and M. E. Patzkowsky. 1991. The Fujita-Ziegler model: A new semi-quantitative technique for estimating paleoclimate from paleogeographic maps. *Palaeogeography, Palaeoclimatology, Palaeoecology* 86:41–66.
Habicht, J. K. A. 1979. *Paleoclimate, paleomagnetism, and continental drift*. American Association of Petroleum Geologists Studies in Geology 9.
Hallam, A. 1975. *Jurassic Environments*. Cambridge: Cambridge University Press.
———. 1977. Jurassic bivalve biogeography. *Paleobiology* 3:58–73.
———. 1983. Early and mid-Jurassic molluscan biogeography and the establishment of the central Atlantic seaway. *Palaeogeography, Palaeoclimatology, Palaeoecology* 43:181–93.
———. 1984a. Continental humid and arid zones during the Jurassic and Cretaceous. *Palaeogeography, Palaeoclimatology, Palaeoecology* 47:195–223.
———. 1984b. Distribution of fossil marine invertebrates in relation to climate. In P. J. Brenchley, ed. *Fossils and Climate*, pp. 107–25. Chichester: John Wiley and Sons.
———. 1987. Mesozoic marine organic-rich shales. In J. Brooks and A. J. Fleet, eds., *Marine Petroleum Source Rocks*, pp. 251–62. Geological Society of London Special Paper 26.
———. 1993. Jurassic climates as inferred from the sedimentary and fossil record. *Philosophical Transactions of the Royal Society, Series B* 341:287–96.
———. 1994. *An Outline of Phanerozoic Biogeography*. Oxford: Oxford University Press.
Hallam, A., ed. 1973. *Atlas of Palaeobiogeography*. Amsterdam: Elsevier Scientific.
Hallam, A., J. A. Grose, and A. H. Ruffell. 1991. Palaeoclimatic significance of changes in clay mineralogy across the Jurassic-Cretaceous boundary in England and France. *Palaeogeography, Palaeoclimatology, Palaeoecology* 81:173–87.
Hallock, P. 1988. The role of nutrient availability in bioerosion: Consequences to carbon buildups. *Palaeogeography, Palaeoclimatology, Palaeoecology* 63:275–91.
Hallock, P., A. C. Hine, G. A. Vargo, J. A. Elrod, and W. C. Jaap. 1988. Platforms of the Nicaraguan rise: Examples of the sensitivity of carbonate sedimentation to excess trophic resources. *Geology* 16:1104–7.
Hallock, P. and W. Schlager. 1986. Nutrient excess

and the demise of coral reefs and carbonate platforms. *Palaois* 1:389–98.

Hambrey, M. J., W. U. Ehrmann, and B. Larsen. 1991. Cenozoic glacial record of the Prydz Bay continental shelf, East Antarctica. In J. Barron, B. Larsen, et al. *Proceedings of the Ocean Drilling Program, Scientific Results, Leg 119*, pp. 77–132. College Station, TX: Ocean Drilling Program.

Hambrey, M. J. and W. B. Harland. 1981a. Criteria for the identification of glacigenic deposits. In M. J. Hambrey and W. B. Harland, eds., *Earth's Pre-Pleistocene Glacial Record*, pp. 14–7. Cambridge: Cambridge University Press.

Hambrey, M. J. and W. B. Harland, eds. 1981b. *Earth's Pre-Pleistocene Glacial Record*. Cambridge: Cambridge University Press.

Handford, C. R. 1991. Marginal marine halites: Sabkhas and salinas. In J. L. Melvin, ed. *Evaporites, Petroleum and Mineral Resources*, pp. 1–66. Developments in Sedimentology 50. Amsterdam: Elsevier.

Hanski, I., J. Kouki, and A. Halkka. 1993. Three explanations of the positive relationship between distribution and abundance of species. In R. E. Ricklefs and D. Schluter, eds., *Species Diversity in Ecological Communities*, pp. 108–16. Chicago: University of Chicago Press.

Haq, B. U. 1978. Silicoflagellates and ebridians. In B. U. Haq and A. Boersma, eds., *Introduction to Marine Micropaleontology*, pp. 267–75. New York: Elsevier.

———. 1982. Climatic acme events in the sea and on land. In W. H. Berger and J. C. Crowell, eds., *Climate in Earth History*, pp. 126–32. Washington, D.C.: National Academy Press.

Haq, B. U. and G. P. Lohmann. 1976. Early Cenozoic calcareous nannoplankton biogeography of the Atlantic Ocean. *Marine Micropaleontology* 1:119–94.

Hardie, L. A., J. P. Smoot, and H. P. Eugster. 1978. Saline lakes and their deposits: A sedimentological approach. In A. Matter and M. E. Tucker, eds., *Modern and Ancient Lake Sediments*, pp. 7–41. International Association of Sedimentologists Special Publication 2. Oxford: Blackwell Scientific.

Harland, W. B., A. V. Cox, P. G. Llewellyn, C. A. G. Pickton, A. G. Smith, and R. Walters. 1982. *A Geologic Time Scale*. Cambridge: Cambridge University Press.

Harms, J. C., J. B. Southard, D. R. Spearing, and R. G. Walker. 1975. *Depositional environments as interpreted from primary sedimentary stuctures and stratification sequences*. Society of Economic Paleontologists and Mineralogists Short Course 2.

Harrison, S. P., J. E. Kutzbach, and P. Behling. 1991. General circulation models, palaeoclimatic data and last interglacial climates. *Quaternary International* 10–12:231–42.

Hartl, P., L. Tauxe, and T. Herbert. 1995. Earliest Oligocene increase in South Atlantic productivity as interpreted from "rock magnetics" at Deep Sea Drilling Project Site 522. *Paleoceanography* 10:311–25.

Harvey, R. D. and J. W. Dillon. 1985. Maceral distributions in Illinois coals and their paleoenvironmental implication. *International Journal of Coal Geology* 5:141–65.

Harwood, G. and A. C. Kendall. 1993. Evaporite sequences as palaeoclimate indicators (abs.). Paper read at *Pangea—Carboniferous to Jurassic*, at Calgary, Alberta.

Hasegawa, J. 1997. Cenomanian-Turonian carbon isotope events recorded in terrestrial organic matter from northern Japan. *Palaeogeography, Palaeoclimatology, Palaeoecology* 130:251–73.

Hasiotis, S. T. and R. F. Dubiel. 1993a. Continental trace fossils of the Upper Triassic Chile Formation, Petrified Forest National Park, Arizona. In S. G. Lucas and M. Morales, eds., *The Nonmarine Triassic*, pp. 175–8. New Mexico Museum of Natural History and Science Bulletin 3.

———. 1993b. Crayfish burrows and their paleohydrologic significance—Upper Triassic Chinle Formation, Fort Wingate, New Mexico. In S. G. Lucas and M. Morales, eds., *The Nonmarine Triassic*, pp. G24-G6. New Mexico Museum of Natural History and Science Bulletin 3.

———. 1994. Ichnofossil tiering in Triassic alluvial paleosols: Implications for Pangean continental rocks and paleoclimate. In B. Beauchamp, A. F. Embry, and D. Glass, eds., *Pangea: Global Environments and Resources*, pp. 311–7. Canadian Society of Petroleum Geologists Memoir 17.

———. 1995. Crayfish fossils and burrows from the Upper Triassic Chinle Formation, Canyonlands National Park, Utah. In V. L. Santucci and L. McClelland, eds., National Park Service Paleontological Research Technical Report, pp. 82–8.

Hasiotis, S. T. and C. E. Mitchell. 1989. Lungfish burrows in the Upper Triassic Chinle and Dolores Formations, Colorado plateau—Discussion: New evidence suggests origin by a burrowing decapod crustacean. *Journal of Sedimentary Petrology* 59:871–5.

———. 1993. A comparison of crayfish burrow morphologies: Triassic and Holocene fossil, paleo- and neo-ichnological evidence, and the identification of their burrowing signatures. *Ichnos* 2:291–314.

Hasle, G. R. 1976. The biogeography of some marine planktonic diatoms. *Deep-Sea Research* 23:319–38.

Hauglustaine, D. A. and J.-C. Gérard. 1992. A sensitivity study of the role of continental location and area on Paleozoic climate. *Palaeogeography, Palaeoclimatology, Palaeoecology* 97:311–23.

Havholm, K. G., R. C. Blakey, M. Capps, L. S. Jones, D. D. King, and G. Kocurek. 1993. Aeolian genetic stratigraphy: An example from the Middle Jurassic Page Sandstone, Colorado Plateau. In K. Pye and N. Lancaster, eds., *Aeolian Sediments, Ancient and Modern*, pp. 87–107. International Association of Sedimentologists Special Publication 16. Oxford: Blackwell Scientific.

Hay, W. W. 1992. Erosion and weathering. *Encyclopedia of Earth System Science* 2:177–85.

Hay, W. W., E. J. Barron, and S. L. Thompson. 1990a. Global atmospheric circulation experiments on an Earth with polar and tropical continents. *Journal of the Geological Society of London* 147:749–57.

———. 1990b. Results of global atmospheric circulation experiments on an Earth with a meridional pole-to-pole continent. *Journal of the Geological Society of London* 147:385–92.

Hay, W. W., J. F. Behensky, Jr., E. J. Barron and J. L. Sloan, II. 1982. Late Triassic-Liassic paleoclimatology of the proto-central North Atlantic rift system. *Palaeogeography, Palaeoclimatology, Palaeoecology* 40:13–30.

Hay, W. W., D. L. Eicher, and R. Diner. 1993. Physical oceanography and water masses in the Cretaceous Western Interior seaway. In W. G. E. Caldwell and E. G. Kauffman, eds., *Evolution of the Western Interior Basin*, pp. 297–318. Geological Association of Canada Special Paper 39.

Hayami, I. 1969. Notes on Mesozoic "planktonic" bivalves (in Japanese with English abstract). *Journal of the Geological Society of Japan* 75:375–85.

Hayden, B. P. 1988. Flood climates. In V. R. Baker, R. C. Kochel, and P. C. Patton, eds., *Flood Geomorphology*, pp. 13–26. New York: John Wiley and Sons.

Haydon, D., R. R. Radtkey, and E. R. Pianka. 1993. Experimental biogeography: Interactions between stochastic, historical, and ecological processes in a model archipelago. In R. E. Ricklefs and D. Schluter, eds., *Species Diversity in Ecological Communities*, pp. 117–30. Chicago: University of Chicago Press.

Hayes, J. M. 1993. Factors controlling ^{13}C contents of sedimentary organic compounds: Principles and evidence. *Marine Geology* 113:111–25.

Hayes, J. M., B. N. Popp, R. Takigiku, and M. W. Johnson. 1989. An isotopic study of biogeochemical relationships between carbonates and organic carbon in the Greenhorn Formation. *Geochimica et Cosmochimica Acta* 53:2961–72.

Hayes, M. O. 1967. Relationship between coastal climate and bottom sediment type on the inner continental shelf. *Marine Geology* 5:111–32.

Hays, J. D., J. Imbrie, and N. J. Shackleton. 1976. Variations in the Earth's orbit: Pacemaker of the Ice Ages. *Science* 194:1121–32.

Hays, P. D. and E. L. Grossman. 1991. Oxygen isotopes in meteoric calcite cements as indicators of continental paleoclimate. *Geology* 19:441–4.

Hays, P. E., N. G. Pisias, and A. K. Roelofs. 1989. Paleoceanography of the eastern equatorial Pacific during the Pliocene: A high-resolution radiolarian study. *Paleoceanography* 4:57–73.

Hazel, J. E. 1988. Determining late Neogene and Quaternary palaeoclimates and palaeotemperature regimes using ostracodes. In P. De Deckker, J.-P. Colin, and J.-P. Peypouquet, eds., *Ostracoda in the Earth Sciences*, pp. 89–101. Amsterdam: Elsevier.

Hecht, A. D., ed. 1985. *Paleoclimate Analysis and Modeling*. New York: John Wiley and Sons.

Heckel, P. H. 1986. Sea-level curve for Pennsylvanian eustatic marine transgressive-regressive depositional cycles along midcontinent outcrop belt, North America. *Geology* 14:330–4.

———. 1995. Glacial-eustatic base-level—Climatic model for late middle to late Pennsylvanian coal-bed formation in the Appalachian Basin. *Journal of Sedimentary Research* B65:348–56.

Heggie, D. T., G. W. Skyring, G. W. O'Brien, C. Reimers, A. Herczeg, D. J. W. Moriarty, W. C. Burnett, and A. R. Milnes. 1990. Organic carbon cycling and modern phosphorite formation on the East Australian margin: A overview. In A. J. G. Notholt and I. Jarvis, eds., *Phosphorite Research and Development*, pp. 87–117. Geological Society of London Special Publication 52.

Hein, J. R., A. O. Allwardt, and G. B. Griggs. 1974. The occurrence of glauconite in Monterey Bay, California, diversity, origins and sedimentary environmental significance. *Journal of Sedimentary Petrology* 44:562–71.

Hein, J. R., R. A. Koski, and H.-W. Yeh. 1987. Chert-hosted manganese deposits in sedimentary sequences of the Franciscan complex, Diablo Range, California. In J. R. Hein, ed. *Siliceous Sedimentary Rock-Hosted Ores and Petroleum*, pp. 206–30. New York: Van Nostrand Reinhold Company.

Hein, J. R. and J. T. Parrish. 1987. Distribution of siliceous deposits in space and time. In J. R. Hein, ed. *Siliceous Sedimentary Rock-Hosted Ores and Petroleum*, pp. 10–57. New York: Van Nostrand Reinhold Company.

Hein, J. R., H.-W. Yeh, S. H. Gunn, W. V. Sliter, L. M. Benninger, and C.-H. Wang. 1993. Two major Cenozoic episodes of phosphogenesis recorded in equatorial Pacific seamount deposits. *Paleoceanography* 8:293–311.

Heins, W. A. 1993. Source rock texture versus climate and topography as controls on the composition of modern, plutoniclastic sand. In M. J. Johnsson and A. Basu, eds., *Processes Controlling the Composition of Clastic Sediments*, pp. 135–46. Geological Society of America Special Paper 284.

Heinze, C. and W. S. Broecker. 1995. Closing off the Southern Ocean surface. *Paleoceanography* 10:49–58.

Heinze, C., E. Maier-Reimer, and K. Winn. 1991. Glacial pCO_2 reduction by the World Ocean: Experiments with the Hamburg carbon cycle model. *Paleoceanography* 6:395–430.

Henderson-Sellers, A. and K. McGuffie. 1987. *A Climate Modelling Primer*. Chichester: John Wiley and Sons.

Henrich, R. 1991. Cycles, rhythms and events on high input and low input glaciated continental margins. In H. Einsele, W. Ricken, and A. Seilacher, eds., *Cycles and Events in Stratigraphy*, pp. 751–72. Berlin: Springer-Verlag.

Herries, R. D. 1993. Contrasting styles of fluvial-aeolian interaction at a downwind erg margin: Jurassic Kayenta—Navajo transition, northeastern Arizona, USA. In C. P. North and D. J. Prosser, eds., *Characterization of Fluvial and Aeolian Reservoirs*, pp. 199–218. Geological Society of London Special Publication 73.

Hilbrecht, H., H.-W. Hubberten, and H. Oberhänsli. 1992. Biogeography of planktonic foraminifera and regional carbon isotope variations: Productivity and water masses in Late Cretaceous Europe. *Palaeogeography, Palaeoclimatology, Palaeoecology* 92:407–21.

Hildebrand, R. S. 1988. Implications of ash dispersal for tectonic models with an example from Wopmay orogen. *Geology* 16:1089–91.

Hill, D. 1959. Distribution and sequence of Silurian coral faunas. *Journal and Proceedings of the Royal Society of New South Wales* 92:151–73.

Hirschboeck, K. K. 1988. Flood hydroclimatology. In V. R. Baker, R. C. Kochel, and P. C. Patton, eds., *Flood Geomorphology*, pp. 27–47. New York: John Wiley and Sons.

Hite, R. J. 1970. Shelf carbonate sedimentation controlled by salinity in the Paradox Basin, southeast Utah. *Third Symposium on Salt* 1:48–66.

Hite, R. J., D. E. Anders, and T. G. Ging. 1984. Organic-rich source rocks of Pennsylvanian age in the Paradox Basin of Utah and Colorado. In J. Woodward, F. F. Meissner, and J. L. Clayton, eds., *Hydrocarbon Source Rocks of the Greater Rocky Mountain Region*, pp. 255–74. Denver: Rocky Mountain Association of Geologists.

Hite, R. J. and D. H. Buckner. 1981. Stratigraphic correlations, facies concepts, and cyclicity in Pennsylvanian rocks of the Paradox Basin. *1981 Field Conference, Rocky Mountain Association of Geologists*, pp. 147–59. Denver: Rocky Mountain Association of Geologists.

Hodell, D. A., G. A. Mead and P. A. Mueller. 1990. Variation in the strontium isotopic composition of seawater (8 Ma to present): Implications for chemical weathering rates and dissolved fluxes to the oceans. *Chemical Geology* 80:291–307.

Holland, H. D. 1984. *The Chemical Evolution of the Atmosphere and Oceans*. Princeton: Princeton University Press.

Holland, M. J., A. B. Cadle, R. Pinheiro, and R. M. S. Falcon. 1989. Depositional environments and coal petrography of the Permian Karoo Sequence: Witbank Coalfield, South Africa. *International Journal of Coal Geology* 11:143–69.

Hollis, C. J., K. A. Rodgers, and R. J. Parker. 1995. Siliceous plankton bloom in the earliest Tertiary of Marlborough, New Zealand. *Geology* 23:835–8.

Hoskins, B. J. 1993. The role of palaeoclimate studies: Modelling. *Philosophical Transactions of the Royal Society, Series B* 341:341–2.

Hostetler, S. W., G. Giorgi, G. T. Bates, and P. J. Bartlein. 1994. Lake-atmosphere feedbacks associated with paleolakes Bonneville and Lahontan. *Science* 263:665–8.

Hottinger, L. 1973. Selected Paleogene larger foraminifera. In A. Hallam, ed. *Atlas of Palaeobiogeography*, pp. 443–52. Amsterdam: Elsevier Scientific.

Hovan, S. A. and D. K. Rea. 1992. The Cenozoic record of continental mineral deposition on Broken and Ninetyeast Ridges, Indian Ocean: Southern African aridity and sediment delivery from the Himalayas. *Paleoceanography* 7:833–60.

Hovan, S. A., D. K. Rea, N. G. Pisias, and N. J. Shackleton. 1989. A direct link between the China loess and marine d18O records: Aeolian flux to the North Pacific. *Nature* 340:296–8.

Howell, D. G., E. R. Schermer, D. L. Jones, A. Ben-Avraham, and E. Scheibner. 1983. Tectonostratigraphic terrane map of the circum-pacific region. *U.S. Geological Survey Open-File Report* 83–716.

Hubbard, R. N. L. B. and M. C. Boulter. 1983. Reconstruction of Palaeogene climate from palynological evidence. *Nature* 301:147–50.

Huber, B. T. 1992. Paleobiogeography of Campanian-Maastrichtian foraminifera in the southern high latitudes. *Palaeogeography, Palaeoclimatology, Palaeoecology* 92:325–60.

Huber, B. T., D. A. Hodell, and C. P. Hamilton. 1995. Middle-Late Cretaceous climate of the southern high latitudes: Stable isotopic evidence for minimal equator-to-pole thermal gradients. *Geological Society of America Bulletin* 107: 1164–91.

Hudson, J. D. 1977. Stable isotopes and limestone lithification. *Journal of the Geological Society of London* 133:637–60.

———. 1994. Introduction: Oxford Clay studies. *Journal of the Geological Society of London* 151:111–2.

Hudson, J. D. and T. F. Anderson. 1989. Ocean temperatures and isotopic compositions through time. *Transactions of the Royal Society of Edinburgh* 80:183–92.

Hudson, J. D. and D. M. Martill. 1991. The Lower Oxford Clay: Production and preservation of organic matter in the Callovian (Jurassic) of central England. In R. V. Tyson and T. H. Pearson, eds., *Modern and Ancient Continental Shelf Anoxia*, pp. 363–79. Geological Society of London Special Publication 58.

———. 1994. The Peterborough Member (Callovian, Middle Jurassic) of the Oxford Clay Formation at Peterborough, UK. *Journal of the Geological Society of London* 151:113–24.

Huff, W. D., S. M. Bergström, and D. R. Kolata. 1992. Gigantic Ordovician volcanic ash fall in North America and Europe: Biological, tectonomagmatic, and event-stratigraphic significance. *Geology* 20:875–8.

———. 1993. Gigantic Ordovician volcanic ash fall in North America and Europe: Biological, tectonomagmatic, and event-stratigraphic significance: Reply. *Geology* 21:382–3.

Huggett, R. J. 1991. *Climate, Earth Processes and Earth History*. Berlin: Springer-Verlag.

Hunt, B. G. 1979. The effects of past variations of the Earth's rotation rate on climate. *Nature* 281:188–91.

Hunter, R. E., B. M. Richmond and T. R. Alpha. 1983. Storm-controlled oblique dunes of the Oregon coast. *Geological Society of America Bulletin* 94:1450–65.

Hunter, R. E. and D. M. Rubin. 1983. Interpreting cyclic crossbedding, with an example from the Navajo Sandstone. In M. E. Brookfield and T. S. Ahlbrandt, eds., *Eolian Sediments and Processes*, pp. 429–54 Developments in Sedimentology 38. Amsterdam: Elsevier.

Hyde, W. T., W.-Y. Kim, and T. J. Crowley. 1990. On the relation between polar continentality and climate: Studies with a nonlinear seasonal energy balance model. *Journal of Geophysical Research* 95:18653–68.

Imbrie, J. and N. G. Kipp. 1971. A new micropaleontological method for quantitative paleoclimatology. In K. Turekian, ed. *The Late Cenozoic Glacial Ages*, pp. 71–181. New Haven: Yale University Press.

Imlay, R. W. 1980. Jurassic paleobiogeography of the conterminous United States in its continental setting. *United States Geological Survey Professional Paper* 1062:134 p.

Ingall, E. D., R. M. Bustin, and P. Van Cappellen. 1993. Influence of water column anoxia on the burial and preservation of carbon and phosphorus in marine shale. *Geochimica et Cosmochimica Acta* 57:303–16.

Ingall, E. D. and R. A. Jahnke. 1994. Evidence for enhanced phosphorus regeneration from marine sediments overlain by oxygen-depleted waters. *Geochimica et Cosmochimica Acta* 58:2571–5.

Ingram, B. L. and D. J. DePaolo. 1993. A 4300 year strontium isotope record of estuarine paleosalinity in San Francisco Bay, California. *Earth and Planetary Science Letters* 119:103–19.

Isaacs, C. M., K. A. Pisciotto, and R. E. Garrison. 1983. Facies and diagenesis of the Miocene Monterey Formation, California: A summary. In A. Iijima, J. R. Hein, and R. Siever, eds., *Siliceous Deposits in the Pacific Region*, pp. 247–82. Amsterdam: Elsevier.

Jacobs, B. F. and A. L. Deino. 1996. Test of climate-leaf physiognomy regression models, their application to two Miocene floras from Kenya, and $^{40}Ar/^{39}Ar$ dating of the Late Miocene Kapturo site. *Palaeogeography, Palaeoclimatology, Palaeoecology* 123:259–71.

Jahnke, R. A., S. R. Emerson, K. K. Roe, and W. C. Burnett. 1983. The present day formation of apatite in Mexican continental margin sediments. *Geochimica et Cosmochimica Acta* 47:259–66.

Jahnke, R. A. and G. B. Shimmield. 1995. Particle flux and its conversion to the sediment record: Coastal ocean upwelling systems. In C. P. Summerhayes, K.-C. Emeis, M. V. Angel, R. L. Smith, and B. Zeitzschel, eds., *Upwelling in the Ocean:*

Modern Processes and Ancient Records, pp. 83–100. Chichester: John Wiley and Sons.

James, N. P., Y. Bone, and T. K. Kyser. 1997. Brachiopod $\delta^{18}O$ values do reflect ambient oceanography: Lacepede Shelf, southern Australia. *Geology* 25:551–4.

James, N. P. and P. W. Choquette. 1984. Diagenesis—9. Limestones—the meteoric diagenetic environment. *Geoscience Canada* 11:161–95.

James, W. C., G. H. Mack, and H. C. Monger. 1993. Classification of paleosols: Reply. *Geological Society of America Bulletin* 105:1637.

Janecek, T. R. and D. K. Rea. 1983. Eolian deposition in the northeast Pacific Ocean: Cenozoic history of atmospheric circulation. *Geological Society of America Bulletin* 94:730–8.

Janis, C. M. 1984. The use of fossil ungulate communities as indicators of climate and environment. In P. J. Brenchley, ed. *Fossils and Climate*, pp. 85–104. Chichester: John Wiley and Sons.

Jansen, J. H. F., C. F. Woensdregt, M. J. Kooistra, and S. J. van der Gaast. 1987. Ikaite pseudomorphs in the Zaire deep-sea fan: An intermediate between calcite and porous calcite. *Geology* 15:245–8.

Janvier, P. 1995. News and Views: Conodonts join the club. *Nature* 374:761–3.

Jarvis, I., G. Carson, M. Hart, P. Leary, and B. Tocher. 1988a. The Cenomanian-Turonian (late Cretaceous) anoxic event in SW England: Evidence from Hooken Cliffs near Beer, SE Devon. *Newsletter Stratigraphy* 18:147–64.

Jarvis, I., G. A. Carson, M. K. E. Cooper, M. B. Hart, P. N. Leary, B. A. Tocher, D. Horne, and A. Rosenfeld. 1988b. Microfossil assemblages and the Cenomanian-Turonian (late Cretaceous) oceanic anoxic event. *Cretaceous Research* 9:3–103.

Jasper, J. P. and R. B. Gagosian. 1990. The sources and deposition of organic matter in the Late Quaternary Pigmy Basin, Gulf of Mexico. *Geochimica et Cosmochimica Acta* 54:1117–32.

Jasper, J. P. and J. M. Hayes. 1990. A carbon isotope record of CO_2 levels during the late Quaternary. *Nature* 347:462–4.

Jefferson, T. H. 1982. Fossil forests from the Lower Cretaceous of Alexander Island, Antarctica. *Palaeontology* 25:681–708.

Jeffries, R. P. S. and P. Minton. 1965. The mode of life of two Jurassic species of "*Posidonia*" (Bivalvia). *Palaeontology* 8:156–85.

Jendrzejewski, J. P. and G. A. Zarillo. 1972. Late Pleistocene paleotemperature oscillations defined by silicoflagellate changes in a subantarctic deep-sea core. *Deep-Sea Research* 19:327–9.

Jenkyns, H. C. and C. J. Clayton. 1986. Black shales and carbon isotopes in pelagic sediments from the Tethyan Lower Jurassic. *Sedimentology* 33:87–106.

Jenkyns, H. C., A. S. Gale, and R. M. Corfield. 1994. Carbon- and oxygen-isotope stratigraphy of the English chalk and Italian Scaglia and its palaeoclimatic significance. *Geological Magazine* 131: 1–34.

Jenkyns, H. C. and E. L. Winterer. 1982. Palaeoceanography of Mesozoic ribbon radiolarites. *Earth and Planetary Science Letters* 60:351–75.

Jerzykiewicz, T. and A. R. Sweet. 1987. Semiarid floodplain as a paleoenvironmental setting of the Upper Cretaceous dinosaurs: Sedimentological evidence from Mongolia and Alberta. Paper read at *Fourth Symposium on Mesozoic Terrestrial Ecosystems, Short Papers*, at Drumheller, Alberta.

Jewell, P. W. 1994. Paleoredox conditions and the origin of bedded barites along the Late Devonian North American continental Margin. *Journal of Geology* 102:151–64.

———. 1995. Geologic consequences of globe-encircling equatorial currents. *Geology* 23:117–20.

Jewell, P. W. and R. F. Stallard. 1991. Geochemistry and paleoceanographic setting of central Nevada bedded barites. *Journal of Geology* 99:151–79.

Johnson, C. C. 1997. Middle Cretaceous reef collapse linked to ocean heat transport: Reply. *Geology* 25:478–9.

Johnson, C. C., E. J. Barron, E. G. Kauffman, M. A. Arthur, P. J. Fawcett, and M. K. Yasuda. 1996. Middle Cretaceous reef collapse linked to ocean heat transport. *Geology* 24:376–80.

Johnson, L. R. 1979. Mineral dispersal patterns of North Atlantic deep-sea sediments with particular reference to eolian dusts. *Marine Geology* 30:335–45.

Johnson, S. Y. 1985. Eocene strike-slip faulting and nonmarine basin formation in Washington. In K. T. Biddle and N. Christie-Blick, eds., *Strike-Slip Deformation, Basin Formation, and Sedimentation*, pp. 283–302. Society of Economic Paleontologists and Mineralogists Special Publication 37.

———. 1989. Significance of loessite in the Maroon Formation (Middle Pennsylvanian to Lower Permian), Eagle Basin, Northwest Colorado. *Journal of Sedimentary Petrology* 59:782–91.

Johnsson, M. J. 1990. Tectonic versus chemical-weathering controls on the composition of fluvial sands in tropical environments. *Sedimentology* 37:713–26.

———. 1992. Chemical weathering controls on sand composition. *Encyclopedia of Earth System Science* 1:455–66.

Johnsson, M. J., R. F. Stallard, and R. H. Meade. 1988. First-cycle quartz arenites in the Orinoco River basin, Venezuela and Colombia. *Journal of Geology* 96:263–77.

Jones, C. E., H. C. Jenkyns, A. L. Coe, and S. P. Hesselbo. 1994. Strontium isotopic variations in Jurassic and Cretaceous seawater. *Geochimica et Cosmochimica Acta* 58:3061–74.

Jones, C. E., H. C. Jenkyns, and S. P. Hesselbo. 1994. Strontium isotopes in Early Jurassic seawater. *Geochimica et Cosmochimica Acta* 58:1285–301.

Jones, D. S. and I. R. Quitmyer. 1996. Marking time with bivalve shells: Oxygen isotopes and season of annual increment formation. *Palaios* 11: 340–6.

Jousé, A. P., O. G. Kozlova, and V. V. Muhina. 1971. Distribution of diatoms in the surface layer of sediment from the Pacific Ocean. In B. M. Funnell and W. R. Riedel, eds., *The Micropalaeontology of Oceans*, pp. 263–9. Cambridge: Cambridge University Press.

Juillet-Leclerc, A. and L. Labeyrie. 1987. Temperature dependence of the oxygen isotopic fractionation between diatom silica and water. *Earth and Planetary Science Letters* 84:69–74.

Juillet-Leclerc, A. and H. Schrader. 1987. Variations of upwelling intensity recorded in varved sediment from the Gulf of California during the past 3,000 years. *Nature* 329:146–9.

Kaiho, K. 1994. Benthic foraminiferal dissolved-oxygen index and dissolved-oxygen levels in the modern ocean. *Geology* 22:719–22.

Kalkreuth, W. D., D. J. McIntyre, and R. J. H. Richardson. 1993. The geology, petrography and palynology of Tertiary coals from the Eureka Sound Group at Strathcona Fiord and Bache Peninsula, Ellesmere Island, Arctic Canada. *International Journal of Coal Geology* 24:75–111.

Kanaya, T. and I. Koizumi. 1966. Interpretation of diatom thanatocoenoses from the North Pacific applied to a study of core V20–130 (Studies of a deep-sea cores V20–130, part IV). *Scientific Reports, Tohoku University* 7:89–130.

Karhu, J. and S. Epstein. 1986. The implication of the oxygen isotope records in coexisting cherts and phosphates. *Geochimica et Cosmochimica Acta* 50:1745–56.

Kauffman, E. G. 1973. Cretaceous Bivalvia. In A. Hallam, ed. *Atlas of Palaeobiogeography*, pp. 353–82. Amsterdam: Elsevier.

———. 1978. Benthic environments and paleoecology of the Posedonienschiefer (Toarcian). *Neues Jahrbuch für Geologie und Paläontologie Abhandlung* 157:18–36.

———. 1981. Ecological reappraisal of the German Posidonienschiefer (Toarcian) and the stagnant basin model. In J. Gray, A. J. Boucot, and W. B. N. Berry, eds., *Communities of the Past*, pp. 311–81. Stroudsburg, PA: Hutchinson Ross.

Keany, J., M. Ledbetter, N. Watkins, and T.-C. Huang. 1976. Diachronous deposition of ice-rafted debris in sub-Antarctic deep-sea sediment. *Geological Society of America Bulletin* 87: 873–82.

Keir, R. S. 1988. On the late Pleistocene ocean geochemistry and circulation. *Paleoceanography* 3:413–45.

Keller, A. M. and M. S. Hendrix. 1997. Paleoclimatologic analysis of a Late Jurassic petrified forest, southeastern Mongolia. *Palaios* 12:282–91.

Keller, G. 1985. Depth stratification of planktonic foraminifers in the Miocene ocean. In J. P. Kennett, ed. *The Miocene Ocean: Paleoceanography and Biogeography*, pp. 177–95. Geological Society of America Memoir 163.

Keller, G. and J. A. Barron. 1983. Paleoceanographic implications of Miocene deep-sea hiatuses. *Geological Society of America Bulletin* 94:590–613.

Keller, G. and M. Lindinger. 1989. Stable isotope, TOC and $CaCO_3$ record across the Cretaceous/Tertiary boundary at El Kef, Tunisia. *Palaeogeography, Palaeoclimatology, Palaeoecology* 73: 243–65.

Keller, G., C. E. Zenker, and S. M. Stone. 1989. Late Neogene history of the Pacific-Caribbean gateway. *Journal of South American Earth Sciences* 2:73–108.

Kelley, P. H., A. Raymond, and C. B. Lutken. 1990. Carboniferous brachiopod migration and latitudinal diversity: A new palaeoclimatic method. In W. S. McKerrow and C. R. Scotese, eds., *Palaeozoic Palaeogeography and Biogeography*, pp. 325–32. Geological Society Memoir 12.

Kellogg, T. B. 1976. Late Quaternary climatic changes: Evidence from deep-sea cores of Norwegian and Greenland Seas. In R. M. Cline and J. D. Hays, eds., *Investigation of Late Quaternary Paleoceanography and Paleoclimatology*, pp. 77–110. Geological Society of America Memoir 145.

Kellogg, W. W. and Z.-C. Zhao. 1988. Sensitivity of soil moisture to doubling of carbon dioxide in climate model experiments. Part I. North America. *Journal of Climate* 1:348–78.

Kelts, K. 1988. Environments of deposition of lacustrine petroleum source rocks: An introduction. In A. J. Fleet, K. Kelts, and M. R. Talbot, eds., *Lacustrine Petroleum Source Rocks*, pp. 3–26. Geological Society of London Special Publication 40.

Kemp, A. E. S. 1995. Laminated sediments from coastal and open ocean upwelling zones: What variability do they record? In C. P. Summerhayes, K.-C. Emeis, M. V. Angel, R. L. Smith, and B. Zeitzschel, eds., *Upwelling in the Ocean: Modern Processes and Ancient Records*, pp. 239–57. Chichester: John Wiley and Sons.

Kemper, E. 1987. Das Klima der Kreide-Zeit. *Geologisches Jahrbuch, Reihe A* 96:5–185.

Kendall, A. C. 1988. Aspects of evaporite basin stratigraphy. In B. C. Schreiber, ed. *Evaporites and Hydrocarbons*, pp. 11–65. New York: Columbia University Press.

Kenig, F., J. M. Hayes, B. N. Popp, and R. E. Summons. 1994. Isotopic biogeochemistry of the Oxford Clay Formation (Jurassic), UK. *Journal of the Geological Society of London* 151:139–52.

Kennedy, G. L., D. M. Hopkins, and W. J. Pickthorn. 1987. Ikaite, the glendonite precursor, in estuarine sediments at Barrow, Arctic Alaska (abs.). *Geological Society of America Abstracts with Programs* 19:735.

Kennett, D. M. and J. P. Kennett. 1990. *Bolboforma* Daniels and Spiegler, from Eocene and lower Oligocene sediments, Maud Rise, Antarctica. In P. F. Barker, J. P. Kennett, et al. *Proceedings of the Ocean Drilling Program, Scientific Results, Leg 113*, pp. 667–73. College Station, TX: Ocean Drilling Program.

Kennett, J. P. 1977. Cenozoic evolution of Antarctic glaciation, the circum-polar Antarctic Ocean, and their impact on global paleoceanography. *Journal of Geophysical Research* 82:3843–60.

———. 1982. *Marine Geology*. Englewood Cliffs, NJ: Prentice-Hall.

Kennett, J. P. and P. F. Barker. 1990. Latest Cretaceous to Cenozoic climate and oceanographic developments in the Weddell Sea, Antarctica: An ocean-drilling perspective. In P. F. Barker, J. P. Kennett, et al., *Proceedings of the Ocean Drilling Program, Scientific Results, Vol. 113*, pp. 937–60.

Kennett, J. P., G. Keller, and M. S. Srinivasan. 1985. Miocene planktonic foraminiferal biogeography and paleoceanographic development of the Indo-Pacific region. In J. P. Kennett, ed. *The Miocene Ocean: Paleoceanography and Biogeography*, pp. 197–236. Geological Society of America Memoir 163.

Kennett, J. P. and N. J. Shackleton. 1976. Oxygen isotopic evidence for the development of the psychrosphere 38 Myr ago. *Nature* 260:513–5.

Kennett, J. P. and L. D. Stott. 1990. Proteus and proto-oceanus: Ancestral Paleogene oceans as revealed from Antarctic stable isotopic results: ODP Leg 113. In P. F. Barker, J. P. Kennett, et al., *Proceedings of the Ocean Drilling Program, Scientific Results, Vol. 113*, pp. 865–80.

———. 1991. Abrupt deep-sea warming, palaeoceanographic changes and benthic extinctions at the end of the Palaeocene. *Nature* 353:225–9.

Kennett, J. P. and K. Venz. 1995. Late Quaternary climatically related planktonic foraminiferal assemblage changes: ODP Hole 893A, Santa Barbara Basin, California. In J. P. Kennett, J. G. Baldauf, and M. Lyle eds., *Proceedings of the Ocean Drilling Program, Scientific Results, Part 2, Leg 146*, pp. 281–325. College Station, TX: Ocean Drilling Program.

Kenny, R. and L. P. Knauth. 1992. Continental paleoclimates from δD and $\delta^{18}O$ of secondary silica in paleokarst chert lags. *Geology* 20:219–22.

Kent, D., N. D. Opdyke, and M. Ewing. 1971. Climate change in the North Pacific using ice-rafted detritus as a climatic indicator. *Geological Society of America Bulletin* 82:2741–54.

Kimura, T. and W. Sekido. 1975. *Nilssoniocladus* n. gen. (Nilssoniaceae n. fam.) newly found from the early Lower Cretaceous of Japan. *Palaeontographica* 153-B:111–8.

Kinsman, D. J. J. 1976. Sabkha facies nodular anhydrite and paleotemperature determination (abs.). *American Association of Petroleum Geologists Bulletin* 60:688.

Kipp, N. G. 1976. New transfer function for estimating past sea-surface conditions from sea-bed distribution of planktonic foraminiferal assemblages in the North Atlantic. In R. M. Cline and J. D. Hays, eds., *Investigation of Late Quaternary Paleoceanography and Paleoclimatology*, pp. 3–41. Geological Society of America Memoir 145.

Klappa, C. F. 1980. Rhizoliths in terrestrial carbonates: Classification, recognition, genesis and significance. *Sedimentology* 27:613–29.

Klapper, G. and J. G. Johnson. 1980. Endemism and dispersal of Devonian conodonts. *Journal of Paleontology* 54:400–55.

Klein, G. D. and D. A. Willard. 1989. Origin of the Pennsylvanian coal-bearing cyclothems of North America. *Geology* 17:152–5.

Klein, G. D. and K. M. Marsaglia. 1987. Hummocky cross-stratification, tropical hurricanes, and intense winter storms—Discussion. *Sedimentology* 34:333–7.

Kling, S. A. 1979. Vertical distribution of polycystine radiolarians in the central North Pacific. *Marine Micropaleontology* 4:295–318.

Knauth, L. P. and S. Epstein. 1976. Hydrogen and oxygen isotope ratios in nodular and bedded cherts. *Geochimica et Cosmochimica Acta* 40: 1095–108.

Knox, J. C. 1993. Large increases in flood magnitude in response to modest changes in climate. *Nature* 361:430–2.

Koç Karpuz, N. and E. Jansen. 1992. A high-resolution diatom record of the last deglaciation from the SE Norwegian Sea: Documentation of rapid climatic changes. *Paleoceanography* 7:499–520.

Koch, P. L., J. C. Zachos, and D. L. Dettman. 1995. Stable isotope stratigraphy and paleoclimatology of the Paleogene Bighorn Basin (Wyoming, USA). *Palaeogeography, Palaeoclimatology, Palaeoecology* 115:61–89.

Koch, P. L., J. C. Zachos, and P. D. Gingerich. 1992. Correlation between isotope records in marine and continental carbon reservoirs near the Palaeocene/Eocene boundary. *Nature* 358:319–22.

Kochel, R. C. 1990. Humid fans of the Appalachian Mountains. In A. H. Rachocki and M. Church, eds., *Alluvial Fans: A Field Approach*, pp. 109–29. Chichester: John Wiley and Sons.

Kocurek, G. 1988. First-order and super bounding surfaces in eolian sequences—bounding surfaces revisited. *Sedimentary Geology* 56:193–206.

———. 1991. Interpretation of ancient eolian sand dunes. *Annual Review of Earth and Planetary Sciences* 19:43–75.

———. 1996. Desert aeolian systems. In H. Reading, ed. *Sedimentary Environments: Processes, Facies and Stratigraphy*, pp. 125–53. Oxford: Blackwell Scientific.

Kocurek, G. and K. G. Havholm. 1993. Eolian sequence stratigraphy—A conceptual framework. In P. Weimer and Posamentier, eds., *Recent Advances in and Applications of Siliciclastic Sequence Stratigraphy*, pp. 393–409. American Association of Petroleum Geologists Memoir 58.

Kocurek, G. and J. Nielson. 1986. Conditions favourable for the formation of warm-climate aeolian sand sheets. *Sedimentology* 33:795–816.

Kolodny, Y. and B. Luz. 1991. Oxygen isotopes in phosphates of fossil fish—Devonian to Recent. In H. P. Taylor, Jr., J. R. O'Neil, and I. R. Kaplan, eds., *Stable Isotope Geochemistry: A Tribute to Samuel Epstein*, pp. 105–19. Geochemical Society Special Publications 3.

Kolodny, Y., B. Luz, and O. Navon. 1983. Oxygen isotope variations in phosphate of biogenic apatites, I. Fish bone apatite—rechecking the rules of the game. *Earth and Planetary Science Letters* 64:398–404.

Kolodny, Y., B. Luz, M. Sander, and W. A. Clemens. 1996. Dinosaur bones: Fossils or pseudomorphs? The pitfalls of physiology reconstruction from apatitic fossils. *Palaeogeography, Palaeoclimatology, Palaeoecology* 126:161–71.

Kolodny, Y. and M. Raab. 1988. Oxygen isotopes in phosphatic fish remains from Israel: Paleothermometry of tropical Cretaceous and Tertiary shelf waters. *Palaeogeography, Palaeoclimatology, Palaeoecology* 64:59–67.

Kominz, M. A. and G. C. Bond. 1990. A new method of testing periodicity in cyclic sediments: Application to the Newark Supergroup. *Earth and Planetary Science Letters* 98:233–44.

Konikow, L. F. and J. D. Bredehoeft. 1992. Groundwater models cannot be validated. *Advances in Water Resources* 15:75–83.

Köppen, A. and A. Wegener. 1924. *Die Klimate der geologischen Vorzeit*. Berlin: Borntraeger.

Kotilainen, A. T. and N. J. Shackleton. 1995. Rapid climate variability in the North Pacific Ocean during the past 95,000 years. *Nature* 377:323–6.

Kovach, W. L. and R. A. Spicer. 1996. Canonical correspondence analysis of leaf physiognomy: A contribution to the development of a new palaeoclimatological tool. *Palaeoclimates: Data and Modelling* 1:125–38.

Krasilov, V. A. 1975. *Paleoecology of Terrestrial Plants*. Translated by Hilary Hardin. Israel Program for Scientific Translations. New York: John Wiley and Sons.

Krassilov, V. A. 1992. Coal-bearing deposits of the Soviet Far East. In P. J. McCabe and J. T. Parrish, eds., *Controls on the Distribution and Quality of Cretaceous Coals*, pp. 263–7. Geological Society of America Special Paper 267.

Kraus, M. J. 1997. Lower Eocene alluvial paleosols: Pedogenic development, stratigraphic relationships, and paleosol/landscape associations. *Palaeogeography, Palaeoclimatology, Palaeoecology* 129:387–406.

Kraus, M. J. and T. M. Bown. 1988. Pedofacies analysis; a new approach to reconstructing ancient fluvial sequences. In J. Reinhardt and W. R.

Sigleo, eds., *Paleosols and Weathering Through Geologic Time: Principles and Applications*, pp. 143–52. Geological Society of America Special Paper 216.

Kreisa, R. D. 1981. Storm-generated sedimentary structures in subtidal marine facies with examples from the Middle and Upper Ordovician of southwestern Virginia. *Journal of Sedimentary Petrology* 51:823–48.

Krinsley, D. and P. Trusty. 1985. Environmental interpretation of quartz grain surface textures. In G. G. Zuffa, ed. *Provenance of Arenites*, pp. 213–29. Dordrecht: D. Reidel.

Krishnamurthy, R. V., S. K. Bhattacharya, and S. Kusumgar. 1986. Palaeoclimatic changes deduced from 13C/12C and C/N ratios of Karewa lake sediments, India. *Nature* 323:150–2.

Kroon, D. and A. J. Nederbragt. 1990. Ecology and paleoecology of triserial planktic foraminifera. *Marine Micropaleontology* 16:25–38.

Kröpelin, S. and I. Soulié-Märsche. 1991. Charophyte remains from Wadi Howar as evidence for deep mid-Holocene freshwater lakes in the eastern Sahara of northwest Sudan. *Quaternary Research* 36:210–23.

Kruijs, E. and E. Barron. 1990. Climate model prediction of paleoproductivity and potential source-rock distribution. In A. Y. Huc, ed. *Deposition of Organic Facies*, pp. 195–216. American Association of Petroleum Geologists Studies in Geology 30.

Kumagai, H., T. Sweda, K. Hayashi, S. Kojima, J. F. Basinger, M. Shibuya, and Y. Fukaoa. 1995. Growth-ring analysis of Early Tertiary conifer woods from the Canadian High Arctic and its paleoclimatic interpretation. *Palaeogeography, Palaeoclimatology, Palaeoecology* 116:247–62.

Kumar, N., R. Gwiazda, R. F. Anderson, and P. N. Froelich. 1993. $^{231}Pa/^{230}Th$ ratios in sediments as a proxy for past changes in Southern Ocean productivity. *Nature* 362:45–8.

Kump, L. R. 1990. Neogene geochemical cycles: Implications concerning phosphogenesis. In W. C. Burnett and S. R. Riggs, eds., *Phophate Deposits of the World. Vol. 3, Neogene to Modern Phosphorites*, pp. 273–82. Cambridge: Cambridge University Press.

———. 1993. The coupling of the carbon and sulfur biogeochemical cycles over Phanerozoic time. In R. Wollast, F. T. Mackenzie, and L. Chou, eds., *Interactions of C, N, P and S Biogeochemical Cycles and Global Change*, pp. 475–90. NATO ASI Series I, Global Environmental Change 4. Berlin: Springer-Verlag.

Kupecz, J. A. 1995. Depositional setting, sequence stratigraphy, diagenesis, and reservoir potential of a mixed-lithology, upwelling deposit: The Upper Triassic Shublik Formation, Prudhoe Bay, Alaska. *American Association of Petroleum Geologists Bulletin* 79:1301–19.

Kutzbach, J. E. 1983. Monsoon rains of the late Pleistocene and early Holocene: Patterns, intensity and possible causes of changes. In F. A. Street-Perrott, M. Beran, and R. Ratcliffe, eds., *Variations in the Global Water Budget*, pp. 371–89. Dordrecht: D. Reidel.

———. 1985. Modeling of paleoclimates. *Advances in Geophysics* 28A:159–96.

Kutzbach, J. E. and R. G. Gallimore. 1989. Pangean climates: Megamonsoons of the megacontinent. *Journal of Geophysical Research* 94:3341–57.

Kutzbach, J. E., R. G. Gallimore, and P. J. Guetter. 1991. Sensitivity experiments on the effect of orbitally caused insolation changes on the interglacial climate of high northern latitudes. *Quaternary International* 10–12:223–9.

Kutzbach, J. E. and P. J. Guetter. 1986. The influence of changing orbital parameters and surface boundary conditions on climate simulations for the past 18 000 years. *Journal of the Atmospheric Sciences* 43:1726–59.

Kutzbach, J. E., P. J. Guetter, and W. M. Washington. 1990. Simulated circulation of an idealized ocean for Pangaean time. *Paleoceanography* 5:299–317.

Kutzbach, J. E. and B. L. Otto-Bliesner. 1982. The sensitivity of African-Asian monsoonal climate to orbital parameter changes for 9000 years B. P. in a low-resolution general circulation model. *Journal of the Atmospheric Sciences* 39:1177–88.

Kutzbach, J. E., W. L. Prell, and W. F. Ruddiman. 1993. Sensitivity of Eurasian climate to surface uplift of the Tibetan Plateau. *Journal of Geology* 101:177–90.

Kutzbach, J. E. and F. A. Street-Perrott. 1985. Milankovitch forcing of fluctuations in the level of tropical lakes from 18 to 0 kyr BP. *Nature* 317: 130–4.

Kutzbach, J. E. and A. M. Ziegler. 1993. Simulation of Late Permian climate and biomes with an atmosphere-ocean model: Comparisons with observations. *Philosophical Transactions of the Royal Society, Series B* 341:327–40.

Kvale, E. P., G. S. Fraser, A. W. Archer, A. Zawistoski, N. Kemp, and P. McGough. 1994. Evidence of seasonal precipitation in Pennsylvanian sediments of the Illinois Basin. *Geology* 22:331–4.

Kyte, F. T., M. Leinen, G. R. Heath, and L. Zhou. 1993. Cenozoic sedimentation history of the central North Pacific: Inferences from the elemental geochemistry of core LL44-GPC3. *Geochimica et Cosmochimica Acta* 57:1719–40.

Labandeira, C. C. and J. J. Sepkoski, Jr. 1993. Insect diversity in the fossil record. *Science* 261:310–5.

Lagoe, M. B. 1987. The stratigraphic record of sea-level and climatic fluctuations in an active-margin basin: The Stevens Sandstone, Coles Levee area, California. *Palaios* 2:48–68.

Lagoe, M. B., C. H. Eyles, N. Eyles, and C. Hale. 1993. Timing of late Cenozoic tidewater glaciation in the far North Pacific. *Geological Society of America Bulletin* 105:1542–60.

Lamberson, M. N., R. M. Bustin, W. D. Kalkreuth, and K. C. Pratt. 1996. The formation of inertinite-rich peats in the mid-Cretaceous Gates Formation: Implications for the interpretation of mid-Albian history of paleowildfire. *Palaeogeography, Palaeoclimatology, Palaeoecology* 120: 235–60.

Lambiase, J. J. 1990. A model for tectonic control of lacustrine stratigraphic sequences in continental rift basins. In B. J. Katz, ed. *Lacustrine Basin Exploration: Case Studies and Modern Analogs*, pp. 265–76. American Association of Petroleum Geologists Memoir 50.

Lancaster, N. 1983. Controls of dune morphology in the Namib Sand Sea. In M. E. Brookfield and T. S. Ahlbrandt, eds., *Eolian Sediments and Processes*, pp. 261–89. Developments in Sedimentology 38. Amsterdam: Elsevier.

———. 1993. Origins and sedimentary features of supersurfaces in the northwestern Gran Desierto Sand Sea. In K. Pye and N. Lancaster, eds., *Aeolian Sediments, Ancient and Modern*, pp. 71–83. Special Publications of the International Association of Sedimentologists 16. Oxford: Blackwell Scientific.

Lancaster, N. and J. T. Teller. 1988. Interdune deposits of the Namib Sand Sea. *Sedimentary Geology* 55:91–107.

Land, L. S. 1995. Comment on "Oxygen and carbon isotopic composition of Ordovician brachiopods: Implications for coeval seawater" by H. Qing and J. Veizer. *Geochimica et Cosmochimica Acta* 59:2843–4.

Land, L. S., J. C. Lang, and D. J. Barnes. 1975. Extension rate: A primary control on the isotopic composition of West Indian (Jamaican) scleractinian reef coral skeletons. *Marine Biology* 33: 221–33.

Langbein, W. B. and S. A. Schumm. 1958. Yield of sediment in relation to mean annual precipitation. *Eos* 39:1076–84.

Langford, R. P. and M. A. Chan. 1989. Fluvial-eolian interactions: Part II, ancient systems. *Sedimentology* 36:1037–51.

Lasaga, A. C., R. A. Berner, and R. M. Garrels. 1985. An improved geochemical model of atmospheric CO_2 fluctuations over the past 100 million years. In E. T. Sundquist and W. S. Broecker, eds., *The Carbon Cycle and Atmospheric CO_2: Natural Variations Archean to Present*, pp. 397–411. Geophysical Monograph 32. Washington, D.C.: American Geophysical Union.

Lawrence, D. T. 1992. Primary controls on total reserves, thickness, geometry, and distribution of coal seams: Upper Cretaceous Adaville Formation, southwestern Wyoming. In P. J. McCabe and J. T. Parrish, eds., *Controls on the Distribution and Quality of Cretaceous Coals*, pp. 69–100. Geological Society of America Special Paper 267.

Lawrence, J. R. and J. W. C. White. 1991. The elusive climate signal in the isotopic composition of precipitation. In H. P. Taylor, J. R. O'Neil, and I. R. Kaplan, eds., *Stable Isotope Geochemistry: A Tribute to Samuel Epstein*, pp. 169–93. Geochemical Society, Special Publication 3.

Lea, D. W. and E. A. Boyle. 1990. Foraminiferal reconstruction of barium distributions in water masses of the glacial oceans. *Paleoceanography* 5:719–42.

———. 1991. Barium in planktonic foraminifera. *Geochimica et Cosmochimica Acta* 55:3321–31.

Lea, D. W. and H. J. Spero. 1994. Assessing the reliability of paleochemical tracers: Barium uptake in the shells of planktonic foraminifera. *Paleoceanography* 9:445–52.

Leckie, R. M. 1985. Foraminifera of the Cenomanian-Turonian boundary interval, Greenhorn Formation, Rock Canyon Anticline, Pueblo, Colorado. In L. M. Pratt, E. G. Kauffman, and F. B. Zelt, eds., *Fine-Grained Deposits and Biofacies of the Cretaceous Western Interior Seaway: Evidence of Cylic Sedimentary Processes*, pp. 139–50. Society of Economic Paleontologists and Mineralogists, Field Trip Guide Book 4.

———. 1989. A paleoceanographic model for the early evolutionary history of planktonic foraminifera. *Palaeogeography, Palaeoclimatology, Palaeoecology* 73:107–38.

Leckie, R. M., M. G. Schmidt, D. Finkelstein, and R. Yuretich. 1991. Paleoceanographic and paleoclimatic interpretations of the Mancos Shale (Upper Cretaceous), Black Mesa Basin, Arizona. In J. D. Nations and J. G. Eaton, eds., *Stratigraphy, Depositional Environments, and Sedimentary Tectonics of the Western Margin, Cretaceous Western Interior Seaway*, pp. 139–52. Geological Society of America Special Paper 260.

Lécuyer, C., P. Grandjean, J. R. O'Neil, H. Cappetta, and F. Martineau. 1993. Thermal excursions in the ocean at the Cretaceous-Tertiary boundary (northern Morocco): $\delta^{18}O$ record of phosphatic fish debris. *Palaeogeography, Palaeoclimatology, Palaeoecology* 105:235–43.

Lécuyer, C., P. Grandjean, F. Paris, M. Robardet, and D. Robineau. 1996. Deciphering "temperature" and "salinity" from biogenic phosphates: The $\delta^{18}O$ of coexisting fishes and mammals of the Middle Miocene sea of western France. *Palaeogeography, Palaeoclimatology, Palaeoecology* 126:61–74.

Lee, T. N., J. A. Yoder, and L. P. Atkinson. 1991. Gulf Stream frontal eddy influence on productivity of the Southeast U.S. continental shelf. *Journal of Geophysical Research* 96:22,191–22,205.

Lees, A. 1975. Possible influences of salinity and temperature on modern shelf carbonate sedimentation. *Marine Geology* 13:M67-M73.

Lees, A. and A. T. Buller. 1972. Modern temperate-water and warm-water shelf carbonate sediments contrasted. *Marine Geology* 13:M67-M73.

Legendre, S. 1986. Analysis of mammalian communities from the late Eocene and Oligocene of southern France. *Palaeovertebrata* 16:191–212.

———. 1987. Mammalian faunas as paleotemperature indicators: Concordance between oceanic and terrestrial paleontological evidence. *Evolutionary Theory* 8:77–86.

Lehman, T. M. 1989. Upper Cretaceous (Maastrichtian) paleosols in Trans-Pecos Texas. *Geological Society of America Bulletin* 101:188–203.

Leinen, M. 1979. Biogenic silica accumulation in the central equatorial Pacific and its implications for Cenozoic paleoceanography: Summary. *Geological Society of America Bulletin* 90:801–3.

Leinen, M. and G. R. Heath. 1981. Sedimentary indicators of atmospheric activity in the Northern Hemisphere during the Cenozoic. *Palaeogeography, Palaeoclimatology, Palaeoecology* 36:1–21.

Leithold, E. L. 1993. Preservation of laminated shale in ancient clinoforms; comparison to modern subaqueous deltas. *Geology* 21:359–62.

Lemoalle, J. and B. Dupont. 1973. Iron-bearing oolites and the present conditions of iron sedimentation in Lake Chad (Africa). In G. C. Amstutz and A. J. Bernard, eds., *Ores in Sediments*, pp. 167–78. International Union of Geological Sciences, Series A(3). Berlin: Springer-Verlag.

Lentz, S. J. 1992. The surface boundary layer in coastal upwelling regions. *Journal of Physical Oceanography* 22:1517–39.

———. 1994. Current dynamics over the northern California inner shelf. *Journal of Physical Oceanography* 24:2461–78.

Leonard, J. E., B. Cameron, O. H. Pilkey, and G. M. Friedman. 1981. Evaluation of cold-water carbonates as a possible paleoclimatic indicator. *Sedimentary Geology* 28:1–28.

Leopold, E. B. and G. Liu. 1994. A long pollen sequence of Neogene age, Alaska Range. *Quaternary International* 22/23:103–40.

Lethiers, F. and R. Whatley. 1994. The use of Ostracoda to reconstruct the oxygen levels of Late Palaeozoic oceans. *Marine Micropaleontology* 24:57–69.

Leventer, A. and R. B. Dunbar. 1988. Recent diatom record of McMurdo Sound, Antarctica: Implications for history of sea ice extent. *Paleoceanography* 3:259–74.

Lieth, H. 1975. Primary production of the major vegetation units of the world. In H. Lieth and R. H. Whittaker, eds., *Primary Productivity of the Biosphere*, pp. 203–16. New York: Springer-Verlag.

Lipps, J. H. and E. Mitchell. 1976. Trophic model for the adaptive radiations and extinctions of pelagic marine mammals. *Paleobiology* 2:147–55.

Lisitsyna, N. A. and G. Y. Butuzova. 1981. Genesis of oceanic glauconites. *Lithology and Mineral Resources* 16:488–92.

Lisitzin, A. P. 1972. *Sedimentation in the World Ocean.* Society of Economic Paleontologists and Mineralogists Special Publication 17.

Littke, R., D. R. Baker, D. Leythaeuser, and J. Rullkötter. 1991. Keys to the depositional history of the Posidonia Shale (Toarcian) in the Hils Syncline, northern Germany. In R. V. Tyson and T. H. Pearson, eds., *Modern and Ancient Continental Shelf Anoxia*, pp. 311–33. Geological Society of London Special Publication 58.

Liu, G. and E. B. Leopold. 1994. Climatic comparison of Miocene pollen floras from northern East-China and south-central Alaska, USA. *Palaeogeography, Palaeoclimatology, Palaeoecology* 108:217–28.

Lloyd, R. M. 1964. Variations in the oxygen and carbon isotope ratios of Florida Bay mollusks and their environmental significance. *Journal of Geology* 72:84–111.

Longinelli, A. 1984. Oxygen isotopes in mammal bone phosphate: A new tool for paleohydrological and paleoclimatological research? *Geochimica et Cosmochimica Acta* 48:385–90.

Lottes, A. L. and A. M. Ziegler. 1994. World peat occurrence and the seasonality of climate and vegetation. *Palaeogeography, Palaeoclimatology, Palaeoecology* 106:23–37.

Loubere, P. 1991. Deep-sea benthic foraminiferal assemblage response to a surface ocean productivity gradient: A test. *Paleoceanography* 6:193–204.

Loutit, T. S., J. Hardenbol, P. R. Vail, and G. R. Baum. 1988. Condensed sections: The key to age dating and correlation of continental margin sequences. In C. K. Wilgus, H. Posamentier, C. A. Ross, and C. G. S. C. Kendall, eds., *Sea-Level Changes: An Integrated Approach*, pp. 183–213. American Association of Petroleum Geologists Memoir 42.

Lowenstam, H. A. 1961. Mineralogy, O^{18}/O^{16} ratios, and strontium and magnesium contents of Recent and fossil brachiopods and their bearing on the history of the oceans. *Journal of Geology* 69:241–60.

———. 1964. Palaeotemperatures of the Permian and Cretaceous periods. In A. E. M. Nairn, ed. *Problems in Palaeoclimatology*, pp. 227–48. London: John Wiley and Sons.

Lowenstam, H. A. and S. Epstein. 1954. Paleotemperatures of the post-Aptian Cretaceous as determined by the oxygen isotope method. *Journal of Geology* 62:207–48.

———. 1959. Cretaceous paleo-temperatures as determined by the oxygen isotope method, their relations to and the nature of rudistid reefs. *El Sistema Cretacico. Congreso Geologica Internacional, 20th, Mexico City, 1956* I:65–76.

Lozano, J. A. and J. D. Hays. 1976. Relationship of radiolarian assemblages to sediment types and physical oceanography in the Atlantic and western Indian Ocean sectors of the Antarctic Ocean. In R. M. Cline and J. D. Hays, eds., *Investigation of Late Quaternary Paleoceanography and Paleoclimatology*, pp. 303–36. Geological Society of America Memoir 145.

Luther, M. E. 1995. Modeling climates and upwelling systems of the past. In C. P. Summerhayes, K.-C. Emeis, M. V. Angel, R. L. Smith, and B. Zeitzschel, eds., *Upwelling in the Ocean: Modern Processes and Ancient Records*, pp. 273–82. Chichester: John Wiley and Sons.

Luz, B. and Y. Kolodny. 1985. Oxygen isotope variation in phosphate of biogenic apatites, IV. Mammal teeth and bones. *Earth and Planetary Science Letters* 75:29–36.

Luz, B., Y. Kolodny, and M. Horowitz. 1984. Fractionation of oxygen isotopes between mammalian bone-phosphate and environmental drinking water. *Geochimica et Cosmochimica Acta* 48:1689–93.

Luz, B., Y. Kolodny, and J. Kovach. 1984. Oxygen isotope variations in phosphate of biogenic apatites, III. Conodonts. *Earth and Planetary Science Letters* 69:255–62.

Lyle, M. 1988. Climatically forced organic carbon burial in equatorial Atlantic and Pacific Oceans. *Nature* 335:529–32.

Lyle, M. W., F. G. Prahl, and M. A. Sparrow. 1992. Upwelling and productivity changes inferred from a temperature record in the central equatorial Pacific. *Nature* 355:812–5.

Lyons, T. W. and R. A. Berner. 1992. Carbon-sulfur-iron systematics of Upper Holocene Black Sea sediments. *Chemical Geology* 99:1–27.

MacFadden, B. J., T. E. Cerling, and J. Prado. 1996. Cenozoic terrestrial ecosystem evolution in Argentina: Evidence from carbon isotopes of fossil mammal teeth. *Palaios* 11:319–27.

MacFadden, B. J., Y. Wang, T. E. Cerling, and F. Anaya. 1994. South American fossil mammals and carbon isotopes: A 25 million-year sequence from the Bolivian Andes. *Palaeogeography, Palaeoclimatology, Palaeoecology* 197:257–68.

Machette, M. N. 1985. Calcic soils of the southwestern United States. In D. L. Weide and M. L. Faber, eds., *Soils and Quaternary Geology of the Southwestern United States*, pp. 1–21. Geological Society of America Special Paper 203.

Mack, G. H., W. C. James, and H. C. Monger. 1993. Classification of paleosols. *Geological Society of America Bulletin* 105:129–36.

MacLeod, K. G. and K. A. Hoppe. 1992. Evidence that inoceramid bivalves were benthic and harbored chemosynthetic symbionts. *Geology* 20:117–20.

Macquaker, J. H. S. 1994. A lithofacies study of the Peterborough Member, Oxford Clay Formation (Jurassic), UK: An example of sediment bypass in a mudstone succession. *Journal of the Geological Society of London* 151:161–72.

Magaritz, M. 1987. A new explanation for cyclic deposition in marine evaporite basins: Meteoric water input. *Chemical Geology* 62:239–50.

Magaritz, M. and J. Heller. 1980. A desert migration indicator—Oxygen isotopic composition of land snail shells. *Palaeogeography, Palaeoclimatology, Palaeoecology* 32:153–62.

Magaritz, M., J. Heller, and M. Volokita. 1981. Land-air boundary environment as recorded by the $^{18}O/^{16}O$ and $^{13}C/^{12}C$ isotope ratios in the shells of land snails. *Earth and Planetary Science Letters* 52:101–6.

Maier-Reimer, E., U. Mikolajewicz, and T. Crowley. 1990. Ocean general circulation model sensitivity experiment with an open Central American isthmus. *Paleoceanography* 5:349–66.

Maier-Reimer, E., U. Mikolajewicz, and K. Hasselmann. 1993. Mean circulation of the Hamburg large-scale geostrophic ocean general circulation model and its sensitivity to the thermohaline surface forcing. *Journal of Physical Oceanography* 23:731–57.

Mainguet, M. 1978. The influence of trade winds, local air-masses and topographic obstacles on the aeolian movement of sand particles and the origin and distribution of dunes and ergs in the Sahara and Australia. *Geoforum* 9:17–28.

Maliva, R. G., A. H. Knoll, and R. Siever. 1989. Secular change in chert distribution: A reflection of evolving biological participation in the silica cycle. *Palaios* 4:519–32.

Marchant, E. R., G. H. Denton, C. C. Swisher, III, and N. Potter, Jr. 1996. Late Cenozoic Antarctic paleoclimate reconstructed from volcanic ashes in the Dry Valleys region of southern Victoria Land. *Geological Society of America Bulletin* 108: 181–94.

Margolis, S. V. and J. P. Kennett. 1971. Cenozoic paleoglacial history of Antarctica recorded in subantarctic deep-sea cores. *American Journal of Science* 271:1–36.

Markwick, P. J. 1994. "Equability," continentality, and Tertiary "climate": The crocodilian perspective. *Geology* 22:613–6.

Marsaglia, K. M. and G. d. Klein. 1983. The paleogeography of Paleozoic and Mesozoic storm depositional systems. *Journal of Geology* 91: 117–42.

Martill, D. M., M. A. Taylor, K. L. Duff, J. B. Riding, and P. R. Bown. 1994. The trophic structure of the biota of the Peterborough Member, Oxford Clay Formation (Jurassic), UK. *Journal of the Geological Society of London* 151:173–94.

Martini, E. 1977. Systematics, distribution and stratigraphical application of silicoflagellates. In A. T. S. Ramsay, ed. *Oceanic Micropaleontology Vol. 2*, pp. 1327–43. London: Academic Press.

Martini, I. P., J. K. Kwong, and S. Sadura. 1993. Sediment ice rafting and cold climate fluvial deposits: Albany River, Ontario, Canada. In M. Marzo and C. Puigdefábregas, eds., *Alluvial Sedimentation*, pp. 63–76. International Association of Sedimentologists Special Publication 17.

Marzolf, J. E. 1983. Changing wind and hydrologic regimes during deposition of the Navajo and Aztec Sandstones, Jurassic (?), southwestern United States. In M. E. Brookfield and T. S. Ahlbrandt, eds., *Eolian Sediments and Processes*, pp. 635–60. Developments in Sedimentology 38. Amsterdam: Elsevier.

Massare, J. A. 1987. Tooth morphology and prey preference of Mesozoic marine reptiles. *Journal of Vertebrate Paleontology* 7:121–37.

Masse, J.-P. and J. Philip. 1981. Cretaceous coral-rudistid buildups of France. In D. F. Toomey, ed. *European Fossil Reef Models*, pp. 399–426. Society of Economic Paleontologists and Mineralogists Special Publication 30.

Matthews, R. K. and R. Z. Poore. 1980. Tertiary $\delta^{18}O$ record and glacio-eustatic sea-level fluctuations. *Geology* 8:501–4.

May, F. E. and J. D. Shane. 1985. An analysis of the Umiat delta using palynologic and other data, North Slope, Alaska. *U.S. Geological Survey Bulletin* 1614, pp. 97–120.

Maynard, J. B. 1982. Extension of Berner's "new geochemical classification of sedimentary environments" to ancient sediments. *Journal of Sedimentary Petrology* 52:1325–31.

Mayr, E. 1963. *Animal Species and Evolution*. Cambridge, MA: Harvard University Press.

McArthur, J. M. and A. Herczeg. 1990. Diagenetic stability of the isotopic composition of phosphate-oxygen: Palaeoenvironmental implications. In A. J. G. Notholt and I. Jarvis, eds., *Phosphorite Research and Development*, pp. 119–24. Geological Society of London Special Publication 52.

McCabe, P. J. 1984. Depositional environments of coal and coal-bearing strata. In R. A. Rahmani and R. M. Flores, eds., *Sedimentology of Coal and Coal-bearing Sequences*, pp. 13–42. International Association of Sedimentologists Special Publication 7. Oxford: Blackwell Scientific.

———. 1991a. Geology of coal: Environments of deposition. In H. J. Gluskoter, D. D. Rice, and R. B. Taylor, eds., *Economic Geology*, pp. 469–82. The Geology of North America P-2. Boulder, Colorado: Geological Society of America.

———. 1991b. Tectonic controls on coal accumulation. *Société Géologique de France Bulletin* 162:277–82.

McCabe, P. J. and J. T. Parrish. 1992. Tectonic and climatic controls on the distribution and quality of Cretaceous coals. In P. J. McCabe and J. T. Parrish, eds., *Controls on the Distribution and Quality of Cretaceous Coals*, pp. 1–15. Geological Society of America Special Paper 267.

McCabe, P. J. and K. W. Shanley. 1992. Organic control on shoreface stacking patterns: Bogged down in the mire. *Geology* 20:741–4.

McCahon, T. J. and K. B. Miller. 1993. Paleosols in the Blue Rapids Shale (Lower Permian) of north-central Kansas. In W. C. Johnson, ed. *Second International Paleopedology Symposium, Field Excursion*, pp. 61–7. Kansas Geological Survey Open-File Report 93-30.

———. 1997. Climatic significance of natric horizons in Permian (Wolfcampian) paleosols of northcentral Kansas, U.S.A. *Sedimentology* 44:113–26.

McCartney, K. 1987. Silicoflagellates. In J. H. Lipps, ed. *Fossil Prokaryotes and Protists*, pp. 143–54. Boston: Blackwell Scientific.

———. 1993. Silicoflagellates. In J. H. Lipps, ed. *Fossil Prokaryotes and Protists*, pp. 143–54. Oxford: Blackwell Scientific.

McCartney, K. and S. W. Wise, Jr. 1990. Cenozoic silicoflagellates and ebridians from DOP Leg 113: Biostratigraphy and notes on morphologic variability. In P. F. Barker, J. P. Kennett, et al. *Proceedings of the Ocean Drilling Program, Scientific Results, Leg 113*, pp. 729–60. College Station, TX: Ocean Drilling Program.

McConnaughey, T. 1989. ^{13}C and ^{18}O isotopic disequilibrium in biological carbonates: I. Patterns. *Geochimica et Cosmochimica Acta* 53:151–62.

McElwain, J. C. and W. G. Chaloner. 1996. Fossil cuticle as a skeletal record of environmental change. *Palaios* 11:376–88.

McFadden, L. D. 1988. Climatic influences on rates and processes of soil development in Quaternary deposits of southern California. In J. Reinhardt and W. R. Sigleo, eds., *Paleosols and Weathering Through Geologic Time: Principles and Applications*, pp. 153–77. Geological Society of America Special Paper 216.

McGowran, B. 1989. Silica burp in the Eocene ocean. *Geology* 17:857–60.

McKelvey, V. E., R. W. Swanson, and R. P. Sheldon. 1952. The Permian phosphorite deposits of western United States. *International Geological Congress, Algiers* 11:45–64.

McKenna, M. C. 1980. Eocene paleolatitude, climate, and mammals of Ellesmere Island. *Palaeogeography, Palaeoclimatology, Palaeoecology* 30: 349–62.

McKenzie, J. A. 1985. Carbon isotopes and productivity in the lacustrine and marine environment. In W. Stumm, ed. *Chemical Processes in Lakes*, pp. 99–118. New York: John Wiley and Sons.

McKerrow, W. S. and C. R. Scotese, eds. 1990. *Palaeozoic Palaeogeography and Biogeography*. Geological Society of London Memoir 12.

Meadows, N. S. and A. Beach. 1993. Structural and climatic controls on facies distribution in a mixed fluvial and aeolian reservoir: The Triassic Sherwood Sandstone in the Irish Sea. In C. P. North and D. J. Prosser, eds., *Characterization of Fluvial and Aeolian Reservoirs*, pp. 247–64. Geological Society of London Special Publication 73.

Mendelson, C. V. 1993. Acritarchs and prasinophytes. In J. H. Lipps, ed. *Fossil Prokaryotes and Protists*, pp. 77–104. Oxford: Blackwell Scientific.

Meyer, H. W. 1992. Lapse rates and other variables applied to estimating paleoaltitudes from fossil floras. *Palaeogeography, Palaeoclimatology, Palaeoecology* 99:71–99.

Meyer, S. C., D. A. Textoris, and J. M. Dennison. 1992. Lithofacies of the Silurian Keefer Sandstone, east-central Appalachian basin, USA. *Sedimentary Geology* 76:187–206.

Meyers, P. A. 1990. Impacts of late Quaternary fluctuations in water level on the accumulation of sedimentary organic matter in Walker Lake, Nevada. *Palaeogeography, Palaeoclimatology, Palaeoecology* 78:229–40.

———. 1994. Preservation of elemental and isotopic source identificaion of sedimentary organic matter. *Chemical Geology* 114:289–302.

Meynadier, L., J.-P. Valet, and F. E. Grousset. 1995. Magnetic properties and origin of upper Quaternary sediments in the Somali Basin, Indian Ocean. *Paleoceanography* 10:459–72.

Miall, A. D. 1977. A review of the braided-river depositional environment. *Earth-Science Reviews* 13:1–62.

———. 1990. *Principles of Sedimentary Basin Analysis*. New York: Springer-Verlag.

———. 1992a. Alluvial deposits. In R. G. Walker and N. P. James, eds., *Facies models: Response to Sea-Level Change*, pp. 119–42. St. John's, Newfoundland: Geological Association of Canada.

———. 1992b. Exxon global cycle chart: An event for every occasion? *Geology* 20:787–90.

Mii, H.-S. and E. L. Grossman. 1994. Late Pennsylvanian seasonality reflected in the ^{18}O and elemental composition of a brachiopod shell. *Geology* 22:661–4.

Mikolajewicz, U., E. Maier-Reimer, T. J. Crowley, and K.-Y. Kim. 1993. Effect of Drake and Panamanian gateways on the circulation of an ocean model. *Paleoceanography* 8:409–26.

Miller, K. B. and R. R. West. 1993. Reevaluation of Wolfcampian cyclothems in northeastern Kansas: Significance of subaerial exposure and flooding surfaces. *Current Research on Kansas Geology, Summer 1993, Bulletin 235, Kansas Geological Survey*:1–26.

Miller, K. G., R. G. Fairbanks, and G. S. Mountain. 1987. Tertiary oxygen isotope synthesis, sea level history, and continental margin erosion. *Paleoceanography* 2:1–19.

Miller, K. G., M. D. Feigenson, J. D. Wright, and B. M. Clement. 1991. Miocene isotope reference section, Deep Sea Drilling Project Site 608: An evaluation of isotope and biostratigraphic resolution. *Paleoceanography* 6:33–52.

Miskell, K. J. 1983. Accumulation of opal in deep sea sediments from the mid-Cretaceous to the Miocene: A paleocirculation indicator. *MS thesis*, University of Miami.

Miskell, K. J., G. W. Brass, and C. G. A. Harrison. 1985. Global patterns in opal deposition from Late Cretaceous to Late Miocene. *American Association of Petroleum Geologists Bulletin* 69: 996–1012.

Mix, A. C. 1989. Influence of productivity variations on long-term atmospheric CO_2. *Nature* 337: 541–4.

Mohr, B. A. R. 1990. Eocene and Oligocene sporomorphs and dinoflagellate cysts from Leg 113 drill sites, Weddell Sea, Antarctica. In P. F. Barker, J. P. Kennett, et al. *Proceedings of the Ocean Drilling Program, Scientific Results, Leg 113*, pp. 595–612. College Station, TX: Ocean Drilling Program.

Molina-Cruz, A. 1984. Radiolaria as indicators of upwelling processes: The Peruvian connection. *Marine Micropaleontology* 9:53–75.

Molina-Cruz, A. and M. Martinez-López. 1994. Oceanography of the Gulf of Tehuantepec, Mexico, indicated by Radiolaria remains. *Palaeogeography, Palaeoclimatology, Palaeoecology* 110:179–95.

Molnar, P. and P. England. 1990. Late Cenozoic uplift of mountain ranges and global climate change: Chicken or egg? *Nature* 346:29–34.

Molnia, B. F. 1996. Forum: Modeling geology—The ideal world vs. the real world. *GSA Today* 6(5):8–14.

Mongenot, T., N.-P. Tribovillard, A. Desprairies, E. Lallier-Vergès, and F. Laggoun-Defarge. 1996. Trace elements as palaeoenvironmental markers in strongly mature hydrocarbon source rocks: The Cretaceous La Luna Formation of Venezuela. *Sedimentary Geology* 103:23–37.

Montuire, S., J. Michaux, S. Legendre, and J.-P. Aguilar. 1997. Rodents and climate. 1. A model for estimating past temperature using arvicolids (Mammalia:Rodentia). *Palaeogeography, Palaeoclimatology, Palaeoecology* 128:187–206.

Moore, G. T., D. N. Hayashida, and C. A. Ross. 1993. Late Early Silurian (Wenlockian) general circulation model—generated upwelling, graptolitic black shales, and organic-rich source rocks—An accident of plate tectonics? *Geology* 21:17–20.

Moore, P. D. 1987. Ecological and hydrological aspects of peat formation. In A. C. Scott, ed. *Coal and Coal-bearing Strata*, pp. 7–15. Geological Society of London Special Publication 32.

———. 1989. The ecology of peat-forming processes: A review. *International Journal of Coal Geology* 12:89–103.

Mora, C. I., S. G. Driese, and P. G. Seager. 1991. Carbon dioxide in the Paleozoic atmosphere: Evidence from carbon-isotope compositions of pedogenic carbonate. *Geology* 19:1017–20.

Morgan, M. E., C. Badgley, G. F. Gunnell, P. D. Gingerich, J. W. Kappelman, and M. C. Maas. 1995. Comparative paleoecology of Paleogene and Neogene mammalian faunas: Body-size structure. *Palaeogeography, Palaeoclimatology, Palaeoecology* 115:287–317.

Mount, J. F. and A. S. Cohen. 1984. Petrology and geochemistry of rhizoliths from Plio-Pleistocene fluvial and marginal lacustrine deposits, east Lake Turkana, Kenya. *Journal of Sedimentary Petrology* 54:263–75.

Mudie, P. J. and F. M. G. McCarthy. 1994. Late Quaternary pollen transport processes, western North Atlantic: Data from box models, cross-margin and N-S transects. *Marine Geology* 118:79–105.

Muehlenbachs, K. and R. N. Clayton. 1976. Oxygen isotope composition of the oceanic crust and its bearing on seawater. *Journal of Geophysical Research* 81:4365–9.

Mull, C. G., I. L. Tailleur, C. F. Mayfield, I. Ellersieck, and S. Curtis. 1982. New upper Paleozoic and lower Mesozoic stratigraphic units, central and western Brooks Range, Alaska. *American Association of Petroleum Geologists* 66:348–62.

Muller, J. 1959. Palynology of recent Orinoco delta and shelf sediments: Reports of the Orinoco shelf expedition, v. 5. *Micropaleontology* 5:1–32.

Müller, P. J. and E. Suess. 1979. Productivity, sedimentation rate and sedimentary organic matter in the oceans. I. Organic carbon preservation. *Deep-Sea Research* 26A:1347–62.

Mullins, H. T., J. B. Thompson, K. McDougall, and T. L. Vercoutere. 1985. Oxygen-minimum zone edge effects: Evidence from the central California upwelling system. *Geology* 13:491–4.

Murchey, B. L., R. J. Madrid, and F. G. Poole. 1987. Paleozoic bedded barite associated with chert in western North America. In J. R. Hein, ed. *Siliceous Sedimentary Rock-Hosted Ores and Petroleum*, pp. 269–83. New York: Van Nostrand Reinhold.

Murnane, R. J. and R. F. Stallard. 1988. Germanium/silicon fractionation during biogenic opal formation. *Paleoceanography* 3:461–9.

Murray, D. R. 1995. Plant responses to carbon dioxide. *American Journal of Botany* 82:690–7.

Murray, R. W., D. L. Jones, and M. R. Buchholtz ten Brink. 1992. Diagenetic formation of bedded chert: Evidence from chemistry of the chert-shale couplet. *Geology* 20:271–4.

Mutti, M. and H. Weissert. 1995. Triassic monsoonal climate and its signature in Ladinian-Carnian carbonate platforms (southern Alps, Italy). *Journal of Sedimentary Research* B65:357–67.

Myers, K. J. 1989. The origin of the Lower Jurassic Cleveland Ironstone Formation of North-East England: Evidence from portable gamma-ray spectrometry. In T. P. Young and W. E. G. Taylor, eds., *Phanerozoic Ironstones*, pp. 221–8. Geological Society of London Special Publication 46.

Nairn, A. E. M. and M. E. Smithwick. 1976. Permian paleogeography and climatology. In H. Falke, ed. *The Continental Permian in Central, West, and South Europe*, pp. 283–312. Boston: D. Reidel.

Nakai, S., A. N. Halliday, and D. K. Rea. 1993. Provenance of dust in the Pacific Ocean. *Earth and Planetary Science Letters* 119:143–57.

Nathan, Y. 1984. The mineralogy and geochemistry of phosphorites. In P. O. Nriagu and P. B. Moore, eds., *Phosphate Minerals*, pp. 275–91. Berlin: Springer-Verlag.

National Research Council. 1975. *Understanding Climatic Change: A Program for Action.* U.S. Committee for GARP, National Academy of Sciences, Washington, D.C.

Nelson, C. S. 1978. Temperate shelf carbonate sediments in the Cenozoic of New Zealand. *Sedimentology* 25:737–71.

Nelson, C. S. 1988. An introductory perspective on non-tropical shelf carbonates. *Sedimentary Geology* 60:3–12.

Nemec, W. 1992. Depositional controls on plant growth and peat accumulation in a braidplain delta environment: Helvetiafjellet Formation (Barremian-Aptian), Svalbard. In P. J. McCabe and J. T. Parrish, eds., *Controls on the Distribution and Quality of Cretaceous Coals*, pp. 209–26. Geological Society of America Special Paper 267.

Neuzil, S. G., Supardi, C. B. Cecil, J. S. Kane, and K. Soedjono. 1993. Inorganic geochemistry of domed peat in Indonesia and its implication for the origin of mineral matter in coal. In J. C. Cobb and C. B. Cecil, eds., *Modern and Ancient Coal-Forming Environments*, pp. 23–44. Geological Society of America Special Paper 286.

Newell, N. D. 1972. The evolution of reefs. *Scientific American* 226(6):54–65.

Newton, C. R. 1988. Significance of "Tethyan" fossils in the American Cordillera. *Science* 242:385–91.

Nicolas, J. and P. Bildgen. 1979. Relations between the location of the karst bauxites in the northern hemisphere, the global tectonics and the climatic variations during geologic time. *Palaeogeography, Palaeoclimatology, Palaeoecology* 28:205–39.

Nielson, J. and G. Kocurek. 1986. Climbing zibars of the Algodones. *Sedimentary Geology* 48:1–15.

Nigrini, C. 1970. Radiolarian assemblages in the North Pacific and their application to a study of Quaternary sediments in Core V20–130. In J. D. Hays, ed. *Geological Investigations of the North Pacific*, pp. 139–83. Geological Society of America Memoir 126.

Norris, G. and A. D. Miall. 1984. Arctic biostratigraphic heterochroneity. *Science* 224:173–6.

Norris, R. D., L. S. Jones, R. M. Corfield, and J. E. Cartlidge. 1996. Skiing in the Eocene Uinta Mountains? Isotopic evidence in the Green River Formation for snow melt and large mountains. *Geology* 24:403–6.

North, G. R. and T. J. Crowley. 1985. Application of a seasonal climate model to Cenozoic glaciation. *Journal of the Geological Society of London* 142:475–82.

Nøttvedt, A. and R. D. Kreisa. 1987. Model for the combined-flow origin of hummocky cross-stratification. *Geology* 15:357–61.

O'Brien, G. S., A. R. Milnes, H. H. Veeh, D. T. Heggie, S. R. Riggs, D. J. Cullen, J. F. Marshall, and P. J. Cook. 1990. Sedimentation dynamics and redox iron-cycling: Controlling factors for the apatite-glauconite association on the East Australia continental margin. In A. J. G. Notholt and

I. Jarvis, eds., *Phosphorite Research and Development*, pp. 61–86. Geological Society of London Special Publication 52.

O'Brien, G. W. and H. H. Veeh. 1983. Are phosphorites reliable indicators of upwelling? In E. Suess and J. Thiede, eds., *Coastal Upwelling: Its Sediment Record, Part A*, pp. 399–419. Mt. Kisco, NY: Plenum.

O'Brien, N. R. 1990. Significance of lamination in Toarcian (Lower Jurassic) shales from Yorkshire, Great Britain. *Sedimentary Geology* 67:25–34.

Odin, G. S. 1973. Répartition, nature minéralogique et genèse des granule verts recueilles dans les sédiments marins actuels. *Sciences de la Terre* 18:79–94.

———. 1988a. Glaucony from the Gulf of Guinea. In G. S. Odin, ed. *Green Marine Clays*, pp. 225–47. Developments in Sedimentology 45. Amsterdam: Elsevier.

———. 1988b. The verdine facies from the lagoon off New Caledonia. In G. S. Odin, ed. *Green Marine Clays*, pp. 57–81. Developments in Sedimentology 45. Amsterdam: Elsevier.

Odin, G. S., ed. 1988c. *Green Marine Clays*. Developments in Sedimentology 45. Amsterdam: Elsevier.

Odin, G. S., R. W. O. Knox, R. A. Gygi, and S. Guerrak. 1988. Green marine clays from the oolitic ironstone facies: Habit, mineralogy, environment. In G. S. Odin, ed. *Green Marine Clays*, pp. 29–52. Developments in Sedimentology 45. Amsterdam: Elsevier.

Odin, G. S. and M. Lamboy. 1988. Glaucony from the margin off northwestern Spain. In G. S. Odin, ed. *Green Marine Clays*, pp. 249–75. Developments in Sedimentology 45. Amsterdam: Elsevier.

Odin, G. S. and R. Létolle. 1980. Glauconitization and phosphatization environments: A tentative comparison. In Y. K. Bentor, ed. *Marine Phosphorites: Geochemistry, Occurrence, Genesis*, pp. 227–38. Society of Economic Paleontologists and Mineralologists, Special Publication 29.

Olsen, P. E. 1985. Constraints on the formation of lacustrine microlaminated sediments. *U.S. Geological Survey Circular* 946, pp. 34–35.

———. 1986. A 40-million-year lake record of early Mesozoic orbital climatic forcing. *Science* 234: 842–8.

———. 1990. Tectonic, climatic, and biotic modulation of lacustrine ecosystems- examples from Newark Supergroup of eastern North America. In B. J. Katz, ed. *Lacustrine Basin Exploration: Case Studies and Modern Analogs*, pp. 209–24. American Association of Petroleum Geologists Memoir 50.

Olsen, P. E. and D. V. Kent. 1996. Milankovitch forcing in the tropics of Pangaea during the Late Triassic. *Palaeogeography, Palaeoclimatology, Palaeoecology* 122:1–26.

Olsen, P. E., R. W. Schlische and P. J. W. Gore, eds. 1989. *Tectonic, Depositional, and Paleoecological History of Early Mesozoic Rift Basins, Eastern North America*. International Geological Congress Field Trip T351. Washington, D.C.: American Geophysical Union.

Oreskes, N., K. Shrader-Frechette, and K. Belitz. 1994. Verification, validation, and confirmation of numerical models in the earth sciences. *Science* 263:641–6.

Ortner, P. B. and M. J. Dagg. 1995. Nutrient-enhanced coastal ocean productivity explored in the Gulf of Mexico. *Eos* 76:97, 109.

Oschmann, W. 1988. Kimmeridge clay sedimentation—a new cyclic model. *Palaeogeography, Palaeoclimatology, Palaeoecology* 65:217–51.

———. 1993. Environmental oxygen fluctuations and the adaptive response of marine benthic organisms. *Journal of the Geological Society of London* 150:193–6.

Ostrom, J. H. 1964. A reconsideration of the paleoecology of hadrosaurian dinosaurs. *American Journal of Science* 262:975–97.

———. 1966. Functional morphology and evolution of the ceratopsian dinosaurs. *Evolution* 20: 290–308.

———. 1970. Terrestrial vertebrates as indicators of Mesozoic climates. *Proceedings of the North American Paleontological Convention: Symposium on Paleoclimatology*, pp. 347–76. Lawrence, KS: Allen Press.

Ottens, J. J. and A. J. Nederbragt. 1992. Planktic foraminiferal diversity as indicator of ocean environments. *Marine Micropaleontology* 19:13–28.

Otto-Bliesner, B. L. 1993. Tropical mountains and coal formation: A climate model study of the Westphalian (306 Ma). *Geophysical Research Letters* 20:1947–50.

———. 1995. Continental drift, runoff, and weathering feedbacks: Implications from climate model experiments. *Journal of Geophysical Research* 100:11537–48.

Otto-Bliesner, B. L. and G. R. Upchurch, Jr. 1997. Vegetation-induced warming of high-latitude regions during the Late Cretaceous period. *Nature* 385:804–7.

Palacios-Fest, M. R., A. S. Cohen, and P. Anadón. 1994. Use of ostracodes as paleoenvironmental tools in the interpretation of ancient lacustrine records. *Revista Española de Paleontología* 9:145–64.

Paladino, F. V., M. P. O'Connor, and J. R. Spotila. 1990. Metabolism of leatherback turtles, gigantothermy, and thermoregulation of dinosaurs. *Nature* 344:858–60.

Pannekoek, A. J. 1965. Shallow-water and deep-water evaporite deposition. *American Journal of Science* 263:284–5.

Paquet, H. 1970. *Évolution géochimique des minéraux argileux dans les altérations et les sols des climats méditerranéens et tropicaux á saisons contrastées*. Mémoires du Service de la Carte géologique d'Alsace et de Lorraine 30.

Parrish, J. M. 1989. Vertebrate paleoecology of the Chinle Formation (Late Triassic) of the southwestern United States. *Palaeogeography, Palaeoclimatology, Palaeoecology* 72:227–47.

Parrish, J. M. and S. C. Good. 1987. Preliminary report on vertebrate and invertebrate fossil occurrences, Chinle Formation (Upper Triassic), southeastern Utah. *Four Corners Geological Society Guidebook 10th Field Conference, Cataract Canyon*, pp. 109–15.

Parrish, J. M. and J. T. Parrish. 1983. Were Mesozoic marine reptiles analogs of whales? (abs.). *Geological Society of America Abstracts with Program* 15:659.

Parrish, J. M., J. T. Parrish, J. H. Hutchison, and R. A. Spicer. 1987. Late Cretaceous vertebrate fossils from the North Slope of Alaska and implications for dinosaur ecology. *Palaios* 2:377–89.

Parrish, J. M., J. T. Parrish, and A. M. Ziegler. 1986. Permian-Triassic paleogeography and paleoclimatology and implications for therapsid distributions. In N. H. Hotton, III, P. D. MacLean, J. J. Roth, and E. C. Roth, eds., *The Ecology and Biology of Mammal-like Reptiles*, pp. 109–32. Washington, D.C.: Smithsonian Institution Press.

Parrish, J. T. 1982. Upwelling and petroleum source beds, with reference to the Paleozoic. *American Association of Petroleum Geologists Bulletin* 66:750–74.

———. 1983. Upwelling deposits: Nature of association of organic-rich rocks, chert, chalk, phosphorite, and glauconite (abs.). *American Association of Petroleum Geologists Bulletin* 67:529.

———. 1985. Latitudinal distribution of land and shelf and absorbed solar radiation during the Phanerozoic. *U.S. Geological Survey Open-File Report* 85–31.

———. 1987a. Lithology, geochemistry, and depositional environment of the Shublik Formation (Triassic), northern Alaska. In I. L. Tailleur and P. Weimer, eds., *Alaskan North Slope Geology*, pp. 391–6 1. Anchorage: Pacific Section, Society of Economic Paleontologists and Mineralogists and the Alaska Geological Society.

———. 1987b. Palaeo-upwelling and the distribution of organic-rich rocks. In J. Brooks and A. J. Fleet, eds., *Marine Petroleum Source Rocks*, pp. 199–205. Geological Society of London Special Paper 26.

———. 1990. Paleogeographic and paleoclimatic setting of the Miocene phosphogenic episode. In W. C. Burnett and S. R. Riggs, eds., *Phosphate Deposits of the World. Vol. 3, Neogene to Modern Phosphorites*, pp. 223–40. Cambridge: Cambridge University Press.

———. 1992. Jurassic climate and oceanography of the circum-Pacific region. In G. E. G. Westermann, ed. *The Jurassic of the Circum-Pacific. IGCP Project 171*, pp. 365–79. Oxford: Oxford University Press.

———. 1993. Climate of the supercontinent Pangea. *Journal of Geology* 101:215–33.

———. 1995. Paleogeography of organic-rich rocks and the preservation versus production controversy. In A.-Y. Huc, ed. *Paleogeography, Paleoclimate, and Source Rocks*, pp. 1–20. American Association of Petroleum Geologists Studies in Geology 40.

Parrish, J. T. and E. J. Barron. 1986. *Paleoclimates and Economic Geology*. Society of Economic Paleontologists and Mineralogists Short Course 18.

Parrish, J. T., M. T. Bradshaw, A. T. Brakel, S. M. Mulholland, J. M. Totterdell, and A. N. Yeates. 1996. Paleoclimatology of Australia during the Pangean interval. *Palaeoclimates—Data and Modelling* 1:23–57.

Parrish, J. T. and R. L. Curtis. 1982. Atmospheric circulation, upwelling, and organic-rich rocks in the Mesozoic and Cenozoic Eras. *Palaeogeography, Palaeoclimatology, Palaeoecology* 40:31–66.

Parrish, J. T., I. L. Daniel, E. M. Kennedy, and R. A. Spicer. 1998. Paleoclimatic significance of mid-Cretaceous floras from the middle Clarence Valley, New Zealand. *Palaios* 13:146–56.

Parrish, J. T., T. M. Demko, and G. S. Tanck. 1993. Sedimentary palaeoclimatic indicators: What they are and what they tell us. *Philosophical Transac-*

tions of the Royal Society of London, Ser. A 344:21–5.

Parrish, J. T. and D. L. Gautier. 1993. Sharon Springs Member of Pierre Shale: Upwelling in the Western Interior Seaway? In W. G. E. Caldwell and E. G. Kauffman, eds., *Evolution of the Western Interior Basin*, pp. 319–32. Geological Association of Canada Special Paper 39.

Parrish, J. T. and F. Peterson. 1988. Wind directions predicted from global circulation models and wind directions determined from eolian sandstones of the western United States—A comparison. *Sedimentary Geology* 56:261–82.

Parrish, J. T. and S. D. Samson. 1988. Possible paleoclimatic constraints on the positions of Laurentia, southern Britain, and Baltica in the Ordovician (abs.). *Geological Society of America Abstracts with Programs* 20:A192.

Parrish, J. T. and R. A. Spicer. 1988a. Middle Cretaceous wood from the Nanushuk Group, central North Slope, Alaska. *Palaeontology* 31:19–34.

———. 1988b. Late Cretaceous terrestrial vegetation: A near-polar temperature curve. *Geology* 16:22–5.

Parrish, J. T., A. M. Ziegler and R. G. Humphreville. 1983. Upwelling in the Paleozoic Era. In J. Thiede and E. Suess, eds., *Coastal Upwelling: Its Sediment Record. Part B.* , pp. 553–78. Mt. Kisco, NY: Plenum.

Parrish, J. T., A. M. Ziegler, and C. R. Scotese. 1982. Rainfall patterns and the distribution of coals and evaporites in the Mesozoic and Cenozoic. *Palaeogeography, Palaeoclimatology, Palaeoecology* 40:67–101.

Parron, C. and D. Nahon. 1980. Red bed genesis by lateritic weathering of glauconitic sediments. *Journal of the Geological Society of London* 137:689–93.

Patterson, W. P., G. R. Smith, and K. C. Kohmann. 1993. Continental paleothermometry and seasonality using the isotopic composition of aragonitic otoliths of freshwater fishes. In P. K. Swart, K. C. Lohmann, J. McKenzie, and S. Savin, eds., *Climate Change in Continental Isotopic Records*, pp. 191–201. Geophysical Monograph 78. Washington, D.C.: American Geophysical Union.

Patterson, W. P. and L. M. Walter. 1994. Depletion of ^{13}C in seawater SCO_2 on modern carbonate platforms: Significance for the carbon isotopic record of carbonates. *Geology* 22:885–8.

Patzkowsky, M. E., L. H. Smith, P. J. Markwick, C. J. Engberts, and E. D. Gyllenhaal. 1991. Application of the Fujita-Ziegler paleoclimate model: Early Permian and Late Cretaceous examples. *Palaeogeography, Palaeoclimatology, Palaeoecology* 86:67–85.

Paytan, A. and M. Kastner. 1996. Benthic Ba fluxes in the central Equatorial Pacific, implications for the oceanic Ba cycle. *Earth and Planetary Science Letters* 142:439–50.

Pedersen, T. F. and S. E. Calvert. 1990. Anoxia vs. productivity: What controls the formation of organic-carbon-rich sediments and sedimentary rocks? *American Association of Petroleum Geologists Bulletin* 74:454–66.

Pedersen, T. F., M. Pickering, J. S. Vogel, J. N. Southon, and D. E. Nelson. 1988. The response of benthic foraminifera to productivity cycles in the eastern equatorial pacific: Faunal and geochemical constraints on glacial bottom water oxygen levels. *Paleoceanography* 3:157–68.

Pelet, R. 1987. A model of organic sedimentation on present-day continental margins. In J. Brooks and A. J. Fleet, eds., *Marine Petroleum Source Rocks*, pp. 167–80. Geological Society of London Special Paper 26.

Peterson, F. 1988. Pennsylvanian to Jurassic eolian transportation systems in the western United States. *Sedimentary Geology* 56:207–60.

Petránek, J. and F. B. Van Houten. 1997. Phanerozoic ooidal ironstones. *Czech Geological Survey Special Papers* 7.

Phillips, R. L. 1987. Late Cretaceous to early Tertiary deltaic to marine sedimentation, North Slope, Alaska (abs.). *American Association of Petroleum Geologists Bulletin* 71:601–2.

———. 1990. Summary of Late Cretaceous environments near Ocean Point, North Slope, Alaska. *Geologic studies in Alaska by the U.S. Geological Survey, 1989. U.S. Geological Survey Bulletin* 1990, pp. 101–106.

Phillips, T. L. and W. A. DiMichele. 1990. From plants to coal: Peat taphonomy of Upper Carboniferous coals. *International Journal of Coal Geology* 16:151–6.

Phillips, T. L. and R. A. Peppers. 1984. Changing patterns of Pennsylvanian coal-swamp vegetation and implications of climatic control on coal occurrence. *International Journal of Coal Geology* 3:205–55.

Pichon, J.-J., L. D. Labeyrie, G. Bareille, M. Labracherie, J. Duprat, and J. Jouzel. 1992. Surface water temperature changes in the high latitudes of the Southern Hemisphere over the last glacial-interglacial cycle. *Paleoceanography* 7:289–318.

Pietrafesa, L. 1990. Upwelling processes associated

with Western Boundary Currents. In W. C. Burnett and S. R. Riggs, eds., *Phosphate Deposits of the World. Vol. 3, Neogene to Modern Phosphorites*, pp. 3–26. Cambridge: Cambridge University Press.

Piper, D. Z. 1994. Seawater as the source of minor elements in black shales, phosphorites and other sedimentary rocks. *Chemical Geology* 114: 95–114.

Pirrie, D. and J. D. Marshall. 1990a. High-paleolatitude Late Cretaceous paleotemperatures: New data from James Ross Island, Antarctica. *Geology* 18:31–4.

———. 1990b. Diagenesis of *Inoceramus* and Late Cretaceous paleoenvironmental geochemistry: A case study from James Ross Island, Antarctica. *Palaios* 5:336–45.

Pisias, N. G., A. Roelofs, and M. Weber. 1997. Radiolarian-based transfer functions for estimating mean surface ocean temperatures and seasonal range. *Paleoceanography* 12:365–79.

Pitman, A. J., A. Henderson-Sellers, and Z.-L. Yang. 1990. Sensitivity of regional climates to localized precipitation in global models. *Nature* 346: 734–7.

Poag, C. W. and A. L. Karowe. 1986. Stratigraphic potential of *Bolboforma* significantly increased by new finds in the North Atlantic and South Pacific. *Palaios* 1:162–71.

Pokras, E. M. and A. C. Mix. 1985. Eolian evidence for spatial variability of late Quaternary climates in tropical Africa. *Quaternary Research* 24:137–49.

Pollard, D. and M. Schulz. 1994. A model for the potential locations of Triassic evaporite basins driven by paleoclimatic GCM simulations. *Global and Planetary Change* 9:233–49.

Pomerol, B. 1983. Geochemistry of the Late Cenomanian-Early Turonian chalks of the Paris Basin: Manganese and carbon isotopes in carbonates as paleooceanographic indicators. *Cretaceous Research* 4:85–93.

Ponel, P. 1995. Rissian, Eemian and Würmian Coleoptera assemblages from La Grande Pile (Vosges, France). *Palaeogeography, Palaeoclimatology, Palaeoecology* 114:1–41.

Poole, I. 1994. "Twig"—Wood anatomical characters as palaeoecological indicators. *Review of Palaeobotany and Palynology* 81:33–52.

Poore, R. Z. and R. K. Matthews. 1984. Oxygen isotope ranking of late Eocene and Oligocene planktonic foraminifers: Implications for Oligocene sea-surface temperatures and global ice-volume. *Marine Micropaleontology* 9:111–34.

Popp, B. N., T. F. Anderson, and P. A. Sandberg. 1986a. Textural, elemental, and isotopic variations among constituents in Middle Devonian limestones, North America. *Journal of Sedimentary Petrology* 56:715–27.

———. 1986b. Brachiopods as indicators of original isotopic compositions in some Paleozoic limestones. *Geological Society of America Bulletin* 97:1262–9.

Popp, B. N., R. Takigiku, J. M. Hayes, J. W. Louda, and E. W. Baker. 1989. The post-Paleozoic chronology and mechanism of ^{13}C depletion in primary marine organic matter. *American Journal of Science* 289:436–54.

Pospichal, J. J. and S. W. Wise, Jr. 1990. Paleocene to middle Eocene calcareous nannofossils of ODP sites 689 and 690, Maud Rise, Weddell Sea. In P. F. Barker, J. P. Kennett, et al. *Proceedings of the Ocean Drilling Program, Scientific Results, Leg 113*, pp. 613–38. College Station, TX: Ocean Drilling Program.

Povey, D. A. R., R. A. Spicer, and P. C. England. 1994. Palaeobotanical investigation of early Tertiary palaeoelevations in northeastern Nevada: Initial results. *Review of Palaeobotany and Palynology* 81:1–10.

Prahl, F. G., L. A. Muehlhausen, and D. L. Zahnle. 1988. Further evaluation of long-chain alkenones as indicators of paleoceanographic conditions. *Geochimica et Cosmochimica Acta* 52:2303–10.

Pratt, L. M. 1984. Influence of paleoenvironmental factors on preservation of organic matter in the Middle Cretaceous Greenhorn Formation of Pueblo, Colorado. *American Association of Petroleum Geologists Bulletin* 68:1146–59.

Pratt, L. M., M. A. Arthur, W. E. Dean, and P. A. Scholle. 1993. Paleo-oceanographic cycles and events during the Late Cretaceous in the Western Interior seaway of North America. In W. G. E. Caldwell and E. G. Kauffman, eds., *Evolution of the Western Interior Basin*, pp. 333–53. Geological Association of Canada Special Paper 39.

Pratt, L. M., E. R. Force, and B. Pomerol. 1991. Coupled manganese and carbon-isotopic events in marine carbonates at the Cenomanian-Turonian boundary. *Journal of Sedimentary Petrology* 61:370–83.

Prell, W. L. and J. E. Kutzbach. 1992. Sensitivity of the Indian monsoon to forcing parameters and implications for its evolution. *Nature* 360:647–52.

Prell, W. L., D. W. Murray, and S. C. Clemens. 1992. Evolution and variability of the Indian Ocean summer monsoon: Evidence from the western

Arabian Sea drilling program. In R. A. Duncan, D. K. Rea, R. B. Kidd, U. von Rad, and J. K. Weissel, eds., *Synthesis of Results from Scientific Drilling in the Indian Ocean*, pp. 447–69. Geophysical Monograph 70. Washington, D.C.: American Geophysical Union.

Prentice, M. L. and R. K. Matthews. 1988. Cenozoic ice-volume history: Development of a composite oxygen isotope record. *Geology* 16:963–6.

Price, G. D. and B. W. Sellwood. 1997. "Warm" palaeotemperatures from high Late Jurassic palaeolatitudes (Falkland Plateau): Ecological, environmental or diagenetic controls? *Palaeogeography, Palaeoclimatology, Palaeoecology* 129:315–27.

Price, G. D., P. J. Valdes, and B. W. Sellwood. 1997. Prediction of modern bauxite occurrence: Implications for climate reconstruction. *Palaeogeography, Palaeoclimatology, Palaeoecology* 131:1–13.

Prospero, J. M. 1981. Arid regions as sources of mineral aerosols in the marine atmosphere. In T. L. Péwé, ed. *Desert Dust: Origin, Characteristics, and Effect on Man*, pp. 71–86. Geological Society of America Special Paper 186.

Pye, K. 1987. *Aeolian Dust and Dust Deposits*. San Diego: Academic Press.

Pye, K. and H. Tsoar. 1987. The mechanics and geological implications of dust transport and deposition in deserts with particular reference to loess formation and dune sand diagenesis in the northern Negev, Israel. In L. E. Frostick and I. Reid, eds., *Desert Sediments: Ancient and Modern*, pp. 139–56. Geological Society of London Special Publication 35.

———. 1990. *Aeolian Sand and Sand Dunes*. London: Unwin Hyman.

Pye, K. and L.-P. Zhou. 1989. Late Pleistocene and Holocene aeolian dust deposition in North China and the Northwest Pacific Ocean. *Palaeogeography, Palaeoclimatology, Palaeoecology* 73:11–23.

Qiu, L., D. F. Williams, A. Gvorzdkov, E. Karabanov, and M. Shimaraeva. 1993. Biogenic silica accumulation and paleoproductivity in the northern basin of Lake Baikal during the Holocene. *Geology* 21:25–8.

Quade, J., J. M. L. Cater, T. P. Ojha, J. Adam, and T. M. Harrison. 1995. Late Miocene environmental change in Nepal and the northern Indian subcontinent: Stable isotopic evidence from paleosols. *Geological Society of America Bulletin* 107:1381–97.

Quade, J. and T. Cerling. 1995. Expansion of C_4 grasses in the Late Miocene of northern Pakistan: Evidence from stable isotopes in paleosols. *Palaeogeography, Palaeoclimatology, Palaeoecology* 115:91–116.

Quade, J., T. E. Cerling, and J. R. Bowman. 1989. Development of Asian monsoon revealed by marked ecological shift during the latest Miocene in northern Pakistan. *Nature* 342:163–6.

Quade, J., L. Roe, P. G. DeCelles, and T. P. Ojha. 1997. The late Neogene $^{87}Sr/^{86}Sr$ record of lowland Himalayan rivers. *Science* 276:1828–31.

Quade, J., N. Solounias, and T. E. Cerling. 1994. Stable isotopic evidence from paleosol carbonates and fossil teeth in Greece for forest or woodlands over the past 11 Ma. *Palaeogeography, Palaeoclimatology, Palaeoecology* 108:41–53.

Railsback, L. B. 1990. Influence of changing deep ocean circulation on the Phanerozoic oxygen isotopic record. *Geochimica et Cosmochimica Acta* 54:1501–9.

———. 1992. A geological numerical model for Paleozoic global evaporite deposition. *Journal of Geology* 100:261–77.

Railsback, L. B., S. C. Ackerly, T. F. Anderson, and J. L. Cisne. 1990. Palaeontological and isotope evidence for warm saline deep waters in Ordovician oceans. *Nature* 343:156–9.

Raiswell, R., F. Buckley, R. A. Berner, and T. F. Anderson. 1988. Degree of pyritization of iron as a paleoenvironmental indicator of bottom-water oxygenation. *Journal of Sedimentary Petrology* 58:812–9.

Ramsay, A. T. S. 1973. A history of organic siliceous sediment in oceans. In N. F. Hughes, ed. *Organisms and Continents Through Time*, pp. 199–234. Special Papers in Palaeontology 12.

Ramstein, G., F. Fluteau, J. Besse, and S. Joussaume. 1997. Effect of orogeny, plate motion and land-sea distribution on Eurasian climate change over the past 30 million years. *Nature* 386:788–95.

Rao, C. P. 1981a. Criteria for recognition of cold-water carbonate sedimentation: Berriedale limestone (Lower Permian), Tasmania, Australia. *Journal of Sedimentary Petrology* 51:491–506.

———. 1981b. Geochemical differences between tropical (Ordovician) and subpolar (Permian) carbonates, Tasmania, Australia. *Geology* 9:205–9.

Rao, C. P. and M. P. J. Jayawardane. 1994. Major minerals, elemental and isotopic composition in modern temperate shelf carbonates, eastern Tasmania, Australia: Implications for the occurrence of extensive ancient non-tropical carbonates. *Palaeogeography, Palaeoclimatology, Palaeoecology* 107:49–63.

Rao, C. P. and C. S. Nelson. 1992. Oxygen and carbon isotope fields for temperate shelf carbonates from Tasmania and New Zealand. *Marine Geology* 103:273–86.

Rau, G. H., M. A. Arthur, and W. E. Dean. 1987. $^{15}N/^{14}N$ variations in Cretaceous Atlantic sedimentary sequences: Implication for past changes in marine nitrogen biogeochemistry. *Earth and Planetary Science Letters* 82:269–79.

Raup, D. M. 1976. Species diversity in the Phanerozoic: An interpretation. *Paleobiology* 2:289–97.

———. 1986. *The Nemesis Affair*. New York: W. W. Norton.

Raup, O. B. and R. J. Hite. 1992. Lithology of evaporite cycles and cycle boundaries in the upper part of the Paradox Formation of the Hermosa Group of Pennsylvanian Age in the Paradox Basin, Utah and Colorado. *U.S. Geological Survey Bulletin* 2000-B, pp. B1-B37.

Raymo, M. E. 1991. Geochemical evidence supporting T. C. Chamberlin's theory of glaciation. *Geology* 19:344–7.

Raymo, M. E. and W. F. Ruddiman. 1992. Tectonic forcing of late Cenozoic climate. *Nature* 359: 117–22.

Raymo, M. E., W. F. Ruddiman, and P. N. Froelich. 1988. Influence of late Cenozoic mountain building on ocean geochemical cycles. *Geology* 16: 649–53.

Raymond, A., P. H. Kelley, and C. B. Lutken. 1989. Polar glaciers and life at the equator: The history of Dinantian and Namurian (Carboniferous) climate. *Geology* 17:408–11.

Raymont, J. E. G. 1963. *Plankton and Productivity in the Oceans*. New York: Macmillan.

Rea, D. K. 1986. Neogene history of the South Pacific tradewinds: Evidence for hemispherical asymmetry of atmospheric circulation. *Palaeogeography, Palaeoclimatology, Palaeoecology* 55: 55–64.

———. 1989. Geologic record of atmospheric circulation on tectonic time scales. In M. Leinen and M. Sarnthein, eds., *Paleoclimatology and Paleometeorology: Modern and Past Patterns of Global Atmospheric Transport*, pp. 841–57. NATO ASI Series C: Mathematical and Physical Sciences 282. Dordrecht: Kluwer Academic.

———. 1993. Geologic records in deep sea muds. *GSA Today* 3:205, 8, 9–10.

———. 1994. The paleoclimatic record provided by eolian deposition in the deep sea: The geologic history of wind. *Reviews of Geophysics* 32: 159–95.

Rea, D. K. and S. A. Hovan. 1995. Grain size distribution and depositional processes of the mineral component of abyssal sediments: Lessons from the North Pacific. *Paleoceanography* 10:251–8.

Rea, D. K. and T. R. Janecek. 1981. Late Cretaceous history of eolian deposition in the Mid-Pacific mountains, central North Pacific Ocean. *Palaeogeography, Palaeoclimatology, Palaeoecology* 36:55–67.

Rea, D. K. and M. Leinen. 1988. Asian aridity and the zonal westerlies: Late Pleistocene and Holocene record of eolian deposition in the Northwest Pacific Ocean. *Palaeogeography, Palaeoclimatology, Palaeoecology* 66:1–8.

Rea, D. K., M. Leinen, and T. R. Janecek. 1985. Geologic approach to the long-term history of atmospheric circulation. *Science* 227:721–5.

Rea, D. K., J. C. Zachos, R. M. Owen, and P. D. Gingerich. 1990. Global change at the Paleocene-Eocene boundary: Climatic and evolutionary consequences of tectonic events. *Palaeogeography, Palaeoclimatology, Palaeoecology* 79:117–28.

Reinhardt, J. and W. R. Sigleo. 1988. *Paleosols and Weathering Through Geologic Time: Principles and Applications*. Geological Society of America Special Paper 216.

Renaut, R. W. and J.-J. Tiercelin. 1994. Lake Bogoria, Kenya Rift Valley—A sedimentological overview. In R. W. Renaut and W. M. Last, eds., *Sedimentology and geochemistry of Modern and Ancient Saline Lakes*, pp. 101–23. SEPM (Society for Sedimentary Geology) Special Publication 50.

Research on Cretaceous Cycles Group. 1986. Rhythmic bedding in Upper Cretaceous pelagic carbonate sequences: Varying sedimentary response to climatic forcing. *Geology* 14:153–6.

Retallack, G. J. 1981. Preliminary observations on fossil soils in the Clarno Formation (Eocene to early Oligocene) near Clarno, Oregon. *Oregon Geology* 43:147–50.

———. 1983. A paleopedological approach to the interpretation of terrestrial sedimentary rocks: The mid-Tertiary fossil soils of Badlands National Park, South Dakota. *Geological Society of America Bulletin* 94:823–40.

———. 1988. Field recognition of paleosols. In J. Reinhardt and W. R. Sigleo, eds., *Paleosols and Weathering Through Geologic Time: Principles and Applications*, pp. 1–20. Geological Society of America Special Paper 216.

———. 1990. *Soils of the Past*. Boston: Unwin Hyman.

———. 1993. Classification of paleosols: Discussion. *Geological Society of America Bulletin* 105:1635–6.

———. 1994. A pedotype approach to latest Cretaceous and earliest Tertiary paleosols in eastern Montana. *Geological Society of America Bulletin* 106:1377–97.

Reverdin, G. 1995. The physical processes of open ocean upwelling systems. In C. P. Summerhayes, K.-C. Emeis, M. V. Angel, R. L. Smith, and B. Zeitzschel, eds., *Upwelling in the Ocean: Modern Processes and Ancient Records*, pp. 125–48. Chichester: John Wiley and Sons.

Rhoads, D. C. and J. W. Morse. 1971. Evolutionary and ecological significance of oxygen-deficient marine basins. *Lethaia* 4:413–28.

Rich, T. H. V. and P. V. Rich. 1989. Polar dinosaurs and biotas of the Early Cretaceous of southeastern Australia. *National Geographic Research* 5:15–53.

Richey, J. E. and R. L. Victoria. 1993. C, N, P export dynamics in the Amazon River. In R. Wollast, F. T. Mackenzie, and L. Chou, eds., *Interactions of C, N, P and S Biogeochemical Cycles and Global Change*, pp. 123–39. NATO ASI Series I, Global Environmental Change 4. Berlin: Springer-Verlag.

Richter, F. M., D. B. Rowley, and D. J. DePaolo. 1992. Sr isotope evolution of seawater: The role of tectonics. *Earth and Planetary Science Letters* 109:11–23.

Rickards, B., S. Rigby, and J. H. Harris. 1990. Graptoloid biogeography: Recent progress, future hopes. In W. S. McKerrow and C. R. Scotese, eds., *Palaeozoic Palaeogeography and Biogeography*, pp. 139–45. Geological Society of London Memoir 12.

Ridgway, K. D. and P. G. DeCelles. 1993. Stream-dominated alluvial fan and lacustrine depositional systems in Cenozoic strike-slip basins, Denali fault system, Yukon Territory, Canada. *Sedimentology* 40:645–66.

Riedel, W. R. and A. Sanfilippo. 1977. Cainozoic radiolaria. In A. T. S. Ramsay, ed. *Oceanic Micropalaeontology*, pp. 847–912 2. London: Academic Press.

Riggs, S. R. 1984. Paleoceanographic model of Neogene phosphorite deposition, U.S. Atlantic continental margin. *Science* 223:123–31.

Robert, C. and H. Chamley. 1987. Cenozoic evolution of continental humidity and paleoenvironment, deduced from the kaolinite content of oceanic sediments. *Palaeogeography, Palaeoclimatology, Palaeoecology* 60:171–87.

———. 1991. Development of early Eocene warm climates, as inferred from clay mineral variations in oceanic sediments. *Palaeogeography, Palaeoclimatology, Palaeoecology (Global and Planetary Change)* 89(3):315–31.

Robert, C. and J. P. Kennett. 1992. Paleocene and Eocene kaolinite distribution in the South Atlantic and Southern Ocean: Antarctic climatic and paleoceanographic implications. *Marine Geology* 103:99–110.

———. 1994. Antarctic subtropical humid episode at the Paleocene-Eocene boundary: Clay-mineral evidence. *Geology* 22:211–4.

———. 1997. Antarctic continental weathering changes during Eocene-Oligocene cryosphere expansion: Clay mineral and oxygen isotope evidence. *Geology* 25:587–90.

Robert, C. and H. Maillot. 1990. Paleoenvironments in the Weddell Sea area and Antarctic climates, as deduced from clay mineral associations and geochemical data, ODP Leg 113. In P. F. Barker, J. P. Kennett, et al. *Proceedings of the Ocean Drilling Program, Scientific Results, Leg 113*, pp. 51–70. College Station, TX: Ocean Drilling Program.

Roberts, L. N. R. and P. J. McCabe. 1992. Peat accumulation in coastal-plain mires: A model for coals of the Fruitland Formation (Upper Cretaceous) of southern Colorado, USA. *International Journal of Coal Geology* 21:115–38.

Robinson, D. and V. P. Wright. 1987. Ordered illite-smectite and kaolinite-smectite: Pedogenic minerals in a Lower Carboniferous paleosol sequence, South Wales? *Clay Minerals* 22:109–18.

Robinson, P. L. 1973. Palaeoclimatology and continental drift. In D. H. Tarling and S. K. Runcorn, eds., *Implications of Continental Drift to the Earth Sciences Vol. I*, pp. 449–76. London: Academic Press.

Rocha-Campos, A. C., P. R. Dos Santos, and J. R. Canuto. 1994. Ice scouring structures in Late Paleozoic rhythmites, Paraná Basin, Brazil. In M. Deynoux, J. M. G. Miller, E. W. Domack, N. Eyles, I. J. Fairchild, and G. M. Young, eds., *Earth's Glacial Record*, pp. 234–40. Cambridge: Cambridge University Press.

Roeschmann, G. 1971. Problems concerning investigations of paleosols in older sedimentary rocks, demonstrated by the example of Wurzelböden of the Carboniferous System. In D. H. Yaalon, ed. *Paleopedology*, pp. 311–20. Jerusalem: International Society of Soil Science and Israel Universities Press.

Romine, K. 1985. Radiolarian biogeography and paleoceanography of the North Pacific at 8 Ma. In

J. P. Kennett, ed. *The Miocene Ocean: Paleoceanography and Biogeography*, pp. 237–72. Geological Society of America Memoir 163.

Ronov, A. B. 1964. Common tendencies in the chemical evolution of the Earth's crust, ocean and atmosphere. *Geochemistry* 8:715–43.

Rosen, B. R. 1984. Reef coral biogeography and climate through the Late Cainozoic. In P. Brenchley, ed. *Fossils and Climate*, pp. 201–62. Chichester: Wiley.

———. 1992. Empiricism and the biogeographical black box: Concepts and methods in marine palaeobiogeography. *Palaeogeography, Palaeoclimatology, Palaeoecology* 92:171–205.

Ross, C. A. 1973. Carboniferous Foraminiferida. In A. Hallam, ed. *Atlas of Palaeobiogeography*, pp. 128–32. Amsterdam: Elsevier Scientific.

Ross, J. R. P. and C. A. Ross. 1990. Late Palaeozoic bryozoan biogeography. In W. S. McKerrow and C. R. Scotese, eds., *Palaeozoic Palaeogeography and Biogeography*, pp. 353–62. Geological Society of London Memoir 12.

Ross, R. J., Jr. 1975. Early Paleozoic trilobites, sedimentary facies, lithospheric plates, and ocean currents. *Fossils and Strata* 4:307–29.

Rossinsky, V., Jr., H. R. Wanless, and P. K. Swart. 1992. Penetrative calcretes and their stratigraphic implications. *Geology* 20:331–4.

Roth, B. and P. K. M. Megaw. 1989. Early Tertiary land mollusks (Gastropoda: Pulmonata) from Sierra Santa Eulalia, Chihuahua, Mexico, and the origins of the North American arid-land mollusk fauna. *Malacological Review* 22:1–16.

Roth, P. H. and K. R. Krumbach. 1986. Middle Cretaceous calcareous nannofossil biogeography and preservation in the Atlantic and Indian Oceans: Implications for paleoceanography. *Marine Micropaleontology* 10:235–66.

Roulier, L. M. and T. M. Quinn. 1995. Seasonal- to decadal-scale climatic variability in southwest Florida during the middle Pliocene: Inferences from a coralline stable isotope record. *Paleoceanography* 10:429–43.

Rowley, D. B., A. Raymond, J. T. Parrish, A. L. Lottes, C. R. Scotese, and A. M. Ziegler. 1985. Carboniferous paleogeographic, phytogeographic, and paleoclimatic reconstructions. *International Journal of Coal Geology* 5:7–42.

Rubin, D. M. 1987. *Cross-Bedding, Bedforms, and Paleocurrents*. Concepts in Sedimentology and Paleontology 1. Tulsa: Society of Economic Paleontologists and Mineralogists.

Rubin, D. M. and R. E. Hunter. 1982. Bedform climbing in theory and nature. *Sedimentology* 29:121–38.

———. 1983. Reconstructing bedform assemblages from compound crossbedding. In M. E. Brookfield and T. S. Ahlbrandt, eds., *Eolian Sediments and Processes*, pp. 407–27. Developments in Sedimentology 38. Amsterdam: Elsevier.

Rubin, D. M. and H. Ikeda. 1990. Flume experiments on the alignment of transverse, oblique, and longitudinal dunes in directionally varying flows. *Sedimentology* 37:673–84.

Ruddiman, W. F. 1977. Investigations of Quaternary climate based on planktonic foraminifera. In A. T. S. Ramsay, ed. *Oceanic Micropalaeontology Vol. 1*, pp. 101–62. London: Academic Press.

Ruddiman, W. F. and J. E. Kutzbach. 1991. Plateau uplift and climatic change. *Scientific American* 264(3):66–75.

Ruddiman, W. F. and A. McIntyre. 1976. Northeast Atlantic paleoclimatic changes over the last 600,000 years. *Geological Society of America Memoir* 145:111–46.

Ruddiman, W. F., W. L. Prell, and M. E. Raymo. 1989. History of late Cenozoic uplift on Southeast Asia and the American Southwest: Rationale for general circulation modeling experiments. *Journal of Geophysical Research* 94:18379–91.

Ruddiman, W. F. and M. E. Raymo. 1988. Northern Hemisphere climate régimes during the past 3 Ma: Possible tectonic connections. *Philosophical Transactions of the Royal Society of London, Series B*:411–30.

Rumney, G. R. 1968. *Climatology and the World's Climates*. New York: Macmillan.

Runnegar, B. 1979. Ecology of *Eurydesma* and the *Eurydesma* fauna, Permian of eastern Australia. *Alcheringa* 3:261–85.

Rust, B. R. 1981. Sedimentation in an arid-zone anastomosing fluvial system: Cooper's Creek, central Australia. *Journal of Sedimentary Petrology* 51:745–55.

Ruttenberg, K. C. 1993. Reassessment of the oceanic residence time of phosphorus. *Chemical Geology* 107:405–9.

Ruttenberg, K. C. and R. A. Berner. 1993. Authigenic apatite formation and burial in sediments from non-upwelling continental margin environments. *Geochimica et Cosmochimica Acta* 57:991–1007.

Ryer, T. A. 1984. Transgressive-regressive cycles and the occurrence of coal in some Upper Cretaceous strata of Utah, U.S.A. In R. A. Rahmani and R. M. Flores, eds., *Sedimentology of Coal and*

Coal-bearing Strata, pp. 217–27. International Association of Sedimentologists Special Publication 7. Oxford: Blackwell Scientific.

Ryer, T. A. and A. W. Langer. 1980. Thickness change involved in the peat-to-coal transformation for a bituminous coal of Cretaceous age in central Utah. *Journal of Sedimentary Petrology* 50:987–92.

Sachs, H. M., T. Webb, III, and D. R. Clark. 1977. Paleoecological transfer functions. *Annual Reviews of Earth and Planetary Science* 5:159–78.

Sadler, P. M. and D. J. Strauss. 1990. Estimation of completeness of stratigraphical sections using empirical data and theoretical methods. *Journal of the Geological Society of London* 147:471–85.

Sælen, G., P. Doyle, and M. R. Talbot. 1996. Stable-isotope analyses of belemnite rostra from the Whitby Mudstone Fm., England: Surface water conditions during deposition of a marine black shale. *Palaios* 11:97–117.

Sageman, B. B. and D. J. Hollander. (In press.) Integration of paleoecological and geochemical proxies: A holistic approach to the study of past global change. In E. Barrera and C. C. Johnson, eds., *The Evolution of Cretaceous Ocean/Climate Systems.* Geological Society of America Special Paper.

Sageman, B. B., P. B. Wignall, and E. G. Kauffman. 1991. Biofacies models for oxygen-deficient facies in epicontinental seas: Tool for paleoenvironmental analysis. In G. Einsele, W. Ricken, and A. Seilacher, eds., *Cycles and Events in Stratigraphy*, pp. 542–64. Berlin: Springer-Verlag.

Salisbury, E. J. 1928. On the causes and ecological significance of stomatal frequency, with special reference to the woodland flora. *Philosophical Transactions of the Royal Society of London, Series B* 216:1–65.

Saltzman, E. S. and E. J. Barron. 1982. Deep circulation in the Late Cretaceous: Oxygen isotope paleotemperatures from *Inoceramus* remains in D. S. D. P. cores. *Palaeogeography, Palaeoclimatology, Palaeoecology* 40:167–81.

Samson, S. D., P. J. Patchett, J. C. Roddick, and R. R. Parrish. 1989. Origin and tectonic setting of Ordovician bentonites in North America: Isotopic and age constraints. *Geological Society of America Bulletin* 101:1175–81.

Sancetta, C., L. Heusser, and M. A. Hall. 1992. Late Pliocene climate in the Southeast Atlantic: Preliminary results from a multi-disciplinary study of DSDP Site 532. *Marine Micropaleontology* 20:59–75.

Sancetta, C. and S. Silvestri. 1986. High-resolution biostratigraphy and oceanographic events in the late Pliocene and Pleistocene North Pacific Ocean. *Paleoceanography* 1:163–80.

Sánchez Chillón, B., M. T. Alberdi, G. Leone, F. P. Bonadonna, B. Stenni, and A. Longinelli. 1994. Oxygen isotopic composition of fossil equid tooth and bone phosphate: An archive of difficult interpretation. *Palaeogeography, Palaeoclimatology, Palaeoecology* 107:317–28.

Sandberg, C. A. 1973. Conodont biofacies of Late Devonian *Polygnathus styriacus* zone in western United States. In C. R. Barnes, ed. *Conodont Paleoecology*, pp. 171–86. Geological Association of Canada Special Paper 15.

Sarmiento, J. L., T. D. Herbert, and J. R. Toggweiler. 1988a. Causes of anoxia in the World Ocean. *Global Biogeochemical Cycles* 2:115–28.

———. 1988b. Mediterranean nutrient balance and episodes of anoxia. *Global Biogeochemical Cycles* 2:427–44.

Sarnthein, M., U. Pflaumann, R. Ross, R. Tiedemann, and K. Winn. 1992. Transfer functions to reconstruct ocean palaeoproductivity: A comparison. In C. P. Summerhayes, W. L. Prell, and K. C. Emeis, eds., *Upwelling Systems: Evolution Since the Early Miocene*, pp. 411–27. Geological Society of London Special Publication 64.

Sarnthein, M., G. Tetzlaff, B. Koopmann, K. Wolter, and U. Pflaumann. 1981. Glacial and interglacial wind regimes over the eastern subtropical Atlantic and North-West Africa. *Nature* 293:193–6.

Sarnthein, M., J. Thiede, U. Pflaumann, H. Erlenkeuser, D. Fütterer, B. Koopmann, H. Lange, and E. Seibold. 1982. Atmospheric and oceanic circulation patterns off Northwest Africa during the past 25 million years. In U. von Rad, K. Hinz, M. Sarnthein, and E. Seibold, eds., *Geology of the Northwest African Continental Margin*, pp. 545–604. Berlin: Springer-Verlag.

Sarnthein, M., K. Winn, J.-C. Duplessy, and M. R. Fontugne. 1988. Global variations of surface ocean productivity in low and mid latitudes: Influence on CO_2 reservoirs of the deep ocean and atmosphere during the last 21,000 years. *Paleoceanography* 3:361–99.

Savin, S. M. 1977. The history of the Earth's surface temperature during the past 100 million years. *Annual Review of Earth and Planetary Sciences* 5:319–56.

Savin, S. M., L. Abel, E. Barrera, D. Hodell, J. P. Kennett, M. Murphy, G. Keller, J. Killingley, and E. Vincent. 1985. The evolution of Miocene surface and near-surface marine temperatures: Oxygen isotopic evidence. In J. P. Kennett, ed. *The Miocene Ocean: Paleoceanography and Biogeog-*

raphy, pp. 49–82. Geological Society of America Memoir 163.

Savrda, C. E. and D. J. Bottjer. 1988. The exaerobic zone, a new oxygen-deficient marine biofacies. *Nature* 327:54–6.

———. 1989. Trace-fossil model for reconstructing oxygenation histories of ancient marine bottom waters: Application to Upper Cretaceous Niobrara Formation, Colorado. *Palaeogeography, Palaeoclimatology, Palaeoecology* 74:49–74.

Savrda, C. E., D. J. Bottjer, and D. S. Gorsline. 1984. Development of a comprehensive oxygen-deficient marine biofacies model: Evidence from Santa Monica, San Pedro, and Santa Barbara Basins, California continental borderland. *American Association of Petroleum Geologists Bulletin* 68:1179–92.

Scherer, M. 1977. Preservation, alteration and multiple cementation of aragonitic skeletons from the Cassian Beds (U. Triassic, southern Alps): Petrographic and geochemical evidence. *Neues Jahrbuch für Geologie und Paläontologie Abhandlungen* 154:213–62.

Schermerhorn, L. J. G. 1974. Late Precambrian mixtites: Glacial and/or nonglacial? *American Journal of Science* 274:673–824.

Schlager, W. 1981. The paradox of drowned reefs and carbonate platforms. *Geological Society of America Bulletin* 92:197–211.

Schlager, W. and J. Philip. 1990. Cretaceous carbonate platforms. In R. N. Ginsburg and B. Beaudoin, eds., *Createceous Resources, Events and Rhythms*, pp. 173–95. Dordrecht: Kluwer Academic.

Schlanger, S. O. 1981. Shallow-water limestones in ocean basins as tectonic and paleoceanic indicators. In J. E. Warme, R. G. Douglas, and E. L. Winterer, eds., *The Deep Sea Drilling Project: A Decade of Progress*, pp. 209–26. Society of Economic Paleontologists and Mineralogists Special Publication 32.

Schlanger, S. O., M. A. Arthur, H. C. Jenkyns, and P. A. Scholle. 1987. The Cenomanian-Turonian oceanic anoxic event, I. Stratigraphy and distribution of organic carbon-rich beds and the marine δ^{13} C excursion. In J. Brooks and A. J. Fleet, eds., *Marine Petroleum Source Rocks*, pp. 371–400. Geological Society of London Special Publication 15.

Schlanger, S. O. and H. C. Jenkyns. 1976. Cretaceous oceanic anoxic events: Causes and consequences. *Geologie en Mijnbouw* 55:179–84.

Schlesinger, M. E. 1984. Mathematical modeling and simulation of climate and climatic change. Report, Institut D'Astronomie et de Geophysique, Univ. Catholique de Louvain, Georges Lemaitre, Belgium.

Schlesinger, M. E. and M. Verbitsky. 1996. Simulation of glacial onset with a coupled atmospheric general circulation/mixed-layer ocean—Ice-sheet/asthenosphere model. *Palaeoclimates: Data and Modelling* 1:179–201.

Schlische, R. W. and P. E. Olsen. 1990. Quantitative filling model for continental extensional basin with applications to early Mesozoic rifts of eastern North America. *Journal of Geology* 98: 135–55.

Schluter, D. and R. E. Ricklefs. 1993. Species diversity, an introduction to the problem. In R. E. Ricklefs and D. Schluter, eds., *Species Diversity in Ecological Communities*, pp. 1–10. Chicago: University of Chicago Press.

Schmalz, R. F. 1969. Deep-water evaporite deposition: A genetic model. *American Association of Petroleum Geologists Bulletin* 53:798–823.

Schmidt, G. A. and L. A. Mysak. 1996. Can increased poleward oceanic heat flux explain the warm Cretaceous climate? *Paleoceanography* 11:579–93.

Schneider, R. R., P. J. Müller, and G. Ruhland. 1995. Late Quaternary surface circulation in the east equatorial South Atlantic: Evidence from alkenone sea surface temperatures. *Paleoceanography* 10:197–219.

Schoell, M., S. Schouten, J. S. Sinninghe Damsté, J. W. de Leeuw, and R. E. Summons. 1994. A molecular organic carbon isotope record of Miocene climate changes. *Science* 263:1122–5.

Scholle, P. A. and M. A. Arthur. 1980. Carbon isotope fluctuations in Cretaceous pelagic limestones: Potential stratigraphic and petroleum exploration tool. *American Association of Petroleum Geologists Bulletin* 64:67–87.

Schopf, J. M. 1973. Coal, climate and global tectonics. In D. H. Tarling and S. K. Runcorn, eds., *Implications of Continental Drift to the Earth Sciences Vol. I*, pp. 609–22. London: Academic Press.

Schopf, T. J. M. 1970. Taxonomic diversity gradients of ectoprocts and bivalves and their geologic implications. *Geological Society of America Bulletin* 81:3765–8.

———. 1979. The role of biogeographic provinces in regulating marine faunal diversity through geologic time. In J. Gray and A. J. Boucot, eds., *Historical Biogeography, Plate Tectonics, and the Changing Environment*, pp. 449–57. Corvallis: Oregon State University Press.

Schrader, H. 1992. Peruvian coastal primary palaeoproductivity during the last 200 000 years. In C. P. Summerhayes, W. L. Prell, and K. C. Emeis, eds., *Upwelling Systems: Evolution Since the Early Miocene*, pp. 391–409. Geological Society of London Special Publication 64.

Schramm, C. T. 1989. Cenozoic climatic variation recorded by quartz and clay minerals in North Pacific sediments. In M. Leinen and M. Sarnthein, eds., *Paleoclimatology and Paleometeorology: Modern and Past Patterns of Global Atmospheric Transport*, pp. 805–39. NATO ASI Series C: Mathematical and Physical Sciences 282. Dordrecht: Kluwer Academic.

Schreiber, B. C. 1988. Introduction. In B. C. Schreiber, ed. *Evaporites and Hydrocarbons*, pp. 1–10. New York: Columbia University Press.

Schroeder, J. O., R. W. Murray, M. Leinen, R. C. Pflaum, and T. R. Janecek. 1997. Barium in equatorial Pacific carbonate sediment: Terrigenous, oxide, and biogenic associations. *Paleoceanography* 12:125–46.

Schumm, S. A. and G. R. Brakenridge. 1987. River responses. In W. F. Ruddiman and H. E. Wright, Jr., eds., *North America and Adjacent Oceans During the Last Deglaciation*, pp. 221–40. The Geology of North America K-3. Boulder: Geological Society of America.

Schumm, S. A., M. P. Mosley, and W. E. Weaver. 1987. *Experimental Fluvial Geomorphology*. Chichester: John Wiley and Sons.

Schwan, J. 1988. The structure and genesis of Weichselian to early Holocene aeolian sand sheets in western Europe. *Sedimentary Geology* 55:197–232.

Schwarzacher, W. and A. G. Fischer. 1982. Limestone-shale bedding and perturbations of the Earth's orbit. In G. Einsele and A. Seilacher, eds., *Cyclic and Event Stratification*, pp. 72–95. Berlin: Springer-Verlag.

Schweitzer, H. J. 1980. Environment and climate in the Early Tertiary of Spitzbergen. *Palaeogeography, Palaeoclimatology, Palaeoecology* 30:297–311.

Sclater, J. C., S. Hellinger, and C. Tapscott. 1977. The paleobathymetry of the Atlantic Ocean from the Jurassic to the present. *Journal of Geology* 85:509–52.

Scotese, C. R., R. K. Bambach, C. Barton, R. Van der Voo, and A. M. Ziegler. 1979. Paleozoic base maps. *Journal of Geology* 87:217–77.

Scotese, C. R. and S. F. Barrett. 1990. Gondwana's movement over the South Pole during the Palaeozoic: Evidence from lithological indicators of climate. In W. S. McKerrow and C. R. Scotese, eds., *Palaeozoic Palaeogeography and Biogeography*, pp. 75–85. Geological Society of London Memoir 12.

Scotese, C. R. and J. Golonka. 1992. *Paleogeographic Atlas*. Arlington: PALEOMAP Project, Dept. of Geology, University of Texas, Arlington.

Scotese, C. R. and W. S. McKerrow. 1990. Revised world maps and introduction. In W. S. McKerrow and C. R. Scotese, eds., *Palaeozoic Palaeogeography and Biogeography*, pp. 1–21. Geological Society of London Memoir 12.

Scotese, C. R. and C. P. Summerhayes. 1986. Computer model of paleoclimate predicts coastal upwelling in the Mesozoic and Cenozoic. *Geobyte* 1:28–42.

Scott, R. W. 1975. Patterns of Early Cretaceous molluscan diversity gradients in south-central United States. *Lethaia* 8:241–52.

———. 1995. Global environmental controls on Cretaceous reefal ecosystems. *Palaeogeography, Palaeoclimatology, Palaeoecology* 119:187–99.

Scrutton, C. T. 1978. Periodic growth features in fossil organisms and the length of the day and month. In P. Brosche and J. Sündermann, eds., *Tidal Friction and the Earth's Rotation*, pp. 154–96. Berlin: Springer-Verlag.

Seilacher, A. 1967. Bathymetry of trace fossils. *Marine Geology* 5:413–28.

———. 1982. Distinctive features of sandy tempestites. In G. Einsele and A. Seilacher, eds., *Cyclic and Event Stratification*, pp. 333–49. Berlin: Springer-Verlag.

Selby, M. J., R. B. Rains, and R. W. P. Palmer. 1974. Eolian deposits of the ice-free Victoria Land, Antarctica. *New Zealand Journal of Geology and Geophysics* 17:543–62.

Sen Gupta, B. K. and M. L. Machain-Castillo. 1993. Benthic foraminifera in oxygen-poor habitats. *Marine Micropaleontology* 20:183–201.

Sepkoski, J. J., Jr. 1982. Flat-pebble conglomerates, storm deposits, and the Cambrian bottom fauna. In G. Einsele and A. Seilacher, eds., *Cyclic and Event Stratification*, pp. 372–85. Berlin: Springer-Verlag.

Shackleton, N. J. 1967. Oxygen isotope analysis and Pleistocene temperatures re-assessed. *Nature* 215: 15–7.

———. 1984. Oxygen isotope evidence for Cenozoic climatic change. In P. J. Brenchley, ed. *Fossils and Climate*, pp. 27–34. Chichester: John Wiley and Sons.

———. 1986. Paleogene stable isotope events. *Palaeogeography, Palaeoclimatology, Palaeoecology* 57:91–102.

Shackleton, N. J. and J. P. Kennett. 1975a. Late Cenozoic oxygen and carbon isotopic changes at DSDP Site 284: Implications for glacial history of the Northern Hemisphere and Antarctica. In J. P. Kennett, R. E. Houtz, et al. *Initial Reports of the Deep-Sea Drilling Project, Leg 29*, pp. 801–7. Washington, D.C.: U.S. Government Printing Office.

———. 1975b. Paleotemperature history of the Cenozoic and the initiation of Antarctic glaciation: Oxygen and carbon isotope analyses in DSDP Sites 277, 279, and 281. In J. P. Kennett, R. E. Houtz, et al. *Initial Reports of the Deep Sea Drilling Project, Leg 29*, pp. 743–55. Washington, D.C.: U.S. Government Printing Office.

Shaeffer, B. 1970. Mesozoic fishes and climate. *Proceedings of the North American Paleontological Convention, Chicago, Section D*:376–88.

Sheard, M. J. 1990. Glendonites from the southern Eromanga Basin in South Australia: Palaeoclimatic indicators for Cretaceous ice. *Quarterly Geological Notes of the Geological Survey of South Australia* 114:17–21.

Shearer, J. C., J. R. Staub, and T. A. Moore. 1994. The conundrum of coal bed thickness: A theory for stacked mire sequences. *Journal of Geology* 102:611–7.

Shearman, D. and A. J. Smith. 1985. Ikaite, the parent mineral of jarrowsite-type pseudomorphs. *Proceedings of the Geologists' Association* 96: 305–14.

Sheehan, P. M. and P. J. Coorough. 1990. Brachiopod zoogeography across the Ordovician-Silurian extinction event. In W. S. McKerrow and C. R. Scotese, eds., *Palaeozoic Palaeogeography and Biogeography*, pp. 181–7. Geological Society of London Memoir 12.

Sheldon, R. P. 1980. Episodicity of phosphate deposition and deep ocean circulation. In Y. K. Bentor, ed. *Marine Phosphorites*, pp. 239–48. Society of Economic Paleontologists and Mineralogists Special Publication 29.

Shemesh, A., L. H. Burckle, and J. D. Hays. 1995. Late Pleistocene oxygen isotope records of biogenic silica from the Atlantic sector of the Southern Ocean. *Paleoceanography* 10:179–96.

Shemesh, A., C. D. Charles, and R. G. Fairbanks. 1992. Oxygen isotopes in biogenic silica: Global changes in ocean temperature and isotopic composition. *Science* 256:1434–6.

Shemesh, A., Y. Kolodny, and B. Luz. 1983. Oxygen isotope variations in phosphate of biogenic apatites, II. Phosphorite rocks. *Earth and Planetary Science Letters* 64.

———. 1988. Isotope geochemistry of oxygen and carbon and phosphate and carbonate of phosphorite francolite. *Geochimica et Cosmochimica Acta* 52:2565–72.

Shemesh, A., S. A. Macko, C. D. Charles, and G. H. Rau. 1993. Isotopic evidence for reduced productivity in the glacial Southern Ocean. *Science* 262:407–10.

Shemesh, A., R. A. Mortlock, and P. N. Froelich. 1989. Late Cenozoic Ge/Si record of marine biogenic opal: Implications for variations of riverine fluxes to the ocean. *Paleoceanography* 4:221–34.

Sheppard, R. A. and A. J. Gude, 3rd. 1968. Distribution and genesis of authigenic silicate minerals in tuffs of Pleistocene Lake Tecopa, Inyo County, California. *U.S. Geological Survey Professional Paper* 597.

Shi, G. R. 1993a. A comparative study of 39 binary similarity coefficients. *Memoir of the Assoication of Australasian Palaeontologists* 15:329–41.

———. 1993b. Multivariate data analysis in palaeoecology and palaeobiogeography—a review. *Palaeogeography, Palaeoclimatology, Palaeoecology* 105: 199–234.

Shimmield, G. B. and R. A. Jahnke. 1995. Particle flux and its conversion to the sediment record: Open ocean upwelling systems. In C. P. Summerhayes, K.-C. Emeis, M. V. Angel, R. L. Smith, and B. Zeitzschel, eds., *Upwelling in the Ocean: Modern Processes and Ancient Records*, pp. 171–91. Chichester: John Wiley and Sons.

Siehl, A. and J. Thein. 1989. Minette-type ironstones. In T. P. Young and W. E. G. Taylor, eds., *Phanerozoic Ironstones*, pp. 175–93. Geological Society of London Special Publication 46.

Siesser, W. G. 1993. Calcareous nannoplankton. In J. H. Lipps, ed. *Fossil Prokaryotes and Protists*, pp. 169–201. Oxford: Blackwell Scientific.

Siesser, W. G., T. J. Bralower, and E. H. De Carlo. 1992. Mid-Tertiary *Braarudosphaera*-rich sediments on the Exmouth Plateau. In U. von Rad, B. U. Haq, et al. *Proceedings of the Ocean Drilling Program, Scientific Results, Leg 122*, pp. 653–63. College Station, TX: Ocean Drilling Program.

Simon-Coinçon, R., M. Thiry, and J.-M. Schmitt. 1997. Variety and relationships of weathering features along the early Tertiary palaeosurface in the southerwestern French Massif Central and the nearby Aquitaine Basin. *Palaeogeography, Palaeoclimatology, Palaeoecology* 129:51–79.

Singer, A. 1984. The paleoclimatic interpretation of clay minerals in sediments—A review. *Earth-Science Reviews* 21:251–93.

Skelton, P. W. and S. K. Donovan. 1997. Middle Cretaceous reef collapse linked to ocean heat transport: Comment. *Geology* 25:477–8.

Slaughter, M. and J. W. Earley. 1965. *Mineralogy and geological significance of the Mowry bentonites.* Geological Society of America Special Paper 83.

Slingerland, R., L. R. Kump, M. A. Arthur, P. J. Fawcett, B. B. Sageman, and E. J. Barron. 1996. Estuarine circulation in the Turonian Western Interior seaway of North America. *Geological Society of America Bulletin* 108:941–52.

Sloan, L. C. 1994. Equable climates during the early Eocene: Significance of regional paleogeography for North American climate. *Geology* 22:881–4.

Sloan, L. C. and E. J. Barron. 1990. "Equable" climates during Earth history? *Geology* 18:489–92.

———. 1991. Reply to comments on " 'Equable' climates during Earth history?." *Geology* 19:540–2.

———. 1992. A comparison of Eocene climate model results to quantified interpretations. *Palaeogeography, Palaeoclimatology, Palaeoecology* 93:183–202.

Sloan, L. C. and D. K. Rea. 1995. Atmospheric carbon dioxide and early Eocene climate: A general circulation modeling sensitivity study. *Palaeogeography, Palaeoclimatology, Palaeoecology* 119: 275–92.

Sloan, L. C., J. C. G. Walker, and T. C. Moore, Jr. 1995. Possible role of oceanic heat transport in early Eocene climate. *Paleoceanography* 10: 347–56.

Sloan, L. C., J. C. G. Walker, T. C. Moore, Jr., D. K. Rea, and J. C. Zachos. 1992. Possible methane-induced polar warming in the early Eocene. *Nature* 357:320–2.

Smalley, I. J. 1966. The properties of glacial loess and the formation of loess deposits. *Journal of Sedimentary Petrology* 36:669–76.

Smart, C. W., S. C. King, A. J. Gooday, J. W. Murray, and E. Thomas. 1994. A benthic foraminiferal proxy of pulsed organic matter paleofluxes. *Marine Micropaleontology* 23:89–99.

Smart, C. W. and J. W. Murray. 1994. An early Miocene Atlantic-wide foraminiferal/palaeoceanographic event. *Palaeogeography, Palaeoclimatology, Palaeoecology* 108:139–48.

Smiley, C. J. 1967. Paleoclimatic interpretations of some Mesozoic floral sequences. *American Association of Petroleum Geologists Bulletin* 51: 849–63.

———. 1969. Floral zones and correlations of Cretaceous Kukpowruk and Corwin Formations, northwestern Alaska. *American Association of Petroleum Geologists Bulletin* 53:2079–93.

Smith, A. G., J. C. Briden, and G. E. Drewry. 1973. Phanerozoic world maps. In N. F. Hughes, ed. *Organisms and Continents Through Time*, pp. 1–42. Special Paper of the Palaeontological Association 12.

Smith, G. A. 1994. Climatic influences on continental deposition during late-stage filling of an extensional basin, southeastern Arizona. *Geological Society of America Bulletin* 106:1212–28.

Smith, G. A., Y. Wang, T. E. Cerling, and J. W. Geissman. 1993. Comparison of a paleosol-carbonate isotope record to other records of Pliocene-early Pleistocene climate in the western United States. *Geology* 21:691–4.

Smith, G. I. 1979. Subsurface stratigraphy and geochemistry of late Quaternary evaporites, Searles Lake, California. *U.S. Geological Survey Professional Paper* 1043.

Smith, G. R. and J. G. Lundberg. 1972. The Sand Draw fish fauna. In M. F. Skinner and C. W. Hibbard, eds., *Early Pleistocene Preglacial and Glacial Rocks and Faunas of North-Central Nebraska*, pp. 40–54. American Museum of Natural History Bulletin 148 (1).

Smith, G. R. and W. P. Patterson. 1994. Mio-Pliocene seasonality on the Snake River plain: Comparison of faunal and oxygen isotopic evidence. *Palaeogeography, Palaeoclimatology, Palaeoecology* 107: 291–302.

Smith, R. L. 1995. The physical processes of coastal ocean upwelling systems. In C. P. Summerhayes, K.-C. Emeis, M. V. Angel, R. L. Smith, and B. Zeitzschel, eds., *Upwelling in the Ocean: Modern Processes and Ancient Records*, pp. 39–64. Chichester: John Wiley and Sons.

Smoot, J. P. 1985. The closed-basin hypothesis and its use in facies analysis of the Newark Supergroup. *U.S. Geological Survey Circular* 946.

Smoot, J. P. and B. Castens-Seidell. 1994. Sedimentary features produced by efflorescent salt crusts, Saline Valley and Death Valley, California. In R. W. Renaut and W. M. Last, eds., *Sedimentology and Geochemistry of Modern and Ancient Saline Lakes*, pp. 73–90. SEPM (Society for Sedimentary Geology) Special Publication 50.

Smoot, J. P. and T. K. Lowenstein. 1991. Depositional environments of non-marine evaporites. In J. L. Melvin, ed. *Evaporite, Petroleum and Mineral Resources*, pp. 189–347. Developments in Sedimentology 50. Amsterdam: Elsevier.

Smoot, J. P. and P. E. Olsen. 1985. Massive mudstones in basin analysis and paleoclimatic interpretation of the Newark Supergroup. *U.S. Geological Survey Circular* 946, pp. 29–33.

———. 1988. Massive mudstones in basin analysis and paleoclimatic interpretation of the Newark Supergroup. In W. Manspeizer, ed. *Triassic-Jurassic Rifting, Part A*, pp. 249–74. Developments in Geotectonics 22. Amsterdam: Elsevier.

Smyth, M. 1989. Organic petrology and clastic depositional environments with special reference to Australian coal basins. *International Journal of Coal Geology* 12:635–56.

Soil Survey Staff. 1975. *Soil Taxonomy*. Soil Conservation Survey Handbook No. 436: U.S. Department of Agriculture.

Soil Survey Staff. 1990. *Keys to Soil Taxonomy*. 4th ed. Soil Management Support Services Technical Monograph 19.

Sonett, C. P., S. A. Finney, and C. R. Williams. 1988. The lunar orbit in the late Precambrian and the Elatina sandstone laminae. *Nature* 335:806–8.

Sonett, C. P. and G. E. Williams. 1985. Solar periodicities expressed in varves from glacial Skilak Lake, southern Alaska. *Journal of Geophysical Research* 90:12019–26.

Sonnenfeld, P. and J.-P. Perthuisot. 1989. *Brines and Evaporites*. 28th International Geological Congress Short Course in Geology 3. Washington, D.C.: American Geophysical Union.

Soreghan, G. S. 1992a. Preservation and paleoclimatic significance of eolian dust in the Ancestral Rocky Mountains province. *Geology* 20:1111–4.

———. 1992b. Sedimentology and process stratigraphy of the Upper Pennsylvanian, Pedregosa (Arizona) and Orogrande (New Mexico) Basins. *Ph.D. thesis,* University of Arizona.

Soulié-Märsche, I. 1993. Apport des charophytes fossiles á la recherche de changements climatiques abrupts. *Bulletin de la Societé Géologique de France* 164:123–30.

Soutar, A., S. R. Johnson, and T. R. Baumgartner. 1981. In search of modern depositional analogs to the Monterey Formation. In R. E. Garrison and R. G. Douglas, eds., *The Monterey Formation and Related Siliceous Rocks of California*, pp. 123–47. Pacific Section, Society of Economic Paleontologists and Mineralogists Special Publication. Los Angeles: Pacific Section, Society of Economic Paleontologists and Mineralogists.

Spaeth, C., J. Hoefs, and U. Vetter. 1971. Some aspects of isotopic composition of belemnites and related paleotemperatures. *Geological Society of America Bulletin* 82:3139–50.

Spero, H. J. 1992. Do planktic foraminifera accurately record shifts in the carbon isotopic composition of seawater ΣCO_2? *Marine Micropaleontology* 19:275–85.

Spezzaferri, S. 1995. Planktonic foraminiferal paleoclimatic implications across the Oligocene-Miocene transition in the oceanic record (Atlantic, Indian and South Pacific). *Palaeogeography, Palaeoclimatology, Palaeoecology* 114:43–74.

Spicer, R. A. 1986a. Comparative leaf architectural analysis of Cretaceous radiating angiosperms. In R. A. Spicer and B. A. Thomas, eds., *Systematic and Taxonomic Approaches in Palaeobotany*, pp. 221–32. Systematics Association Special Volume 31.

———. 1986b. Pectinal veins: A new concept in terminology for the description of dicotyledonous leaf venation patterns. *Botanical Journal of the Linnean Society* 93:379–88.

———. 1989a. Physiological characteristics of land plants in relation to environment through time. *Transactions of the Royal Society of Edinburgh: Earth Sciences* 80:321–9.

———. 1989b. The formation and interpretation of plant fossil assemblages. *Advances in Botanical Research* 16:96–191.

———. 1991. Plant taphonomic processes. In P. A. Allison and D. E. G. Briggs, eds., *Taphonomy: Releasing the Data Locked in the Fossil Record*, pp. 72–113. Mt. Kisco, NY: Plenum.

Spicer, R. A. and A. G. Greer. 1986. Plant taphonomy in fluvial and lacustrine systems. In R. A. Gastaldo, ed. *Land Plants*, pp. 10–26. University of Tennessee Department of Geological Sciences Studies in Geology 15.

Spicer, R. A. and J. T. Parrish. 1986. Paleobotanical evidence for cool North Polar climates in middle Cretaceous (Albian-Cenomanian) time. *Geology* 14:703–6.

———. 1990a. Late Cretaceous-early Tertiary palaeoclimates of northern high latitudes: A quantitative view. *Journal of the Geological Society, London* 147:329–41.

———. 1990b. Latest Cretaceous woods of the central North Slope, Alaska. *Palaeontology* 33:225–42.

Spicer, R. A. and J. A. Wolfe. 1987. Plant taphonomy of late Holocene deposits in Trinity (Clair Engle) Lake, northern California. *Paleobiology* 13: 227–45.

Stach, E., M.-T. Mackowsky, M. Teichmüller, G. H. Taylor, D. Chandra, and R. Teichmüller. 1982. *Coal Petrology*. 3rd ed. Berlin: Gebrüder Borntraeger.

Stanistreet, I. G. and T. S. McCarthy. 1993. The

Okavango Fan and the classification of subaerial fan systems. *Sedimentary Geology* 85:115–33.

Stanton, R. W., T. A. Moore, P. D. Warwick, S. S. Crowley, and R. M. Flores. 1989. Comparative facies formation in selected coal beds of the Powder River Basin. *Coal and Hydrocarbon Resources of North America*, pp. T132:19–27. 28th International Geological Congress 1. Washington, D.C.: American Geophysical Union.

Steers, J. A. 1977. Physiography. In V. J. Chapman, ed. *Wet Coastal Ecosystems*, pp. 31–60. Ecosystems of the World 1. Amsterdam: Elsevier.

Stehli, F. G. 1970. A test of the Earth's magnetic field during Permian time. *Journal of Geophysical Research* 75:3325–42.

Stein, C. and A. J. Smith. 1986. Authigenic carbonate nodules in the Nankai Trough, Site 583. In H. Kagami, D. H. Karig, et al. *Initial Reports of the Deep Sea Drilling Project, Leg 87*, pp. 659–068. Washington, D.C.: U.S. Government Printing Office.

Steuber, T. 1996. Stable isotope sclerochronology of rudist bivalves: Growth rates and Late Cretaceous seasonality. *Geology* 24:315–8.

Stevens, G. R. 1971. Relationship of isotopic temperatures and faunal realms to Jurassic and Cretaceous palaeogeography, particularly of the Southwest Pacific. *Journal of the Royal Society of New Zealand* 1:145–58.

———. 1980. Southwest Pacific faunal palaeobiogeography in Mesozoic and Cenozoic times: A review. *Palaeogeography, Palaeoclimatology, Palaeoecology* 31:153–96.

Stevens, G. R. and R. N. Clayton. 1971. Oxygen isotope studies on Jurassic and Cretaceous belemnites from New Zealand and their biogeographic significance. *New Zealand Journal of Geology and Geophysics* 14:829–97.

Stihler, S. D., D. B. Stone, and J. E. Beget. 1992. "Varve" counting vs. tephrochronology and ^{137}Cs and ^{210}Pb dating: A comparative test at Skilak Lake, Alaska. *Geology* 20:1019–22.

Stoddart, D. R. and T. P. Scottin. 1983. Phosphate rock on coral reef islands. In A. S. Goudie and K. Pye, eds., *Chemical Sediments and Geomorphology*, pp. 369–400. London: Academic Press.

Stow, D. A. V. and W. E. Dean. 1984. Middle Cretaceous black shales at Site 530 in the southeastern Angola Basin. In W. W. Hay, J.-C. Sibuet, et al. *Initial Reports of the Deep Sea Drilling Project, Leg 75*, pp. 809–17. Washington, D.C.: U.S. Government Printing Office.

Stranks, L. and P. England. 1997. The use of a resemblance function in the measurement of climatic parameters from the physiognomy of woody dicotyledons. *Palaeogeography, Palaeoclimatology, Palaeoecology* 131:15–28.

Street-Perrott, F. A. and S. P. Harrison. 1984. Temporal variations in lake levels since 30,000 yr bp—an index of the global hydrological cycle. In J. E. Hansen and T. Takahashi, eds., *Climate Processes and Climate Sensitivity*, pp. 118–29. Geophysical Monograph 29. Washington, D.C.: American Geophysical Union.

———. 1985. Lake levels and climate reconstruction. In A. D. Hecht, ed. *Paleoclimate Analysis and Modeling*, pp. 291–340. New York: John Wiley and Sons.

Suchecki, R. K., J. F. Hubert and C. C. Birney de Wet. 1988. Isotopic imprint of climate and hydrogeochemistry on terrestrial strata of the Triassic-Jurassic Hartford and Fundy rift basins. *Journal of Sedimentary Petrology* 58:801–11.

Suess, E., W. Balzer, K.-F. Hesse, P. J. Müller, and G. Wefer. 1982. Calcium carbonate hexahydrate from organic-rich sediments of the Antarctic shelf: Precursors of glendonites. *Science* 216:1128–31.

Summerfield, M. A. 1983a. Silcrete. In A. S. Goudie and K. Pye, eds., *Chemical Sediments and Geomorphology*, pp. 59–91. London: Academic Press.

———. 1983b. Silcrete as a palaeoclimatic indicator: Evidence from southern Africa. *Palaeogeography, Palaeoclimatology, Palaeoecology* 41:65–79.

Summerhayes, C. P., K.-C. Emeis, M. V. Angel, R. L. Smith, and B. Zeitzschel, eds. 1995. *Upwelling in the Ocean: Modern Processes and Ancient Records*. Chichester: John Wiley and Sons.

Surdam, R. C. and R. A. Sheppard. 1978. Zeolites and saline, alkaline-lake deposits. In L. B. Sand and F. A. Mumpton, eds., *Natural Zeolites: Occurrence, Properties, Use*, pp. 145–74. Elmsford, NY: Pergamon Press.

Suttner, L. J. and P. K. Dutta. 1986. Alluvial sandstone composition and paleoclimate, I. Framework mineralogy. *Journal of Sedimentary Petrology* 56:329–45.

Sverdrup, H. U., M. W. Johnson, and R. H. Fleming. 1942. *The Oceans: Their Physics, Chemistry, and General Biology*. Englewood Cliffs, NJ: Prentice-Hall.

Sweet, M. L. and G. Kocurek. 1990. An empirical model of aeolian dune lee-face flow. *Sedimentology* 37:1023–38.

Sweet, M. L., J. Nielson, K. Havholm, and J. Farrelley. 1988. Algodones dune field of southeastern

California: Case history of a migrating modern dune field. *Sedimentology* 35:939–52.

Sweet, W. C. 1988. *The Conodonta*. Oxford: Oxford University Press.

Sweet, W. C. and S. M. Bergström. 1984. Conodont provinces and biofacies of the Late Ordovician. In D. L. Clark, ed. *Conodont Biofacies and Provincialism*, pp. 69–87. Geological Society of America Special Paper 196.

Swift, D. P. and D. Nummedal. 1987. Hummocky cross-stratification, tropical hurricanes, and intense winter storms—Discussion. *Sedimentology* 34:338–44.

Talbot, M. R. 1985. Major bounding surfaces in aeolian sandstones—a climatic model. *Sedimentology* 32:257–65.

———. 1988. The origns of lacustrine oil source rocks: Evidence from the lakes of tropical Africa. In A. J. Fleet, K. Kelts, and M. R. Talbot, eds., *Lacustrine Petroleum Source Rocks*, pp. 29–43. Geological Society of London Special Paper 40.

———. 1990. A review of the palaeohydrological interpretation of carbon and oxygen isotopic ratios in primary lacustrine carbonates. *Chemical Geology* 80:261–79.

Talbot, M. R. and K. Kelts. 1986. Primary and diagenetic carbonates in the anoxic sediments of Lake Bosumtwi, Ghana. *Geology* 14:912–5.

———. 1990. Paleolimnological signatures from carbon and oxygen isotopic ratios in carbonates from organic carbon-rich lacustrine sediments. In B. J. Katz, ed. *Lacustrine Basin Exploration: Case Studies and Modern Analogs*, pp. 99–112. American Association of Petroleum Geologists Memoir 50.

Talbot, M. R. and D. A. Livingstone. 1989. Hydrogen index and carbon isotopes of lacustrine organic matter as lake level indicators. *Palaeogeography, Palaeoclimatology, Palaeoecology* 70: 121–37.

Tanck, G. S. 1997. Distribution and origin of organic carbon in the Upper Cretaceous Niobrara Formation and Sharon Springs Member of the Pierre Shale, Western Interior, United States. *Ph.D. thesis,* University of Arizona.

Tandon, S. K. and M. R. Gibling. 1994. Calcrete and coal in late Carboniferous cyclothems of Nova Scotia, Canada: Climate and sea-level changes linked. *Geology* 22:755–8.

Tappan, H. 1980. *The Paleobiology of Plant Protists*. San Francisco: W. H. Freeman.

Tardy, Y. and D. Nahon. 1985. Geochemistry of laterite, stability of Al-goethite, Al-hematite, and Fe+3-kaolinite in bauxites and ferricretes: An approach to the mechanism of concretion formation. *American Journal of Science* 285:865–903.

Taylor, D. G., J. H. Callomon, R. Hall, P. L. Smith, H. W. Tipper, and G. E. G. Westermann. 1984. Jurassic ammonite biogeography of western North America: The tectonic implications. In G. E. G. Westermann, ed. *Jurassic Cretaceous Biochronology and Paleogeography of North America*, pp. 121–41. Geological Association of Canada Special Paper, I. G. C. P. Project 171.

Taylor, F., R. A. Eggleton, C. C. Holzhauer, L. A. Maconachie, M. Gordon, M. C. Brown, and K. G. McQueen. 1992. Cool climate lateritic and bauxitic weathering. *Journal of Geology* 100: 669–77.

Taylor, G. H., S. Y. Liu, and C. F. K. Diessel. 1989. The cold-climate origin of inertinite-rich Gondwana coals. *International Journal of Coal Geology* 11:1–22.

Taylor, M. E. and R. M. Forester. 1979. Distributional model for marine isopod crustaceans and its bearing on early Paleozoic paleozoogeography and continental drift. *Geological Society of America Bulletin* 90:405–13.

Teichmüller, M. 1989. The genesis of coal from the view point of coal petrology. *International Journal of Coal Geology* 12:1–87.

Thickpenny, A. and J. K. Leggett. 1987. Stratigraphic distribution and paleo-oceanographic significance of European early Palaeozoic organic-sediments. In J. Brooks and A. J. Fleet, eds., *Marine Petroleum Source Rocks*, pp. 231–48. Geological Society of London Special Paper 26.

Thierstein, H. R. 1979. Paleoceanographic implications of organic carbon and carbonate distribution in Mesozoic deepsea sediments. In M. Talwani, W. Hay, and W. B. F. Ryan, eds., *Deep Drilling Results in the Atlantic Ocean: Continental Margins and Paleoenvironment*, pp. 249–74. Maurice Ewing Series 3. Washington, D.C.: American Geophysical Union.

Thomas, E. and A. J. Godday. 1996. Cenozoic deepsea benthic foraminifers: Tracers for changes in oceanic productivity? *Geology* 24:355–8.

Thomas, R. D. K. and E. C. Olson. 1980. *A Cold Look at the Warm-Blooded Dinosaurs*. American Association for the Advancement of Science Selected Symposium 28.

Thomasson, J. R., M. E. Nelson, and R. J. Zakrzewski. 1986. A fossil grass (Graminae: Chloridoideae) from the Miocene with Kranz anatomy. *Sciences* 233:876–8.

Thompson, E. I. and B. Schmitz. 1997. Barium and the late Paleocene $\delta^{13}C$ maximum: Evidence of increased marine surface productivity. *Paleoceanography* 12:239–54.

Thompson, J. B., H. T. Mullins, C. R. Newton, and T. L. Vercoutere. 1985. Alternative biofacies model for dysaerobic communities. *Lethaia* 18:167–80.

Thomson, J., S. E. Calvert, S. Mukherjee, W. C. Burnett, and J. M. Bremner. 1984. Further studies of the nature, composition and ages of contemporary phosphorite from the Namibian Shelf. *Earth and Planetary Science Letters* 69:341–53.

Thunell, R. and L. R. Sautter. 1992. Planktonic foraminiferal faunal and stable isotopic indices of upwelling: A sediment trap study in the San Pedro Basin, southern California bight. In C. P. Summerhayes, W. L. Prell and K. C. Emeis, eds., *Upwelling Systems: Evolution Since the Early Miocene*, pp. 77–91. Geological Society of London Special Publication 64.

Thunell, R. C., W. B. Curry, and S. Honjo. 1983. Seasonal variation in the flux of planktonic foraminifera: Time series sediment trap results from the Panama Basin. *Earth and Planetary Science Letters* 64:44–55.

Thunell, R. C., C. J. Pride, E. Tappa, and F. E. Muller-Karger. 1994. Biogenic silica fluxes and accumulation rates in the Gulf of California. *Geology* 22:303–6.

Thurow, J., H.-J. Brumsack, J. Rullkötter, R. Littke, and P. Meyers. 1992. The Cenomanian-Turonian boundary event in the Indian Ocean—A key to understand the global picture. In R. A. Duncan, D. K. Rea, R. B. Kidd, U. von Rad, and J. K. Weissel, eds., *Synthesis of Results from Scientific Drilling in the Indian Ocean*, pp. 253–73. Geophysical Monograph 70. Washington, D.C.: American Geophysical Union.

Tissot, B., G. Demaison, P. Masson, J. R. Delteil, and A. Combaz. 1980. Paleoenvironment and petroleum potential of middle Cretaceous black shales in Atlantic basins. *American Association of Petroleum Geologists Bulletin* 64:2051–63.

Tissot, B. P. and D. H. Welte. 1984. *Petroleum Formation and Occurrence*. 2nd ed. Berlin: Springer-Verlag.

Tourtelot, H. A. and R. O. Rye. 1969. Distribution of oxygen and carbon isotope fossils of Late Cretaceous age, Western Interior region of North America. *Geological Society of America Bulletin* 80:1903–22.

Tozer, E. T. 1982. Marine Triassic faunas of North America: Their significance for assessing plate and terrane movements. *Geologische Rundschau* 71: 1077–104.

Tracy, C. R., J. S. Turner, and R. B. Huey. 1986. A biophysical analysis of possible thermoregulatory adaptation in sailed pelycosaurs. In N. H. Hotton, III, P. D. MacLean, J. J. Roth, and E. C. Roth, eds., *The Ecology and Biology of Mammal-like Reptiles*, pp. 195–206. Washington, D.C.: Smithsonian Institution Press.

Traverse, A. and R. N. Ginsburg. 1967. Pollen and associated microfossils in the marine surface sediments of the Great Bahama Bank. *Review of Palaeobotany and Palynology* 3:243–54.

Tribovillard, N.-P., A. Desprairies, E. Lallier-Vergès, and P. Bertrand. 1994a. Vulcanization of lipidic organic matter in reactive-iron deficient environments: A possible enhancement for the storage of hydrogen-rich organic matter. *Comptes Rendus Académie des Sciences Paris* 319, série II:1199–206.

Tribovillard, N.-P., A. Desprairies, E. Lallier-Vergès, P. Bertrand, N. Moureau, A. Ramdani, and L. Ramanampisoa. 1994b. Geochemical study of organic-matter rich cycles from the Kimmeridge Clay Formation of Yorkshire (UK): productivity versus anoxia. *Palaeogeography, Palaeoclimatology, Palaeoecology* 108:165–81.

Trolard, F. and Y. Tardy. 1987. The stabilities of gibbsite, boehmite, aluminous goethites and aluminous hematites in bauxites, ferricretes and laterites as a function of water activity, temperature and particle size. *Geochimica et Cosmochimica Acta* 51:945–57.

Tsoar, H. and K. Pye. 1987. Dust transport and the question of desert loess formation. *Sedimentology* 34:139–53.

Tucker, M. E. and M. J. Benton. 1982. Triassic environments, climates and reptile evolution. *Palaeogeography, Palaeoclimatology, Palaeoecology* 40: 361–79.

Turner, C. E. and N. S. Fishman. 1991. Jurassic Late T'oo'dichi': A large alkaline, saline lake, Morrison Formation, eastern Colorado Plateau. *Geological Society of America Bulletin* 103:538–58.

Turner, P. 1980. *Continental Red Beds*. Developments in Sedimentology 29. Amsterdam: Elsevier.

Turner, S. and D. H. Tarling. 1982. Thelodont and other Agnathan distributions as tests of Lower Palaeozoic continental reconstructions. *Palaeogeography, Palaeoclimatology, Palaeoecology* 39: 295–311.

Tyson, R. V. 1987. The genesis and palynofacies characteristics of marine petroleum source rocks. In J. Brooks and A. J. Fleet, eds., *Marine Petro-*

leum Source Rocks, pp. 47–67. Geological Society of London Special Publication 26.

Upchurch, G. R., Jr. 1995. Dispersed angiosperm cuticles: Their history, preparation, and application to the rise of angiosperms in Cretaceous and Paleocene coals, southern western interior of North America. *International Journal of Coal Geology* 28:161–227.

Upchurch, G. R. and J. A. Wolfe. 1987. Mid-Cretaceous to early Tertiary vegetation and climate: Evidence from fossil leaves and woods. In E. M. Friis, W. G. Chaloner, and P. R. Crane, eds., *The Origins of Angiosperms and Their Biological Consequences*, pp. 75–105. Cambridge: Cambridge University Press.

Vail, P. R., J. Mitchum, R. M., and I. Thompson, S. 1977. Global cycles of relative changes of sea level. *Seismic Stratigraphy—Applications to Hydrocarbon Exploration* 26:83–97.

Vakhrameyev, V. A. 1982. *Classopollis* pollen as an indicator of Jurassic and Cretaceous climate. *International Geology Review* 24:1190–6.

Valdes, P. 1993. Atmospheric general circulation models of the Jurassic. *Philosophical Transactions of the Royal Society, Series B* 341:317–26.

Valdes, P. J. and B. W. Sellwood. 1992. A palaeoclimate model for the Kimmeridgian. *Palaeogeography, Palaeoclimatology, Palaeoecology* 95:45–72.

Valdes, P. J., B. W. Sellwood, and G. D. Price. 1996. Evaluating concepts of Cretaceous equability. *Palaeoclimates: Data and Modelling* 1:139–58.

Valentine, J. W. and E. M. Moores. 1970. Plate-tectonic regulation of faunal diversity. *Nature* 228:657–9.

van Andel, T. H. 1975. Mesozoic/Cenozoic calcite compensation depth and the global distribution of calcareous sediment. *Earth and Planetary Science Letters* 26:187–94.

Van Cappellen, P. and E. D. Ingall. 1994. Benthic phosphorus regeneration, net primary production, and ocean anoxia: A model of the coupled marine biogeochemical cycles of carbon and phosphorus. *Paleoceanography* 9:677–92.

Van Cappellen, P. and Y. Wang. 1995. Metal cycling in surface sediments: Modeling the interplay of transport and reaction. In H. E. Allen, ed. *Metal Speciation and Contamination of Aquatic Sediments*, pp. 21–64. Chelsea, Michigan: Ann Arbor Press.

Van de Water, P. K., S. W. Leavitt, and J. L. Betancourt. 1994. Trends in stomatal density and $^{13}C/^{12}C$ ratios of *Pinus flexilis* needles during last glacial-interglacial cycle. *Science* 264:239–43.

Van Der Burgh, J., H. Visscher, D. L. Dilcher, and W. M. Kürschner. 1993. Paleoatmospheric signatures in Neogene fossil leaves. *Science* 260:1788–90.

Van der Voo, R. 1988. Paleozoic paleogeography of North America, Gondwana, and intervening displaced terranes: Comparison of paleomagnetism with paleoclimatology and biogeographical patterns. *Geological Society of America Bulletin* 100:311–24.

van der Zwan, C. J., M. C. Boulter, and R. N. L. B. Hubbard. 1985. Climatic change during the Lower Carboniferous in Euramerica, based on multivariate statistical analysis of palynological data. *Palaeogeography, Palaeoclimatology, Palaeoecology* 52:1–20.

van Eijden, A. J. M. 1995. Morphology and relative frequency of planktic foraminiferal species in relation to oxygen isotopically inferred depth habitats. *Palaeogeography, Palaeoclimatology, Palaeoecology* 113:267–301.

van Geen, A., D. C. McCorkle, and G. P. Klinkhammer. 1995. Sensitivity of the phosphate-cadmium-carbon isotope relation in the ocean to cadmium removal by suboxic sediments. *Paleoceanography* 10:159–69.

Van Houten, F. B. 1964. Cyclic lacustrine sedimentation, Upper Triassic Lockatong Formation, central New Jersey and adjacent Pennsylvania. *Kansas Geological Survey Bulletin* 169:497–531.

———. 1982a. Redbeds. *McGraw-Hill Encyclopedia of Science and Technology* 5/e:441–2.

———. 1982b. Ancient soils and ancient climates. In W. H. Berger and J. C. Crowell, eds., *Climate in Earth History*, pp. 112–7. Washington, D.C.: National Academy of Sciences.

Van Houten, F. B. and M. E. Purucker. 1984. Glauconitic peloids and chamositic ooids—Favorable factors, constraints, and problems. *Earth-Science Reviews* 20:211–43.

Van Os, B. J. H., L. J. Lourens, F. J. Hilgen, G. J. De Lange, and L. Beaufort. 1994. The formation of Pliocene sapropels and carbonate cycles in the Mediterranean: Diagenesis, dilution, and productivity. *Paleoceanography* 9:601–17.

Veeh, H. H., S. E. Calvert, and N. B. Price. 1974. Accumulation of uranium in sediment and phosphorites on the South West Africa Shelf. *Marine Chemistry* 2:189–202.

Veevers, J. J. and C. M. Powell. 1987. Late Paleozoic glacial episodes in Gondwanaland reflected in transgressive-regressive depositional sequences in Euramerica. *Geological Society of America Bulletin* 98:475–87.

Veizer, J., P. Fritz, and B. Jones. 1986. Geochemistry of brachiopods: Oxygen and carbon isotopic records of Paleozoic oceans. *Geochimica et Cosmochimica Acta* 50:1679–96.

Verardo, D. J. and A. McIntyre. 1994. Production and destruction: Control of biogenous sedimentation in the tropical Atlantic 0–300,000 years B. P. *Paleoceanography* 9:63–86.

Vermeij, G. J. 1978. *Biogeography and Adaptation.* Cambridge, MA: Harvard University Press.

———. 1990. Tropical Pacific pelecypods and productivity: A hypothesis. *Bulletin of Marine Science* 47:62–7.

Veron, J. E. N. 1995. *Corals in Space and Time.* New York: Cornell University Press.

Vincent, E. and W. H. Berger. 1985. Carbon dioxide and polar cooling in the Miocene: The Monterey hypothesis. In E. T. Sundquist and W. S. Broecker, eds., *The Carbon Cycle and Atmospheric CO_2: Natural Variations Archean to Present*, pp. 455–68. Geophysical Monograph 32. Washington, D.C.: American Geophysical Union.

Visscher, H. and C. J. van der Zwan. 1981. Palynology of the circum-Mediterranean Triassic: Phytogeographical and palaeoclimatological implications. *Geologische Rundschau* 70:625–34.

Visser, J. N. J. 1983. Glacial-marine sedimentation in the late Paleozoic Karoo Basin, southern Africa. In B. F. Molnia, ed. *Glacial-Marine Sedimentation*, pp. 667–701. Mt. Kisco, NY: Plenum.

von Hillebrandt, A., G. E. G. Westermann, J. H. Callomon, and R. L. Detterman. 1992. Ammonites of the circum-Pacific region. In G. E. G. Westermann, ed. *The Jurassic of the Circum-Pacific*, pp. 342–59. Cambridge: Cambridge University Press.

Walker, J. C. G. and B. C. Opdyke. 1995. Influence of variable rates of neritic carbonate deposition on atmospheric carbon dioxide and pelagic sediments. *Paleoceanography* 10:415–27.

Walker, T. R. 1967. Formation of red beds in modern and ancient deserts. *Geological Society of America Bulletin* 78:353–68.

———. 1976. Diagenetic origin of continental red beds. In H. Falke, ed. *The Continental Permian in Central, West, and South Europe*, pp. 240–82. Dordrecht-Holland: D. Reidel.

Walker, T. R., B. Waugh, and A. J. Crone. 1978. Diagenesis in first-cycle desert alluvium of Cenozoic age, southwestern United States and northwestern Mexico. *Geological Society of America Bulletin* 89:19–32.

Wall, D., B. Dale, G. P. Lohmann, and W. K. Smith. 1977. The environmental and climatic distribution of dinoflagellate cysts in modern marine sediments from regions in the North and South Atlantic Oceans and adjacent seas. *Marine Micropaleontology* 2:121–200.

Walter, H. 1977. Climate. In V. J. Chapman, ed. *Wet Coastal Ecosystems*, pp. 61–7. Ecosystems of the World 1. Amsterdam: Elsevier.

Walther, M. 1982. A contribution to the origin of limestone-shale sequences. In G. Einsele and A. Seilacher, eds., *Cyclic and Event Stratification*, pp. 113–20. Berlin: Springer-Verlag.

Wang, Y., T. E. Cerling, and B. J. MacFadden. 1994. Fossil horses and carbon isotopes: New evidence for Cenozoic dietary, habitat, and ecosystem changes in North America. *Palaeogeography, Palaeoclimatology, Palaeoecology* 107:269–79.

Warren, J. K. 1989. *Evaporite Sedimentology.* Englewood Cliffs, NJ: Prentice-Hall.

Waterhouse, J. B. 1969. The palaeoclimatic significance of Permian Productacea from Queensland. In K. S. W. Campbell, ed. *Stratigraphy and Palaeontology*, pp. 226–35. Canberra: Australian National University Press.

Waterhouse, J. B. and G. F. Bonham-Carter. 1975. Global distribution and character of Permian biomes based on brachiopod assemblages. *Canadian Journal of Earth Sciences* 12:1085–146.

Weaver, J. N. and R. M. Flores. 1989. A summary description of synorogenic conglomerate in the Fort Union Formation: Alluvial fan facies, Mowry Basin, Wyoming. *Coal and Hydrocarbon Resources of North America*, pp. T132:11–14. 28th International Geological Congress 1. Washington, D.C.: American Geophysical Union.

Weber, E. T., II, R. M. Owen, G. R. Dickens, A. N. Halliday, C. E. Jones, and D. K. Rea. 1996. Quantitative resolution of eolian continental crustal material and volcanic detritus in North Pacific surface sediment. *Paleoceanography* 11:115–27.

Weber, J. N. and D. M. Raup. 1966. Fractionation of the stable isotopes of carbon and oxygen in marine calcareous organisms—The Echinoidea. 1. Variation of ^{13}C and ^{18}O content within individuals. *Geochimica et Cosmochimica Acta* 30:681– 703.

Weber, J. N. and P. M. J. Woodhead. 1970. Carbon and oxygen isotope fractionation in the skeletal carbonates of reef-building corals. *Chemical Geology* 6:93–117.

———. 1972. Temperature dependence of oxygen-18 concentration in reef coral carbonates. *Journal of Geophysical Research* 77:463–73.

Weber, M. E., M. Wiedicke, and V. Riech. 1995. Carbonate preservation history in the Peru Basins: Paleoceanographic implications. *Paleoceanography* 10:775–800.

Webster, P. J. 1987. The elementary monsoon. In J. S. Fein and P. L. Stephens, eds., *Monsoons*, pp. 3–32. New York: John Wiley and Sons.

Weddige, K. and W. Ziegler. 1987. Lithic and faunistic ratios of conodont sample data as facies indicators. In R. L. Austin, ed. *Conodonts: Investigative Techniques and Applications*, pp. 333–40. Chichester: Ellis Horwood Ltd.

Wei, W., G. Villa, and S. W. Wise, Jr. 1992. Paleoceanographic implications of Eocene-Oligocene calcareous nannofossils from Sites 711 and 748 in the Indian Ocean. In S. W. Wise, Jr., R. Schlich, et al. *Proceedings of the Ocean Drilling Program, Scientific Results, Leg 120*, pp. 979–99. College Station, TX: Ocean Drilling Program.

Weigelt, J. 1989. *Recent Vertebrate Carcasses and Their Paleobiological Implications*. Translated by Schaefer, J. Chicago: University of Chicago Press.

Weissert, H. and A. Lini. 1991. Ice Age interludes during the time of Cretaceous greenhouse climate? In D. W. Müller, J. A. McKenzie, and H. Weissert, eds., *Controversies in Modern Geology*, pp. 173–91. London: Academic Press.

Wells, S. G., L. D. McFadden, and J. C. Dohrenwend. 1987. Influence of late Quaternary climatic changes on geomorphic and pedogenic processes on a desert piedmont, eastern Mojave Desert, California. *Quaternary Research* 27:130–46.

West, R. M., M. R. Dawson, and J. H. Hutchison. 1977. Fossils from the Paleogene Eureka Sound Formation, N.W.T., Canada: Occurrence, climatic and paleogeographic implications. *Paleontology and Plate Tectonics* 2:77–93.

West, R. R., A. W. Archer, and K. B. Miller. 1997. The role of climate in stratigraphic patterns exhibited by late Palaeozoic rocks exposed in Kansas. *Palaeogeography, Palaeoclimatology, Palaeoecology* 128:1–16.

Westermann, G. E. G. 1981. Ammonite biochronology and biogeography of the circum-Pacific Middle Jurassic. In M. R. House and J. R. Senior, eds., *The Ammonoidea*, pp. 459–98. Systematics Association Special Volume 18. London: Academic Press.

———. 1985. Comments on the biogeographical papers: A plate-tectonic alternative. In G. E. G. Westermann, ed. *Jurassic Biogeography and Stratigraphy of East U.S.S.R.*, pp. 28–9. I.G.C.P. Project 171, Circum-Pacific Jurassic, Special Paper 10.

Westgate, J. W. and C. T. Gee. 1990. Paleoecology of a middle Eocene mangrove biota (vertebrates, plants, and invertebrates) from southwest Texas. *Palaeogeography, Palaeoclimatology, Palaeoecology* 78:163–77.

Wheeler, E. A. and P. Baas. 1991. A survey of the fossil record for dicotyledonous wood and its significance for evolutionary and ecological wood anatomy. *International Association of Wood Anatomists Bulletin* 12:275–332.

———. 1993. The potentials and limitations of dicotyledonous wood anatomy for climate reconstructions. *Paleobiology* 19:487–98.

Wheeler, E. A., J. McClammer, and C. A. LaPasha. 1995. Similarities and differences in dicotyledonous woods of the Cretaceous and Paleocene, San Jan Basin, New Mexico, USA. *IAWA Journal* 16:223–54.

White, J. M., T. A. Ager, D. P. Adam, E. B. Leopold, G. Liu, H. Jetté, and C. E. Schweger. 1997. An 18 million year record of vegetation and climate change in northwestern Canada and Alaska: Tectonic and global climatic correlates. *Palaeogeography, Palaeoclimatology, Palaeoecology* 130: 293–306.

Wignall, P. B. 1993. Distinguishing between oxygen and substrate control in fossil benthic assemblages. *Journal of the Geological Society of London* 150:193–6.

Wignall, P. B. and K. J. Myers. 1988. Interpreting benthic oxygen levels in mudrocks: A new approach. *Geology* 25:452–5.

Wignall, P. B. and A. H. Ruffell. 1990. The influence of a sudden climatic change on marine deposition in the Kimmeridgian of northwest Europe. *Journal of the Geological Society, London* 147: 365–71.

Wignall, P. B. and M. J. Simms. 1990. Pseudoplankton. *Palaeontology* 33:359–78.

Wilde, P. and W. B. N. Berry. 1982. Progressive ventilation of the oceans—Potential for return to anoxic conditions in the post-Paleozoic. In S. L. Schlanger and M. B. Cita, eds., *Nature and Origin of Cretaceous Carbon-rich Facies*, pp. 209–24. London: Academic Press.

Wilf, P. 1997. When are leaves good thermometers? A new case for Leaf Margin Analysis. *Paleobiology* 23:373–90.

Wilf, P., S. L. Wing, D. R. Greenwood, and C. L. Greenwood. 1997. Using fossil leaves as paleorain gauges—an Eocene example (abstract). Annual meeting of the Geological Society of America. *Abstracts and Programs* 29:A-432.

Williams, G. E. 1975. Possible relation between periodic glaciation and the flexure of the galaxy. *Earth and Planetary Science Letters* 26:361–9.

———. 1981. Sunspot periods in the late Precambrian glacial climate and solar-planetary relations. *Nature* 291:624–8.

———. 1988. Cyclicity in the late Precambrian Elatina Formation, South Australia: Solar or tidal signature? *Climatic Change* 13:117–28.

———. 1989a. Late Precambrian tidal rhythmites in South Australia and the history of the Earth's rotation. *Journal of the Geological Society of London* 146:97–111.

———. 1989b. Precambrian tidal sedimentary cycles and earth's paleorotation. *Eos* 70:33,40–1.

Williams, G. E. and C. P. Sonett. 1985. Solar signature in sedimentary cycles from the late Precambrian Elatina Formation, Australia. *Nature* 318:523–7.

Williams, G. L. 1977. Dinocysts. In A. T. S. Ramsay, ed. *Oceanic Micropalaeontology Vol. 2*, pp. 1231–325. London: Academic Press.

———. 1978. Dinoflagellates, acritarchs, and tasmanitids. In B. U. Haq and A. Boersma, eds., *Introduction to Marine Micropaleontology*, pp. 293–326. New York: Elsevier.

Williams, J. R. and T. J. Bralower. 1995. Nannofossil assemblages, fine fraction stable isotopes, and the paleoceanography of the Valanginian-Barremian (Early Cretaceous) North Sea Basin. *Paleoceanography* 10:815–39.

Wilson, A. T. 1979. Geochemical problems of the Antarctic dry areas. *Nature* 280:205–8.

Wilson, I. G. 1973. Ergs. *Sedimentary Geology* 10:77–106.

Wilson, K. M., D. Pollard, W. W. Hay, S. L. Thompson, and C. N. Wold. 1994. General circulation model simulations of Triassic climates: Preliminary results. In G. D. Klein, ed. *Pangea: Paleoclimate, Tectonics, and Sedimentation During Accretion, Zenith, and Breakup of a Supercontinent*, pp. 91–116. Geological Society of America Special Paper 288.

Wilson, P. A. and B. N. Opdyke. 1996. Equatorial sea-surface temperatures for the Maastrichtian revealed through remarkable preservation of metastable carbonate. *Geology* 24:555–8.

Wing, S. L. 1991. Comment on " `Equable' climates during Earth history?. " *Geology* 19:539–40.

Wing, S. L., J. Alroy, and L. J. Hickey. 1995. Plant and mammal diversity in the Paleocene to Early Eocene of the Bighorn Basin. *Palaeogeography, Palaeoclimatology, Palaeoecology* 115:117–55.

Wing, S. L. and D. R. Greenwood. 1993. Fossils and fossil climate: The case for equable continental interiors in the Eocene. *Philosophical Transactions of the Royal Society, Series B* 341:243–52.

Winston, R. B. 1986. Characteristic features and compaction of plant tissues traced from permineralized peat to coal in Pennsylvanian coals (Desmoinesian) from the Illinois Basin. *International Journal of Coal Geology* 6:21–41.

———. 1988. Paleoecology of Middle Pennsylvanian-age peat-swamp plants in Herrin Coal, Kentucky, U.S.A. *International Journal of Coal Geology* 10:203–38.

Winston, R. B. and R. W. Stanton. 1989. Plants, coal, and climate in the Pennsylvanian of the central Appalachians. *Coal and Hydrocarbon Resources of North America*, pp. T143:18–26. 28th International Geological Congress 2. Washington, D.C.: American Geophysical Union.

Wise, S. W., Jr., J. R. Breza, D. M. Harwood, W. Wei, and J. C. Zachos. 1992. Paleogene glacial history of Antarctica in light of Leg 120 drilling results. In S. W. Wise, Jr., R. Schlich, et al. *Proceedings of the Ocean Drilling Program, Scientific Results, Leg 120*, pp. 1001–30. College Station, TX: Ocean Drilling Program.

Wise, S. W., Jr. and K. J. Hsü. 1971. Genesis and lithification of a deep sea chalk. *Eclogae Geologicae Helvetiae* 64:273–8.

Witzke, B. J. 1990. Palaeoclimatic constraints for Palaeozoic palaeolatitudes of Laurentia and Euramerica. In W. S. McKerrow and C. R. Scotese, eds., *Palaeozoic Palaeogeography and Biogeography*, pp. 57–73. The Geological Society of London Memoir 12.

Wnuk, C. and H. W. Pfefferkorn. 1987. A Pennsylvanian-age terrestrial storm deposit: Using plant fossils to characterize the history and process of sedimentation. *Journal of Sedimentary Petrology* 57:212–21.

Wolfe, J. A. 1971. Tertiary climatic fluctuations and methods of analysis of Tertiary floras. *Palaeogeography, Palaeoclimatology, Palaeoecology* 9:27–57.

———. 1979. Temperature parameters of humid to mesic forests of eastern Asia and relation to forests of other regions of the northern hemisphere and Australasia. *U.S. Geological Survey Professional Paper* 1106.

———. 1980. Tertiary climates and floristic relationships at high latitudes in the northern hemisphere. *Palaeogeography, Palaeoclimatology, Palaeoecology* 30:313–23.

———. 1985. Distribution of major vegetational

types during the Tertiary. In E. T. Sundquist and W. S. Broecker, eds., *The Carbon Cycle and Atmospheric CO_2: Natural Variations Archean to Present*, pp. 357–75. Geophysical Monograph 32. Washington, D.C.: American Geophysical Union.

———. 1990. Palaeobotanical evidence for a marked temperature increase following the Cretaceous/Tertiary boundary. *Nature* 343:153–6.

———. 1991. Palaeobotanical evidence for a June "impact winter" at the Cretaceous/Tertiary boundary. *Nature* 352:420–3.

———. 1992. An analysis of present-day terrestrial lapse rates in the western conterminous United States and their significance to paleoaltitudinal estimates. *U.S. Geological Survey Bulletin* 1964.

———. 1993. A method of obtaining climatic parameters from leaf assemblages. *U.S. Geological Survey Bulletin* 2040.

———. 1994a. Tertiary climatic changes at middle latitudes of western North America. *Palaeogeography, Palaeoclimatology, Palaeoecology* 108: 195–205.

———. 1994b. Alaskan Palaeogene climates as inferred from the CLAMP database. In M. C. Boulter and H. C. Fisher, eds., *Cenozoic Plants and Climates of the Arctic*, pp. 223–37. NATO ASI Series I 27. Berlin: Springer-Verlag.

———. 1995. Paleoclimatic estimates from Tertiary leaf assemblages. *Annual Review of Earth and Planetary Sciences* 23:119–42.

Wolfe, J. A., H. E. Schorn, C. E. Forest, and P. Molnar. 1997. Paleobotanical evidence for high altitudes in Nevada during the Miocene. *Science* 276:1672–5.

Wolfe, J. A. and G. R. Upchurch, Jr. 1986. Vegetation, climatic and floral changes at the Cretaceous-Tertiary boundary. *Nature* 324:148–52.

———. 1987a. North American nonmarine climates and vegetation during the Late Cretaceous. *Palaeogeography, Palaeoclimatology, Palaeoecology* 61:33–77.

———. 1987b. Leaf assemblages across the Cretaceous-Tertiary boundary in the Raton Basin, New Mexico and Colorado. *Proceedings of the National Academy of Sciences* 84:5096–100.

Woodcock, D. W. and C. M. Ignas. 1994. Prevalence of wood characters in eastern North America: What characters are most promising for interpreting climates from fossil wood? *American Journal of Botany* 81:1243–51.

Woodruff, F. 1985. Changes in Miocene deep-sea benthic foraminiferal distribution in the Pacific Ocean: Relationship to paleoceanography. In J. P. Kennett, ed. *The Miocene Ocean: Paleoceanography and Biogeography*, pp. 131–75. Geological Society of America Memoir 163.

Woodruff, F. and S. M. Savin. 1989. Miocene deepwater oceanography. *Paleoceanography* 4:87–140.

Woodworth-Lynas, C. M. T. and J. A. Dowdeswell. 1994. Soft-sediment striated surfaces and massive diamicton facies produced by floating ice. In M. Deynoux, J. M. G. Miller, E. W. Domack, N. Eyles, I. J. Fairchild, and G. M. Young, eds., *Earth's Glacial Record*, pp. 241–59. Cambridge: Cambridge University Press.

Wray, J. L. 1978. Calcareous algae. In B. U. Haq and A. Boersma, eds., *Introduction to Marine Micropaleontology*, pp. 171–87. New York: Elsevier.

Wright, E. K. 1987. Stratification and paleocirculation of the Late Cretaceous Western Interior Seaway of North America. *Geological Society of America Bulletin* 99:480–90.

Wright, J. D., K. Miller, G., and R. G. Fairbanks. 1992. Early and middle Miocene stable isotopes: Implications for deepwater circulation and climate. *Paleoceanography* 7:357–89.

Wright, V. P. 1988. Paleokarsts and paleosols as indicators of paleoclimate and porosity evolution: A case study from the Carboniferous of South Wales. In N. P. James and P. W. Choquette, eds., *Paleokarst*, pp. 329–41. New York: Springer-Verlag.

Wright, V. P., N. H. Platt, and W. A. Wimbledon. 1988. Biogenic laminar calcretes: Evidence of calcified root-mat horizons in paleosols. *Sedimentology* 35:603–20.

Wright, V. P. and S. D. Vanstone. 1991. Assessing the carbon dioxide continent of ancient atmospheres using palaeo-calcretes: Theoretical and empirical constraints. *Journal of the Geological Society, London* 148:945–7.

Wu, G., J. C. Herguera, and W. H. Berger. 1990. Differential dissolution: Modification of late Pleistocene oxygen isotope records in the western equatorial Pacific. *Paleoceanography* 5:581–94.

Yaalon, D. H., ed. 1971. *Paleopedology: Origin, Nature, and Dating of Paleosols*. Jerusalem: International Society of Soil Science and Israel University Press.

Yanshin, A. L. and M. A. Zharkov. 1986. *Phosphorus and Potassium in Nature (in Russian)*. Novosibirsk: Nauka.

Yapp, C. J. 1979. Oxygen and carbon isotope measurements of land snail shell carbonate. *Geochimica et Cosmochimica Acta* 43:629–35.

———. 1993. The stable isotope geochemistry of low temperature Fe(III) and Al "oxides" with implications for continental paleoclimates. In P. K. Swart, K. C. Lohmann, J. McKenzie, and S. Savin, eds., *Climate Change in Continental Isotopic Records*, pp. 285–94. Geophysical Monograph 78. Washington, D.C.: American Geophysical Union.

Yapp, C. J. and H. Poths. 1992. Ancient atmospheric CO_2 pressures inferred from natural goethites. *Nature* 355:342–4.

Yemane, K. 1993. Contribution of Late Permian palaeogeography in maintaining a temperate climate in Gondwana. *Nature* 361:51–4.

Yemane, K., C. Siegenthaler, and K. Kelts. 1989. Lacustrine environment during Lower Beaufort (Upper Permian) Karoo deposition in northern Malawai. *Palaeogeography, Palaeoclimatology, Palaeoecology* 70:165–78.

Yentsch, C. S. 1974. The influence of geostrophy on primary production. *Tethys* 6:111–8.

Young, G. C. 1990. Devonian vertebrate distribution patterns and cladistic analysis of palaeogeographic hypotheses. In W. S. McKerrow and C. R. Scotese, eds., *Palaeozoic Palaeogeography and Biogeography*, pp. 243–55. Geological Society of London Memoir 12.

Young, J. A. 1987. Physics of monsoons: The current view. In J. S. Fein and P. L. Stephens, eds., *Monsoons*, pp. 211–43. New York: John Wiley and Sons.

Young, T. P. 1989. Phanerozoic ironstones: An introduction and review. In T. P. Young and W. E. G. Taylor, eds., *Phanerozoic Ironstones*, pp. ix-xxv. Geological Society of London Special Publication 46.

Yurtesever, Y. 1975. Worldwide survey of stable isotopes in precipitation. *Report of the Section on Isotope Hydrology, IAEA* November 1975:40.

Zachos, J. C., M. A. Arthur, and W. E. Dean. 1989. Geochemical evidence for suppression of pelagic marine productivity at the Cretaceous/Tertiary boundary. *Nature* 337:61–4.

Zachos, J. C., W. A. Berggren, M.-P. Aubry, and A. Mackensen. 1992. Isotope and trace element geochemistry of Eocene and Oligocene foraminifers from Site 748, Kerguelen Plateau. In S. W. Wise, Jr., R. Schlich, et al. *Proceedings of the Ocean Drilling Program, Scientific Results, Leg 120*, pp. 839–54. College Station, TX: Ocean Drilling Program.

Zachos, J. C., L. D. Stott, and K. C. Lohmann. 1994. Evolution of early Cenozoic marine temperatures. *Paleoceanography* 9:353–87.

Zahn R. K. Winn, and M. Sarnthein. 1986. Benthic foraminiferal δ ^{13}C and accumulation rates of organic carbon: *Uvigerina peregrina* group and *Cibicidoides wuellerstorfi*. *Paleoceanograpy* 1:27–42.

Ziegler, A. M. 1965. Silurian marine communities and their environmental significance. *Nature* 207:270–2.

———. 1990. Phytogeographic patterns and continental configurations during the Permian period. In W. S. McKerrow and C. R. Scotese, eds., *Palaeozoic Palaeogeography and Biogeography*, pp. 363–77. Geological Society of London Memoir 12.

Ziegler, A. M., R. K. Bambach, J. T. Parrish, S. F. Barrett, E. H. Gierlowski, W. C. Parker, A. Raymond, and J. J. Sepkoski, Jr. 1981. Paleozoic biogeography and climatology. In K. J. Niklas, ed. *Paleobotany, Paleoecology, and Evolution Vol. 2*, pp. 231–66 2. New York: Praeger.

Ziegler, A. M., L. R. M. C. Cocks, and R. K. Bambach. 1968. The composition and structure of Lower Silurian marine communities. *Lethaia* 1:1–27.

Ziegler, A. M., K. S. Hansen, M. E. Johnson, M. A. Kelly, C. R. Scotese, and R. Van der Voo. 1977. Silurian continental distributions, paleogeography, climatology, and biogeography. *Tectonophysics* 40:13–51.

Ziegler, A. M., M. L. Hulver, A. L. Lottes, and W. F. Schmachtenberg. 1984. Uniformitarianism and palaeoclimates: Inferences from the distribution of carbonate rocks. In P. J. Brenchley, ed. *Fossils and Climate*, pp. 3–25. Chichester: John Wiley and Sons.

Ziegler, A. M., A. L. Raymond, T. C. Gierlowski, M. A. Horrell, D. B. Rowley, and A. L. Lottes. 1987. Coal, climate and terrestrial productivity: The present and Early Cretaceous compared. In A. C. Scott, ed. *Coal and Coal-Bearing Strata*, pp. 25–49. Geological Society of London Special Publication 32.

Ziegler, A. M., D. B. Rowley, A. L. Lottes, D. L. Sahagian, M. L. Hulver, and T. C. Gierlowski. 1985. Paleogeographic interpretation: With an example from the mid-Cretaceous of North America. *Annual Reviews of Earth and Planetary Sciences*.

Ziegler, A. M., C. R. Scotese, W. S. McKerrow, M. E. Johnson, and R. K. Bambach. 1979. Paleozoic paleogeography. *Annual Review of Earth and Planetary Sciences* 7:473–502.

Index